En

Ro

N

N

Ar

o

o

E.

o

I

Bet

U

n

Gar

te

vi

Other

Stuar
EN

Vene
ICS

OF

Jame
PLA

Davi
ISM

Forthc

Morri
GEN

GAN

J. Gre
BIOT

For mor
call 1-80
countrie

Microbial
Mineral
Recovery

Henry L. Ehrlich

Department of Biology
Rensselaer Polytechnic Institute
Troy, New York

Corale L. Brierley

VistaTech Partnership, Ltd.
Salt Lake City, Utah

McGraw-Hill Publishing Company

New York St. Louis San Francisco Auckland Bogotá
Caracas Hamburg Lisbon London Madrid Mexico
Milan Montreal New Delhi Oklahoma City
Paris San Juan São Paulo Singapore
Sydney Tokyo Toronto

Library of Congress Cataloging-in-Publication Data

Microbial mineral recovery / edited by Henry L. Ehrlich, Corale L.
  Brierley.

    p.   cm.
    Includes bibliographical references.
    ISBN 0-07-007781-9
    1. Bacterial leaching. I. Ehrlich, Henry Lutz.
  II. Brierley, Corale L.
  TN688.3.B33M53   1990
  669'.028'3—dc20   89-13007

1234567890 DOC/DOC 9543210

ISBN 0-07-007781-9

*The editors for this book were Jennifer Mitchell, Gretlyn Cline, and
Stephen M. Smith, the designer was Naomi Auerbach, and the production supervisor was Dianne L. Walber. It was set in Century School-
book. It was composed by the McGraw-Hill Publishing Company Pro-
fessional and Reference Division composition unit.*

Printed and bound by R. R. Donnelley & Sons Company.

Cover art adapted from photograph courtesy of Biorecovery
Systems, Inc.

*For more information about other McGraw-Hill materials,
call 1-800-2-MCGRAW in the United States. In other
countries, call your nearest McGraw-Hill office.*

Contents

Contributors

Terry J. Beveridge   *Department of Microbiology, University of Guelph, Guelph, Ontario, Canada*

Pieter Bos   *Department of Microbiology and Enzymology, Kluyver Laboratory of Biotechnology, Delft University of Technology, Delft, The Netherlands*

Corale L. Brierley   *VistaTech Partnership, Ltd., Salt Lake City, Utah*

Dennis W. Darnall   *Biorecovery Systems, Inc., Las Cruces, New Mexico, and Department of Chemistry, New Mexico State University, Las Cruces, New Mexico*

Henry L. Ehrlich   *Department of Biology, Rensselaer Polytechnic Institute, Troy, New York*

Phillip M. Fedorak   *Department of Microbiology, University of Alberta, Edmonton, Alberta, Canada*

Julia M. Foght   *Department of Microbiology, University of Alberta, Edmonton, Alberta, Canada*

Geoffrey Michael Gadd   *Department of Biological Sciences, University of Dundee, Dundee, Scotland, U.K.*

Gill Geesey   *Department of Microbiology, California State University, Long Beach, California*

W. D. Gould   *CANMET, Energy, Mines and Resources, Canada, Ottawa, Ontario, Canada*

Murray R. Gray   *Department of Chemical Engineering, University of Alberta, Edmonton, Alberta, Canada*

Benjamin Greene   *Lockheed-EMSCO, Las Cruces, New Mexico*

David S. Holmes   *Department of Biology, Rensselaer Polytechnic Institute, Troy, New York*

Larry Jang *Department of Microbiology, California State University, Long Beach, California*

J. Gijs Kuenen *Department of Microbiology and Enzymology, Kluyver Laboratory of Biotechnology, Delft University of Technology, Delft, The Netherlands*

Richard W. Lawrence *Coastech Research, Inc., North Vancouver, British Columbia, Canada*

R. G. L. McCready *CANMET, Energy, Mines and Resources, Canada, Ottawa, Ontario, Canada*

Michael J. McInerney *Department of Botany and Microbiology, University of Oklahoma, Norman, Oklahoma*

Robert J. C. McLean *Department of Urology and Department of Microbiology and Immunology, Queen's University, Kingston, Ontario, Canada*

T. P. McNulty *T. P. McNulty and Associates, Inc., Evergreen, Colorado*

K. A. Natarajan *Department of Metallurgy, Indian Institute of Science, Bangalore, India*

Paul R. Norris *Department of Biological Sciences, University of Warwick, Coventry, U.K.*

Giovanni Rossi *Dipartimento di Ingegneria Mineraria e Mineralurgica, Universita degli Studi di Cagliari, Cagliari, Italy*

David L. Thompson *Hazen International, Inc., Golden, Colorado*

Marios Tsezos *Chemical Engineering, McMaster University, Hamilton, Ontario, Canada*

Olli H. Tuovinen *Department of Microbiology, The Ohio State University, Columbus, Ohio*

Donald W. S. Westlake *Department of Microbiology, University of Alberta, Edmonton, Alberta, Canada*

James R. Yates *Research and Development Center, General Electric Company, Schenectady, New York*

Preface

A major objective of this book is to present the scientific basis (1) for microbial leaching by which metal values can be recovered from ores or by which interfering mineral constituents can be removed from ores or fossil fuels and (2) for using microbial biomass for removing metals from solution. A second important objective is to give an overview of the current status of available and potentially available microbial technology in the areas of bioleaching of ores, biobeneficiation of ores and fossil fuels, and metal recovery from solution, and to provide some scientific ground rules by which some microorganisms used in these technologies might be improved through genetic engineering.

The book is aimed at a readership which includes scientists and engineers specializing in mining, ore dressing, metallurgy, fossil-fuel processing, industrial-waste treatment, and water-pollution abatement. The book should also be of interest to management from industries concerned with one or more of the above topics.

Each chapter was written by one or more experts and is presented at a level which is comprehensible by nonspecialists. The references at the end of each chapter are intended to provide guidance to literature sources for clarification and/or for more in-depth study of any topic.

It will be clear after reading this book, that the biotechnologies that are sufficiently developed for immediate industrial exploitation are those involving heap, dump, and in situ leaching of copper and uranium ores and the biobeneficiation of gold ores and pyritic coal. For several years the biobeneficiation of gold ores and concentrates has been carried out in various pilot scale programs. These tests confirmed that bioleaching of pyrite and arsenopyrite from gold ores and concentrates is competitive from both a capital and operating standpoint with conventional hydrometallurgical and pyrometallurgical technologies. Biobeneficiation of arsenopyrite-containing gold concentrate is now being conducted on a commercial scale and one commercial scale reactor for bioleaching of sulfide-containing gold ore is under construction.

More progress needs to be made in reactor bioleaching of pyritic coal before that approach becomes competitive with more conventional hydrometallurgical and pyrometallurgical technologies. Some of the biotechnologies involving the use of biomass in removing dissolved metal species from solution are also sufficiently developed for industrial application. Other extractive or beneficiating biotechnologies appear to be rapidly reaching a state of development that suggests a possibility for industrial exploitation in the very near or, at the very least, not too distant future. In any event, they hold real promise.

The editors are indebted to all authors for the thoroughness with which they prepared their chapters. Readers expert in biohydrometallurgy, biosorption, or fossil-fuel desulfurization may not agree with all assessments of biotechnological promise or lack thereof that are to be found in this book, but such disagreement should be a stimulus to work toward further exploration and development of this biotechnology. With ever more stringent environmental constraints on conventional nonmicrobiological extractive metallurgy, ore and fossil-fuel beneficiation, and bioremediation of metal-containing industrial wastewater, the technologies described herein deserve to be closely examined for their industrial applicability.

The editors of the book wish to thank Jennifer Mitchell from McGraw-Hill for her patience and advice in bringing this book to fruition.

Henry L. Ehrlich
Corale L. Brierley

Bioleaching and Biobeneficiation

Some microbes have been shown to be able to extract metal values from ores on a practical scale through metal solubilization. Although leaching of copper-containing, low-grade ores and ore tailings has been a long-standing practice going back at least as far as the 1700s and possibly to the times of some ancient civilizations, the involvement of microbes in this leaching was not experimentally demonstrated until the 1950s.

Biobeneficiation of ores is a technology that has emerged relatively recently. It involves the removal by microbial leaching of ore constituents which interfere with the winning of metal values from the ore. At the moment, this technology has taken on special significance in the processing of some gold ores.

The chapters in this section examine the scientific and, in a few instances, the economic basis for these technologies, and survey the range of processes which are presently practicable, potentially practicable, or, at this time, do not hold much promise.

Acidophilic Bacteria and Their Activity in Mineral Sulfide Oxidation

Paul R. Norris

Introduction

A variety of acidophilic bacteria can participate in the oxidation of mineral sulfides with the consequent solubilization of some metals. The bacteria that are most active in the degradation of a range of mineral sulfides oxidize both ferrous iron and sulfur. The bacteria that oxidize only iron or sulfur appear to have a limited capacity in pure culture to degrade most mineral sulfides but could nevertheless contribute significantly to mineral leaching in industrial operations where pure cultures are unlikely to be found and where the aggregate of the different oxidation capacities of the various bacteria is likely to be important. The taxonomy of some of the bacteria is currently unresolved and the limited substrate range of many of them precludes any simple, practical scheme of classification on nutritional grounds. The morphologies of some of the iron-oxidizing bacteria are distinctive, but different genera are being revealed within some groups of bacteria of similar appearance. The bacteria are introduced in this account, therefore, under headings of their temperature ranges for growth, which allows the description of the bacteria with reference to one of the major factors that influences their activity and the rate of metal extraction from mineral sulfides. The bacteria are divided into mesophiles, moderate thermophiles, and extreme thermophiles. The borders between these categories are not always sharply defined, but, considering the bacteria that have been most studied, the respective optimum temperatures for those in these three groups can be taken as 30, 50, and 70°C, respectively. There are exceptions to this general pattern, with some strains of mesophiles and moderate thermophiles, the latter more correctly described as thermotolerant, being more active toward 40 than 30 or 50°C, respectively. In addition, some of the recently isolated extreme thermophiles grow well above 70°C.

Iron- and Sulfur-Oxidizing, Acidophilic Bacteria

Morphologies and optimum temperatures for growth

The mesophiles

Thiobacillus ferrooxidans.   This appears to be the dominant organism in the oxidation of mineral sulfides in most acidic environments if the temperature is below 40°C. The gram-negative, rod-shaped bacteria are usually 0.5 by 1 to 1.5 μm, occurring singly or in pairs (Fig. 1.1), and sometimes possessing a polar flagellum. The optimum temperature for most strains is probably about 30 to 35°C (Fig. 1.2), but the upper and lower temperature limits for growth may depend to some extent on the strain and other growth conditions. The growth rate has been found approximately to halve with each 6°C fall in temperature from 25 to 2°C, with some isolates better able to adapt to low temperatures (McCready, 1988). One isolate with an optimum temperature of about 20°C for growth on ferrous iron at pH 1.6 (*T. ferrooxidans* Hardanger, Fig. 1.2) was found to oxidize pyrite one and one-half times more rapidly than *T. ferrooxidans* DSM583 during growth on the mineral at 10°C (Norris and Kelly, 1982). The upper temperature limit for the growth of both of these strains was raised with a decrease in acidity. For example, the *T. ferrooxidans* Hardanger strain did not grow at 30°C on ferrous iron at pH 1.6 (Fig. 1.2), whereas at pH 2 it grew most rapidly at this temperature. In contrast to *T. ferrooxidans* DSM583, it did not grow at 37°C.

Some variations, attributable to two major causes, have been noted in the reported characteristics of *T. ferrooxidans*. First, cultures that oxidize iron or mineral sulfides in acid media at about 30°C have often been labeled *T. ferrooxidans* without sufficiently close scrutiny to ensure culture purity, with the result that the activity of heterogeneous cultures has sometimes been inadvertently ascribed to *T. ferrooxidans*. Second, there are different strains correctly classified as *T. ferrooxidans* that show some reproducible differences in behavior, as

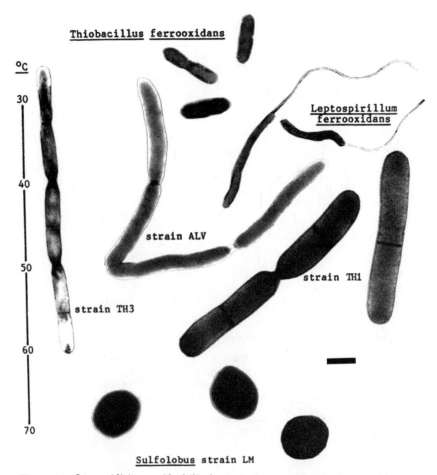

Figure 1.1 Iron-oxidizing, acidophilic bacteria in approximate juxtaposition to their optimum growth temperatures. Bar marker equivalent to 1 μm.

noted above with reference to their optimum growth temperatures. Where large numbers of isolates have been compared, a range of iron and sulfur oxidation rates have been found and correlated with the activity of the different strains in the oxidation of mineral sulfides (Groudev, 1985). However, physiological diversity has not been examined in the context of taxonomy except where different temperature tolerances have been associated with groups of strains as defined on the basis of their DNA:DNA homologies (Harrison, 1982). Twenty-three strains of *"T. ferrooxidans"* were divided into seven groups on the basis of their DNA:DNA homology. However, group 1 has been shown to comprise strains of *Leptospirillum ferrooxidans* (see later). Groups 4, 5, 6, and 7 contained only single isolates but at least some of

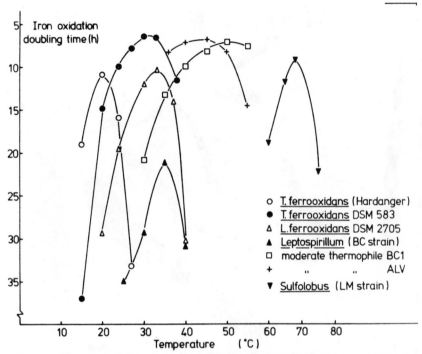

Figure 1.2 The effect of the temperature on the growth-associated iron oxidation by acidophilic bacteria that are described in the text. All bacteria were grown in shaken flasks in a medium containing $(g \cdot L^{-1})$ $(NH_4)_2SO_4$ (0.2), $MgSO_4 \cdot 7H_2O$ (0.4), K_2HPO_4 (0.1), and $FeSO_4 \cdot 7H_2O$ (13.9; 50 mM) at pH 1.6. For growth of strain ALV, strain BC1, and *Sulfolobus*, the medium was gassed with 1% (v/v) CO_2 in air, and for growth of strain BC1 and *Sulfolobus* it was supplemented with thiosulfate (0.5 mM) as a source of reduced sulfur (see text).

these would be disqualified from inclusion in the species *T. ferrooxidans* (iron and sulfur-oxidizing, mole percent GC content usually 57–59) on the basis of a lack of a sulfur-oxidizing capacity and mole percent GC contents as high as 65 (e.g., strain m-1). The most studied strains of *T. ferrooxidans* include American Type Culture Collection (ATCC) 19859, Deutsche Sammlung von Mikroorganismen (DSM) 583, and ATCC 13661 which, like the type strain ATCC 23270, all belong to the same DNA homology group 3a (Harrison, 1982). Strains of this group are widespread in environments of mineral oxidation as shown by the similarity of the electrophoretic protein profiles of ATCC 19859 and of DSM 583 and isolates from metal and coal mine waters of many countries (Jerez et al., 1986; P. Norris, unpublished data). Furthermore, the strains in homology group 3a were reported to grow at 40°C, whereas those in group 3b (38 to 73 percent homology with group 3a strains) were inhibited. Strains in group 2 (18 to 47 percent homology with groups 3a and 3b strains) grew best at 30°C, and in comparison with group 3b strains were inhibited at 35°C (Harrison, 1982). Other phenotypic traits have not been correlated with the position of strains in these major homology groups, and it might prove difficult to establish any subspecies level of classification based on physiological or biochemical characteristics.

The growth of heterotrophic (*Acidiphilium* species) or facultatively autotrophic (*Thiobacillus acidophilus*) contaminants of cultures of *T. ferrooxidans* when iron has been replaced by organic substrates in growth media has resulted in some false claims for heterotrophic growth of the iron-oxidizing bacteria. The utilization of glucose during iron oxidation with a consequent increase in the cell yield has been confirmed with one strain which showed 39% DNA:DNA homology with ATCC 19859 (Barros et al., 1984), but most strains of *T. ferrooxidans* are considered to be obligate autotrophs obtaining their carbon from carbon dioxide primarily via the Calvin cycle.

Leptospirillum ferrooxidans.   The vibrioid cells of *L. ferrooxidans,* which often develop into spiral forms of varying length (Pivovarova et al., 1981; Fig. 1.1), are readily distinguishable from *T. ferrooxidans*; the cells are slightly thinner and are much more frequently motile with a long polar flagellum. In contrast to *T. ferrooxidans,* no strains of *L. ferrooxidans* have been reported to oxidize sulfur. *L. ferrooxidans* does, however, grow on and efficiently degrade pyrite in pure culture (Norris and Kelly, 1982), but the lack of a sulfur-oxidizing capacity restricts its growth on some other mineral sulfides, including chalcopyrite concentrates, except in mixed culture with sulfur-oxidizing bacteria (Balashova et al., 1974; Norris, 1983). The maximum growth rate in shaken, batch culture on ferrous iron of perhaps the most studied strain of *L. ferrooxidans* is about half that of *T. ferrooxidans* (Fig. 1.2). The growth of other strains (e.g., strain BC, Fig. 1.2; Harrison and Norris, 1985) has been found to be even slower on the transfer of bacteria from pyrite- to ferrous-iron-containing media. On serial subculture, growth at rates similar to that of the well-studied strain have been obtained on ferrous iron when a specific requirement for zinc was met (Norris, 1989). The growth on iron usually proceeds with macroscopic aggregations of cells embedded in slime. The aggregations are most evident with growth at low temperatures (e.g., 15 to 20°C), with several strains reproducibly forming a variety of forms, including ribbons and almost spherical pellets. The optimum temperature has not been determined for many strains, but with some it is clearly above 30°C (e.g., strain BC, Fig. 1.2). A strain from a copper leach dump has shown activity at 45°C on pyrite (Norris, 1983) and on chalcopyrite in mixed culture with sulfur-oxidizing bacteria (Fig. 1.3).

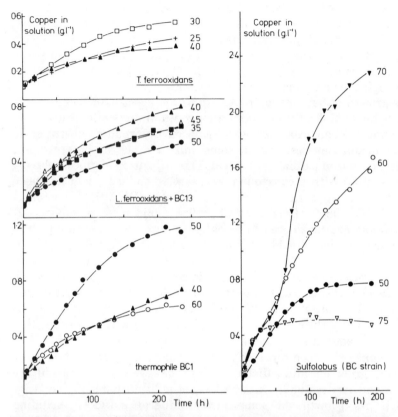

Figure 1.3 The solubilization of copper from a copper concentrate (primarily chalcopyrite) in stirred reactors during the growth of different bacteria at the indicated temperatures (°C). The medium, initially at pH 2, was gassed with 1% (v/v) CO_2 in air and contained (g · L^{-1}) (NH$_4$)$_2$SO$_4$ (0.4), MgSO$_4$ · 7H$_2$O (0.5), K$_2$HPO$_4$ (0.2), and the mineral [20 g · L^{-1}; 20% (w/w) Cu, 75–250 μm particle size].

Thiobacillus thiooxidans. *T. thiooxidans* resembles *T. ferrooxidans* in its acidophily, in its growth on sulfur with an optimum temperature of about 30°C, and in its morphology, although the rods usually possess a polar flagellum giving consistent motility in comparison with *T. ferrooxidans*. *T. thiooxidans* differs from *T. ferrooxidans* in the lower GC content of its DNA (see Table 1.1) and in its inability to oxidize iron. In pure culture, *T. thiooxidans* is unable to effectively degrade pyrite and chalcopyrite concentrates, but it does influence the leaching of some minerals, particularly zinc sulfides (Khalid and Ralph, 1977; Lizama and Suzuki, 1988) and cadmium sulfide, where the leaching depends on the oxidation of elemental sulfur (see Kelly et al., 1979). In mixed culture with iron-oxidizing bacteria, *T. thiooxidans*

TABLE 1.1   Characteristics of Iron- and/or Sulfur-Oxidizing, Acidophilic Bacteria

Bacteria	DNA mol% GC	Temp[a] (°C)	Growth via oxidation of					Carbon source	
			Fe	S^0	FeS_2	$CuFeS_2$	S_4O_6[b]	CO_2	Yeast extract[c]
T. ferrooxidans	58	30	+	+	+	+	+	+	−
T. thiooxidans	52	30	−	+	−/+[d]	−/+	+	+	−
L. ferrooxidans	51–55	35	+	−	+	−/+	−	+	−
Strain BC13	61	45	−	+	−/+	−/+	+	+	−
Strain TH1/BC1[e]	50	50	+	+	+	+	−	+	+
S. thermo-sulfidooxidans	45–49	50	+	+	+	+	−	+	+
Strain ALV	57	50	+	+	+/−[f]	+	−	+	+
Strain LM2	60	50	+	+	−	ND[g]	−	+	+
Strain TH3	69	50	+	+	+	+	−	−	+
Sulfolobus BC[h]	ND	70	+	+	+	+	+	+	+
Sulfolobus B6-2	ND	ND[i]	−	+	ND	ND	+	+	+

[a]Approximate optimum temperatures for growth.
[b]The bacteria growing on tetrathionate also utilize thiosulfate.
[c] + indicates yeast extract supports heterotrophic growth, but, apart from strain TH3, is not required for growth on iron or sulfur.
[d] − / + indicates growth in mixed culture (see text).
[e]Strains TH1 and BC1 share over 90% DNA:DNA homology and represent the same species.
[f] + / − indicates limited growth relative to +.
[g]ND = not determined.
[h]37 mole percent GC in *Sulfolobus acidocaldarius*.
[i]Data from growth at 70°C.

can contribute to the dissolution of chalcopyrite concentrates through the oxidation of sulfur, which could otherwise protect the mineral surface from attack by ferric iron or by bacteria such as *L. ferrooxidans*, which apparently cannot remove the sulfur.

Other mesophilic, acidophilic bacteria.   The comparison of *T. ferrooxidans* strains discussed above revealed iron-oxidizing acidophiles, such as strain m-1, which do not fit the taxonomic criteria for inclusion in the genus *Thiobacillus* (Harrison, 1982). Other iron- and mineral-oxidizing mesophiles which are clearly distinct from *T. ferrooxidans* and *L. ferrooxidans* have also been isolated (Huber et al., 1986) but so far have received relatively little study. The isolation of *Thiobacillus albertis* (Bryant et al., 1983), a sulfur-oxidizing acidophile with a higher DNA GC content (61.5 mole percent) and slightly less tolerance of acidity than *T. thiooxidans* (usually 52 to 53 mole percent GC), increased the number of acidophiles known to be active in sulfur oxidation at about 30°C. The sulfur-oxidizing *Thiobacillus acidophilus* shares a capacity for heterotrophic growth with species of *Acidiphilium* (see Harrison, 1984). In the absence of any added organic nutrients to inorganic media with *T. ferrooxidans*, *Acidiphilium*, and *T. acidophilus* can coexist in mixed culture with *T. ferrooxidans* by utilizing organic matter excreted from the autotroph.

Moderate thermophiles. Among the isolates of bacteria which have been shown to oxidize iron and mineral sulfides at about 50°C are types which are sufficiently different to warrant classification in different genera (see mole percent GC values, Table 1.1; Harrison, 1986). They all appear to be gram-positive (A. P. Harrison, personal communication), and they also have in common a wider nutritional versatility than the obligately autotrophic, gram-negative *T. ferrooxidans* and *L. ferrooxidans*.

Sulfobacillus thermosulfidooxidans and strain TH1/BC1. *S. thermosulfidooxidans* and closely related subspecies have been isolated from geothermal environments, mineral sulfide ore, and coal dumps across eastern Europe (see Karavaiko et al., 1988). Strain TH1 (Fig. 1.1) was originally isolated from an Icelandic hot spring. Comparison with bacteria subsequently isolated from self-heating coal spoils, copper leach dumps, and metal mine waters from several countries has shown that this type is also widespread. *S. thermosulfidooxidans* and strain TH1 have not been directly compared. The former has been described as nonmotile and spore-forming. Sporulation has not been observed in strain TH1, but motility has been observed, although not consistently. It seems likely, however, that being similar in other respects, they are representatives of the same species. Strain BC1, the activity of which is shown in Figs. 1.2 and 1.3, has over 90% DNA:DNA homology with strain TH1 (A. P. Harrison, personal communication) and is therefore another isolate of the same species. The cell shape can vary, with rods of strain TH1 grown on ferrous iron and yeast extract having dimensions of about 0.8 by 6 μm (Fig. 1.1, where pairs of cells are also shown within this length without constriction at the cross walls). The bacteria are generally smaller when grown autotrophically on iron, in the absence of yeast extract (Norris et al., 1986a). More elongated forms have been observed during growth on pyrite, and even greater morphological variation, including branched forms, has been described for *S. thermosulfidooxidans* (Karavaiko et al., 1988). The bacteria appear unable to utilize sulfate as a source of the sulfur required in biosynthesis; consequently, unlike *T. ferrooxidans,* they require a source of reduced sulfur in growth media (Norris and Barr, 1985). Sulfur-containing amino acids such as cysteine can be utilized to fulfill this requirement, but inorganic reduced sulfur compounds, including mineral sulfides, are also adequate.

The optimum temperature of growth-associated iron oxidation of the moderate thermophiles is generally 45 to 50°C (Fig. 1.2), but they are active over a wide temperature range, with growth of some isolates on ferrous iron occurring at 60°C and the oxidation of minerals at 35°C occurring as rapidly as with *T. ferrooxidans*. The maximum rate of autotrophic growth by these moderate thermophiles on ferrous iron is slightly slower than that of *T. ferrooxidans* (Fig. 1.2), but the growth is more rapid on iron in the presence of some sugars (glucose, fructose, and sucrose; Wood and Kelly, 1984) or yeast extract (Marsh and Norris, 1983a). The sugars are utilized simultaneously with carbon dioxide during mixotrophic growth on ferrous iron. The growth yield under autotrophic growth conditions on iron in batch culture is similar to that of *T. ferrooxidans* or *L. ferrooxidans*. It amounts to about 0.3 g (dry weight) per mole of ferrous iron oxidized, but is slightly higher during mixotrophic growth (Wood and Kelly, 1985). Growth of the thermophiles on iron in the presence of yeast extract is chemolithoheterotrophic, proceeding rapidly without significant utilization of carbon dioxide (Wood and Kelly, 1983). The growth on sugars in a mineral salts medium in the absence of iron is slow, but the rate of purely heterotrophic growth on yeast extract is similar to that of autotrophic growth on ferrous iron.

Other iron-oxidizing moderate thermophiles.   Enrichment cultures from mine water or leach dump samples which are maintained at 50°C with mineral sulfides usually become dominated by bacteria resembling *S. thermosulfidooxidans* and strain TH1. Other bacteria that oxidize iron and mineral sulfides at 50°C have been isolated, but generally only on single occasions or from single sites. This could indicate that they are not widespread or that particular enrichment conditions are required for their isolation.

Strain ALV (Figs. 1.1 and 1.2) does not require a source of reduced sulfur when growing autotrophically on ferrous iron, in contrast to the other moderately thermophilic strains so far studied (Norris and Barr, 1985). It has been isolated from the same coal spoil tip on two occasions 9 years apart when samples were placed in ferrous-iron enrichment cultures that did not include any reduced sulfur. The inclusion of yeast extract in the isolation medium or the use of pyrite as the substrate resulted instead in growth of strain-TH1-type bacteria from the same samples. Strain ALV was not isolated from another spoil heap about 10 miles away where the TH1-type was also abundant, but the real extent of its distribution remains to be examined. The original isolate of strain ALV appeared almost filamentous in comparison to strain TH1 when grown autotrophically on iron (Fig. 1.1; Harrison, 1986), but it can more closely resemble strain TH1 when grown on iron plus yeast extract. Its metabolism of sugars has revealed oxidation of glucose mainly by the oxidative pentose-phosphate pathway during mixotrophic growth (Wood and Kelly, 1984).

The isolation of the iron-, sulfur-, and mineral-oxidizing strain TH3 from samples from a copper leach dump (Chino Mine, Hurley, New Mexico) was described in 1978 (Brierley, J. A., 1978) and again several years later (Norris and Barr, 1985). As with strain ALV, strain TH3 appears to have a limited distribution in comparison with strain TH1, since its presence has not been confirmed in enrichment cultures of mine water, leach dump, or coal spoil samples from elsewhere, even though its distinctive morphology would aid in its recognition. The cell width of strain TH3 is similar to that of *T. ferrooxidans* (Fig. 1.1), but growth tends to be filamentous. Motility has been observed, particularly during heterotrophic growth on yeast extract. Autotrophic growth on iron has been obtained with one isolate described as TH3 (see Brierley and Brierley, 1986), but growth on iron of the TH3 isolate described herein (Fig. 1.1, Table 1.1) has not been obtained in the absence of yeast extract (Norris and Barr, 1985).

The growth of strain LM2 (Table 1.1) on iron (Norris and Barr, 1985) and on sulfur (Norris et al., 1986b) has been described, but its growth on pyrite is extremely poor. This strain was isolated from a hot-spring sample via a sulfur enrichment culture at 50°C, while enrichment cultures of the sample containing iron or pyrite became dominated by strain-TH1-type bacteria.

Thermotolerant, sulfur-oxidizing acidophiles. The presence in geothermal habitats of acidophilic bacteria which resemble *T. thiooxidans* in their appearance, in their capacity for sulfur oxidation, and in their lack of iron oxidation, but which grow at higher temperatures than the mesophile, has been known for some time (e.g., Schwartz and Schwartz, 1965). Similar strains have been isolated from hot springs, coal spoil tips, and mineral sulfide mine waters. A typical strain, BC13, is gram-negative and should be classified in the genus *Thiobacillus* but, in view of its thermotolerance and higher DNA GC content (Table 1.1), should be assigned to a different species from *T. thiooxidans*. Strain BC13 oxidizes sulfur over a wide temperature range and more rapidly at 45 to 50°C (Norris et al., 1986b) than do the iron- and sulfur-oxidizing, gram-positive strains such as LM2 and ALV.

The extreme thermophiles. The similar morphologies of the approximately spherical, extremely thermophilic, iron- and/or sulfur-oxidizing bacteria that have been studied conceal a diversity of types in terms of their taxonomy and of their activity in mineral oxidation. The sulfur-oxidizing strains were initially classified as *Sulfolobus acidocaldarius* (Brock et al., 1972). The diversity of types was recognized with the division of the genus into three species, *S. acido-*

caldarius, S. solfataricus, and *S. brierleyi* (Zillig et al., 1980), and by the subsequent classification of *S. brierleyi* as *Acidianus brierleyi* together with *A. infernus* in a genus of bacteria capable of anaerobic sulfur reduction as well as aerobic sulfur oxidation (Segerer et al., 1986). Few characterized strains of these thermophiles appear to have been used in mineral oxidation studies, so there is little information with which to compare the activity of particular strains in this context.

Some culture collection strains designated *S. acidocaldarius* were found to grow only heterotrophically and to be unable to oxidize sulfur or iron (Marsh et al., 1983). An optimum temperature of about 70°C for growth on iron (Fig. 1.2) and mineral sulfides (Fig. 1.3) has been found with isolates that have been referred to as *Sulfolobus* strains BC and LM (Fig. 1.1). These strains generally conform to the original description of *S. acidocaldarius,* but have not been directly compared with a sulfur-oxidizing type strain. The study of other isolates would most likely reveal various optimum temperatures for mineral oxidation. Optimum temperatures for growth on sulfur of 65° and 80°C have been described for different isolates from hot springs (Mosser et al., 1974). At least one isolate active in mineral oxidation at 85°C has a DNA GC content of 45 mole percent (Huber and Stetter, 1986) and could therefore be classified separately from *S. acidocaldarius* (mole percent GC 37) and *A. brierleyi* (mole percent GC 31). Yeast extract was found to be required for the growth of *S. brierleyi* (*A. brierleyi*) on iron and pyrite, and the oxidation of the mineral was slower than during the autotrophic growth of *Sulfolobus* strains LM and BC (Marsh et al., 1983). The capacity for autotrophic growth by *S. brierleyi* has been demonstrated with sulfur as the substrate, but was inhibited by agitation which did not prevent the growth of other *Sulfolobus* strains (Norris et al., 1986b). A strain of *Sulfolobus,* designated strain B6-2 (Konig et al., 1982), which does not appear to be closely related to the other species on the basis of quite different electrophoretic protein profiles (Norris et al., 1986b), is unable to oxidize iron, which probably restricts its capacity to degrade mineral sulfides in pure culture, making it similar to that of *T. thiooxidans* in comparison with *T. ferrooxidans* at lower temperatures.

The Microbiology of Mineral Oxidation

The maximum rates of growth-associated iron oxidation by different bacteria are not necessarily reflected in the relative rates of mineral oxidation during their growth. The growth rates on minerals are slower, sometimes limited by the availability of iron from the solid substrate. Thus, *T. ferrooxidans* may grow twice as fast as *L. ferrooxidans* on ferrous iron (Fig. 1.2), but these bacteria degrade pyrite at

similar rates (Norris et al., 1988). Similarly, the moderate thermophiles do not grow more rapidly autotrophically on ferrous iron than
does *T. ferrooxidans* (Fig. 1.2), but their rate of pyrite oxidation during autotrophic growth at 50°C is clearly more rapid than that by the
mesophile at 30°C (Marsh and Norris, 1983b) through the influence of
the temperature on the rate of mineral dissolution. The efficiency of
mineral leaching by bacteria of inherently similar activity can differ
therefore through the capacity of some of the bacteria to remain active
under different conditions which alter the rate or extent of mineral
dissolution. This is illustrated with the solubilization of copper from a
concentrate comprising primarily chalcopyrite by acidophilic, iron-
and sulfur-oxidizing bacteria at different temperatures (Fig. 1.3). The
incomplete oxidation of chalcopyrite by bacteria is well-documented
(e.g., Brierley, C. L., 1978) and attributed to the deposition of secondary minerals or precipitates on the mineral surface. Increasing the
temperature of the leaching might result in the slower formation of
such a barrier to continued dissolution of the mineral or in the development of a more porous layer, which would allow the most extensive
leaching by the extreme thermophiles. As noted earlier, the lack of
iron oxidation by *T. thiooxidans* and the lack of sulfur oxidation by *L.
ferrooxidans* restrict their activity in the dissolution of many mineral
sulfides unless they are grown in mixed cultures that have the oxidative range of *T. ferrooxidans*. The activity of the thermotolerant,
sulfur-oxidizing strain BC13 is similarly restricted, but it can usefully
complement that of *L. ferrooxidans* in chalcopyrite oxidation (Fig. 1.3)
at temperatures which exceed the range of *T. thiooxidans*.

Some basic characteristics of the acidophiles and their oxidation capacities in relation to growth on minerals are summarized in Table
1.1. The poor growth on pyrite of some bacteria which can oxidize iron
and sulfur, for example, strain ALV (Table 1.1), is discussed later in
relation to the characteristics of iron oxidation.

The bacterial oxidation of thiosalts has application in biological oxidation ponds for the treatment of ore mill process waters (Silver,
1985), but the growth of bacteria on the soluble sulfur compounds
tetrathionate and thiosulfate or heterotrophically on yeast extract
(Table 1.1) is not directly relevant to application of the bacteria in
mineral treatment. However, these substrates could be useful in facilitating the study of aspects of metabolism that are concerned with
mineral oxidation. For example, the study of the mechanism or control
of iron oxidation may benefit from the examination of clones in which
the process has been altered. The use of a mixture of thiosulfate and
ferrous iron in a solid medium for the growth of *T. ferrooxidans* has
enabled the selection of variants deficient in iron oxidation which utilize the sulfur compound for growth to give colonies that are white in

comparison to the brown, oxidized-iron-encrusted colonies of iron-oxidizing bacteria (Holmes et al., 1988). Similarly, a different combination of substrates, ferrous iron and yeast extract, should allow the selection of iron-deficient mutants of moderate thermophiles such as strain TH1 which could not utilize thiosulfate if it were provided as the alternative substrate to iron.

The following discussion of the bacterial characteristics influencing growth on mineral sulfides, in the context of mixed cultures, follows the description of three major areas of metabolism involved in the growth on minerals: the oxidations of iron and sulfur, and the fixation of carbon dioxide.

Iron oxidation

The oxidation of iron, which provides energy for the growth of acidophilic bacteria, can be described at two levels. First, as a superficially similar process in the various bacteria, the rate of which is affected by factors such as the temperature and the concentrations of ferrous and ferric iron. Second, as the mechanism or mechanisms of iron oxidation in the different bacteria.

Observations with whole cells.   The rate of iron oxidation by different isolates of *T. ferrooxidans* can be quite different at a given temperature, as noted earlier (Fig. 1.2). The comparison of a large number (370) of isolates of *T. ferrooxidans* showed that some oxidized iron up to one and one-half times more rapidly than others (Groudev, 1985). The kinetics of iron oxidation by *T. ferrooxidans* DSM 583 have been determined with resting cells, using oxygen electrodes to monitor oxidation rates, and with cells growing in batch and continuous cultures (Kelly et al., 1977; Kelly and Jones, 1978; Jones and Kelly, 1983). The effects of the absolute and relative concentrations of ferrous and ferric iron were complex in continuous culture in which substrate inhibition by high concentrations of ferrous iron and different forms of product inhibition by ferric iron occurred (Jones and Kelly, 1983). The inhibition by ferric iron was influenced by factors which included the pH and the potassium ion concentration, and could be either competitive or noncompetitive. Purely competitive inhibition of ferrous-iron oxidation by ferric iron was observed with washed cell suspensions of *T. ferrooxidans*. Similar values of the affinity for iron in the oxidation process were obtained with resting and growing cells. An apparent Michaelis constant (K_m, equal to the concentration of substrate at which the reaction rate is half the maximum) in the range of 0.43 to 0.9 mM ferrous iron was found with cell suspensions; a substrate saturation coefficient (K_s) in the range 0.7 to 2.4 mM ferrous iron was obtained from chemostat work.

L. ferrooxidans has been found to possess a high affinity for ferrous iron (apparent K_m, 0.25 mM ferrous iron versus 1.34 mM for *T. ferrooxidans*) and the least sensitivity among the iron-oxidizing bacteria to ferric iron as an end-product, competitive inhibitor of iron oxidation (Norris et al., 1988). Competition experiments between *T. ferrooxidans* and *L. ferrooxidans* in chemostats in which ferrous iron was the growth-limiting substrate resulted in *L. ferrooxidans* becoming dominant (Norris et al., 1988). At the other extreme, the rate of oxidation of ferrous iron by strain ALV was found to be that most readily lowered with a decrease in ferrous-iron concentration and was the most severely lowered in the presence of ferric iron. In these examples at least, the relative affinities for iron appear to reflect to some extent the capacity of the bacteria for growth on pyrite. Although other factors also affect the growth of bacteria on the mineral, it seems likely that the relatively low concentration of available ferrous iron against a background of an increasing concentration of ferric iron during the mineral dissolution favors *L. ferrooxidans,* potentially the best adapted for growth with such a ferrous-ferric-iron ratio. In comparison with strain TH1/BC1, strain ALV, which would seem to be the least suited for growth with a low ferrous-ferric-iron ratio, was inhibited by lower concentrations of ferric iron (Norris et al., 1988) and exhibited initially similar, but then relatively restricted, growth on pyrite (Norris et al., 1986a).

The mechanism of iron oxidation. The process of iron oxidation in *T. ferrooxidans,* the only bacterium in which it has received much study, has been thoroughly reviewed (Ingledew, 1982). Some aspects of the process, including possible key components, have still not been resolved, whereas others have been relatively thoroughly characterized. Ferrous ions are oxidized at the cell surface, with electron transfer ultimately to a terminal cytochrome oxidase in the cytoplasmic membrane (Fig. 1.4). Almost 25 years ago, Dugan and Lundgren (1965) proposed a reaction involving a lattice of iron and sulfate ions and an iron oxidase in the cell envelope of *T. ferrooxidans.* More recently, the possible transfer of electrons through a polynuclear Fe(III) layer to the periplasm has been suggested (Ingledew, 1986, Fig. 1.2). Rusticyanin, a small blue copper protein (Cox and Boxer, 1978), and cytochrome c on the reducing side of rusticyanin (Blake et al., 1988) have also been suggested as potential initial acceptors of electrons from ferrous iron. More complexity is indicated by recent descriptions of the simultaneous induction by iron of an outer-membrane glycoprotein (M_r 92,000) and rusticyanin (Mjoli and Kulpa, 1988), by the purifica-

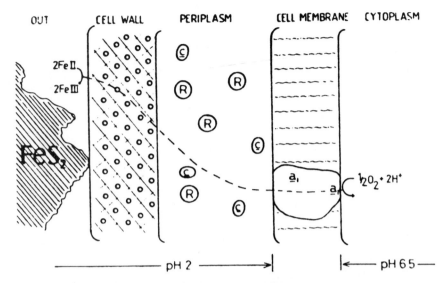

Figure 1.4 A diagrammatic representation of iron oxidation in *T. ferrooxidans* illustrating a possible route of electrons via cell-surface-bound iron (small circles in the cell wall) and the periplasmic electron carriers rusticyanin (R) and cytochrome c. (*From Ingledew, Ferrous iron oxidation by Thiobacillus ferrooxidans, in H. L. Ehrlich and D. S. Holmes [eds.], Workshop of Biotechnology for the Mining, Metal-Refining and Fossil Fuel Processing Industries, Wiley, New York, 1986, pp. 23–33.*)

tion of a protein (M_r 63,000) which apparently reduced cytochrome c_{552} in the presence of Fe(II) (Fukumori et al., 1988), and by the presence of an iron-sulfur protein which might act with sulfate to facilitate electron transfer to rusticyanin or a cytochrome (Fry et al., 1986).

The nature of the substrate probably ensures an essentially similar process in other iron-oxidizing acidophiles with iron oxidized extracytoplasmically but the nature of the components catalyzing the oxidation and electron transfer could be different in taxonomically unrelated groups of bacteria. A comparison by optical spectroscopy of some of the bacteria noted above (Table 1.1) has revealed differences in the respiratory chains (Fig. 1.5; D. W. Barr, W. J. Ingledew, and P. Norris, unpublished data). The cytochromes of *T. ferrooxidans* have been reviewed (Ingledew, 1982); the Soret peak at 419 nm and the alpha peaks around 551 nm have been attributed to at least two c-type cytochromes, and the peaks at 442 and 595 nm to cytochrome a_1. The prominent peaks in the spectrum of *L. ferrooxidans* have been attributed to an acid-stable cytochrome which contains iron and zinc, the latter possibly accounting for the specific zinc requirement for growth of the bacteria (see Norris, 1989). This cytochrome could occupy a position analogous to that of the rusticyanin or a cytochrome c in *T. ferrooxidans*.

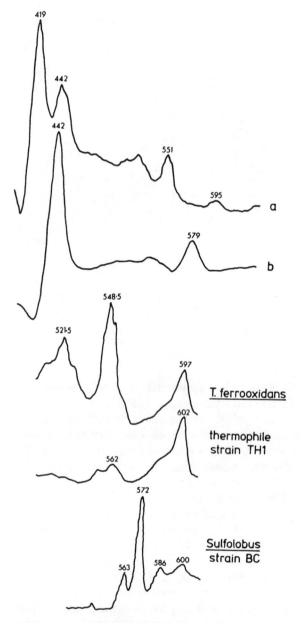

Figure 1.5 A comparison of difference spectra (dithionite-reduced minus oxidized) at room temperature of whole cells of (a) *T. ferrooxidans* and (b) *L. ferrooxidans* grown on ferrous iron and at liquid nitrogen temperature, of *T. ferrooxidans* and thermophilic iron-oxidizing acidophiles grown on iron or (*Sulfolobus* only) pyrite (unpublished work of D. W. Barr, W. J. Ingledew, and P. Norris).

The optical spectrum of strain TH1 (Fig. 1.5), and a similar spectrum of strain ALV (not shown), indicated respiratory chain components that have yet to be characterized but which are different from those in the mesophiles. With *Sulfolobus* strain BC, the absorption at 572 nm (Fig. 1.5) was observed in cells grown on pyrite but not in cells grown on thiosulfate, which suggests that, as with the other iron-oxidizing bacteria, this thermophile might possess an electron carrier involved specifically in the process of iron oxidation.

Sulfur oxidation

Studies of the mechanism of reduced sulfur oxidation by bacteria have generally concentrated on the metabolism of soluble substrates rather than elemental sulfur or mineral sulfides, and much of the work has involved the nonacidophilic thiobacilli. The oxidation of thiosulfate by *Thiobacillus versutus* is catalyzed, without any detectable free intermediates, by a periplasmic, multienzyme complex (see Kelly, 1988), but other thiobacilli, including the obligately autotrophic acidophiles, convert thiosulfate initially to tetrathionate. Enzymes catalyzing the tetrathionate formation and the final step of reduced sulfur compound oxidation, sulfite to sulfate, have been characterized in various thiobacilli. In *T. ferrooxidans,* elemental sulfur could be an intermediate in thiosalt oxidation (Hazeu et al., 1988) as well as the first product of the oxidation of the sulfide moiety of mineral sulfides through the action of a sulfide oxidase or chemically with ferric iron. The oxidation pathway of elemental sulfur formed as an intermediate or utilized as a primary substrate has not been completely resolved. The accumulation of sulfur in the cell wall, in the periplasm, and on the cytoplasmic membrane has been observed (Gromova et al., 1983) and might be facilitated by association with phospholipids from the bacteria. An alternative feature of the interaction with elemental sulfur has been suggested by the formation of sulfur globules from tetrathionate by *T. ferrooxidans*; a coat of long-chain polythionates over a hydrophobic nucleus of predominantly S_8 molecules would give an overall hydrophilic character to the particles (Steudel et al., 1987). Enzyme action on a form of colloidal sulfur or polysulfide sol, possibly in the form of a glutathione polysulfide has been envisaged (Suzuki, 1974). It is not clear, however, whether the formation of sulfite might involve oxygenase activity, which has been found in vitro in various thiobacilli but which would be non-energy-conserving, or possibly a sulfite reductase (as found in *Thiobacillus denitrificans*) operating in the direction of sulfite formation (see Kelly, 1982). The oxidation of sulfite to sulfate in the thiobacilli can occur via the adenosine phosphosulfate (APS) pathway or via a sulfite-cytochrome c oxido-reductase, the former producing ATP by substrate-level phosphorylation and the latter by an electron transport chain and oxidative phosphorylation (see Kelly, 1982, 1988). The sulfite oxidase of *T. ferrooxidans* has been described (Vestal and Lundgren, 1971).

The couplings of sulfur and sulfite oxidations to ferric-ion reduction have been proposed as alternative routes of reduced sulfur oxidation in *T. ferrooxidans* with growth dependent on the oxidation of the resultant ferrous ions. A sulfur-ferric-ion oxidoreductase has been purified (Sugio et al., 1987), and the activity of a sulfite oxidase which utilizes ferric ion as the electron acceptor in place of cytochrome c has been described (Sugio et al., 1988). The physiological significance of these reactions in *T. ferrooxidans* is not clear in comparison with the system previously outlined that would be expected to operate without ferric-ferrous-ion cycling as in the non-iron- but sulfur-oxidizing *T. thiooxidans*. Ferric-iron reduction occurs during the oxidation of sulfur by *T. thiooxidans* and *Sulfolobus* as well as by *T. ferrooxidans* (Brock and Gustafson, 1976).

As suggested earlier for iron oxidation, it is possible that sulfur oxidation systems which differ in detail, but which share common intermediates such as sulfite, have evolved in phylogenetically distinct acidophiles. The process of sulfur oxidation in the gram-positive, iron-oxidizing, moderate thermophiles has not been described, but with differences in cell surface layers as one of the features that distinguish these bacteria from the gram-negative thiobacilli some differences in their interaction with sulfur are likely. Growth of the gram-positive moderate thermophiles on tetrathionate has not been obtained (Table 1.1), although thiosulfate has been cited as a substrate for *S. thermosulfidooxidans* (Karavaiko et al., 1988). The growth of various species of *Sulfolobus* on tetrathionate has been described (Wood et al., 1986). The mechanism of thiosalt metabolism is unknown except that thiosulfate, as in most thiobacilli, is metabolized via tetrathionate (Norris et al., 1989). A sulfur oxygenase with a pH optimum of 7 has been purified from *S. brierleyi* in the first study of the oxidation process in the thermophilic archaebacteria (Emmel et al., 1986).

Carbon dioxide fixation

Carbon dioxide is assimilated by *T. ferrooxidans,* as in many other autotrophic eubacteria, primarily via the Calvin reductive pentose phosphate pathway. The enzyme catalyzing the carbon dioxide fixation, ribulose 1,5-bisphosphate carboxylase/oxygenase, has been purified from *T. ferrooxidans* (Holuigue et al., 1987). Activity of the same enzyme has been found in the moderately thermophilic iron-oxidizing bacteria (Wood and Kelly, 1985). However, in comparison with *T. ferrooxidans,* the autotrophic growth of aerated cultures of the moderate thermophiles is severely restricted unless the carbon dioxide concentration in the gas supply is increased above that in air. In the absence of such carbon dioxide enrichment, the rate of pyrite oxidation during growth of *T. ferrooxidans* and moderate thermophile strain BC1 on the mineral was found to be 15 and 86 percent reduced, respectively (Norris et al., 1989). In the same experiment, the activity of *Sulfolobus* strain BC was reduced by 26 percent in a culture gassed with air. The capacity of *Sulfolobus* to respond to a limiting carbon dioxide concentration appeared to reside in the increased production of an enzyme catalyzing the carboxylation of acetyl CoA. The Calvin cycle does not operate in *Sulfolobus* and archaebacteria in general.

Mixed cultures and factors affecting bacterial activity

The heterogeneous microenvironments and temperature gradients which exist in some mineral sulfide mine waters, ore leaching dumps, acid coal spoils, and geothermal habitats ensure that a variety of microorganisms will be present and active in ferrous-iron, sulfur, and mineral sulfide oxidation at such sites. For example, bacteria that require different conditions for optimum activity, such as *Leptospirillum* strain BC, the thermotolerant or moderately thermophilic strains BC13 and BC1, and the thermophile *Sulfolobus* BC, which were noted in this chapter, as well as *T. ferrooxidans*, were all isolated via enrichment cultures of a single small sample from a drainage channel of a pyrite-bearing spoil heap at a colliery. In contrast to such "natural" sites of mineral sulfide oxidation, bioreactors for industrial ore or concentrate treatment present a relatively homogeneous environment for bacterial growth. The nature of the substrate and the process conditions would determine which bacteria would be most active in the mixed, rather than pure, cultures which might be expected in industrial mineral processing. The predominance of either mesophiles or thermophiles would depend on the temperature. Considering the possible competition between bacteria that grow at similar temperatures, differences in their behavior that would affect their relative activity in mineral oxidation can be obvious, such as sulfur oxidation by *T. ferrooxidans* but not by *L. ferrooxidans*, or more subtle, as with different characteristics of iron oxidation and carbon dioxide fixation among the iron-oxidizing autotrophs.

The difference in the affinities of *T. ferrooxidans* and *L. ferrooxidans* for ferrous and ferric iron has been discussed and could be one of the factors favoring the growth of *L. ferrooxidans* when initially mixed cultures are serially cultured on pyrite. The growth of *L. ferrooxidans* on pyrite at an acidity, pH 1.3, which inhibited *T. ferrooxidans* (Norris, 1983) indicated another factor which can influence the relative activity of these bacteria. During continuous leaching of pyrite at pH 1.5, the displacement of *T. ferrooxidans* by *L. ferrooxidans* has been observed, whereas *T. ferrooxidans* was dominant at higher pH values (Helle and Onken, 1988). The relative sensitivity of these bacteria to potentially toxic metal cations and anions varies with different metals (Norris et al., 1986a, 1988). The apparently greater resistance of *T. ferrooxidans* to several divalent cations, however, including copper, would make it less likely to be inhibited than *L. ferrooxidans* during the leaching of some mineral concentrates where there would be high concentrations of the metals in solution.

Indirectly cooperative growth on mineral sulfides can occur with combinations of species such as *L. ferrooxidans* and *T. thiooxidans* or strain BC13. Mixed cultures might also contain different strains of a single iron- and sulfur-oxidizing species, such as *T. ferrooxidans* strains, which might have different and complementary capacities for iron and sulfur oxidation. The enhanced growth of *T. ferrooxidans* which has been observed in mixed culture with heterotrophic acidophiles such as *Acidiphilium* (Harrison, 1984; Wichlacz and Thompson, 1988) most likely results from the utilization by the heterotrophs of organic compounds which could otherwise inhibit the iron-oxidizing autotroph. Some observations of better mineral leaching by uncharacterized mixed cultures in comparison with pure cultures could be explained by the activity of bacteria which have yet to be isolated and studied. More extensive solubilization of pyrite in comparison with that by pure cultures of *T. ferrooxidans* was explained when the presence and activity of strains of *L. ferrooxidans* in some mixed cultures was established (Norris and Kelly, 1982). The use of a combination of immunofluorescence and DNA fluorescence staining techniques has indicated that *T. ferrooxidans* may be displaced by other bacteria, yet to be characterized, during growth of mixed cultures on pyrite in coal (Muyzer et al., 1987).

It seems likely that at the overlap of the temperature ranges for growth of the mesophilic and the moderately thermophilic iron-oxidizing bacteria, the mesophiles would dominate aerated, mixed cultures unless the carbon dioxide concentration was increased to meet the requirement of the thermophiles. Different characteristics of some of the iron-oxidizing strains, such as the sensitivity of strain ALV to ferric iron and the requirement of TH3 for an organic nutrient or growth factor, would preclude their application in industrial mineral treatment. However, these strains serve to illustrate that a variety of moderately thermophilic, mineral-oxidizing bacteria does exist. Other strains as useful as *S. thermosulfidooxidans* and strain TH1/BC1 could be isolated in the future. Although strain BC13 is a readily iso-lated, thermotolerant "counterpart" of *T. thiooxidans,* a thermotolerant gram-negative equivalent of *T. ferrooxidans* has yet to be described.

This account of the acidophilic bacteria which have been most stud-ied in the context of the oxidation of mineral sulfides has been limited to outlining some of their characteristics in relation to factors affect-ing their activity. Explanations of some of the observed differences in bacterial behavior, for example the responses to the ferrous- and ferric-iron concentrations, await detailed studies of their iron oxida-tion systems, perhaps in relation to their particular cell surface char-acteristics. Possible interactions of bacterial iron and sulfur metabo-lism (as noted earlier, Sugio et al., 1987) and the interfacial aspects of bacteria-mineral interactions (Rodriguez-Leiva and Tributsch, 1988) are being investigated with reference mostly to the activity of *T. ferrooxidans.* The diversity of taxonomically unrelated, mineral-oxidizing bacteria with some quite different characteristics provides the option of a comparative route by which the role and importance of particular features of metabolism in mineral oxidation might be fur-ther established. This diversity also ensures bacterial activity over a range of process conditions and allows, therefore, some selection of bacteria or operating parameters, particularly in bioreactors, to give the most efficient mineral treatment.

References

Balashova, V. V., Vedinina, I. Ya., Markosyan, G. E., and Zarvarzin, G. A., The autotrophic growth of *Leptospirillum ferrooxidans, Microbiology* **43**:491–494 (1974).

Barros, M. E. C., Rawlings, D. E., and Woods, D. R., Mixotrophic growth of a *Thiobacillus ferrooxidans* strain, *Appl. Environ. Microbiol.* **47**:593–595 (1984).

Blake, R. C., White, K. J., and Shute, E. A., Electron transfer from iron to rusticyanin is catalyzed by an acid-stable cytochrome, in P. R. Norris and D. P. Kelly (eds.), *Biohydrometallurgy, Proc. Intern. Symp., Warwick,* Science and Technology Letters, Kew, U.K., 1988, pp. 103–110.

Brierley, C. L., Bacterial leaching, *Crit. Rev. Microbiol.* **6**:207–262 (1978).

Brierley, J. A., Thermophilic iron-oxidizing bacteria found in copper leaching dumps, *Appl. Environ. Microbiol.* **36**:523–525 (1978).

Brierley, J. A., and Brierley, C. L., Microbial mining using thermophilic microorgan-isms, in T. D. Brock (ed.), *Thermophiles: General, Molecular and Applied Microbiology,* Wiley, New York, 1986, pp. 279–305.

Brock, T. D., Brock, K. M, Belly, R. T., and Weiss, R. L., *Sulfolobus:* A new genus of sulfur-oxidizing bacteria living at low pH and high temperature, *Arch. Microbiol.* **84**:54–68 (1972).

Brock, T. D., and Gustafson, J., Ferric iron reduction by sulfur- and iron-oxidizing bac-teria, *Appl. Environ. Microbiol.* **32**:567–571 (1976).

Bryant, R. D., McGroarty, K. M., Costerton, J. W., and Laishley, E. J., Isolation and characterization of a new acidophilic *Thiobacillus* species (*T. albertis*), *Can. J. Microbiol.* **29**:1159–1170 (1983).

Cox, J. C., and Boxer, D. H., The purification and some properties of rusticyanin, a blue copper protein involved in iron(II) oxidation from *Thiobacillus ferrooxidans, Biochem. J.* **174**:497–502 (1978).

Dugan, P. R., and Lundgren, D. G., Energy supply for the chemoautotroph *Ferrobacillus ferrooxidans, J. Bacteriol.* **89**:825–834 (1965).

Emmel, T., Sand, W., Konig, W. A., and Bock, E., Evidence for the existence of a sulphur oxygenase in *Sulfolobus brierleyi, J. Gen. Microbiol.* **132**:3415–3420 (1986).

Fry, I. V., Lazaroff, N., and Packer, L., Sulfate-dependent iron oxidation by *Thiobacillus ferrooxidans:* Characterization of a new EPR detectable electron transport component on the reducing side of rusticyanin, *Arch. Biochem. Biophys.* **246**:650–654 (1986).

Fukumori, Y., Yano, T., Sato, A., and Yamanaka, T., Fe(II)-oxidizing enzyme purified from *Thiobacillus ferrooxidans, FEMS Microbiol. Lett.* **50**:169–172 (1988).

Gromova, L. A., Karavaiko, G. I., Sevtsov, A. V., and Pereverzev, N. A., Identification and distribution of elemental sulfur in *Thiobacillus ferrooxidans* cells, *Microbiology* **52**:357–363 (1983).

Groudev, S. N., Differences between strains of *Thiobacillus ferrooxidans* with respect to their ability to oxidize sulphide minerals, in G. I. Karavaiko and S. N. Groudev (eds.), *Biogeotechnology of Metals,* UNEP, Centre of International Projects GKNT, Moscow, 1985, pp. 83–96.

Harrison, A. P., Genomic and physiological diversity amongst strains of *Thiobacillus ferrooxidans* and genomic comparison with *Thiobacillus thiooxidans, Arch. Microbiol.* **131**:68–76 (1982).

Harrison, A. P., The acidophilic thiobacilli and other acidophilic bacteria that share their habitat, *Ann. Rev. Microbiol.* **38**:265–292 (1984).

Harrison, A. P., Characteristics of *Thiobacillus ferrooxidans* and other iron-oxidizing bacteria, with emphasis on nucleic acid analyses, *Biotechnol. Appl. Biochem.* **8**:249–257 (1986).

Harrison, A. P., and Norris, P. R., *Leptospirillum ferrooxidans* and similar bacteria: Some characteristics and genomic diversity, *FEMS Microbiol. Lett.* **30**:99–102 (1985).

Hazeu, W., Steudel, R., Batenburg-van de Vegte, W. H., Bos, P., and Kuenen, J. G., Elemental sulfur as an intermediate in the oxidation of reduced sulfur compounds by *Thiobacillus ferrooxidans:* Localization and characterization, in P. R. Norris and D. P. Kelly (eds.), *Biohydrometallurgy, Proc. Intern. Symp., Warwick,* Science and Technology Letters, Kew, U.K., 1988, pp. 111–117.

Helle, U., and Onken, U., Continuous microbial leaching of a pyritic concentrate by *Leptospirillum*-like bacteria, *Appl. Microbiol. Biotechnol.* **28**:553–558 (1988).

Holmes, D. S., Yates, J., and Schrader, J., Mobile, repeated DNA sequences in *Thiobacillus ferrooxidans* and their significance for biomining, in P. R. Norris and D. P. Kelly (eds.), *Biohydrometallurgy, Proc. Intern. Symp., Warwick,* Science and Technology Letters, Kew, U.K., 1988, pp. 153–160.

Holuigue, L., Herrera, L., Phillips, O. M., Young, M., and Allende, J. G., CO_2 fixation by mineral-leaching bacteria: characteristics of the ribulose bisphosphate carboxylase-oxygenase of *Thiobacillus ferrooxidans, Biotechnol. Appl. Biochem.* **9**:497–505 (1987).

Huber, G., Huber, H., and Stetter, K. O., Isolation and characterization of new metal-mobilizing bacteria, in H. L. Ehrlich and D. S. Holmes (eds.), *Workshop on Biotechnology for the Mining, Metal-Refining and Fossil Fuel Processing Industries,* Wiley, New York, 1986, pp. 239–251.

Huber, G., and Stetter, K. O., Properties of newly isolated metal-mobilizing bacteria, in O. Kandler and W. Zillig (eds.), *Archaebacteria '85,* Fischer Verlag, Stuttgart, 1986, p. 413.

Ingledew, W. J., *Thiobacillus ferrooxidans:* The bioenergetics of an acidophilic chemolithotroph, *Biochem. Biophys. Acta.* **683**:89–117 (1982).

Ingledew, W. J., Ferrous iron oxidation by *Thiobacillus ferrooxidans,* in H. L. Ehrlich and D. S. Holmes (eds.), *Workshop on Biotechnology for the Mining, Metal-Refining and Fossil Fuel Processing Industries,* Wiley, New York, 1986, pp. 23–33.

Jerez, C. A., Peirano, I., Chamorro, D., and Campos, G., Immunological and electrophoretic differentiation of *Thiobacillus ferrooxidans* strains, in R. W. Lawrence, R. M. R. Branion and H. G. Ebner (eds.), *Fundamental and Applied Biohydrometallurgy,* Elsevier, Amsterdam, 1986, pp. 443–456.

Jones, C. A., and Kelly, D. P., Growth of *Thiobacillus ferrooxidans* on ferrous iron in chemostat culture: Influence of product and substrate inhibition, *J. Chem. Tech. Biotechnol.* **33B**:241–261 (1983).

Karavaiko, G. I., Golovacheva, R. S., Pivovarova, T. A., Tzaplina, I. A., and Vartanjan, N. S., Thermophilic bacteria of the genus *Sulfobacillus*, in P. R. Norris and D. P. Kelly (eds.), *Biohydrometallurgy, Proc. Intern. Symp., Warwick*, Science and Technology Letters, Kew, U.K., 1988, pp. 29–41.

Kelly, D. P., Biochemistry of the chemolithotrophic oxidation of inorganic sulphur, *Phil. Trans. R. Soc. Lond. B* **298**:499–528 (1982).

Kelly, D. P., Oxidation of sulphur compounds, in J. A. Cole and S. J. Ferguson (eds.), *The Nitrogen and Sulphur Cycles*, Cambridge University Press, Cambridge, 1988, pp. 65–98.

Kelly, D. P., Eccleston, M., and Jones, C. A., Evaluation of continuous cultivation of *Thiobacillus ferrooxidans* on ferrous iron or tetrathionate, in W. Schwartz (ed.), *Conference: Bacteria Leaching*, Verlag Chemie, Weinheim, 1977, pp. 1–7.

Kelly, D. P., and Jones, C. A., Factors affecting metabolism and ferrous iron oxidation in suspensions and batch cultures of *Thiobacillus ferrooxidans:* Relevance to ferric iron leach solution regeneration, in L. E. Murr, A. E. Torma, and J. A. Brierley (eds.), *Metallurgical Applications of Bacterial Leaching and Related Microbiological Phenomena*, Academic Press, New York, 1978, pp. 19–44.

Kelly, D. P., Norris, P. R., and Brierley, C. L., Microbiological methods for the extraction and recovery of metals, in A. T. Bull, C. Ratledge, and D. C. Ellwood (eds.), *Microbial Technology: Current State, Future Prospects*, Cambridge University Press, Cambridge, 1979, pp. 263–308.

Khalid, A. M., and Ralph, B. J., The leaching behaviour of various zinc sulphide minerals with three *Thiobacillus* species, in W. Schwartz (ed.), *Conference: Bacterial Leaching*, Verlag Chemie, Stuttgart, 1977, pp. 261–270.

Konig, H., Skorko, R., Zillig, W., and Reiter, W.-D., Glycogen in thermoacidophilic archaebacteria of the genera *Sulfolobus, Thermoproteus, Desulfurococcus* and *Thermococcus, Arch. Microbiol.* **132**:297–303 (1982).

Lizama, H. M., and Suzuki, I., Bacterial leaching of a sulphide ore by *Thiobacillus ferrooxidans* and *Thiobacillus thiooxidans:* 1. Shake flask studies, *Biotechnol. Bioeng.* **32**:110–116 (1988).

Marsh, R. M., and Norris, P. R., The isolation of some thermophilic, autotrophic, iron- and sulphur-oxidizing bacteria, *FEMS Microbiol. Lett.* **17**:311–315 (1983a).

Marsh, R. M., and Norris, P. R., Mineral sulphide oxidation by moderately thermophilic acidophilic bacteria, *Biotechnol. Lett.* **5**:585–590 (1983b).

Marsh, R. M., Norris, P. R., and Le Roux, N. W., Growth and mineral oxidation studies with *Sulfolobus*, in G. Rossi and A. E. Torma (eds.), *Recent Progress in Biohydrometallurgy*, Associazione Mineraria Sarda, Iglesias, 1983, pp. 71–81.

McCready, R. G. L., Progress in the bacterial leaching of metals in Canada, in P. R. Norris and D. P. Kelly (eds.), *Biohydrometallurgy, Proc. Intern. Symp., Warwick*, Science and Technology Letters, Kew, U.K., 1988, pp. 177–195.

Mjoli, N., and Kulpa, C. F., Identification of a unique outer membrane protein required for iron oxidation in *Thiobacillus ferrooxidans*, in P. R. Norris and D. P. Kelly (eds.), *Biohydrometallurgy, Proc. Intern. Symp., Warwick*, Science and Technology Letters, Kew, U.K., 1988, pp. 89–102.

Mosser, J. L., Mosser, A. G., and Brock, T. D., Population ecology of *Sulfolobus acidocaldarius* I. Temperature strains, *Arch. Microbiol.* **97**:169–179 (1974).

Muyzer, G., De Bruyn, A. C., Schmedding, D. J. M., Bos, P., Westbroek, P., and Kuenen, G. J., A combined immunofluorescence-DNA-fluorescence staining technique for enumeration of *Thiobacillus ferrooxidans* in a population of acidophilic bacteria, *Appl. Environ. Microbiol.* **53**:660–664 (1987).

Norris, P. R., Iron and mineral oxidation with *Leptospirillum*-like bacteria, in G. Rossi and A. E. Torma (eds.), *Recent Progress in Biohydrometallurgy*, Associazione Mineraria Sarda, Iglesias, 1983, pp. 83–96.

Norris, P. R., Mineral-oxidizing bacteria: Metal-organism interactions, in R. K. Poole and G. M. Gadd (eds.), *Metal-Microbe Interactions*, IRL Press, Oxford, 1989, pp. 99–117.

Norris, P. R., and Barr, D. B., Growth and iron oxidation by acidophilic moderate thermophiles, *FEMS Microbiol. Lett.* **28**:221–224 (1985).

Norris, P. R., and Kelly, D. P., The use of mixed microbial cultures in metal recovery, in A. T. Bull and J. H. Slater (eds.), *Microbial Interactions and Communities*, Academic Press, London, 1982, pp. 443–474.

Norris, P. R., Barr, D. W., and Hinson, D., Iron and mineral oxidation by acidophilic bacteria: Affinities for iron and attachment to pyrite, in P. R. Norris and D. P. Kelly (eds.), *Biohydrometallurgy, Proc. Intern. Symp., Warwick*, Science and Technology Letters, Kew, U.K., 1988, pp. 43–59.

Norris, P. R., Marsh, R. M., and Linstrom, E. B., Growth of mesophilic and thermophilic acidophilic bacteria on sulfur and tetrathionate, *Biotechnol. Appl. Biochem.* **8**:318–329 (1986b).

Norris, P. R., Nixon, A., and Hart, A., Acidophilic, mineral-oxidizing bacteria: The utilization of carbon dioxide with particular reference to autotrophy in *Sulfolobus*, in M. S. de Costa, J. C. Duarte, and R. A. D. Williams (eds.), *Microbiology of Extreme Environments and Its Potential for Biotechnology*, Elsevier, London, 1989, pp. 24–43.

Norris, P. R., Parrott, L., and Marsh, R. M., Moderately thermophilic mineral-oxidizing bacteria, in H. L. Ehrlich and D. S. Holmes (eds.), *Workshop on Biotechnology for the Mining, Metal-Refining and Fossil Fuel Processing Industries*, Wiley, New York, 1986a, pp. 253–262.

Pivovarova, T. A., Markosyan, G. E., and Karavaiko, G. I., Morphogenesis and fine structure of *Leptospirillum ferrooxidans*, *Microbiology* **50**:339–344 (1981).

Rodriguez-Leiva, M., and Tributsch, H., Morphology of bacterial leaching patterns by *Thiobacillus ferrooxidans* on synthetic pyrite, *Arch. Microbiol.* **149**:401–405 (1988).

Schwartz, A., and Schwartz, W., Geomikrobiologische Untersuchungen VII. Uber das Vorkommen von Mikroorganismen in solfataran und heissen Quellen, *Z. Allg. Mikrobiol.* **5**:395–405 (1965).

Segerer, A., Neuner, A., Kristjansson, J. K., and Stetter, K. O., *Acidianus infernus* gen. nov., sp. nov.: Facultatively aerobic, extremely acidophilic thermophilic sulfur-metabolizing archaebacteria, *Int. J. Syst. Bacteriol.* **36**:559–664 (1986).

Silver, M., Parameters for the operation of bacterial thiosalt oxidation ponds, *Appl. Environ. Microbiol.* **50**:663–669 (1985).

Steudel, R., Holdt, G., Gobel, T., and Hazeu, W., Chromatographic separation of higher polythionates $S_nO_6^{2-}$ ($n = 3,...,22$) and their detection in cultures of *Thiobacillus ferrooxidans;* molecular composition of bacterial sulfur secretions, *Angew. Chem. Int. Ed. Engl.* **26**:151–153 (1987).

Sugio, T., Katagiri, T., Moriyama, M., Zhen, Y. L., Inagaki, K., and Tano, T., Existence of a new type of sulfite oxidase which utilizes ferric ions as an electron acceptor in *Thiobacillus ferrooxidans, Appl. Environ. Microbiol.* **54**:153–157 (1988).

Sugio, T., Mizunashi, W., Inagaki, K., and Tano, T., Purification and properties of sulfur:ferric ion oxidoreductase from *Thiobacillus ferrooxidans, J. Bacteriol.* **169**:4916–4922 (1987).

Suzuki, I., Mechanisms of inorganic oxidation and energy coupling, *Ann. Rev. Microbiol.* **28**:85–101 (1974).

Vestal, J. R., and Lundgren, D. G., The sulfite oxidase of *Thiobacillus ferrooxidans* (*Ferrobacillus ferrooxidans*), *Can. J. Biochem.* **49**:1125–1130 (1971).

Wichlacz, P. L., and Thompson, D. L., The effect of acidophilic bacteria on the leaching of cobalt by *Thiobacillus ferrooxidans*, in P. R. Norris and D. P. Kelly (eds.), *Biohydrometallurgy, Proc. Intern. Symp., Warwick*, Science and Technology Letters, Kew, U.K., 1988, pp. 77–86.

Wood, A. P., and Kelly, D. P., Autotrophic and mixotrophic growth of three thermoacidophilic iron-oxidizing bacteria, *FEMS Microbiol. Lett.* **20**:102–112 (1983).

Wood, A. P., and Kelly, D. P., Growth and sugar metabolism of a thermoacidophilic iron-oxidizing mixotrophic bacterium, *J. Gen. Microbiol.* **130**:1337–1349 (1984).

Wood, A. P., and Kelly, D. P., Autotrophic and mixotrophic growth and metabolism of some moderately thermoacidophilic iron-oxidizing bacteria, in D. E. Caldwell, J. A. Brierley, and C. L. Brierley (eds.), *Planetary Ecology*, Van Nostrand Reinhold, New York, 1985, pp. 251–262.

Wood, A. P., Kelly, D. P., and Norris, P. R., Autotrophic growth of four *Sulfolobus* strains on tetrathionate and the effect of organic nutrients, *Arch. Microbiol.* **146:**382–389 (1986).

Zillig, W., Stetter, K. O., Wunderl, A., Schulz, W., Priess, H., and Scholz, I., The *Sulfolobus-"Caldariella"* group: Taxonomy on the basis of the structure of DNA-dependent RNA polymerases, *Arch. Microbiol.* **125:**259–269 (1980).

Basic Principles of Genetic Manipulation of *Thiobacillus ferrooxidans* for Biohydrometallurgical Applications

David S. Holmes

James R. Yates

Introduction

The objective of this chapter is to describe in simple terms the fundamental principles of classical genetics and genetic engineering so that the nonspecialist can understand recent progress in the genetic manipulation of *Thiobacillus ferrooxidans*. This chapter is not a review of the literature. We have focused attention on *T. ferrooxidans* because of its significance for the mineral recovery industry and because more is known about it, even though this is embarrassingly little, than about other pertinent microorganisms.

We first describe certain selected topics of classical genetics. These include how genes work and how they control the metabolism of a cell. The important role played by mutations in providing the geneticist with experimental material is emphasized. We also point out several serious deficiencies of classical genetic approaches for bacterial strain improvement and describe how genetic engineering can overcome these deficiencies. Finally, we describe recent progress in the genetic manipulation and genetic engineering of *T. ferrooxidans* in the context of the simple framework developed in this chapter.

Principles of Classical Genetics

Genes and biochemical pathways

Organisms are capable of breaking down complex organic molecules that might occur in their food to simpler compounds and then using these metabolites to build other complex molecules that they might need. Organisms are also able to take up some simple compounds and assemble them into more complex molecules. All these chemical conversions take place in a stepwise, sequential, and controlled fashion. This stepwise sequence is called a *biochemical pathway,* and each step is catalyzed by a different enzyme. Enzymes are complex protein molecules which speed up the biochemical conversions, hence the term *catalyst.* Altogether, there are about 3000 or so different enzymes in a bacterium like *Escherichia coli.* Each enzyme requires a gene, and therefore *E. coli* has about 3000 genes for enzymes. Actually, *E. coli* has close to 4000 genes. Some of these additional genes are responsible for making proteins that are not enzymes, such as structural proteins that are part of the architecture of the cell.

How a gene makes an enzyme will be discussed in the next section. Here we want to focus on the power of genetics as an experimental tool to help us understand these chemical conversions. In order to develop superior strains of microorganisms for metal recovery, we must understand how these microorganisms accomplish what they do. Anybody who has stood in awe at the wall charts illustrating the incredibly detailed and intricate interrelationship of known biochemical pathways must have wondered how such information was derived. To a large extent it was done using genetic analysis, and one of the key concepts to understanding genetic analysis is the concept of a mutation.

A *mutation* is a mistake in a gene so that it can no longer produce a functional enzyme. Since the enzyme activity is missing, the chemical conversion that was catalyzed by the enzyme will not take place and that particular biochemical pathway becomes blocked.

Mutations have been very useful in helping to deduce the order of chemical conversions in a biochemical pathway. For example, consider the conversion of a cellular biochemical component (e.g., the product of a metabolized food source) to a required growth compound (e.g., an amino acid, termed Z for discussion). This bioconversion proceeds along the following pathway in bacteria:

$$\text{METABOLITE}$$
$$\downarrow \text{A}$$
$$\text{INTERMEDIATE 1}$$
$$\downarrow \text{B}$$
$$\text{INTERMEDIATE 2}$$
$$\downarrow \text{C}$$
$$\text{Z}$$

where A, B, and C represent specific enzymes catalyzing the conversion of a metabolite and intermediates 1 and 2, respectively. If a strain of bacteria (called strain I in the following discussion) has a mutation in the gene making enzyme B, then a nonfunctional enzyme (or no enzyme) will be produced. This blocks the conversion of intermediate 1 to 2, resulting in the accumulation of 1 (see Fig. 2.1). This strain will not be able to grow because it requires the small precursor molecule Z, which it can now no longer make. Normally, the level of intermediate 1 in the cell is fairly low because it is efficiently converted to 2. However, the mutation in strain I will cause the accumulation of intermediate 1 to levels where it can be isolated in quantities sufficient for identification and other biochemical analyses.

If another bacterial strain (called strain II for the purposes of this discussion) has a mutation resulting in a nonfunctional enzyme C,

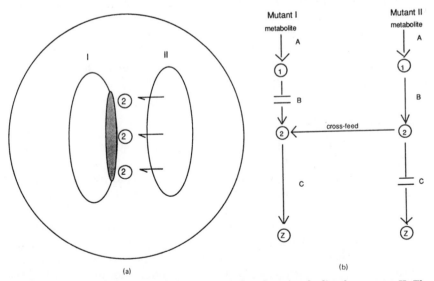

Figure 2.1 (*a*) Growth of mutant I on a petri plate by cross-feeding by mutant II. The solid region indicates growth of mutant I cells. (*b*) An explanation of the results observed in (*a*).

then intermediate 2 will accumulate. This strain could be used to produce intermediate 2 in quantities sufficient for detailed analysis. A biochemist could now begin to predict the order of conversion of intermediates in this pathway by comparing their biochemical properties and structures. A simple biological test can be devised to indicate the probable order of bioconversions in this pathway (see Fig. 2.1). Massive inocula of two complementary bacterial strains are placed side by side on a solid growth medium in a petri plate. The growth medium contains all the compounds required for growth except for the required precursor molecule Z. Neither mutant I nor mutant II grows by itself on this medium because each requires Z, but mutant II is able to convert the precursor metabolite as far as intermediate 2 before it is blocked. The accumulated intermediate 2 happens to diffuse out of the mutant II cells in this example and can be taken up by the nearby mutant I cells. The mutant I cells retain the capability to convert intermediate 2 to the precursor of Z and subsequently to molecule Z and, hence, are able to grow. Mutant II in this example is said to be able to *cross-feed* mutant I.

Intermediate 1, accumulated by mutant I, can similarly diffuse to mutant II, which can convert it to intermediate 2. However, mutant II cannot convert intermediate 2 to the required product Z; therefore mutant I cannot cross-feed mutant II. Information from this simple type of experiment together with an analysis of the nature of the accumulated intermediates can be used to unravel biochemical pathways. Note that this powerful approach would not have been possible without prior isolation of the two genetic mutants.

A second important application of mutations that block bioconversions is to generate strains of microorganisms that can be used in industrial processes to produce valuable intermediates, which might otherwise be converted to less valuable end products.

How a gene works

A gene consists of two adjoining regions of double-stranded DNA (see Fig. 2.2). The principal region of a gene has the genetic code that specifies the composition of a particular enzyme, and is termed the *coding region*. The second region is a segment of DNA that controls the activity of the coding region. Hence it is termed the *control region* or, in technical parlance, the *promoter*. Under certain conditions of growth a bacterial cell may not require the use of a particular enzyme. Under these circumstances, the promoter switches off the activity of the gene encoding that specific enzyme. Conversely, under another condition of growth, the bacterial cell may require the enzyme, and the promoter switches back on the appropriate gene. Some promoters are capable of

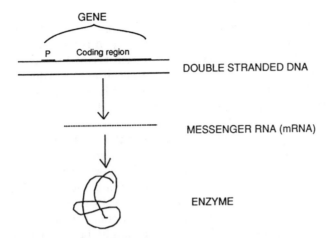

GENE

P   Coding region

DOUBLE STRANDED DNA

MESSENGER RNA (mRNA)

ENZYME

Figure 2.2 Schematic representation of a gene illustrating the flow of genetic information from the coding region to a messenger RNA and ultimately to an enzyme. P = promoter or control region.

regulating the level of activity of the coding region. Such fine-tuning permits the cell to make a little or a lot of the enzyme as dictated by its needs.

In natural bacterial populations the evolution of promoters has occurred over billions of years to keep the complex biochemical pathways under control. However, in the laboratory it is possible to mutate (alter) specific promoters so that genes can become more active (or less active, if desired). The value of such a mutation is obvious. It can permit the geneticist to obtain more of a certain enzyme which could, in turn, speed up the rate of a particular biochemical conversion. This would be a type of mutation that might speed up the rate of iron oxidation in *T. ferrooxidans*.

Both the promoter and coding region of a gene consist of double-stranded DNA, made famous by Watson and Crick. The genetic information found in the coding region consists of a sequence of four different monomers called *bases*. The four bases are abbreviated by A, C, G, and T. A typical coding region contains about 1000 of these sequential monomers. Since the number of ways these four monomers could be arranged in a sequence of 1000 is enormous (4^{1000}), the potential coding information is virtually unlimited.

The conversion of the information on the coding region of the gene to an actual structure of an enzyme involves two steps. First, a complementary copy of the bases on the coding regions is made. This copy is called *messenger RNA* (mRNA). Second, the sequence of bases on the mRNA is read sequentially, three at a time, by complex cellular machines (ribosomes) that join monomers called *amino acids* to form the enzyme (Fig. 2.3). There are 20 different amino acids in proteins, and each one has its own special three-base combination termed a *codon*. Thus the order of different codons on the mRNA specifies the order of assembly of different amino acids in the enzyme, and it is the order of the different amino acids that dictates the structure. The structure of the enzyme, in turn, dictates its function.

Many mutations result from the loss of a base or the replacement of one base by another, different, one in a codon. It can result in an inactive enzyme or sometimes no enzyme at all. Mutations occur naturally, but at very low frequencies and randomly along the gene. Of these rare mutations, only a very small percentage result in a more active enzyme. Thus, using natural mutation rates alone, it could be a very long time before a superior microorganism might be developed for an industrial application.

This has been a very simplistic description of how a gene works and what a mutation is. Interested readers can consult a genetic textbook (e.g., Goodenough, 1984) for further information.

Naturally occurring mutations were the only available mutations for strain improvement before 1920. With an increased understanding of genetics in general, and the role of DNA as the carrier of genetic information in particular, techniques were developed to enhance the rate at which mutations occurred. These techniques included the use of x-rays, ultraviolet light, and certain chemicals (*mutagens*) that cause changes in DNA. The rate of mutation could be increased several hundred- to several thousandfold by using these techniques. However, these techniques still only result in random mutations, and the rate of deleterious mutations still overwhelmed the beneficial ones. Beneficial mutations could only be detected by employing tedious selection procedures on all the mutants that were produced by these

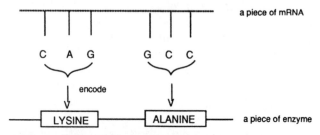

Figure 2.3   Representation of two adjacent codons on the mRNA encoding the amino acids lysine and alanine. Left out of the diagram for clarity are the complex cell machines called ribosomes that translate the code into the order of amino acids. A typical coding region has approximately 1000 bases or 333 codons. Only two codons are shown here for simplicity.

techniques. It remained essentially a matter of extreme perseverance and luck to obtain a mutation that improved the usefulness of a microorganism for industrial applications. A good example of what could be achieved by using these mutagenic techniques, with perseverance and an enormous research effort, was the enhancement of penicillin production from the *Penicillium* mold. A major effort to obtain higher producing strains was initiated at the beginning of World War II. During the next 20 years the yield of penicillin was increased several thousandfold by using the techniques of x-ray, ultraviolet light, and chemical mutagenesis.

Developing new industrial microorganisms by conjugation

Conjugation is another technique that geneticists use to obtain new strains of microorganisms. Conjugation has many similarities to sexual recombination of genes of higher organisms.

Most of the 4000 or so genes of a bacterium are joined to form a single, circular chromosome. Some bacteria have a few (e.g., 5 or 10) genes on another very small circular DNA molecule called a *plasmid*. In conjugation, one bacterium, termed the *male* bacterium, forms a conjugation tube which attaches to another bacterium, the *female*. A part of the male chromosome passes down the conjugation tube, enters the female cell, and replaces the equivalent part of the female chromosome. The conjugation tube breaks, and the cells separate. The female cell now contains integrated into its chromosome a segment of the male chromosome, which is passed down to all daughter cells when the cell divides. If the integrated male part of the chromosome contains some genetic differences from the female, then a new "recombinant" strain has been produced, which might exhibit interesting new properties.

For conjugation to be detected in the laboratory, it is necessary that the progeny of the conjugation be distinguished from either of their parents. This is usually done by using parents that exhibit two different properties. For example, the male bacterium might be able to grow in medium lacking the amino acid arginine, because it has a normal gene for making arginine (arg+). However, it might be sensitive to the antibiotic ampicillin. The female, on the other hand, is the converse. It has a mutation that requires the addition of arginine (arg−) but is resistant to ampicillin.

After allowing conjugation to occur, the bacteria can be spread on a medium lacking arginine and containing ampicillin. Male parents die because they are sensitive to the ampicillin, and female parents do not grow because there is no arginine in the medium. Only progeny of female parents which have received a piece of male DNA that includes the arg+ gene will survive and grow on the medium. This is termed a *selection procedure*. Selection procedures such as this are a common requirement for carrying out many classical procedures of genetic manipulation, and in most cases the prior existence of characterized mutants is a prerequisite.

During conjugation, plasmids can also be readily transferred from a male donor to a female recipient (see Fig. 2.4). Indeed, some plasmids even carry genes that direct the formation of the conjugation tube. Conjugative transfer is one of the most prevalent mechanisms that

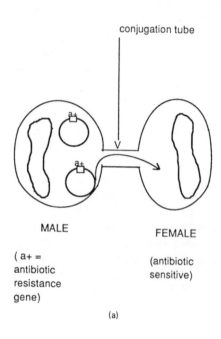

conjugation tube

MALE   FEMALE

(a+ =
antibiotic
resistance
gene)

(antibiotic
sensitive)

(a)

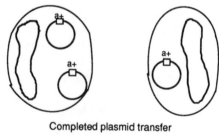

Completed plasmid transfer

(b)

Figure 2.4   (*a*) A male and a female bacterium joined by a conjugation tube. A plasmid carrying a gene (a+) that confers resistance to the antibiotic ampicillin can be seen about to enter the conjugation tube and pass into the female cell. (*b*) On uptake of the plasmid, the female cell and all progeny would become antibiotic resistant.

permits plasmid-encoded antibiotic resistances to be passed among microorganisms around the world.

Conjugation is thought to be an important mechanism in generating new strains of bacteria in laboratory chemostat selection experiments, which have been very successfully employed in creating new strains for toxic waste degradation (Harder, 1981). In a typical chemostat, a collection of different strains of microorganisms, often isolated from a toxic waste site, are grown initially in a rich medium. Gradually the rich medium is replaced by a recalcitrant (hard-to-metabolize) toxic material. Eventually the only source of carbon is provided by the toxic material, and only those microorganisms that can degrade the toxic material to obtain carbon will grow. This is a very powerful selection procedure. Frequently, none of the microorganisms present at the inception of the experiment can degrade the toxic material. It is thought that the degradation ability arises in one or a few microorganisms as they transfer genes from one to another by conjugation, allowing them to create entirely new biochemical pathways composed of different elements of several existing pathways.

Despite its success, chemostat selection has two severe limitations. First, it depends on the preexistence of appropriate biochemical pathway elements in the initial starting bacterial population. This can be optimized by starting with strains of microorganisms from a toxic waste site which have a higher than average likelihood of being able to degrade a range of organic compounds. However, if the right genes are not present in the first place, and this often happens, then virtually no amount of conjugation can come up with the right new combination of genes to specify the required new biochemical pathway. Second, conjugation generally occurs only between related species, so that if the correct genes are present in two or more strains that will not conjugate, then the required recombination will not take place. As will be described later in this chapter, genetic engineering can overcome both of these deficiencies.

Transformation and transduction

Two other classical techniques exist which are used primarily as tools for genetic analysis but which could, in principle, be used to generate new strains of microorganisms for industrial applications.

The first is called *transformation*. If certain strains of laboratory microorganisms, such as *E. coli*, are treated with calcium, they gain the ability to take up DNA (genes) from the surrounding solution. The DNA taken up could be a plasmid, which once inside the cell can replicate or it could be a fragment of DNA containing one or a few genes.

In the latter case, the DNA can occasionally become incorporated into the bacterial chromosome. If it doesn't become incorporated, then it is rapidly degraded with the cell.

In nature, transformation is probably not a common way to obtain new strains because the concentration of free DNA in the environment is usually very low and most microorganisms do not readily take up DNA. In the laboratory, transformation is a very important step in genetic engineering strategies because it represents the way that genetically manipulated DNA can be introduced into bacterial cells.

The second technique, *transduction,* involves the introduction of DNA into a bacterial cell by means of bacterial virus. A typical bacterial virus infects a cell by first attaching to a specific site on the bacterial membrane. It then injects its own DNA into the bacterial cell by a mechanism not unlike the operation of a hypodermic syringe. It is possible to incorporate a small piece of foreign DNA into the viral DNA without affecting the ability of the virus to inject DNA into a bacterial cell. When such a virus injects its own DNA into a bacterial cell, the foreign DNA is carried along with the viral DNA. Once inside the cell the foreign DNA may be incorporated into the recipient bacterial chromosome, or else it is rapidly degraded.

Principles of Genetic Engineering

Advantages of genetic engineering

Natural mutations are very rare and are random. Therefore, it can be extremely difficult or impossible to obtain a beneficial mutation by using classical genetic approaches. In contrast, genetic engineering techniques, developed during the last 16 years, have enabled scientists to mutate a gene at precisely the desired position. This is termed *site-directed mutagenesis.*

There is a second major advantage to genetic engineering. The probability of obtaining one beneficial mutation by classical procedures is very low, as mentioned. The probability of obtaining two simultaneous mutations is the product of the probability of the two independent mutations, which in practical terms is essentially zero. However, genetic engineering permits any number of desired mutations to be made simultaneously.

A third key advantage to genetic engineering concerns its use in transferring DNA across species boundaries. For example, the human insulin gene has been cloned into the microorganism *E. coli.* There is still debate about the extent to which DNA is exchanged between different species in nature, but it is safe to say that the use of genetic engineering has greatly facilitated the exchange of genetic material between nonrelated species. Thus one can build a new biochemical pathway from components taken from two or more nonrelated organisms. Genes from any organism in the world can be used to modify a microorganism. This represents an enormous repository of genetic information.

These three advantages have revolutionized the ability to create new genes, and many of these changes have resulted in organisms with superior biochemical properties for specific industrial applications.

A typical genetic engineering strategy

A typical genetic engineering strategy is illustrated in Fig. 2.5. The strategy can be divided into 10 steps.

1. It starts with the isolation of DNA by standard procedures from the experimental organisms, for example, *T. ferrooxidans.*

2. The DNA is then broken down to gene-size pieces by using *restriction enzymes* that cut DNA.

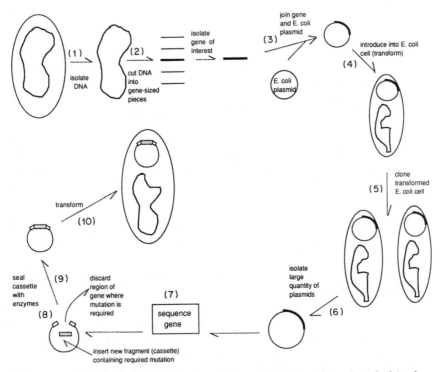

Figure 2.5   A typical genetic engineering strategy. Steps 1 to 10 are described in the text.

3. A single gene-size piece is then isolated from the complex of gene pieces. Usually this is some particular gene of interest, for example, a gene that codes for an enzyme under study. It is not always easy to identify and isolate, since an average bacterium contains about 4000 different genes. However, there are several powerful procedures of molecular biology that can be used for this step. These procedures are too technical for discussion here, but the interested reader will find them described in standard textbooks of molecular biology (e.g., Freifelder, 1983, Chap. 20). Once the gene of interest has been isolated, it is incorporated into an *E. coli* plasmid. This procedure is done entirely in the test tube and is the heart of genetic engineering. The *E. coli* plasmid is a small circle of DNA, and it is cut once with a restriction enzyme to form a linear molecule. The gene of interest is then added to the linear plasmid. The end of the gene attaches to the linear plasmid to form a recombinant linear molecule consisting of part plasmid DNA and part gene DNA, hence the term *recombinant DNA*. The recombinant linear molecule then closes on itself to reform a circle. The junctions between the plasmid and the inserted gene are sealed with an enzyme called ligase.

4. The recombinant plasmid is then put back into *E. coli*. This process is called *transformation*.

5. The *E. coli* cell is then allowed to divide many times in culture, producing billions of bacteria, each identical to the starting bacterium. Thus they represent a clone of the original bacterium. Since the recombinant plasmid is also formed in all the progeny bacteria, the inserted gene of interest is said to "have been cloned."

6. Recombinant plasmids are now isolated from the host *E. coli*, giving billions of copies of a single gene of interest. This provides enough material for geneticists to study and manipulate, as shown for example in the next two steps.

7. The gene of interest is then sequenced to deduce the order of bases along the coding and control regions. Because the genetic code is known, it is possible to predict the order of amino acids in the enzyme by reading the triplet codons in the coding regions.

In early experiments, an enormous effort was required to sequence 100 bases of DNA (G. Khorana received the Nobel prize in 1968 for this effort). Since a typical gene contains 1000 bases of DNA, progress was slow. In the 1970s, two new techniques were developed which greatly facilitated DNA sequencing. (W. Gilbert and F. Sanger shared the Nobel prize in 1980 for these inventions.) Today, well over 1 million of the 4 million bases of *E. coli* DNA have been sequenced. Altogether over 10 million bases of DNA have been sequenced from various organisms, including bacteria, viruses, fungi, plants, and animals. These sequences have been deposited in computer banks around the world. Also, there are now several first-generation automatic DNA sequencing machines that have just emerged on the market, and these will simplify even further our ability to sequence genes. This is an excellent example of how advances in science can often be driven by advances in supporting technology and instrumentation.

8. The region on the gene of the bacterium where the desired mutation is required is identified, cut out using restriction enzymes, and is then discarded. A new segment of DNA is chemically synthesized with a machine. This new segment contains a programmed change in the order of bases on the coding region that will result in the desired mutation in the enzyme. The new segment is then used to replace the discarded segment and "glued" into place by ligase enzymes. Because a DNA segment can be "plugged" into a gene, it is called a *cassette*.

Machines are on the market that can chemically synthesize about 100 bases of a gene. The desired sequence of bases is programmed in the DNA by a computer. Complete genes of about 1000 bases have now been made by "gluing" together several such individual 100-base units of DNA. It is hoped that future machines will become available to chemically synthesize whole genes of 1000 or more bases without recourse to having to "glue" pieces together.

9. The mutated gene is now amplified billions of times by repeating steps 4 through 6.

10. The final step is to return the plasmid containing the mutated gene back to the microorganism which donated the gene in the first place. Similar to step 4, this is called transformation. Four requirements for successful transformation must be met. First, the plasmid must enter the bacterial cell. Second, it must be stably replicated once inside the cell so that all daughter cells receive a copy of the plasmid. Third, the host cell must be able to recognize the control region of the genetically engineered gene and thus be able to convert the code of the coding region into an enzyme. This usually does not represent a problem if one is returning a genetically engineered gene to the microorganism that donated the gene in the first place, because the organism recognizes its "own" control region. In contrast, the control region of a gene from another microorganism might not be correctly recognized and, as a consequence, the genetically engineered gene is not copied into mRNA and therefore no enzyme is made. This can be a problem even between related species of microorganisms. For example, it has been found that some *Pseudomonas* genes do not work in *E. coli,* and vice versa, whereas genes from some other kinds of bacteria do. Fourth, the bacterium that successfully takes up the recombinant

plasmid, successfully replicates it, and successfully expresses the genetically engineered gene must be identifiable. The reason is that transformation, even in the most favorable circumstances, is a rare event. For example, only about one out of 10,000 apparently identical cells of *E. coli* takes up a plasmid. Therefore, it is necessary to identify this 1-in-10,000 cell. Usually this is done by putting a gene on the recombinant plasmid that confers resistance to a drug, such as ampicillin. Under these circumstances the 1-in-10,000 cell that takes up the recombinant plasmid becomes resistant to the drug and can be selectively grown in a medium containing the drug.

A rather striking example of the use to which the repository of genetic information can be put is the introduction of the light gene (luciferase gene) of the firefly into plants. The luciferase gene is responsible for the enzyme that creates the light flashes of the firefly. Genetic engineers have isolated this gene from the firefly and have introduced it into plants, thus making parts of the plant glow in the dark. Clearly, this could not have been achieved by natural processes. Before the reader gets carried away imagining glow-in-the-dark *T. ferrooxidans* for underground mining, it can be estimated that the number of *T. ferrooxidans* cells in natural habitats is insufficient to create much light. Perhaps that is just as well because we do not want our mines to fill up with lovesick fireflies!

Another example of the power of genetic engineering to mutate a gene at a desired location is provided by the work of Estell et al. (1984). These researchers were interested in the structure and function of the protease enzyme subtilisin. The enzyme is called a protease because its function is to degrade other proteins. It does this by cutting the target protein into small pieces like a pair of molecular scissors. However, it cuts at only certain sites in the target protein. These sites have hydrophobic (water-disliking) amino acids. The part of the subtilisin that "cuts" the target protein, the so-called active site, is actually embedded in a deep crevice in the enzyme. It has been postulated that the hydrophobic amino-acid region of the target protein likes to enter this crevice to escape from the surrounding water. Once inside the crevice, the active site cuts the target protein. This process is illustrated diagrammatically in Fig. 2.6.

Estell and coworkers (1984) wanted to know whether they could genetically engineer subtilisin so that it could target proteins at sites other than those with hydrophobic amino acids. To do this, they first isolated the gene for subtilisin and determined the nucleotide sequence of the entire gene. From the sequence, they were able to deduce the section of the gene coding for the deep crevice (the active cutting site) of the enzyme; they cut out this section of the gene and

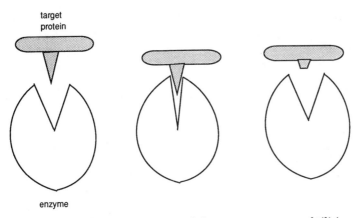

target
protein

enzyme

Figure 2.6 The mechanism of action of the protease enzyme subtilisin.

discarded it. Next they replaced the removed section sequentially with 19 fragments (cassettes) of DNA, chemically synthesized by a machine, in 19 separate experiments. Each cassette was made so that it encoded a different amino acid. In this way, they were able to test the effect of replacing one of the natural amino acids found in the crevice of subtilisin with all the remaining amino acids available. They found that if a charged amino acid, such as arginine, was introduced into the crevice, it would now accept a hydrophilic (water-loving) region of the target protein and cut it successfully. This occurred because a hydrophilic region of the target protein "likes" a charged molecule such as arginine. Thus, Estell et al. were able to dramatically change the substrate specificity of the enzyme. To change 19 amino acids sequentially in exactly the same position of an enzyme by classical genetic techniques is inconceivable and illustrates the power of genetic engineering.

Application of Classical Genetics to *T. ferrooxidans*

Special problems

In the section "Principles of Classical Genetics," some of the basic concepts of classical genetics have been described with special reference to how these concepts have been utilized to generate new strains of microorganisms. In this section, we review how these concepts have been applied to *T. ferrooxidans*.

Although *T. ferrooxidans* is a gram-negative bacterium like *E. coli*, it differs from *E. coli* in many important respects. Some of these differences make *T. ferrooxidans* harder to work with than *E. coli*. For example, it grows in the laboratory to cell densities that are typically three orders of magnitude less than can be achieved with *E. coli,* and its rate of cell division is about 10 times slower than that of *E. coli.* This results in 1/10,000 of the cell yield per unit volume per unit time which can be obtained with *E. coli.*

A particularly vexing problem that has impeded progress has been the difficulty encountered in trying to obtain colonies of *T. ferrooxidans* on a solid medium. The ability to isolate colonies is a requirement for many mutagenic experiments with bacteria. If bacteria, such as *E. coli,* are spread on a suitable nutrient medium gelled with agar, individual bacterial cells will divide repeatedly. The cells will remain localized until they form visible colonies (each containing more than a million cells) as shown in Fig. 2.7. Ideally each colony arises from a single cell and is genetically pure. A colony is a clone of bacteria. Spontaneous mutants that arise as the colony develops are very rare and therefore make up an exceedingly small part of the whole. In contrast, if one of the original cells spread on the solid me-

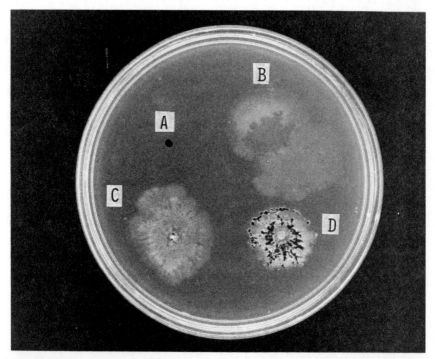

Figure 2.7 *T. ferrooxidans* colonies grown on solid media containing a mixture of Fe^{2+} and thiosulfate. A = normal colony. B–D = mutant colonies.

dium was a mutant, then all its descendent cells that form the colony will also be mutants. With this technique, rare mutant colonies that are present in a background of normal ones can be chosen for further study. The problem has been that *T. ferrooxidans* cells did not readily form colonies on standard agar media. The reason for this is that *T. ferrooxidans* is inhibited by some of the organic compounds found in unpurified agar. Several media formulations have now been devised that overcome this problem (e.g., Tuovinen and Kelly, 1973; Manning 1975). One of the simplest is to substitute very pure (e.g., electrophoresis-grade) agarose for the agar (Holmes et al., 1983).

The ability to form colonies on a solid medium is also necessary for strain purification. One can apply a mixed population of bacterial cells derived, for example, from a bioleaching operation, to the solid medium and pick out individual species as they grow into colonies. One can imagine the confusion that could result if a mixed culture of bacteria is not purified prior to mutagenic treatment. Unusual colonies that grow up would be incorrectly identified as mutants, whereas they might represent a novel species of naturally occurring bacteria.

The difficulty encountered in trying to grow *T. ferrooxidans* on a solid medium is typical of the problems that investigators had to solve before genetic studies on this organism could proceed. Other such problems will become apparent later in this chapter.

Developing special strains of
T. ferrooxidans by mutation

Groudeva et al. (1980) mutagenized *T. ferrooxidans* using a chemical mutagen, nitrosoguanidine. Treated cells were applied to solid media containing either Fe^{2+} or S^0 as an energy source. Colonies that grew larger or faster than usual were picked and subcultured in liquid medium for further tests of Fe^{2+}- and S^0-oxidizing activity. The rationale was that larger than usual colonies might represent mutants with an enhanced ability to oxidize either Fe^{2+} or S^0. Some additional work was reported by Groudeva et al. (1981a,b), but detailed descriptions of the nature of the putative mutations were not published.

More recently, Cox and Boxer (1986) isolated putative mutants of *T. ferrooxidans* that could grow on a solid thiosulfate medium but not on a medium containing Fe^{2+}. These investigators used a common and important technique to isolate mutants. After treatment with a mutagenic agent, they applied cells to a solid thiosulfate medium in a petri plate. They transferred colonies which developed in a replica pattern to plates containing the solid medium with Fe^{2+} or thiosulfate, as shown in Fig. 2.8. All the sampled colonies were expected to grow on the thiosulfate medium since that was the medium from which they were picked. However, any cell that was mutated in the biochemical

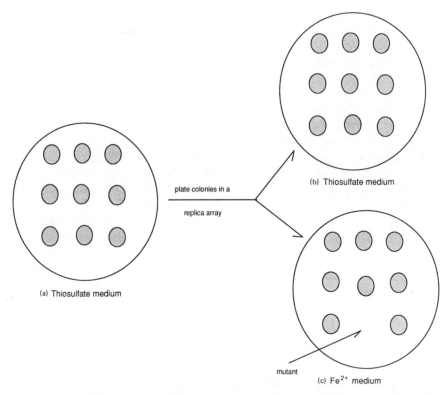

Figure 2.8 Replica plating of *T. ferrooxidans* as a technique to detect mutants unable to grow on Fe^{2+}. Cells are first treated with a chemical agent to enhance the rate of normal mutation and spread on a petri plate containing growth medium with thiosulfate (*a*) as an energy source. Colonies grow up where individual cells were laid down. Individual colonies are transferred in a replica pattern to plates containing either thiosulfate (*b*) or Fe^{2+} (*c*) as an energy source. All the colonies should grow successfully on plate (*b*) because they previously grew on a similar plate [plate (*a*)]. However, if a colony on plate (*a*) grew up from a mutant cell which cannot utilize Fe^{2+} as an energy source, it will not grow on plate (*c*) [e.g., middle of the bottom row, plate (*c*)]. This colony can be picked from the equivalent position on plate (*b*) for further study.

pathway for iron oxidation should grow on a solid medium containing Fe^{2+} as an energy source. Therefore, colonies derived from such mutants will not grow when transferred to an Fe^{2+}-containing solid medium. This can be readily detected because an absent colony results in a conspicuous gap in the orderly array of replica colonies (see Fig. 2.8). The mutant can be preserved by picking the equivalent colony from the plate containing the thiosulfate medium. Candidate mutant colonies are always retested to ensure that they lack the ability to grow on the Fe^{2+}-containing medium, and are not merely experimental artifacts resulting from things like failure to transfer a colony from the source plate to the replica plate.

Using this technique, Cox and Boxer (1986) found an unexpectedly large number of potential mutant *T. ferrooxidans* colonies which grew on the thiosulfate but not on the Fe^{2+} medium. Two of these prospective mutants lacked rusticyanin, one of the enzymes involved in iron oxidation. However, after this promising beginning, they carried out no further analysis (Cox, personal communication).

A novel approach for the isolation of *T. ferrooxidans* mutants has recently been described (Holmes et al., 1988; Schrader and Holmes, 1988). In this technique, cells were first mutagenized and spread on a medium containing a mixture of Fe^{2+} and thiosulfate. Normal cells gave rise to small rusty colonies, whereas those unable to utilize Fe^{2+} (potential mutants) resulted in large clear colonies, as shown in Fig. 2.7. An advantage of this technique is that prospective mutant colonies can be immediately identified without recourse to the tedious replica plating technique illustrated in Fig. 2.8.

During the course of the above experiments it was observed, unexpectedly, that even *T. ferrooxidans* cells which had not been subjected to prior chemical mutation frequently gave rise to colonies that appeared to be mutant. An occasional such colony is not unexpected since natural mutations occur at a low frequency. However, the frequency of appearance of the mutants exceeded normal mutation rates by several orders of magnitude. It was discovered that those mutants also reverted back to forming the normal colony type at an unusually high frequency. Preliminary evidence indicates that this high frequency of forward and backward mutation is caused by special gene segments called *insertion sequences* (IS), which appear to move from position to position around the chromosome of *T. ferrooxidans* (Yates and Holmes, 1987; Yates et al., 1988). These investigators speculated that if one of these IS jumps into and inactivates a region of the chromosome encoding enzymes involved in iron metabolism, then a mutant unable to oxidize iron will arise. Similarly, normal metabolism will be restored when the IS jumps out again.

The discovery of the phenomenon of natural high-frequency mutations in *T. ferrooxidans* has several important implications. For example, such mutations can occur spontaneously at such high frequency that they numerically dominate any mutation that might arise after treatment with a chemical mutagen. This makes chemically derived mutants unusually difficult to detect. Clearly, strains of *T. ferrooxidans* that do not give rise to high-frequency spontaneous mutants will have to be utilized in the future before classical techniques of mutation, described in the subsection "Developing New Industrial Microorganisms by Conjugation" can be used to develop new stable strains of the organism.

High-frequency mutation has been found in several laboratory strains of *T. ferrooxidans* and several strains isolated from the environment (Schrader and Holmes, 1988). The occurrence of high-frequency mutation in natural populations could account, at least in some instances, for the commonly observed ability of *T. ferrooxidans* to adapt to specific laboratory culture conditions (e.g., increasing levels of copper concentration) and to adapt to particular conditions in mining operations. If this explanation is correct, then it suggests that strains of *T. ferrooxidans* exhibiting high-frequency spontaneous mutation might be good candidates for chemostat selection to generate new strains. Clearly the issue of the stability of such strains in a bioleaching operation would have to be addressed. However, a particular advantage of such "naturally" engineered strains is that they would probably not be subject to guidelines or regulations that govern the deliberate use of genetically engineered microorganisms in the environment.

Another important consideration regarding the appearance of *T. ferrooxidans* variants at high frequencies concerns the genetic engineering of superior strains. Clearly, it will not be desirable to genetically engineer an organism that is genetically unstable, and which could easily lose its engineered properties by the same mechanism that causes spontaneous variability. Strains of *T. ferrooxidans* which lack this genetic instability for use as recipients in genetic engineering experiments must be found, or else one must learn to regulate this instability. A definite advantage to the latter approach is that by learning more about the mechanism that controls the natural genetic instability, one may be able to manipulate the mechanism to create desired mutations.

In conclusion, no detailed analysis of any *T. ferrooxidans* mutant has been published. However, there are promising commercial applications and research directions for strains of *T. ferrooxidans* that exhibit high-frequency spontaneous mutation. In the section "Principles of Classical Genetics" a heavy emphasis was placed on the importance of mutations as a tool for genetic analysis. In *E. coli* thousands of different mutations are available to researchers, whereas in *T. ferrooxidans* none is available. It is critical that this situation be remedied.

Developing special strains of
T. ferrooxidans by conjugation

There is no published work describing conjugation experiments with *T. ferrooxidans*. The major problem is that there are no mutants available to permit selection of the progeny from conjugation experiments between two *T. ferrooxidans*. In theory it might be feasible to conjugate another species of bacterium carrying, for example, an antibiotic resistance gene with *T. ferrooxidans* and to select for progeny cells that can grow on thiosulfate in the presence of the antibiotic. One of the impediments to this approach is the difference in pH requirements for *T. ferrooxidans* (pH 4.5 on thiosulfate) and most other available bacteria (pH 6 to 7). One possible solution would be to conjugate a gene down a "pH ladder" using organisms having progressively lower pH optima. This might make it possible to take a gene all the way from *E. coli* to *T. ferrooxidans*. With this type of approach in mind, Kulpa et al. (1983) have successfully conjugated a known plasmid into *Thiobacillus neopolitans*.

One piece of evidence that suggests that *T. ferrooxidans* is capable of conjugation is the discovery of a gene on a *T. ferrooxidans* plasmid that, when introduced into *E. coli,* assists in the transfer of plasmids between *E. coli* strains (Rawlings et al., 1986). Since such a gene is present in *T. ferrooxidans,* it must surely assist in *Thiobacillus* conjugation. Hopefully further experiments will clarify this point.

Application of Genetic Engineering to
T. ferrooxidans

Has *T. ferrooxidans* been modified yet by genetic engineering? Unfortunately, the answer is "no." *T. ferrooxidans* cells contain DNA that can be isolated quite easily, cloned into *E. coli* plasmids, and introduced (transformed) into *E. coli* cells (Holmes et al., 1983; Rawlings et al., 1983). That is, steps 1 to 6 of the typical genetic engineering strategy shown in Fig. 2.5 can be readily accomplished. Inside the *E. coli* cells, the *T. ferrooxidans* DNA is replicated as part of the *E. coli* plasmid. In fact, the *E. coli* cells do not "know" that they are replicating *T. ferrooxidans* DNA. This means that genetic engineers can use *E. coli* to manufacture large quantities of *T. ferrooxidans* genes for analysis. This has been put to good use in studying *T. ferrooxidans* genes that encode enzymes used in nitrogen fixation and metabolism (Pretorius et al., 1986) and to learn about the mobile insertion sequences that probably cause the genetic variability of *T. ferrooxidans* (Yates and Holmes, 1987; Yates et al., 1988). In several instances it has been shown that some *T. ferrooxidans* genes can actually function in *E. coli* (Rawlings and Woods, 1985; Barros et al., 1986; Kulpa et al., 1986). This is very encouraging because it means that the regulatory regions that govern the expression of these genes also work in *E. coli*. The possibility of learning how the *T. ferrooxidans* regulatory regions work is thus opened up, and this, as will be discussed shortly, is going to be very important.

The bad news is that no one has successfully returned functioning genes from *E. coli* to *T. ferrooxidans* (i.e., step 10 of Fig. 2.5). This means that, in principle, genes can be extracted from *T. ferrooxidans* and genetically modified in the test tube by the techniques of genetic engineering but cannot yet be returned to *T. ferrooxidans* to create an improved organism. Recombinant plasmids containing *E. coli* DNA encoding resistance to various metals and drugs have been constructed. Unsuccessful attempts have been made to introduce these constructions into *T. ferrooxidans* or other acidophilic microorganisms (Holmes et al., 1984, 1986; Rawlings et al., 1984). The rationale is that if the *T. ferrooxidans* cells take up and use the recombinant DNA constructions, then they will become resistant to toxic levels of the metals or drugs encoded by the introduced genes. Once this has been accomplished, any gene can be introduced into the construction and returned to the *T. ferrooxidans* cell.

The failure to get recombinant DNA genes to be expressed in *T. ferrooxidans* could be due either to the inability of the recombinant plasmid DNA to penetrate into the cells (perhaps due to peculiarities of the *T. ferrooxidans* membrane) or to the inability of *T. ferrooxidans* cells to utilize the *E. coli* drug and metal-resistant genes once the plasmid has entered the cell. This is a definite possibility. It is known that some *E. coli* genes will not work in some other bacteria due to the inability of the other bacteria to recognize the promoter (control region) of the incoming gene. There are several potential solutions to this problem, which fall into three categories: (1) Devise alternative ways to introduce DNA into *T. ferrooxidans* cells; (2) place the *E. coli* gene for metal or drug resistance next to a *T. ferrooxidans* promoter; (3) introduce a *T. ferrooxidans* gene back into *T. ferrooxidans*, thereby leaving no doubt about its ability to function in *T. ferrooxidans*. Solutions (1) and (2) are technically straightforward, but whether they will work remains to be seen. Solution (3) raises another issue, namely what gene should be chosen for introduction into *T. ferrooxidans*? A requirement is that the return of the gene must be an identifiable event. There are two possible sources of genes whose introduction would result in an identifiable event. The first source is from natural variants of *T. ferrooxidans*. For example, different strains of *T. ferrooxidans* vary with regard to their resistance to toxic metals (Tuovinen et al., 1971). Therefore, one could introduce the gene(s) for a particular metal resistance derived from a metal-resistant strain into a *T. ferrooxidans* plasmid by genetic engineering. Then one could introduce the genetically engineered plasmid (recombinant DNA plasmid) into a strain that is sensitive to the metal. Those cells that take up the plasmid will be able to survive in an otherwise toxic concentration of the metal, and cells that do not take up the plasmid will die. This is called a *selective procedure*. One possible problem with this approach is that the metal-resistant trait (or any other selectable trait) might be encoded by several genes, only one of which was introduced into the recombinant DNA plasmid. This pitfall could be avoided by utilizing a selective trait encoded by only one gene. However, our knowledge of the genetics of *T. ferrooxidans* is so limited that we do not have even this simple kind of information.

Another way in which genes could be introduced into *T. ferrooxidans* and result in an identifiable event would be to clone genes into a mutant strain. For example, suppose a cell lacks the ability to oxidize iron due to a specific mutation in a gene encoding an enzyme involved in iron oxidation. If genes were randomly cloned from a normal strain into a plasmid, then the recombinant plasmid containing the specific iron-oxidizing gene could enter the mutant cell and correct (complement) the deficiency caused by the mutation. This cell would then be able to grow. The specific iron-oxidizing gene would be continually replicated into the introduced plasmid and would be available for isolation and detailed characterization. There are currently no mutations of *T. ferrooxidans* that have been characterized in sufficient detail to act as recipients for recombinant DNA plasmids. It is clearly important to develop such mutants.

An interesting variation on this idea is to complement *E. coli* mutants with genes derived form *T. ferrooxidans* (Kulpa et al., 1986). For example, Barros et al., (1986) have prepared recombinant DNA plasmids containing *T. ferrooxidans* DNA and have then introduced these plasmids into mutant *E. coli* cells that lack the ability to synthesize glutamine. *E. coli* cells that took up the recombinant DNA plasmid that contained the equivalent *T. ferrooxidans* glutamine gene were able to grow. These investigators now had a source of the *T. ferrooxidans* glutamine gene which could be examined in detail. The logical step now is to return the recombinant plasmid to a glutamine mutant of *T. ferrooxidans*. Unfortunately, such a mutant has not yet been isolated. Clearly, this represents another good reason to develop mutants of *T. ferrooxidans*.

In summary, it has proved feasible to remove genes from *T. ferrooxidans* and to clone them in *E. coli*. Details of the structure of these genes, including their DNA sequence, are beginning to emerge. This will soon generate the kind of information required in order to genetically engineer them. It has also proved possible to isolate plasmids from *T. ferrooxidans*. These plasmids could, in principle, serve as vectors to carry genes back into *T. ferrooxidans* cells. However, methods to return such recombinant DNA plasmids and to permit the expression of the genes they carry in *T. ferrooxidans* have not yet been developed. This is a major roadblock to the development of genetic engineering of *T. ferrooxidans*. Conjugation and transduction could also be of use, in principle, to introduce genetically engineered genes back into *T. ferrooxidans*. But, as has already been pointed out, a conjugation system has not yet been established in *T. ferrooxidans* because of a lack of mutational markers. Also, no viruses specific for *T. ferrooxidans* have been described in the literature. Such viruses almost certainly exist. It is just a question of finding them. At present, too few laboratories in the world are working on these sorts of problems to cover all the options adequately.

Further Discussion

Once a method is developed for introducing genes back into *T. ferrooxidans,* our understanding of its genetics and biochemistry will proceed rapidly. Also, the powerful technique of site-directed mutagenesis can then be applied to improving the capabilities of *T. ferrooxidans* for bioleaching applications.

What characteristics of *T. ferrooxidans* might one manipulate in order to enhance their usefulness? The answer is not simple. The physical and biochemical parameters that govern the rate and extent of biooxidation are extremely complex, and the rate-limiting step has not yet been identified. Until these parameters have been identified, it is only possible to speculate on what modifications might be desirable for strain improvement. An enhancement of growth rate might be attractive in certain circumstances, as would an enhancement of the rates of iron and sulfur oxidation. Other characteristics that may prove to be important and worth modifying by genetic engineering include enhanced ability to oxidize nonferrous metals, reduced sensitivity to organic compounds, better attachment properties to solid substances, increased metal tolerance, tolerance to cyanide, surfactants, and chloride ions, enhanced activity in the cold, secretion or intracellular production of specific metal-binding proteins, resistance to phage infection, reduced tendency to undergo genetic changes, enhanced ability to fix nitrogen, and so on.

Whatever improvements are ultimately made either directly by genetic manipulation or indirectly by laboratory selection of naturally occurring strains, the end result will be a strain of bacteria that must compete and retain the activity of desirable traits in the presence of native microorganisms. The question of the survivability of a laboratory strain in a bioleaching operation, whether in an open dump or a bioreactor, has not been investigated. A useful objective of genetic engineering would be to introduce a suitable genetic marker into a laboratory strain to facilitate strain identification and to improve a selection mechanism to promote the maintenance of a laboratory strain in a bioleaching operation.

We reemphasized that a major objective of research laboratories is to utilize genetics and genetic engineering to develop the necessary understanding of the biochemistry of relevant microorganisms. Without a thorough understanding of this biochemistry, it would be difficult at best, impossible at worst, to improve strains for practical metal recovery. However, the prospects for strain improvement by genetic engineering will be exciting to contemplate, once such fundamental biochemical studies have been completed.

Acknowledgments

We are very grateful to Stacey Hood for her expert typing. We would also like to thank Dr. Henry L. Ehrlich for preparing the figures.

References

Barros, M. E. C., Rawlings, D. E., and Woods, D. R., Purification and regulation of a cloned *Thiobacillus ferrooxidans* glutamine synthetase, *J. Gen. Microbiol.* **132**:1989–1995 (1986).

Cox, J. C., and Boxer, D. H., The role of rusticyanin, a blue-copper protein, in the electron transport cahin of *Thiobacillus ferrooxidans* grown on iron or thiosulfate, *Biotech. Appl. Biochem.* **8**:269–275 (1986).

Estell, D. A., Miller, J. V., Graycar, T. P., Powers, D. B., and Wells, J. A., Site-directed mutagenesis of the active site of subtilisin BPN, in *Biotech. 84 USA*, Online Publishers, Pinnar, U.K., 1984, pp. 181–187.

Freifelder, D., *Molecular Biology*, Jones and Bartlett, Boston, 1983.

Goodenough, U., *Genetics*, 3d ed., Saunders, Philadelphia, 1984.

Groudeva, V. I., Groudev, S. N., and Markov, K. I., Nitrosoguanidine mutagenesis of *Thiobacillus ferrooxidans* in relation to the levels of its oxidizing activity, *Bulgar. Acad. Sci.* **33**:1401–1404 (1980).

Groudeva, V. I., Groudev, S. N., and Markov, K. I., Different types of *Thiobacillus ferrooxidans* mutants possessing high sulfur-oxidizing activity, *Bulgar. Acad. Sci.* **34**:1594–1551 (1981a).

Groudeva, V. I., Groudev, S. N., and Markov, K. I., Biological bases of increased ferrous-oxidizing activity of *Thiobacillus ferrooxidans*, *Bulgar. Acad. Sci.* **35**:371–373 (1981b).

Harder, W., Enrichment and characterization of degrading organisms, in T. Lesinger, A. M. Cook, R., Hunter, and J. Neusch (eds.), *Microbial Degradation of Xenobiotic and Recalcitrant Compounds*, Academic Press, London, 1981.

Holmes, D. S., Lobos, J. H., Bopp, L. H., and Welch, G. C., Setting up a genetic system de novo for studying the acidophilic thiobacillus *T. ferrooxidans*, in G. Rossi and A. E. Torma (eds.), *Recent Progress in Biohydrometallurgy*, Associazione Mineroria, Sarda, Iglesias, Italy, 1983, pp. 541–554.

Holmes, D. S., Yates, J. R., Lobos, J. H., and Doyle, M. V., Genetic engineering of biomining organisms, in *Biotech. 84*, Online Publishing, London, 1984, pp. A64–A81.

Holmes, D. S., Yates, J. R., Lobos, J. H., and Doyle, M. V., Strategy for the establishment of a genetic system for studying the novel heterotroph, *Acidiphilium organovorum*, *Biotech. Appl. Biochem.* **8**:258–268 (1986).

Holmes, D. S., Yates, J. R., and Schrader, J., Mobile repeated DNA sequences in *Thiobacillus ferrooxidans* and their significance for biomining, in P. R. Norris and D. P. Kelly (eds.), *Biohydrometallurgy*, Science and Technology Letters, Kew, U.K., 1988, pp. 153–160.

Kulpa, C. F., Roskey, M. T., and Mjoli, N., Construction of genomic libraries and introduction of iron-oxidation in *Thiobacillus ferrooxidans*, *Biotech. Appl. Biochem.* **8**:330–341 (1986).

Kulpa, C. F., Roskey, M. T., and Travis, M. T., Transfer of plasmid RPI into chemolithotropic *Thiobacillus neopolitanus, J. Bacteriol.* **156**:434–436 (1983).

Manning, H. L., New medium for isolating iron-oxidizing and heterotrophic bacteria from acid mine drainage, *Appl. Microbiol.* **30**:1010–1016 (1975).

Pretorius, I. M., Rawlings, D. E., and Woods, D. R., Identification and change of *Thiobacillus ferrooxidans* structural genes in *Escherichia coli, Gene* **45**:59–65 (1986).

Rawlings, D. E., Gawith, C., Petersen, A., and Woods, D. R., Characterization of plasmids and potential genetic markers in *Thiobacillus ferrooxidans,* in G. Rossi and A. E. Torma (eds.), *Recent Progress in Biohydrometallurgy,* Associazione Mineroria, Sarda, 1983, pp. 555–570.

Rawlings, D. E., Pretorius, I., and Woods, D. R., Expression of a *Thiobacillus ferrooxidans* origin of replication in *Escherichia coli, J. Bacteriol.,* **158**:737–738 (1984a).

Rawlings, D. E., Pretorius, I. M., and Woods, D. R., Construction of arsenic-resistant *Thiobacillus ferrooxidans* recombinant plasmids and the expression of autotrophic plasmid genes in a heterotrophic cell-free system, *J. Biotech.* **1**:129–133 (1984b).

Rawlings, D. E., and Woods, D. R., Mobilization of *T. ferrooxidans* plasmids among *Escherichia coli* strains, *Appl. Environ. Microbiol.* **49**:1323–1325 (1985).

Rawlings, D. E., Sewcharan, R., and Woods, D. R., in R. W. Lawrence, R. M. R. Branion, and H. G. Ebner (eds.), *Fundamental and Applied Biohydrometallurgy,* Elsevier, Amsterdam, 1986, pp. 419–427.

Schrader, J., and Holmes, D. S., Phenotypic switching of *Thiobacillus ferrooxidans, J. Bacteriol.* **170**:3915–3923 (1988).

Tuovinen, O. H., and Kelly, D. P., Studies on the growth *Thiobacillus ferrooxidans:* 1. Use of membrane filters and ferrous iron agar to determine viable numbers, and comparison with $^{14}CO_2$ fixation and iron oxidation as measures of growth. *Arch. Microbiol.* **88**:285–296 (1973).

Tuovinen, O. H., Niemela, S. I., and Gyllenberg, H. G., Tolerance of *T. ferrooxidans* to some metals, *Antonie van Leeuwenhoek* **37**:489–496 (1971).

Yates, J. R., Cunningham, R. R., and Holmes, D. S., IST2: A new insertion sequence in *Thiobacillus ferrooxidans, Proc. Nat. Acad., Sci. USA,* **85**:7284–7287 (1988).

Yates, J. R., and Holmes, D. S., Two families of repeated DNA sequences in *Thiobacillus ferrooxidans, J. Bacteriol.* **169**:1861–1870 (1987).

Biological Fundamentals of Mineral Leaching Processes

Olli H. Tuovinen

Introduction

Microbiological leaching of metals from ore materials is presently being practiced in dump and underground uranium and copper leaching operations, as outlined by McCready and Gould (1990) for uranium mines in the Elliot Lake area in Ontario, Canada, and by Ralph (1985) for copper leach mines in many different regions of the world. There also appear to be economically attractive prospects for commercialization of the biological treatment of precious-metal-containing ore materials for gold recovery (Lawrence, 1990).

Biological leaching processes have been evaluated on a bench scale with numerous samples of different sulfide mineralizations from many geographical regions. No two ores are identical, and within each ore deposit the mineralogical composition and the concentration of metals display heterogeneity. These variations necessitate experimental evaluation of multiple samples even from a single ore body because mineralogical and geochemical characteristics have a major influence on the microbiological leaching of the material. Biological leaching profiles of mineral samples from each mineralization are unique because of variation in inclusions and mineralogical composition, acid demand characteristics, and concentration of metals as well as in intermediates and by-products of oxidative dissolution.

This chapter focuses on biological leaching studies and laboratory evaluations of sulfide minerals that have been performed with acidophilic *Thiobacillus* type or related bacteria. Besides ferrous sulfate, elemental sulfur, and soluble inorganic sulfur compounds, these bacteria can oxidize many sulfide minerals as electron donors while utilizing CO_2 as the carbon source. The main physiological characteristics of the bacteria are described in an accompanying chapter (Norris, 1990). Solubilization of metals from mineral materials has also been tested with varying success employing chemoorganotrophic bacteria and fungi utilizing extraneously added organic carbon and energy substrates, but this topic is not within the scope of this review.

Laboratory-Scale Reactors for Bioleaching

Experimental approaches to bioleaching evaluations of ore materials have usually involved at least one of the following techniques: (1) shake flasks, (2) columns or percolators, and (3) stirred-tank reactors or pachucas. Each technique has specific, unique characteristics with respect to experimental variables such as aeration, particle size, pulp density, or solids-to-liquid ratio.

The shake-flask technique has major limitations because of continuously changing conditions. Steady states cannot be reached, and it is difficult to control experimental variables such as pH or dissolved oxygen. The lack of steady-state operation makes it difficult to examine effects of experimental factors because they may be amplified or negated by continuously changing conditions. While suitable for screening and preliminary testing, shake-flask experiments should be accompanied by other approaches before extrapolating experimental results to pilot-scale or commercial-scale systems.

Unlike the shake-flask leaching technique, column or percolation leaching utilizes coarser ore material which, depending on the experimental scale, may be more representative of the material in dump or underground leaching applications. Column leaching studies are of long duration, lasting several months, sometimes years, thereby imposing various constraints which may be difficult to resolve. A feature characteristic of column or percolation leaching is that it appears to result in the formation of different zones within the sample ore material that have distinct chemical and physical gradients. These zones may exhibit differences in redox potential, iron precipitation, and elemental sulfur formation even in relatively small, laboratory-scale columns. Although not examined in detail, these zones may have some analogy with those in commercial-scale dump and heap leaching operations. Some examples of differences in the chemical composition of the ore material, based on postleaching examination, are given in Table 3.1. In large-scale commercial applications, such zones have not been characterized at mine sites but are likely to also include temperature gradients.

Stirred-tank reactors or other experimental vessels, such as pachucas, that allow aeration, complete mixing of suspended solids, and control of various parameters can provide most useful information

TABLE 3.1   Partial Chemical Composition of Leach Residues from Biological Leaching Systems, Demonstrating Variation in Metal Dissolution and Formation of Elemental Sulfur

Sample	\multicolumn{8}{c}{Chemical composition (% wt/wt)}
	Fe	Cu	Co	Zn	Ni	S^0	SO_4^{2-}	S_{total}
Purified pyrite (-100 to $+200$ mesh); shake-flask leaching for 87 d								
Initial	45.0	0.05	—[a]	0.08	—	—	—	52.3
Final	45.5	—	—	—	—	0.6	1.6	50.3
Purified pyrrhotite (-100 to $+200$ mesh); shake-flask leaching for 87 d								
Initial	57.5	0.24	—	1.1	—	—	—	35.8
Final	6.3	—	—	—	—	78.0	12.8	81.3
Purified chalcopyrite (-100 to $+200$ mesh); shake-flask leaching for 87 d								
Initial	30.0	32.5	—	0.9	—	—	—	33.5
Final	30.5	32.5	—	—	—	0.6	1.4	33.7
Sulfide ore I (1.68 to 5.0 mm); percolator leaching for 457 d								
Initial	12.2	0.23	0.058	0.31	0.17	—	—	7.4
Final	7.4	0.24	0.032	0.039	0.053	—	—	7.1
Sulfide ore II (1.68 to 5.0 mm); percolator leaching for 650 d								
Initial	8.3	0.64	0.08	0.07	0.10	—	—	10.1
Final	4.4	0.37	0.04	0.002	0.012	—	—	4.4
Jarosite sample from percolator after 650 d of leaching of ore II[b]								
	28.0	0.16	—	—	—	—	35.7	—

[a]—, not determined.
[b]The sample also contained 2% K, 1.46% Na, and trace amounts (< 50 ppm) of Zn, Ni, and Co.
SOURCE: From L. Ahonen and O. H. Tuovinen, unpublished results.

on the various factors that influence the kinetics of biological leaching of minerals. An experimental system utilizing a rotating barrel-type reactor and inoculated with *Thiobacillus ferrooxidans* was suggested by Edvardsson (1988) to improve the rate of arsenopyrite leaching compared with the stirred-tank technique, but it needs further evaluation.

Very few experimental continuous leaching systems have been reported. They appear to entail several technical problems due to corrosiveness of the leach solution, high attrition associated with mineral particles, technical difficulties in feeding mineral suspensions, and ensuring sample homogeneity. A semicontinuous biological leaching approach based on stirred-tank reactors was reported by Lindström and Gunneriusson (1988), and has been successfully utilized in arsenopyrite leaching studies with *Sulfolobus* spp.

Other reactor types have also been developed for various bioleaching applications based on rapid microbiological regeneration of ferric-sulfate solutions. Many of these are related to fixed-film systems employed as either rotating biological contactors (Nakamura et al., 1986; Nikolov et al., 1988), packed-bed bioreactors (Grishin and Tuovinen, 1988; Nikolov et al., 1988), or fluidized-bed bioreactors (Nikolov and Karamanev, 1987; Karamanev and Nikolov, 1988). Iron-oxidizing thiobacilli in fixed-film samples taken from bioreactors designed for rapid ferrous-sulfate oxidation are displayed in Figure 3.1. A large-scale iron oxidation process employing diatomaceous earth charged with *T. ferrooxidans* was described by Murayama et al. (1987) for biological treatment of acid mine water.

Nutrients and Metals in Biological Leaching

The essential major elements in cell biomass and growth are carbon, oxygen, hydrogen, nitrogen, phosphorus, and sulfur. Carbon assimilation in acidophilic thiobacilli is based on the reductive fixation of carbon dioxide (Holuigue et al., 1987). In principle, it is not necessary to amend biological leaching systems with organic carbon sources as an alternative to carbon dioxide. Organic carbon amendment may have some benefit for the bacteria if there are toxic metals in solution. Complex material such as yeast extract provided at relatively low concentrations have been shown to alleviate silver-ion toxicity to bacterial ferrous-sulfate oxidation (Tuovinen et al., 1985). Similar effects due to complex formation can be demonstrated with the use of chelating agents for divalent cations (Mahapatra and Mishra, 1984).

Organic waste such as sewage sludge or process effluent from bakers' yeast production has been shown to stimulate the biological leaching of a multimetallic sulfide ore material (Puhakka and Tuovinen, 1987). This observation may be related either to a nutrient or to a metal chelation effect, but the exact cause is not known. In general, sugars and other low-molecular-weight organic compounds are potentially inhibitory to acidophilic thiobacilli. Several explanations have been suggested for the adverse influence, including repression of carbon dioxide fixation associated with the inability to utilize organic carbon for growth. It has also been recognized that organic acids can accumulate in the cytoplasm of the bacteria, thereby causing an intracellular pH change which ultimately causes inhibition (Alexander et al., 1987). It should also be noted that some bacteria, particularly some thermophilic acidophiles, appear to have a requirement for yeast extract or other complex organic material (Norris, 1990), presumably to satisfy various trace nutrient requirements or as a carbon source to support a chemolithoheterotrophic mode of growth.

Oxygen in cellular compounds derives from the elemental content of oxygen in the various intermediary metabolites involved in biosyn-

Figure 3.1 Scanning electron micrographs of iron-oxidizing bacteria employed in bioreactors for ferrous sulfate oxidation (S. I. Grishin and O. H. Tuovinen, unpublished). (a) Exterior surface colonization of activated carbon particles by *T. ferrooxidans* in a packed-bed reactor after 12 weeks of operation (bar = 5 μm). (b) Iron precipitation and bacteria on an activated carbon sample collected near the exit port of a packed-bed reactor after 12 weeks of operation (bar = 5 μm). (c) Surface opening of activated carbon displaying *T. ferrooxidans* cells in a fluidized bed reactor after five weeks of operation (bar = 5 μm). (d) Colonization of anion-exchange resin beads (Amberlite IR-45) by *T. ferrooxidans* in a packed-bed reactor after 11 weeks of operation (bar = 1μm). The performance data of these bioreactors have been published (Grishin et al., 1988; Grishin and Tuovinen, 1988).

thetic pathways which are associated with carbon and nitrogen assimilation and with oxyanions of sulfur and phosphorus. Oxygenase-type enzymes, incorporating molecular oxygen into acceptor molecules, have been reported only for elemental sulfur oxidation in acidophilic thiobacilli (Hooper and DiSpirito, 1985; Kelly, 1985) and in a

Sulfolobus sp. (Emmel et al., 1986). However, the acidophiles described in this chapter all require oxygen as the terminal electron acceptor in order to couple energy metabolism with growth, although limited uncoupled substrate oxidation activities may occur in the absence of dissolved oxygen.

Hydrogen is required as a reductant for $NAD(P)^+$ in many assimilatory and biosynthetic pathways, and it is also necessary in the terminal reduction of oxygen in the electron transport chain described by the anodic half-reaction in Table 3.2. In addition, the transmembrane movement of hydrogen and electrons is essential for establishing the proton motive force in acidophilic bacteria (Hooper and DiSpirito, 1985).

Ammonia appears to be the preferred nitrogen source, but some strains of *T. ferrooxidans* can also fix dinitrogen (Stevens et al., 1986). The significance of nitrogen fixation in biological leaching systems is not known. Various organic nitrogen compounds have been tested either as N sources in leaching situations or as hydrometallurgical reagents in metal extraction, but their stabilities in acid leach solutions have not been systematically determined. Their efficacy as a nutrient source is based on direct assimilation or on ammonification by satellite microbes to provide ammonium for cellular assimilation.

Phosphate- or sulfate-limited growth conditions cause serious meta-

TABLE 3.2 Examples of Anodic Half-Cell Reactions Associated with Oxidative Leaching of Sulfide Minerals

Half-cell reactions of mineral oxidations are determined by thermodynamic phase boundary conditions.

Mineral (substrate)	Anodic reaction*
Arsenopyrite	$FeAsS + 8H_2O \rightarrow Fe^{3+} + AsO_4^{3-} + SO_4^{2-} + 16H^+ + 14e^-$
Chalcopyrite	$2CuFeS_2 \rightarrow Cu^+ + Fe^{2+} + 2S^0 + CuFeS_2 + 3e^-$
	$CuFeS_2 \rightarrow Cu^{2+} + Fe^{2+} + 2S^0 + 4e^-$
	$CuFeS_2 + 8H_2O \rightarrow Cu^{2+} + Fe^{2+} + 2SO_4^{2-} + 16H^+ + 16e^-$
	$CuFeS_2 + 8H_2O \rightarrow Cu^{2+} + Fe^{3+} + 2SO_4^{2-} + 16H^+ + 17e^-$
Elemental sulfur	$S^0 + 4H_2O \rightarrow SO_4^{2-} + 8H^+ + 6e^-$
Ferrous ion	$Fe^{2+} \rightarrow Fe^{3+} + e^-$
Pentlandite	$Ni_9S_8 \rightarrow 9Ni^{2+} + 8S^0 + 18e^-$
Pyrite	$FeS_2 \rightarrow Fe^{2+} + 2S^0 + 2e^-$
	$FeS_2 + 8H_2O \rightarrow Fe^{2+} + 2SO_4^{2-} + 16H^+ + 14e^-$
	$FeS_2 + 8H_2O \rightarrow Fe^{3+} + 2SO_4^{2-} + 16H^+ + 15e^-$
Pyrrhotite	$Fe_{0.95}S \rightarrow 0.95Fe^{2+} + S^0 + 1.9e^-$
Sphalerite	$ZnS \rightarrow Zn^{2+} + S^0 + 2e^-$
	$ZnS + 4H_2O \rightarrow Zn^{2+} + SO_4^{2-} + 8H^+ + 8e^-$
Uraninite	$UO_2 \rightarrow UO_2^{2+} + 2e^-$

*The cathodic reaction can be represented by the following equations with either oxygen or ferric iron as the electron acceptor:

$$0.25xO_2 + xH^+ + xe^- \rightarrow 0.5xH_2O$$

$$xFe^{3+} + xe^- \rightarrow xFe^{2+}$$

where x is the number of electrons available from the respective anodic half-cell reaction.

bolic imbalance and restricted growth because both nutrients are required by growing cells. Phosphates produce insoluble complexes with Fe(III) (Hoffmann et al., 1985) that may effectively scavenge available phosphate from leach solutions. Phosphate may thus be effectively in limited supply, depending on its concentration in ore material. Besides being an essential nutrient, the sulfate ion is also required in ferrous-iron oxidation by *T. ferrooxidans* (Ingledew, 1982). Sulfate is likely to occur in excess in biological leaching processes because it is a major product of oxidative leaching. Some acidophiles are known to require a reduced sulfur source in the form of thiosulfate or yeast extract (Norris, 1990), indicating deficiency in the reductive pathway of sulfate assimilation.

Trace metals and other trace nutrient requirements of acidophilic thiobacilli have not been quantitatively characterized. *T. ferrooxidans* is known to have distinct cationic requirements (e.g., Cu for rusticyanin, Fe for cytochromes, Mg for energy-transducing enzyme activities), but these are difficult to measure because the commonly used substrates, especially ferrous sulfate and various sulfide minerals, are known to contain metallic impurities whose further separation is difficult.

Metal toxicity for acidophilic thiobacilli has been evaluated in numerous laboratory studies using synthetic mineral-salts solutions. Silver and molybdenum are some of the metals that are particularly toxic to iron-oxidizing thiobacilli. However, the toxic threshold concentrations in biological leaching systems have not been established. These may be difficult to determine because of the formation of various soluble and insoluble complexes minimizing the inhibition.

Precious-metal concentrations in particular are extremely low in sulfate-containing environments. For example, Bruynesteyn et al. (1986) reported that silver and gold concentrations in biological leach solutions never exceeded 0.1 mg Ag/L and 0.1 μg Au/L, respectively, in biological leaching experiments with sulfidic (arsenopyritic) silver and gold concentrates. Molybdenum, toxic at high levels, was found to have a beneficial effect on sulfur compound oxidation by both mesophilic thiobacilli (*T. thiooxidans*) (Takakuwa et al., 1977) and thermoacidophiles such as *Sulfolobus* spp. (Buckingham et al., 1989).

For base metals (Co, Cu, Mn, Ni, Zn), the toxic threshold concentrations are high, at least several grams of metal ion per liter. Base metals have not been reported to reach toxic levels in leach solutions. For uranium, iron-oxidizing bacteria can be resistant to at least 10 mM UO_2^{2+}, which is within the target concentration in uranium leaching in mines.

The biochemical and genetic basis of resistance to transition metals in iron-oxidizing thiobacilli is not well understood, although it has been recognized for a long time that these bacteria can be relatively readily adapted to increasingly higher concentrations of metals. The resistance to mercury in *T. ferrooxidans* is conferred by mercuric reductase enzyme, similar in structural and functional properties to those described in other gram-negative bacteria (Booth and Williams, 1984). Arsenic toxicity and development of resistance to it are areas in need of research because soluble arsenic species may potentially limit the biological leaching application of arsenopyritic ore materials for gold recovery.

In other bacterial systems, arsenite is usually more toxic than arsenate. The toxicity is based on several metabolic effects, with arsenate competing with phosphate in uptake and incorporation being one of the better-characterized mechanisms of toxicity in biological systems. The speciation and stoichiometry of soluble and insoluble products of arsenopyrite leaching have not been reported. Arsenic resistance was reported to reach levels as high as 10 to 20 g As/L (Karavaiko et al., 1986), but experimental data have not been presented to support these claims. The oxidation of arsenite to arsenate occurs in biological leaching systems since the bulk of As may occur in the oxidized form, but whether the oxidation is chemical or possibly microbiological remains to be elucidated.

Initial attempts to improve the genetic basis of As resistance in *T. ferrooxidans* were not particularly successful, but showed that plasmid DNA from *T. ferrooxidans* was expressed in *Escherichia coli* transformants (Rawlings et al., 1984a,b). This was an important demonstration because of the need to study genetic determinants and regulation of acidophilic thiobacilli by plasmid transfer and recombination in better-characterized heterotrophic bacteria such as *E. coli,* whose transcription and translation systems are much better understood than those of the thiobacilli. Several vectors have since been constructed for developing genetic systems for *T. ferrooxidans* (Rawlings and Woods, 1988), but successful transformation or conjugation of *T. ferrooxidans* is yet to be demonstrated.

Biological Leaching in Acid Ferric-Sulfate Solutions

The ferrous-ferric-ion couple in acid solutions is an important redox couple because the ferric ion acts as an electron acceptor for sulfide and uranium minerals, and the ferrous ion thus formed can be subsequently reoxidized to ferric iron by iron-oxidizing thiobacilli (*T. ferrooxidans*) or by bacteria of the *Leptospirillum ferrooxidans* type. Thus a prerequisite for acid leaching is the regeneration of ferric-sulfate solutions, which can be accomplished either by chemical means with various nonrecoverable chemical oxidants (e.g., $KMnO_4$,

MnO_2, H_2O_2) or by the microbiological route by promoting bacterial oxidation of ferrous iron in sulfate environments. In the absence of chemical oxidants, a high ratio of Fe^{3+}/Fe^{2+} indicates microbiological activity. In the absence of biological and chemical oxidizing agents, Fe^{2+} oxidation rates at the acid pH values prevailing in sulfide leaching are so slow that they make essentially no practical contribution toward ferric-iron regeneration. The ferric-ferrous couple is a major redox component in leach solutions, and their redox potential sensitively reflects biological activities of iron oxidation.

Although oxidation quotients for bacterial Fe^{2+} oxidation may be high (Table 3.3), these values should be contrasted with relatively low biomass yields on ferrous sulfate. Biomass yields depend heavily on the number of electrons available from substrate oxidation, varying from one electron for the oxidation of Fe^{2+} to 15 for the complete oxidation of FeS_2. Consequently, relative to ferrous iron, biomass yields are greater on substrates such as thiosulfate and tetrathionate (Eccleston and Kelly, 1978), or pyrite (Basaran and Tuovinen, 1987), or other sulfide minerals that can donate multiple electrons to the electron transport system.

Ferric-chloride leaching usually results in faster chemical leaching rates of various sulfide minerals because the activation energies are lower in comparison with ferric-sulfate leaching media. These differences can be illustrated by an activation energy of 42 kJ/mol for

TABLE 3.3 Oxidation Quotients for Ferrous Ion Oxidation by *T. ferrooxidans* and the Corresponding Productivity Estimates Obtained with Various Experimental Techniques, as Summarized by Grishin and Tuovinen (1988)

	Oxidation quotient		Productivity
Experimental system	mmol Fe^{2+}/(mg protein · h)	mg Fe^{2+}/(mg protein · h)	g Fe^{3+}/(L · h)
Packed-bed reactor (fixed-film)			
Glass beads	0.25	14.0	1.3
Ion-exchange resin	0.08	4.5	9.5
Activated carbon	0.02	1.1	16
Polyvinylchloride	Nd*	Nd	1.8
Alginate bead-entrapped cells in packed-bed column	0.04	0.002	0.03
Rotating biological contactor	Nd	Nd	1.46
Fixed-film fiber reactor	90	5.0	1.0–1.2
Bacfox process	Nd	Nd	0.75
Fluidized bed (fixed film)			
Activated carbon	0.14	7.8	0.9–1.6
Polystyrene	1.61	0.09	0.49–1.68
Batch culture	514	28.7	0.1–0.2
Chemostat	152–1580	8.5–88	0.11–0.22

*Nd = not determined.

chalcopyrite leaching in ferric chloride as opposed to 75 kJ/mol in ferric-sulfate systems (Dutrizac, 1981). Chloride leaching systems appear to be difficult to develop for bioleaching applications because the bacteria potentially useful for acid leaching applications display a sulfate requirement (Ingledew, 1982; Fry et al., 1986) and are sensitive to the chloride ion (Cameron et al., 1984). In some dry, desert areas of sulfide mineralizations, freshwater resources may be limited while brine or seawater may be readily available to provide for water balance in leaching processes. Halotolerant, halophilic, or marine species of iron-oxidizing acidophiles have not been described in the literature. At present, therefore, there is no recognized microbiological basis for developing biohydrometallurgical applications for ferric-chloride leaching in chloride environments.

Ferrous iron is a ubiquitous product in leach solutions in biological mineral leaching processes. It is derived from the solubilization of ferrous minerals (e.g., pyrite, pyrrhotite, chalcopyrite) and is regenerated by bacterial iron oxidation:

$$4Fe^{2+} + O_2 + 4H^+ \rightarrow 4Fe^{3+} + 2H_2O$$

In acid solutions ferric iron is subject to hydrolysis to yield various hexaaquocoordinated hydroxycomplexes (Brown and Kester, 1980), simplified as $Fe(OH)^{2+}$ and $Fe(OH)_2{}^+$. The hydrolytic reactions are acid-yielding reactions:

$$Fe^{3+} + H_2O \rightleftharpoons Fe(OH)^{2+} + H^+$$

$$Fe(OH)^{2+} + H_2O \rightleftharpoons Fe(OH)_2{}^+ + H^+$$

Smith et al. (1988) presented also the following principal ferric-sulfate complexes for acidic ferric-sulfate solutions:

$$Fe^{3+} + SO_4{}^{2-} \rightleftharpoons FeSO_4{}^+$$

$$Fe^{3+} + 2SO_4{}^{2-} \rightleftharpoons Fe(SO_4)_2{}^-$$

The main solid-phase products forming in ferric-sulfate systems can potentially include phosphates, Fe(III) oxyhydroxides, and jarosites. Secondary, ordered Fe(III) mineral products in bacterial iron oxidation systems are primarily jarosite-type compounds which have been found to occur at pH values as low as pH 1.5 (Grishin et al., 1988). These have a general formula of $XFe_3(SO_4)_2(OH)_6$ (where X = Na^+, K^+, $NH_4{}^+$, or H_3O^+). The net reactions of jarosite formation are acid-yielding and may have a strong influence on the pH characteristics of the leaching system, as shown for hydronium jarosite at pH 1.9 ($HSO_4{}^- \rightleftharpoons SO_4{}^{2-} + H^+$; pK_a 1.91):

$$3Fe^{3+} + SO_4{}^{2-} + HSO_4{}^- + 7H_2O \rightarrow H_3OFe_3(SO_4)_2(OH)_6 + 6H^+$$

It is desirable to prevent jarosite formation in bacterial leaching systems because it may form diffusion barriers on mineral surfaces and scavenge metals or K^+, Na^+, or NH_4^+ from the leach solution. The formation of diffusion barriers due to insoluble complexes of Fe(III) is presumed to create potentially serious problems in heap leaching, but it has not been possible to adequately mimic the problem in experimental systems on a laboratory scale. Secondary Fe(III)-oxide zones have been found on mineral particles in laboratory studies, as shown in Fig. 3.2, but they do not appear to form tenacious layers. The monovalent cation entity in jarosite can be replaced with silver and other monovalent metals, or with divalent ions such as zinc (Dutrizac, 1984).

Bacterial Leaching of Iron Sulfides

Three types of biological leaching mechanisms for a pyrite-chalcopyrite system are schematically presented in Fig. 3.3. Half-cell reactions for these sulfides are presented in Table 3.1. Pyrite is the most abundant sulfide mineral and is associated in varying amounts with other sulfide minerals, such as those of copper, nickel, and zinc, or with uranium ores (e.g., uraninite, uranothorite). Pyrite and arsenopyrite may also occur in the same mineralizations, as in precious-metal-containing sulfidic ore materials (Lawrence, 1990).

Pyrite oxidation produces sulfuric acid which helps to neutralize acid demand caused by the dissolution of various sulfide and nonsulfide minerals:

$$2FeS_2 + 7O_2 + 2H_2O \rightarrow 2Fe^{2+} + 4SO_4^{2-} + 4H^+$$

Pyrite oxidation results in the dissolution of iron which is maintained in the ferric form in the presence of iron-oxidizing bacteria:

$$2FeS_2 + 7.5O_2 + H_2O \rightarrow 2Fe^{3+} + 4SO_4^{2-} + 2H^+$$

Ferric iron thus produced can act as a redox carrier to accelerate the chemical leaching of pyrite:

$$FeS_2 + 14Fe^{3+} + 8H_2O \rightarrow 15Fe^{2+} + 2SO_4^{2-} + 16H^+$$

Pyrite occurrences in ore materials often have inclusions of other sulfides, and thus partial oxidation of pyrite is a prerequisite for the contact of bacteria and leach solution with inclusions.

Pyrrhotite ($Fe_{1-x}S$; or FeS in a simplified formula) is another common iron sulfide in sulfide mineralizations, but its chemical and microbiological oxidation is not well characterized. The oxidation of this mineral has been reported to be relatively faster than that of pyrite

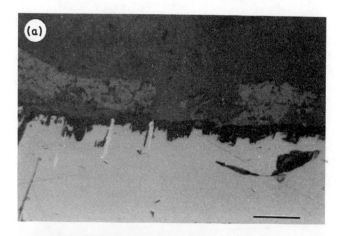

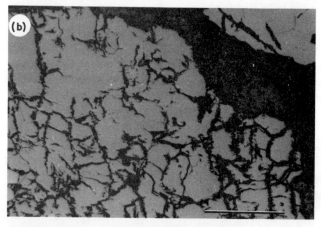

Figure 3.2 Post-leaching examination of solid residue samples from a column containing complex sulfide ore material (L. Ahonen and O. H. Tuovinen, unpublished results). (a) A loosely associated, dark gray layer of Fe(III)-oxide, possibly goethite, is associated with pyrrhotite surface but not as a tenacious boundary layer (bar = 0.1 mm). Two small inclusions of pentlandite are protruding from the pyrrhotite phase, suggesting that pentlandite was not leached as fast as pyrrhotite was. (b) The pyrrhotite phase displays considerable disintegration and a dark gray Fe(III)-oxide zone (goethite) appears between the two pyrrhotite particles (bar = 0.5 mm). The black layers between the pyrrhotite and goethite phases did not display reflectance in these polished specimens and may be composed of elemental sulfur, which is a recognized intermediate of the microbiological leaching of pyrrhotite.

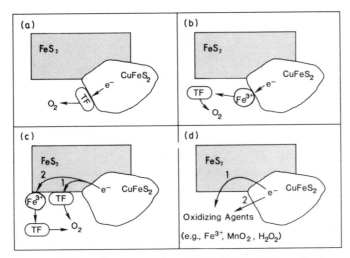

Figure 3.3 Schematic presentation of biological leaching mechanisms for a pyrite/chalcopyrite ($FeS_2/CuFeS_2$) system. (*a*) Direct leaching involving biologically mediated electron transport to oxygen; (*b*) indirect leaching involving Fe^{3+} as the primary electron acceptor, coupled with its reduction to Fe^{2+} and followed by its biological oxidation; (*c*) galvanic coupling involving bacteria as oxidizing agents either directly (1) or indirectly (2) with Fe^{3+}/Fe^{2+} as the redox carrier; (*d*) oxidative leaching mediated by chemical oxidizing agents with (1) or without (2) a galvanic coupling effect. TF = *T. ferrooxidans*; e^- = electrons.

(Ahonen et al., 1986) and its microbiological oxidation at elevated temperatures has also been reported (Norris and Parrott, 1986), but strictly comparable experimental data relative to other sulfide minerals are not available. Pyrrhotite oxidation is an acid-demanding reaction that also produces major amounts of elemental sulfur as a by-product (Fig. 3.4). Ferric iron can again act as a chemical oxidant and is regenerated via bacterial oxidation:

$$2FeS + 4.5O_2 + 2H^+ \rightarrow 2Fe^{3+} + 2SO_4^{2-} + H_2O$$

$$2FeS + 1.5O_2 + 6H^+ \rightarrow 2Fe^{3+} + 2S^0 + 3H_2O$$

$$FeS + 8Fe^{3+} + 4H_2O \rightarrow 9Fe^{2+} + SO_4^{2-} + 8H^+$$

The microbiological leaching of chalcopyrite is a relatively slow reaction by comparison with biological leaching rates of secondary Cu-oxide and Cu-sulfide minerals. Complete oxidation of chalcopyrite can be represented by the following equation:

$$2CuFeS_2 + 8.5O_2 + 2H^+ \rightarrow 2Cu^{2+} + 2Fe^{3+} + 4SO_4^{2-} + H_2O$$

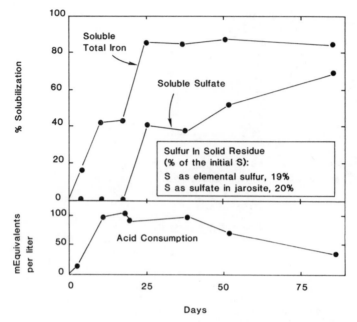

Figure 3.4 Oxidation of pyrrhotite by *T. ferrooxidans,* based on data presented by Ahonen et al. (1986). For the initial two weeks, the reaction was acid-consuming. After 40 days of incubation it changed to net acid production, suggesting gradual oxidation of the intermediate elemental sulfur. The mass balance of sulfur in the solid residue is also indicated.

Elemental sulfur formation

With chalcopyrite, ferric-iron leaching in acid solutions produces elemental sulfur, which tends to coat mineral particles and may thereby interfere with electron transport and fluxes of reactants and products between the solid and solution phases:

$$CuFeS_2 + 4Fe^{3+} \rightarrow Cu^{2+} + 5Fe^{2+} + 2S$$

The sulfur rim is therefore an undesired product because it results in a diffusion barrier on the mineral surface. It should be recognized, on the other hand, that partial oxidation of the sulfur entity to the level of elemental sulfur requires less oxygen (aeration) if compared with complete oxidation to the level of sulfate. Chalcopyrite oxidation may eventually become diffusion-limited due to the formation of a sulfur coating. Because of its poor conductivity, sulfur also prevents electron transport processes on mineral surfaces.

The formation of the sulfur rim is a well-recognized problem both in the chemical (mediated by ferric iron) and biological (mediated by *T.*

ferrooxidans) leaching of chalcopyrite. Sulfur-oxidizing bacteria are presumed to have a beneficial effect on the leaching rate by oxidizing the sulfur layer to soluble products (i.e., sulfate). Current evidence for this beneficial effect is ambiguous, and culture manipulation to promote sulfur-oxidizing thiobacilli has not been shown to decrease sulfur formation.

Silver catalysis

Because it has not been possible to alleviate the formation of intermediate sulfur or to rapidly oxidize it via inoculation of appropriately selected thiobacilli, other attempts have been made to prevent, remove, or modify the sulfur coating of chalcopyrite surfaces. These have included (1) attrition grinding to very small particle size; (2) regrinding of chalcopyrite leach residue to reexpose chalcopyrite surfaces to leach solution and bacteria; and (3) use of catalysts to alter the sulfur chemistry and thus the pathway of elemental sulfur formation. The first two physical methods have shown success but have limited application because of the high cost associated with particle size diminution.

The addition of a silver salt displays an accelerating effect on chalcopyrite leaching in chemical (Miller et al., 1981; Miller and Portillo, 1981) and microbiological (Ahonen and Tuovinen, 1990) leaching systems, as shown in Fig. 3.5 for a mixed sulfide ore sample. The exact mechanism has not been unequivocally elucidated. It has been proposed that the silver effect is based on transient formation of Ag_2S, which is oxidized to Ag^+ and S^0 by excess Fe^{3+}:

$$CuFeS_2 + 4Ag^+ \rightarrow 2Ag_2S + Cu^{2+} + Fe^{2+}$$

$$Ag_2S + 2Fe^{3+} \rightarrow 2Ag^+ + 2Fe^{2+} + S^0$$

The elemental sulfur thus produced appears to form a surface layer which is less tenacious and causes less diffusional resistance than the elemental sulfur layer formed without the intermediate silver sulfide. With excess silver addition, precipitates may include elemental silver and silver sulfate on mineral surfaces. These are not desirable in the leaching process because they contribute to silver loss without alleviating the formation of elemental sulfur barriers on chalcopyrite surfaces.

Because *T. ferrooxidans* regenerates ferric iron, the silver effect is amplified in the presence of iron-oxidizing bacteria due to a favorable ratio of Fe^{3+}/Fe^{2+}. In the absence of biological oxidation of ferrous iron, the silver amendment has a limited effect because the rate of the chemical oxidation of Fe^{2+} at low pH values is negligible and cannot therefore maintain a favorable ratio of oxidized to reduced Fe species.

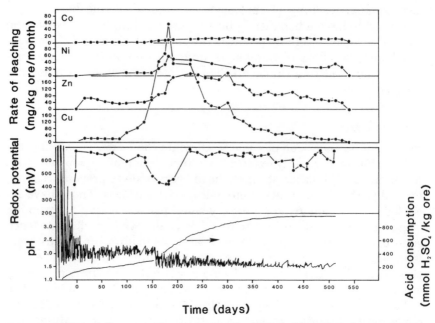

Figure 3.5 Effect of silver amendment on the microbiological leaching of a complex ore sample containing chalcopyrite (Cu), sphalerite (Zn), pentlandite (Ni and Co), and pyrrhotite as the main sulfide minerals (L. Ahonen and O. H. Tuovinen, unpublished results). The column contained 12 kg of the ore material (5–20 mm particle size fraction), 3 L of a leach solution, and an inoculum of acidophilic thiobacilli. After 122 days of leaching, silver sulfate was added to a final concentration of 30 mg Ag/L leach solution (7.5 mg Ag/kg ore). Subsequently the soluble silver concentration remained below 0.2 mg/L. The leaching rate determinations were based on the method of least squares with five data points each. Where possible, these included three data points preceding the sampling data. Therefore, an early change in the rate preceding the silver amendment is apparent. The corresponding changes in redox potential, pH, and acid (H⁺) consumption are also indicated for the same time course.

The rate of the chemical oxidation of elemental sulfur is also insignificant under these conditions, and therefore the role of the thiobacilli becomes important in sulfur oxidation:

$$2S + 3O_2 + 2H_2O \rightarrow 4H^+ + 2SO_4^{2-}$$

A silver-catalyzed pathway appears to accelerate chalcopyrite leaching rates considerably (Ahonen and Tuovinen, 1990), suggesting that it can alleviate the diffusion problem due to elemental sulfur zones. The microbiological leaching of other sulfide minerals may be adversely influenced by Ag amendment (Ahonen and Tuovinen, 1990), which may reflect silver toxicity on surfaces where silver sulfide is not formed. It is a common observation, however, that the dissolved silver concentration is below 0.5 to 0.1 mg/L due to the formation of insoluble complexes such as Ag_2S. A considerable amount of silver may actually be incorporated into jarosite precipitates (Table 3.4), and thus its catalytic effect is diminished.

Bruynesteyn et al. (1983) and Lawrence et al. (1985) described a silver-catalyzed process which required copper sulfate and thiosulfate as activating reagents before Ag catalyzed the biological leaching of chalcopyrite. The reaction mechanism for silver catalysis in this scheme has not been elucidated, but is presumed to involve copper reduction to form cuprous-thiosulfate complexes which subsequently decompose to Cu sulfides. Silver also forms a complex with thiosulfate, which decomposes to Ag_2S when the excess thiosulfate is consumed via unstable complex formation with the added copper ion. This Ag_2S may then act on the chalcopyrite surface via cyclic oxidation-precipitation to catalyze chalcopyrite dissolution. Upon oxidation of the secondary Cu sulfides, chalcopyrite is presumed to dissolve to replace the copper sulfide:

$$CuS + 0.5O_2 + 2H^+ \rightarrow Cu^{2+} + S^0 + H_2O$$

$$CuFeS_2 + Cu^{2+} \rightarrow 2CuS + Fe^{2+}$$

Besides these chemical pathways and formation of various complexes, it can be postulated that Ag_2S on the chalcopyrite surface has a catalytic effect due to its cathodic behavior with respect to anodic chalcopyrite. This electrochemical coupling may compete with cathodic pyrite and might in part explain why pyrite oxidation was inhibited upon Ag addition (Ahonen and Tuovinen, 1990).

While the Ag catalysis is unlikely to be a commercially significant approach, an understanding of the electrochemical changes brought about by Ag catalysis may help in developing other catalyst approaches for the biological leaching of chalcopyrite. The introduction of carbon (graphite) into chemical leaching systems has been tested (Wan et al.,

TABLE 3.4   Distribution of Silver in Jarosite Fractions Recovered in Solid Residue after the Microbiological Leaching of Complex Sulfide Ore Samples (A and B) in Columns for 117 Days Following the Addition, as Silver Sulfate, of 30 mg of Ag/Kg Ore*

| Sample | Concentration of silver (ppm) | | Distribution of silver |
	Jarosite	Solution	$Ag_{jarosite}/Ag_{solution}$
A	50	0.36	139
B	45	0.31	145

*L. Ahonen and O. H. Tuovinen, unpublished results.

1984, 1985) in attempts to modify the electrochemical environment of the leaching, but the graphite effect on biological leaching of chalcopyrite has not been reported.

Pyrite-pyrrhotite-chalcopyrite leaching

A positive effect on chalcopyrite leaching has been demonstrated with pyrite (FeS_2) (Mehta and Murr, 1982; Ahonen et al., 1986). In theory, these two minerals establish a galvanic couple (Fig. 3.3) where chalcopyrite, which has a lower rest potential, appears to behave anodically relative to pyrite, which has a higher rest potential, as summarized by Natarajan (1990). Pyrrhotite ($Fe_{1-x}S$) lacks a positive effect on chalcopyrite leaching, which may be related to its lower rest potential effectively preventing its cathodic behavior. Pyrrhotite does not appear to promote the microbiological leaching of chalcopyrite, and in fact an inhibitory effect has been noted (Rossi et al., 1983). An additional feature with pyrrhotite is that elemental sulfur is typically produced during its microbiological leaching (Ahonen et al., 1986) which may adversely influence the fluxes of dissolved and gaseous reactants and dissolution products.

Leaching based on galvanic coupling offers an avenue to rate-controlling mechanisms, particularly for improving the kinetics of the biological leaching of recalcitrant sulfide minerals. In large-scale operations it may not be possible to differentiate between the various mechanisms and their respective roles in rate limitation.

Oxidation of nonferrous minerals

Direct oxidation is best exemplified in the leaching of nonferrous sulfide minerals. It should be recognized that iron is an abundant element and invariably present in natural mineral deposits and ore bodies. In leaching, the direct mechanism is never the exclusive leaching process because of the ubiquitous presence and concurrent dissolution of iron and its bacterial oxidation.

In laboratory studies it has been shown that the direct leaching mechanism entails the bacterial oxidation of the sulfur entity in nonferrous sulfides to sulfate, which results in the concomitant dissolution of the accompanying metal entity. Examples of the direct mechanism include the bacterial oxidation of synthetic base-metal sulfides in laboratory studies, suggesting the following stoichiometry and end products:

$$ZnS + 2.5O_2 + 2H^+ \rightarrow Zn^{2+} + SO_4^{2-} + H_2O$$

$$NiS + 2.5O_2 + 2H^+ \rightarrow Ni^{2+} + SO_4{}^{2-} + H_2O$$

$$CoS + 2.5O_2 + 2H^+ \rightarrow Co^{2+} + SO_4{}^{2-} + H_2O$$

$$Cu_2S + 2.5O_2 + 2H^+ \rightarrow 2Cu^{2+} + SO_4{}^{2-} + H_2O$$

$$Cu_2S + S^0 \rightarrow 2CuS$$

$$CuS + 2.5O_2 + 2H^+ \rightarrow Cu^{2+} + SO_4{}^{2-} + H_2O$$

A direct oxidation mechanism has also been elucidated for the tetravalent uranous ion which can be directly oxidized by *T. ferrooxidans* in the absence of dissolved iron (DiSpirito and Tuovinen, 1982a,b). The oxidation yields hexavalent uranium which hydrolyzes in acid solutions to uranyl ion:

$$2U^{4+} + O_2 + 4H^+ \rightarrow [2U^{6+} + 2H_2O] \rightarrow 2UO_2{}^{2+} + 4H^+$$

$$2U(SO_4)_2 + O_2 + 2H_2SO_4 \rightarrow [2U(SO_4)_3 + 2H_2O] \rightarrow$$

$$2UO_2(SO_4)_2 + 2H_2SO_4$$

The rates of direct biological oxidation of uranium are slower than those of the indirect system in which the primary oxidant is ferric iron. Uranium ores that are amenable to acid leaching processes typically contain either iron or are associated with iron sulfides such as pyrite or pyrrhotite. The direct oxidation mechanism of U(IV) may have little significance in biological leaching situations because of the overriding kinetic effect by the Fe^{3+}/Fe^{2+} couple recycled by *T. ferrooxidans*. DiSpirito and Tuovinen (1984) also reviewed evidence that has been reported for the direct bacterial oxidation of Cu(I). In ionic form in oxygenated solutions, the cuprous copper has an extremely short half-life because of rapid chemical oxidation even at a pH regime relevant to acid leaching.

Anaerobic oxidation

Although it has been recognized that *T. ferrooxidans* is an aerobic bacterium, utilizing oxygen as the terminal electron acceptor for the oxidation of inorganic substrates, this bacterium displays the ability to oxidize sulfur under anaerobic conditions, whereas ferric iron serves as the reductant (Sugio et al., 1987, 1989). The anaerobic oxidation of elemental sulfur can also be coupled with manganese-oxide (MnO_2) and molybdic-ion (Mo^{6+}) reduction (Sugio et al., 1988a,b). These anaerobic activities do not appear to be coupled with energy-

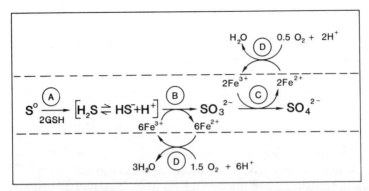

Figure 3.6 Scheme of the anaerobic oxidation of inorganic sulfur coupled with ferric-ion reduction by *T. ferrooxidans*, based on data presented by Sugio et al. (1985, 1988c, 1989). Reactions A–C inside the broken line are indicated for anaerobic conditions as follows. (A) Nonenzymatic reduction of sulfur with reduced glutathione as electron donor. Reduced glutathione may be regenerated with NADH as the electron donor (glutathione reductase). (B) Sulfur-ferric ion oxidoreductase. The hexavalent molybdenum, as molybdate MoO_4^{2-}, can replace ferric ion in this reaction and be reduced to a molybdenum blue complex with a molar stoichiometry between 4.6 to 5.4 molybdenum blue per one sulfite produced (Sugio et al., 1988b). (C) Sulfite oxidase with ferric ion as the electron donor and nonenzymatic oxidation of sulfite to sulfate. (D) Aerobic oxidation of ferrous ion by *T. ferrooxidans*. In this scheme, only the aerobic activity is coupled with energy transduction in this bacterium.

transducing systems for cell growth (Sugio et al., 1988c). The anaerobic reduction of hexavalent molybdenum has also been shown to occur in *Sulfolobus* spp. (Brierley and Brierley, 1982). The anaerobic pathway of sulfur oxidation is summarized in Fig. 3.6.

References

Ahonen, L., Hiltunen, P., and Tuovinen, O. H., The role of pyrrhotite and pyrite in the bacterial leaching of chalcopyrite ores, in R. W. Lawrence, R. M. R. Branion, and H. G. Ebner (eds.), *Fundamental and Applied Biohydrometallurgy*, Elsevier, Amsterdam, 1986, pp. 13–20.

Ahonen, L., and Tuovinen, O. H., Catalytic effects of silver in the microbiological leaching of finely ground chalcopyrite-containing ore materials in shake flasks, *Hydrometallurgy* (1990).

Alexander, B., Leach, S., and Ingledew, W. J., The relationship between chemiosmotic parameters and sensitivity to anions and organic acids in the acidophile *Thiobacillus ferrooxidans*, *J. Gen. Microbiol.* **133**:1171–1179 (1987).

Basaran, A., and Tuovinen, O. H., Iron pyrite oxidation by *Thiobacillus ferrooxidans*: Sulfur intermediates, soluble end products, and changes in biomass, *Coal Prepr.* **5**:39–55 (1987).

Berry, V. K., Murr, L. E., and Hiskey, J. B., Galvanic interaction between chalcopyrite and pyrite during bacterial leaching of low-grade waste, *Hydrometallurgy* **3**:309–326 (1978).

Booth, J. E., and Williams, J. W., The isolation of a mercuric ion-reducing flavoprotein from *Thiobacillus ferrooxidans, J. Gen. Microbiol.* **130**:725–730 (1984).

Brierley, C. L., and Brierley, J. A., Anaerobic reduction of molybdenum by *Sulfolobus* species, *Zbl. Bakteriol. Hyg., I Abt. Orig.* **C3**:289–294 (1982).

Brown, M., and Kester, D. R., Ultraviolet spectroscopic study of ferric iron solutions, *Appl. Spectrosc.* **34**:377–380 (1980).

Bruynesteyn, A., Hackl, R. P., and Wright, F., The BIOTANKLEACH process, in *Gold 100—Proc. Intern. Conf. on Gold.* Vol. 2: *Extractive Metallurgy of Gold*, South African Institute of Mining and Metallurgy, Johannesburg, 1986, pp. 353–365.

Bruynesteyn, A., Lawrence, R. W., Vizsolyi, A., and Hackl, R., An elemental sulphur producing biohydrometallurgical process for treating sulphide concentrates, in G. Rossi and A. E. Torma (eds.), *Recent Progress in Biohydrometallurgy*, Associazione Mineraria Sarda, Cagliari, Italy, 1983, pp. 151–168.

Buckingham, J. A., Stanbury, P. F., LeRoux, N. W., Effect of molybdenum of the efficiency of tetrathionate utilisation by chemostat cultures of *Sulfolobus* BC, *Biotechnol. Lett.* **11**:99–104 (1989).

Cameron, F. J., Jones, M. V., and Edwards, C., Effects of salinity on bacterial iron oxidation, *Curr. Microbiol.* **10**:353–356 (1984).

DiSpirito, A. A., and Tuovinen, O. H., Uranous ion oxidation and carbon dioxide fixation by *Thiobacillus ferrooxidans, Arch. Microbiol.* **133**:28–32 (1982a).

DiSpirito, A. A., and Tuovinen, O. H., Kinetics of uranous ion and ferrous iron oxidation by *Thiobacillus ferrooxidans, Arch. Microbiol.* **133**:33–37 (1982b).

DiSpirito, A. A., and Tuovinen, O. H., Oxidations of nonferrous metals by thiobacilli, in W. R. Strohl and O. H. Tuovinen (eds.), *Microbial Chemoautotrophy*, Ohio State University Press, Columbus, 1984, pp. 11–29.

Dutrizac, J. E., The dissolution of chalcopyrite in ferric sulfate and ferric chloride media, *Metall. Trans.* **12B**:371–378 (1981).

Dutrizac, J. E., The behavior of impurities during jarosite precipitation, in R. G. Bautista (ed.), *Hydrometallurgical Process Fundamentals*, Plenum Press, New York, 1984, pp. 125–169.

Eccleston, M., and Kelly, D. P., Oxidation kinetics and chemostat growth kinetics of *Thiobacillus ferrooxidans* on tetrathionate and thiosulfate, *J. Bacteriol.* **134**:718–727 (1978).

Edvardsson, U., The use of a rotating barrel to determine bacterial leaching of arsenopyrite/pyrite concentrates, *Biorecovery* **1**:43–50 (1988).

Emmel, T., Sand, W., König, W. A., and Bock, E., Evidence for the existence of a sulphur oxygenase in *Sulfolobus brierleyi, J. Gen. Microbiol.* **132**:3415–3420 (1986).

Fry, I. V., Lazaroff, N., and Packer, L., Sulfate-dependent iron oxidation by *Thiobacillus ferrooxidans:* Characterization of a new EPR detectable electron transport component on the reducing side of rusticyanin, *Arch. Biochem. Biophys.* **246**:650–654 (1986).

Grishin, S. I., Bigham, J. M., and Tuovinen, O. H., Characterization of jarosite formed upon bacterial oxidation of ferrous sulfate in a packed-bed reactor, *Appl. Environ. Microbiol.* **54**:3101–3106 (1988).

Grishin, S. I., and Tuovinen, O. H., Fast kinetics of Fe^{2+} oxidation in packed-bed reactors, *Appl. Environ. Microbiol.* **54**:3092–3100 (1988).

Hoffmann, M. R., Hiltunen, P., and Tuovinen, O. H., Inhibition of ferrous ion oxidation by *Thiobacillus ferrooxidans* in the presence of oxyanions of sulfur and phosphorus, in Y. A. Attia (ed.), *Processing and Utilization of High Sulfur Coals*, Elsevier, Amsterdam, 1985, pp. 683–698.

Holuigue, L., Herrera, L., Phillips, O. M., Young, M., and Allende, J. E., CO_2 fixation by mineral-leaching bacteria: Characteristics of the ribulose bisphosphate carboxylase-oxygenase of *Thiobacillus ferrooxidans, Biotechnol. Appl. Biochem.* **9**:497–505 (1987).

Hooper, A. B., and DiSpirito, A. A., In bacteria which grow on simple reductants, generation of a proton gradient involves extracytoplasmic oxidation of substrate. *Microbiol. Rev.* **49**:140–157 (1985).

Ingledew, W. J., *Thiobacillus ferrooxidans:* The bioenergetics of an acidophilic chemolithotroph, *Biochim. Biophys. Acta* **683**:89–117 (1982).

Karamanev, D. G., and Nikolov, L., Influence of some physicochemical parameters on bacterial activity of biofilm: Ferrous iron oxidation by *Thiobacillus ferrooxidans*, *Biotechnol. Bioeng.* **31**:295–299 (1988).

Karavaiko, G. I., Chuchalin, L. K., Pivovarova, T. A., Yemelyanov, B. A., and Dorofeyev, A. G., Microbiological leaching of metals from arsenopyrite containing concentrates, in R. W. Lawrence, R. M. R. Branion, and H. G. Ebner (eds.), *Fundamental and Applied Biohydrometallurgy*, Elsevier, Amsterdam, 1986, pp. 115–126.

Kelly, D. P., Physiology of the thiobacilli: Elucidating the sulphur oxidation pathway, *Microbiol. Sci.* **2**:105–109 (1985).

Lawrence, R. W., Biotreatment of gold ores, in H. L. Ehrlich and C. L. Brierley (eds.), *Microbial Mineral Recovery*, McGraw-Hill, New York, 1990, pp. 127–148.

Lawrence, R. W., Vizsolyi, A., and Vos, R. J., The silver catalyzed bioleach process for copper concentrate, in L. Haas and J. Clum (eds.), *Microbiological Effects on Metallurgical Processes*, AIME, New York, 1985, pp. 65–82.

Lindström, E. B., and Gunneriusson, L., Semi-continuous leaching of arsenopyrite with a *Sulfolobus* strain at 70°C, in P. R. Norris and D. P. Kelly (eds.), *Biohydrometallurgy*, Science and Technology Letters, Kew, U.K., 1988, pp. 525–527.

Mahapatra, S. S. R., and Mishra, A. K., Inhibition of iron oxidation in *Thiobacillus ferrooxidans* by toxic metals and its alleviation by EDTA, *Curr. Microbiol.* **11**:1–6 (1984).

McCready, R. G. L., and Gould, W. D., Bioleaching of uranium, in H. L. Ehrlich and C. L. Brierley (eds.), *Microbial Mineral Recovery*, McGraw-Hill, New York, 1990, pp. 107–125.

Mehta, A. P., and Murr, L. E., Kinetic study of sulfide leaching by galvanic interaction between chalcopyrite, pyrite, and sphalerite in the presence of *T. ferrooxidans* (30°C) and a thermophilic microorganism (55°C), *Biotechnol. Bioeng.* **24**:919–940 (1982).

Miller, J. D., McDonough, P. J., and Portillo, H. Q., Electrochemistry in silver catalysed ferric sulfate leaching of chalcopyrite, in M. C. Kuhn (ed.), *Process and Fundamental Considerations of Selected Hydrometallurgical Systems*, AIME, New York, 1981, pp. 327–344.

Miller, J. D., and Portillo, H. Q., Silver catalysis in ferric sulfate leaching of chalcopyrite, in J. Laskowski (ed.), *Developments in Mineral Processing*, Elsevier, Amsterdam, 1981, pp. 851–897.

Murayama, T., Konno, Y., Sakata, T., and Imaizumi, T., Application of immobilized *Thiobacillus ferrooxidans* for large-scale treatment of acid mine drainage, *Meth. Enzymol.* **136**:530–540 (1987).

Nakamura, K., Noike, T., and Matsumoto, J., Effect of operation conditions on biological Fe^{2+} oxidation with rotating biological contactors, *Water Res.* **20**:73–77 (1986).

Natarajan, K. A., Electrochemical aspects of bioleaching of base metal sulfides, in H. L. Ehrlich and C. L. Brierley (eds.), *Microbial Mineral Recovery*, McGraw-Hill, New York, 1990, pp. 79–106.

Nikolov, L., and Karamanev, D., Experimental study of the inverse fluidized bed biofilm reactor, *Can. J. Chem. Eng.* **65**:214–217 (1987).

Nikolov, L., Mehochev, D., and Dimitrov, D., Continuous bacterial ferrous iron oxidation by *Thiobacillus ferrooxidans* in rotating biological contactors, *Biotechnol. Lett.* **8**:707–710 (1986).

Nikolov, L., Valkova-Vachanova, M., and Mehochev, D., Oxidation of high ferrous iron concentrations by chemolithotrophic *Thiobacillus ferrooxidans* in packed bed bioreactors, *J. Biotechnol* **7**:87–94 (1988).

Norris, P. R., Acidophilic bacteria and their activity in mineral sulfide oxidation, in H. L. Ehrlich and C. L. Brierley (eds.), *Microbial Mineral Recovery*, McGraw-Hill, New York, 1990, pp. 3–27.

Norris, P. R., and Parrott, L., High temperature mineral concentrate dissolution with *Sulfolobus*, in R. W. Lawrence, R. M. R. Branion, and H. G. Ebner (eds.), *Fundamental and Applied Biohydrometallurgy*, Elsevier, Amsterdam, 1986, pp. 355–365.

Puhakka, J. P., and Tuovinen, O. H., Effect of organic compounds on the microbiological leaching of a complex sulphide ore material, *MIRCEN J. Appl. Microbiol. Biotechnol.* **3**:436–442 (1987).

Ralph, B. J., Biotechnology applied to raw minerals processing, in C. W. Robinson and J. A. Howell (eds.), *Comprehensive Biotechnology*, Vol. 4, Pergamon Press, Oxford, 1985, pp. 201–234.

Rawlings, D. E., Pretorius, I., and Woods, D. R., Expression of a *Thiobacillus ferrooxidans* origin of replication in *Escherichia coli*, *J. Bacteriol.* **158**:737–739 (1984a).

Rawlings, D. E., Pretorius, I.-M., and Woods, D. R., Construction of arsenic-resistant *Thiobacillus ferrooxidans* recombinant plasmids and the expression of autotrophic plasmid genes in a heterotrophic cell-free system, *J. Biotechnol.* **1**:129–133 (1984b).

Rawlings, D. E., and Woods, D. R., Recombinant DNA techniques and the genetic manipulation of *Thiobacillus ferrooxidans*, in J. A. Thomson (ed.), *Recombinant DNA and Bacterial Fermentation*, CRC Press, Boca Raton, Fla., 1988, pp. 277–296.

Rossi, G., Torma, A. E., and Trois, P., Bacteria-mediated copper recovery from a cupriferous pyrrhotite ore: chalcopyrite/pyrrhotite interactions, in G. Rossi and A. E. Torma (eds.), *Recent Progress in Biohydrometallurgy*, Associazione Mineraria Sarda, Cagliari, 1983, pp. 185–200.

Smith, J. R., Luthy, R. G., and Middleton, A. C., Microbial ferrous iron oxidation in acidic solution, *J. Water Pollut. Contr. Feder.* **60**:518–530 (1988).

Stevens, C. J., Dugan, P. R., and Tuovinen, O. H., Acetylene reduction (nitrogen fixation) by *Thiobacillus ferrooxidans*, *Biotechnol. Appl. Biochem.* **8**:351–359 (1986).

Sugio, T., Domatsu, C., Munakata, O., Tano, T., and Imai, K., Role of a ferric iron-reducing system in sulfur oxidation of *Thiobacillus ferrooxidans*, *Appl., Environ. Microbiol.* **49**:1401–1406 (1985).

Sugio, T., Katagiri, T., Inagaki, K., and Tano, T., Actual substrate for elemental sulfur oxidation by sulfur:ferric ion oxidoreductase purified from *Thiobacillus ferrooxidans*, *Biochim. Biophys. Acta* **973**:250–256 (1989).

Sugio, T., Mizunashi, W., Inagaki, K., and Tano, T., Purification and some properties of sulfur:ferric ion oxidoreductase from *Thiobacillus ferrooxidans*, *J. Bacteriol.* **169**:4916–4922 (1987).

Sugio, T., Tsujita, Y., Hirayama, K., Inagaki, K., and Tano, T., Mechanism of tetravalent manganese reduction with elemental sulfur by *Thiobacillus ferrooxidans*, *Agric. Biol. Chem.* **52**:185–190 (1988a).

Sugio, T., Tsujita, Y., Katagiri, T., Inagaki, K., and Tano, T., Reduction of Mo^{6+} with elemental sulfur by *Thiobacillus ferrooxidans*, *J. Bacteriol.* **170**:5956–5959 (1988b).

Sugio, T., Wada, K., Mori, M., Inagaki, K., and Tano, T., Synthesis of an iron-oxidizing system during growth of *Thiobacillus ferrooxidans* on sulfur-basal salts medium, *Appl. Environ. Microbiol.* **54**:150–152 (1988c).

Takakuwa, S., Nishiwaki, T., Hosoda, K., Tominaga, N., and Iwasaki, H., Promoting effect of molybdate on the growth of a sulfur-oxidizing bacterium, *Thiobacillus ferrooxidans*, *J. Gen. Appl. Microbiol.* **23**:163–173 (1977).

Tuovinen, O. H., Puhakka, J., Hiltunen, P., and Dolan, K. H., Silver toxicity to ferrous iron and pyrite oxidation and its alleviation by yeast extract in cultures of *Thiobacillus ferrooxidans*, *Biotechnol. Lett.* **7**:389–394 (1985).

Wan, R. Y., Miller, J. D., Foley, J., and Pons, S., Electrochemical features of the ferric sulfate leaching of $CuFeS_2/C$ aggregates, in P. E. Richardson, S. Srinivasan, and R. Woods (eds.), *Electrochemistry in Mineral and Metal Processing*, Electrochemical Society, Pennington, N.J., 1984, pp. 391–416.

Wan, R. Y., Miller, J. D., and Simkovich, G., Enhanced ferric sulphate leaching of copper from $CuFeS_2$ and C particulate aggregates, *Metall. Trans.* **16B**:575–588 (1985).

Electrochemical Aspects of Bioleaching of Base-Metal Sulfides

K. A. Natarajan

Introduction

The mechanisms involved in the bioleaching of multisulfides are rather complex. When living organisms interact with mineral substrates in an acid medium, a number of mutually complementary, as well as divergent, processes come into play. Any study of microbe-mineral interactions should therefore aim at understanding the various chemical, biochemical, and electrochemical factors, all of which influence the rate of mineral dissolution. The role of galvanic interactions in the leaching of multisulfides has been recognized of late. However, the manner in which bacteria of the type *Thiobacillus ferrooxidans* accelerate galvanic dissolution processes is not well understood.

Electrochemical interactions in the bioleaching of complex sulfides could be taken advantage of in achieving selective dissolution of the desired mineral. Preferential attachment of bacteria onto selective sulfide mineral substrates would catalyze the dissolution process. In the development of an appropriate bioleaching process to treat complex sulfides, it is also essential to consider the use of bacterial strains with high metal tolerance.

In this chapter, various electrochemical principles governing the bioleaching of multisulfides are examined. The utility of electrochemical data based on measurements of rest potentials, combination potentials, galvanic current, and anodic and cathodic polarization in the interpretation of mineral dissolution behavior and bacterial activity are illustrated. Electrochemical models for the biooxidation of multisulfide combinations are proposed. Factors affecting galvanic interactions in a multisulfide bioleaching system are discussed with respect to enhancement of selective dissolution of the desired mineral. The influence of an applied dc potential on bacterial activity and on selective mineral dissolution is also discussed. A laboratory process for the electrochemical bioleaching of sphalerite is outlined, and the various mechanisms underlying such a procedure are examined in the light of increase in biomass, control of ferric-ferrous ratio, and acceleration of zinc dissolution rates.

Electrochemistry of Sulfide Minerals

Many natural minerals are good conductors and semiconductors. They behave as electrodes in the presence of a leaching medium. Since mineral dissolution in leaching is essentially a corrosion process, principles of electrochemistry apply. In leaching systems containing more than one mineral, galvanic interactions necessarily come into play. Various sulfide minerals can be arranged in the form of a galvanic series with regard to their relative electrochemical activity (Table 4.1). Depending on the nature, type, and amount of the impurities present in a sulfide mineral, the measured rest potential may vary from sample to sample, and thus one can expect a shift in position for a sulfide mineral in the galvanic series. For example, the rest potential of sphalerite was found to be influenced by its iron content (Dutrizac and MacDonald, 1973; Rossi, 1975).

Pyrite and chalcopyrite, being nobler sulfides, are difficult to oxidize in an acid medium. On the other hand, sphalerite and pyrrhotite are active minerals and hence can be easily oxidized. When two sulfide minerals establish contact in a leaching medium, a galvanic cell will be formed; the more active mineral in the couple will undergo corrosion while the nobler (less active) one is cathodically protected. The

TABLE 4.1 Galvanic Series of Some Base-Metal Sulfides in a Bioleaching Medium

Pyrite	Noble
Chalcopyrite	
Pentlandite	↑
Galena	↓
Pyrrhotite	
Sphalerite	Active

driving force for the galvanic dissolution process is the potential difference existing in the couple, while the rate of anodic oxidation is the galvanic current flowing in the circuit. Based on the galvanic series, it is possible to predict the selective dissolution behavior of a sulfide mineral in a binary or multiple combination. For example, in a pyrite-chalcopyrite combination, chalcopyrite would be expected to be selectively leached in preference to pyrite since it is anodic. Similarly, pentlandite would be selectively oxidized when contacted with chalcopyrite. Prediction of the electrochemical behavior of sulfide minerals in multiple combinations is more difficult.

According to Tomashov (1966), the potential of each individual electrode in a short-circuited multielectrode galvanic system approaches some overall potential E_x, and the sum of all cathodic currents exactly equals the sum of all anodic currents in the system. The operation of components of a multielectrode system depends on their specific polarization, area relationships among electrodes, and relative spatial distribution of electrodes in the electrolyte. It also depends on the specific resistivity of the electrolyte.

1. An electrode with the most negative initial potential will always be anodic; the most positive, always cathodic.

2. If the intermediate electrodes are placed nearer the principal anodes, their tendency to become cathodes increases; if placed near the principal cathodes, their tendency to become anodes will prevail.

Thus, in all types of possible binary, ternary, or quarternary combinations involving the sulfide minerals listed above, pyrite could be expected to behave cathodically and sphalerite anodically, since pyrite is the noblest and sphalerite is the most active. However, the electrochemical dissolution behavior of the sulfide minerals having intermediate rest potential values, such as chalcopyrite, pentlandite, galena, and pyrrhotite, depends on the relative activity of the mineral with which they are in close contact. Thus, chalcopyrite, while behaving anodically in contact with pyrite, acts as a cathode if placed in the vicinity of pentlandite, galena, pyrrhotite, or sphalerite. Similarly, galena is anodic to chalcopyrite but cathodic to sphalerite. The schematic in Fig. 4.1 shows the behavior of a multimineral combination.

The following factors influence the rate of galvanic interactions among sulfide minerals in a bioleaching system.

1. Rest potential difference as well as the galvanic current existing in a combination

2. Relative surface areas between anode and cathode (a smaller anode

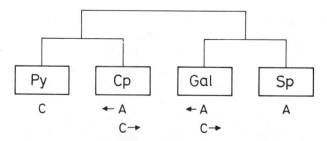

Figure 4.1 Multimineral electrode cell consisting of pyrite (Py), chalcopyrite (Cp), galena (Gal), and sphalerite (Sp) (A = anode, C = cathode, arrow marks indicate relative anodic and cathodic behavior of intermediate minerals with respect to the mineral in the vicinity).

in contact with a larger cathode facilitating enhanced anodic disso-lution)—*area effect*

3. Interelectrode distance—*distance effect*

4. Nature and duration of contact

5. Conductivity of the mineral and the electrolyte

6. Electrolyte properties such as pH, dissolved salts, presence or absence of oxygen, and other redox couples

7. Presence or absence of microorganisms

Among the sulfide minerals listed above, sphalerite, unlike chalco-pyrite and pyrite, exhibits poor conductivity.

E_h and pH as environmental parameters

In addition to the rest potential of the sulfide mineral, the E_h (oxidation potential) of the leaching system is equally important. A large difference between E_h of the medium and the potential of the sulfides must exist for efficient mineral dissolution to proceed. The two most important environmental parameters governing a bioleaching reaction are E_h and pH, since they also influence the metabolic activity of the bacteria. Available energy to the bacteria is in the form of electrons. During the growth of *T. ferrooxidans,* E_h is from 700 to 800 mV and depends on pH level, pO_2 level, and the concentrations of redox species in the system. For the reaction

$$Fe^{2+} = Fe^{3+} + e \qquad (4.1)$$

$$E_h = 0.771 + 0.059 \log \frac{[Fe^{3+}]}{[Fe^{2+}]} \qquad (4.2)$$

The presence of bacteria further enhances the E_h of the system containing ferrous and ferric ions. The important electrochemical criterion for efficient bacterial leaching would therefore be the establishment of an optimum E_h value both for bacterial activity and the oxidation of the desired sulfide mineral.

E_h-pH diagrams can readily be constructed to portray the stability and optimum activity limits for *T. ferrooxidans* (Baas Becking et al., 1968; Natarajan, 1976). The stability regions for the various dissolved and precipitated species in a sulfide mineral system can be incorporated in Fig. 4.2. The stability limits for the important species involved in a chalcopyrite bioleaching system are represented in the simplified E_h-pH diagram in Fig. 4.2. The activity region for *T. ferrooxidans* is confined to the window marked in the diagram within an E_h range of about 400 to 800 mV in the pH region of 1.5 to 3.0. Measured E_h values from a chalcopyrite leaching flask in the presence and absence of *T. ferrooxidans* (Jyothi, 1988) are also indicated in the diagram. The E_h values are higher for chalcopyrite leaching in the presence of the bacteria (700 to 760 mV), indicating the contribution of the microorganisms toward maintaining a higher oxidation potential to facilitate faster mineral dissolution. Cu^{2+}, Fe^{2+}, and SO_4^{2-} are stable in the leaching region marked on the diagram.

Combination potential and galvanic current in mineral combinations

Measured rest potential values for pyrite, chalcopyrite, galena, and sphalerite in a bioleaching medium in the presence and absence of *T. ferrooxidans* are given in Table 4.2. Pyrite exhibits the most positive rest potential under all conditions, closely followed by chalcopyrite. Both galena and sphalerite are active relative to pyrite and chalcopyrite. The presence of *T. ferrooxidans* in the medium shifts the measured potentials in a positive direction for all the sulfides. In a bacterial culture, the potential readings are still higher, owing to the added presence of ferric and ferrous ions in the medium.

TABLE 4.2 Measured Rest Potentials for Some Base-Metal Sulfides in 0.9 K Fe Medium

	Rest potential, E_{SCE}, mV		
Mineral	Uninoculated	Inoculated iron-free medium	Inoculated iron-containing medium
Pyrite	300	340	450
Chalcopyrite	200	240	310
Galena	10	20	175
Spalerite	-35	-20	-10

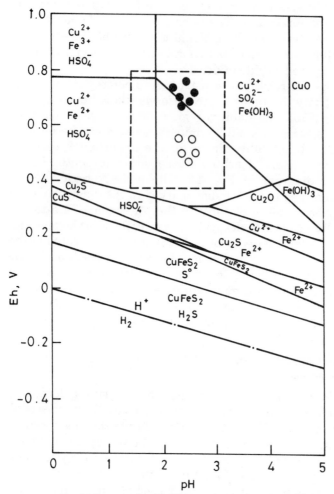

Figure 4.2 E_h-pH diagram for the chalcopyrite-water system incorporating stability and activity limits for *Thiobacillus ferrooxidans* (filled circles indicate measured E_h values during chalcopyrite leaching in the presence of bacteria, open circles in the absence of bacteria).

    The variation of combination potential and galvanic current with time in pyrite-sphalerite, chalcopyrite-pyrite, and chalcopyrite-sphalerite couples in a 0.9 K Fe medium in the presence and absence of *T. ferrooxidans* is illustrated in Fig. 4.3. The behavior of a sulfide mineral in these couples differs from that of the same mineral when present alone. The combination potential measured in a binary mineral combination tends to attain a steady-state value with time and a value between the individual rest potentials of the sulfides in the cou-

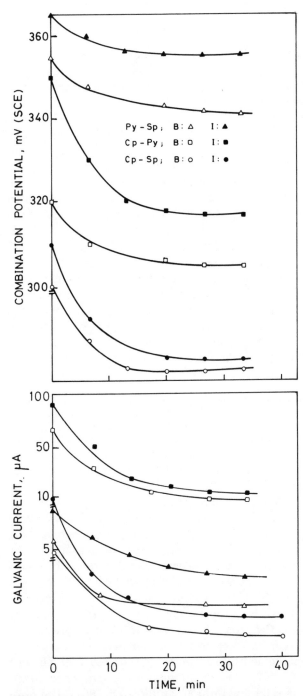

Figure 4.3 Variation of combination potential and galvanic current with time in some bimineral combinations involving pyrite (Py), chalcopyrite (Cp), and sphalerite (Sp) in the presence (I) and absence (B) of *T. ferrooxidans* in 0.9 K Fe medium. [The 0.9 K medium is that of Silverman and Lundgren, (1959), but containing only 0.9 instead of 9 g Fe per liter.]

ple. A pyrite-sphalerite couple exhibits the highest combination potential, while the chalcopyrite-sphalerite couple has the lowest. On the other hand, a chalcopyrite-pyrite couple exhibits the highest galvanic current compared with binary couples involving sphalerite. Lower galvanic currents monitored in the presence of sphalerite could be attributed to the poor conductivity of the sphalerite mineral. In the presence of added graphite powder, the galvanic current in sphalerite involving combinations was found to increase significantly. The presence of T. ferrooxidans enhances the electrochemical activity in the various mineral combinations as substantiated by the increase in combination potentials and galvanic currents in the presence of the bacteria. As the leaching proceeds, the combination potential and galvanic current shifts, resulting in the selective dissolution of the most active mineral; the presence of bacteria further accelerates such a preferential galvanic oxidation process. In ternary and quarternary combinations involving the above sulfides, the magnitude and sign of the galvanic current depend on the activity of the mineral which is connected as the anode in the measuring circuit. It is also essential to monitor the currents at the various possible binary contact points available in the multimineral combination.

Role of the Galvanic Effect in Sulfide Dissolution

The influence of the galvanic effect in sulfide mineral dissolution has been well documented. The rate of oxidation of certain natural sulfides is greatly increased by the presence of either marcasite or pyrite (Gottschalk and Buehler, 1910). The rate of chalcopyrite dissolution was found to be accelerated in the presence of pyrite, molybdenite, and stibnite, but was retarded in the presence of galena (Dutrizac and MacDonald, 1973). In their study of galvanic interaction between Cu and chalcopyrite, Hiskey and Wadsworth (1975) demonstrated the rapid conversion of chalcopyrite to chalcocite. Similar galvanic effects were observed during the reduction of chalcopyrite by Fe, Cu, or Pb in acid solutions (Nicol, 1975). Galvanic interaction in a chalcopyrite-pyrite couple caused the Cu mineral to corrode more rapidly than pyrite, and the presence of T. ferrooxidans was found to catalyze the reaction (Berry et al., 1978). In the presence of pyrite and bacteria, the dissolution rate of molybdenite increased by almost 350 percent (Bryner and Anderson, 1957). Bacterial oxidation of sphalerite mixed with pyrite was increased 30-fold (Malouf and Prater, 1961). The preferential attack of sphalerite over chalcopyrite in the bioleaching of a Cu-bearing ore was also observed (Bruynesteyn and Duncan, 1974). Arsenopyrite, which was found to be more active than chalcopyrite, was selectively leached using T. ferrooxidans (Polkin et al., 1975). The breakdown of arsenopyrite takes place first; that of chalcopyrite starts only after the former reaction is complete. The contribution of the galvanic effect to the bioleaching of mixed sulfides has been systematically studied in the presence and absence of bacteria (Mehta and Murr, 1982, 1983). Coupled chalcopyrite-pyrite, sphalerite-pyrite, and chalcopyrite-pyrite-sphalerite systems showed improved metal dissolution compared with leaching of single minerals. Cu dissolution was found to increase by a factor of 4.6, when the galvanic leaching of chalcopyrite-pyrite was compared with chalcopyrite leaching. When bacteria were present, Cu dissolution increased by an additional factor of 2.1. Similarly, for the sphalerite-pyrite system, Zn dissolution was faster and the presence of bacteria enhanced it further. From a complex ore containing about 8% Zn, 6% Pb, and 1% Cu, almost 99.5 percent of the Zn was selectively bioleached, while only 2 percent of the Cu was solubilized (Natarajan and Iwasaki, 1985).

Laboratory studies have indicated that $T.$ $ferrooxidans$ can utilize sulfides even in the absence of Fe (Imai, 1978; Dave et al., 1979). Rest potentials of chemically prepared CuS and ZnS were measured as 340 mV and 140 mV, respectively, and biooxidation of ZnS was found to be enhanced in the presence of CuS, while Cu dissolution decreased significantly with the addition of ZnS (Dave, 1980). Enhanced dissolution of pyrrhotite after contacting chalcopyrite has also been observed (Natarajan et al., 1982, 1983; Natarajan and Iwasaki, 1983). As the proportion of pyrrhotite in the mixture was increased, Cu dissolution decreased rapidly. From an ore containing 0.8% Cu, 0.2% Ni, 11.9% Fe, and 2% S as chalcopyrite-cubanite, pentlandite, and pyrrhotite, less than about 2% Cu was solubilized while about 25% Ni was bioleached after 40 days. Ni release into the solution was faster and in all cases took place before Cu appeared in the solution; Ni concentrations in solution were several times higher than those of Cu. Cu concentration in solution was found to increase only after most of the Ni had been solubilized (Natarajan and Iwasaki, 1983). Rest potential measurements indicated that pentlandite (E_{SCE} = 100 to 180 mV)* and pyrrhotite (E_{SCE} = 110 to 120 mV) were anodic to chalcopyrite-cubanite (E_{SCE} = 250 mV); hence selective Ni dissolution in preference to Cu was expected.

Effect of Mineral-to-Mineral Ratio in the Bioleaching of Mixed Sulfides

By controlling the ratio of different minerals in a mixture, one can control the dissolution rate of the desired mineral component, as suggested by the surface area effect. The effect of the ratios of mixing of

---

*SCE = saturated calomel electrode.

chalcopyrite and sphalerite concentrates on copper and zinc dissolution is illustrated in Table 4.3 (Ashokkumar, 1980; Natarajan, 1985). Zinc extraction increases as chalcopyrite content increases in the mixture, whereas copper dissolution decreases as the amount of sphalerite increases. Although the presence of *T. ferrooxidans* enhances the dissolution of both copper and zinc, zinc dissolution is more favored galvanically. An optimum ratio of mixing exists for two concentrates, one of copper ore and one of zinc ore, at which zinc extraction would be the highest and copper dissolution would be negligible. Similar experiments with respect to the bioleaching of chalcopyrite-pyrrhotite concentrate mixtures have indicated enhanced dissolution of pyrrhotite with increasing chalcopyrite content, while copper dissolution decreases as the pyrrhotite content increases (Natarajan et al., 1982). A study of the interactions between pyrrhotite and chalcopyrite in a bioleaching process indicated that a critical ratio among the minerals exists at which the dissolution of copper is promoted (Rossi et al., 1983). Copper dissolution from pyrite-chalcopyrite was faster than that from pyrrhotite-chalcopyrite or from pure chalcopyrite, since the galvanic effect is favorable for copper dissolution in the former couple (Ahonen et al., 1986). The leaching behavior of sphalerite in binary combination with pyrite, chalcopyrite, and galena at two different ratios of mixing of the pure mineral samples, namely, 5:5 and 5:3 by weight, is illustrated in Fig. 4.4 (Jyothi et al., 1988). Sphalerite dissolution from the various binary combinations follows the galvanic principle. The percent zinc dissolved was the highest from the sphalerite-pyrite couple followed by that from the sphalerite-chalcopyrite mixture. While 70 percent of zinc was leached in 100 days in the presence of *T. ferrooxidans* from a 5:5 mixture of sphalerite and pyrite, only about 35 percent of zinc was so leached from sphalerite contacted with chalcopyrite under the same conditions. Only about 4.5 percent of zinc was leached from the sphalerite-galena couple under similar

TABLE 4.3 Copper and Zinc Dissolution from Chalcopyrite-Sphalerite Concentrate Mixtures

(9 K, pH 2.3, shake-flask leaching at 10% pulp density for 41 days)

		Percent Zn solubilized		Percent Cu solubilized	
Sphalerite, g	Chalcopyrite, g	Uninoculated control	Inoculated	Uninoculated control	Inoculated
1	9	10.2	90.5	6.8	18.6
2	8		90.1		16.6
4	6		88.7		11.6
5	5	10.7	88.9	6.9	4.8
8	2		88.0		4.6

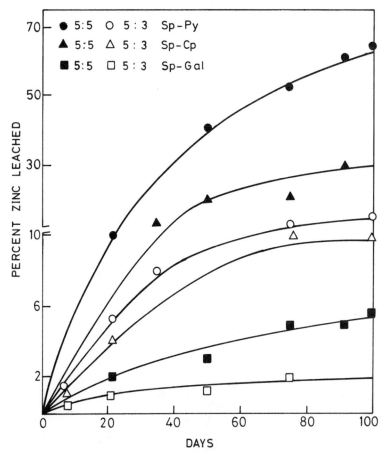

Figure 4.4   Zinc dissolution from bimineral couples involving sphalerite at different mineral-to-mineral ratios in the presence of *T. ferrooxidans*.

conditions. The presence of bacteria as well as the ratio of mixing influenced the galvanic dissolution of sphalerite from the various binary mineral mixtures. Zinc dissolution was higher and faster from a 5:5 mixture than from a 5:3 mixture in all the binary combinations.

Galvanic effects in the bioleaching of chalcopyrite when mixed with pyrite, galena, and sphalerite in a 1:1 ratio by weight are illustrated in Fig. 4.5 (Jyothi et al., 1988). Copper dissolution from a chalcopyrite-pyrite couple was found to be the highest (almost 90 percent extraction after 80 days of bioleaching in shake flasks), because chalcopyrite is anodic to pyrite. Copper extraction from chalcopyrite-galena and chalcopyrite-sphalerite couples under similar leaching conditions was very low, only about 5 percent and 3.5 percent, respectively. Copper dissolution from the above two couples is not galvani-

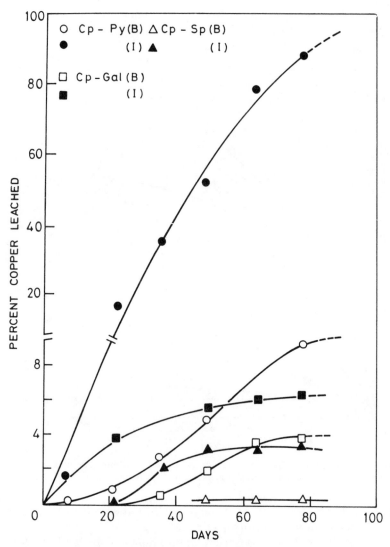

Figure 4.5 Copper dissolution from bimineral couples involving chalcopyrite in the presence (I) and absence (B) of *T. ferrooxidans.*

cally favored, since both galena and sphalerite are anodic to chalcopyrite.

For sustaining the galvanic interactions in a multisulfide leaching system, intimate and continuous mineral-mineral contacts must be established. In agitation leaching of fine particles, making and breaking of such contacts take place continuously. The need to maintain proper mineral-mineral contacts to ensure enhanced selective dissolution is demonstrated through the experimental results presented in Fig. 4.6. As can be seen, a continuously higher rate of bioleaching for copper from a chalcopyrite-pyrite couple was ensured only when the two minerals were in intimate contact throughout the duration of the test, as in a packed bed of the mineral mixture through which the lixiviant

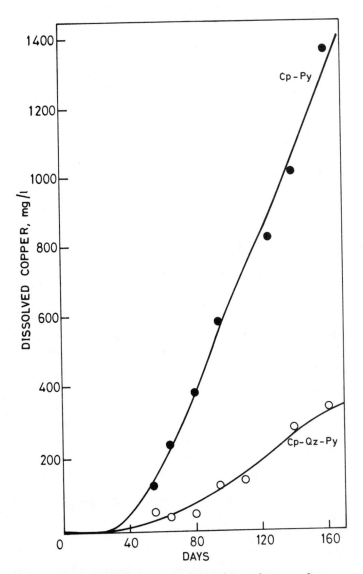

Figure 4.6 Galvanic effect in the bioleaching of copper from a chalcopyrite-pyrite couple in the presence and absence of an intermediate quartz layer.

was continuously percolated. On the other hand, chalcopyrite dissolution from a chalcopyrite-quartz-pyrite mixture, in which a centrally placed insulating quartz layer physically separated the two sulfide minerals, was limited and slower. After about 5 months of leaching in the presence of *T. ferrooxidans,* almost 4½ to 5 times higher copper dissolution was achieved from the chalcopyrite-pyrite mixture than from the same mixture separated by the quartz layer.

Electrobiochemical Reaction Mechanisms

Probable electrochemical reactions involved in the bioleaching of the two types of complex sulfides discussed above are given next.

1. Lead-zinc-copper sulfides

$$ZnS = Zn^{2+} + 2e + S^0 \text{ (anodic oxidation of sphalerite)} \qquad (4.3)$$

$$PbS = Pb^{2+} + 2e + S^0 \text{ (anodic oxidation of galena)} \qquad (4.4)$$

($PbSO_4$ will be precipitated in the bioleaching medium)

$$CuFeS_2 = Cu^{2+} + Fe^{2+} + 4e + 2S^0 \qquad (4.5)$$

(anodic oxidation of chalcopyrite)

$$O_2 + 4H^+ + 4e = 2H_2O \quad \text{(cathodic reduction of oxygen)} \quad (4.6)$$

(at the surfaces of nobler minerals such as pyrite or chalcopyrite)

Oxidation of sphalerite will be the prominent anodic reaction. Anodic oxidation of chalcopyrite takes place only if it comes in contact with pyrite.

2. Copper-nickel-iron sulfides

$$FeS = Fe^{2+} + S^0 + 2e \quad \text{(anodic oxidation of pyrrhotite)} \quad (4.7)$$

$$(Fe_xNi_{1-x})_9S_8 = 9xFe^{2+} + 9(1 - x)Ni^{2+} + 8S^0 + 18e \qquad (4.8)$$

(anodic oxidation of pentlandite)

Cathodic reduction of oxygen as in Eq. (4.6) takes place at chalcopyrite surfaces.

The net electrochemical reaction could be generalized as

$$2MS + O_2 + 4H^+ = 2M^{2+} + 2S^0 + 2H_2O \qquad (4.9)$$

Elemental sulfur, the product of anodic oxidation would accumulate on the leached surfaces and, if not removed, could act as a barrier for further leaching. In the presence of *T. ferrooxidans,* both the Fe^{2+} and elemental sulfur are subsequently oxidized to Fe^{3+} and sulfate ions, leading to the formation of acidic ferric sulfate. The presence of bacteria in the system helps in the efficient removal of the sulfur layer from the leached mineral surfaces, thereby exposing the sulfides for continued bio- and electrochemical oxidation. As pointed out before, the bacteria also help in maintaining a higher oxidation potential in the system through the oxidation of Fe^{2+}. A schematic model illustrating the combined bacterial-galvanic oxidation of a binary mineral combination is given in Fig. 4.7.

Galvanic Displacement of Dissolved Metal Ions

Another important aspect of galvanic effect in the bioleaching of mixed sulfides is the removal of nobler metal ions from an acid leach solution by the presence of active mineral sulfides. As Table 4.3 shows, the dissolved copper concentration in solution decreases as the sphalerite content in the pulp increases. The amount of copper in the leach residues after bioleaching different chalcopyrite-sphalerite concentrate mixtures is tabulated in Table 4.4. As the sphalerite in the concentrate mixture increased, the amount of copper reported in the leach residue also increased. Galvanic cementation of dissolved copper by active ZnS in an acid solution is possible.

$$Cu^{2+} + ZnS = Cu^0 + Zn^{2+} + S^0 \qquad (4.10)$$

A decrease in dissolved copper in the leach liquor with time has also been observed in several other tests involving chalcopyrite-sphalerite. Jyothi et al. (1988) monitored the decrease in copper concentration and the increase in zinc concentration in solution with time when 10 g of

TABLE 4.4 Copper Content in Leach Residues after Bioleaching of Chalcopyrite-Sphalerite Concentrate Mixtures

Ratio of chalcopyrite to sphalerite (weight in grams)	Percent copper reported in the leach residue	
	Uninoculated control	Inoculated
9:1	0.02	1.4
8:2	—	1.9
6:4	—	2.2
5:5	0.02	4.8
2:8	—	8.2

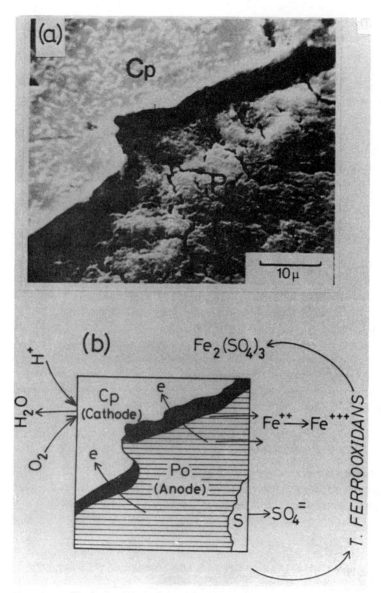

Figure 4.7 Bioelectrochemical model illustrating combined bacterial-galvanic oxidation in a chalcopyrite (Cp)-pyrrhotite (Po) couple: (*a*) Scanning electron micrograph depicting preferential corrosion of active pyrrhotite; (*b*) model showing anodic, cathodic, and biooxidation reactions.

sphalerite mineral sample were bioleached in 0.9 K Fe medium containing up to 17 g of copper per liter as copper sulfate. According to Fig. 4.8, a sharp decrease in dissolved copper occurred over time, and after about 2 weeks no copper was detected in the medium. On the other hand, a substantial increase in dissolved zinc occurred over time, indicating that sphalerite dissolution was enhanced in the process. In the absence of added copper sulfate, bioleaching of sphalerite under similar conditions yielded only about one-sixth of the dissolved zinc recovered in the previous case. X-ray photoelectron spectroscopic analysis of the leach residue revealed significant amounts of copper on the sphalerite surfaces.

A similar observation has been reported in the bioleaching of a chalcopyrite-pyrrhotite mixture. The precipitation of copper from solution due to galvanic displacement by pyrrhotite has been implied (Rossi et al., 1983).

In the bioleaching of copper-nickel sulfides containing chalcopyrite, pentlandite, and pyrrhotite, it was observed that the reaction of dissolved cupric ions with pyrrhotite and pentlandite influenced the copper-nickel ratios in solution. In addition to the galvanic displacement of dissolved copper by the active minerals, exchange reactions involving pyrrhotite and cupric ions may also take place (Natarajan et al., 1983; Natarajan and Iwasaki, 1983).

$$Cu^{2+} + FeS = CuS + Fe^{2+} \qquad (4.11)$$

Bacterial Attachment and Strain Improvement

It is clear from the foregoing discussion that galvanic interactions could be used to advantage in the bioleaching of mixed sulfides. Two approaches are possible in this regard:

1. Direct bioleaching of complex sulfide ores

2. Bioleaching of their bulk flotation concentrates

Selective dissolution of the desired sulfide mineral is the key to the successful application of this biotechnology. Such selectivity also opens the door for application of bioleaching in the beneficiation (upgrading) of sulfide mattes and multimetal ores in that undesirable contaminants can be effectively removed or lowered. The following important factors also need to be considered in the development of an efficient bioleaching process to treat complex sulfides:

1. Bacterial attachment to selected sulfide mineral sites and the role of direct attack

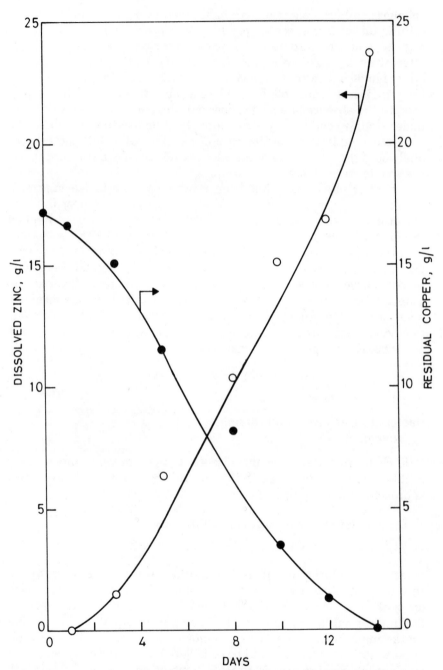

Figure 4.8 Displacement of dissolved copper by sphalerite in a bioleaching medium.

2. Use of specially adapted metal-tolerant strains that are specific to the desired mineral

3. Use of genetically engineered bacterial strains

It is not clear as to whether bacterial attachment is an essential pre-requisite for mineral solubilization. Bacterial attachment to mineral substrates appears to be preferential and may be influenced by the potential difference existing in mineral-mineral contacts. Bacteria tend to change the sulfide potential and bring about depolarization of the mineral surfaces through the oxidation of elemental sulfur and ferrous iron.

Scanning electron micrographs depicting attachment of *T. ferrooxidans* onto a sphalerite mineral surface and the resulting surface corrosion are shown in Fig. 4.9. Extensive surface damage in the form of pits and deep fissures can be seen on sphalerite surfaces that had previously been in contact with pyrite or chalcopyrite in the presence of bacteria. The microorganisms were found to be tenaciously attached. Neither vigorous shaking nor washing could dislodge them. The bacterial leaching mechanism is a heterogeneous process in which the bacterial cell attaches itself to the sulfide mineral, and corrosion occurs in a thin film between the bacterial outer membrane and the sulfide interface (Rodriguez-Leiva and Tributsch, 1988). Bacteria can attach and detach from the mineral surfaces. The organism appears to detach and multiply in the fluid media after a period of corrosion activity on the surface (Wakao et al., 1984).

Distribution of *T. ferrooxidans* between the solution and the ore particles has also been studied (Espejo and Ruiz, 1987). Attached bacteria can help in the transport of electrons from the anodic area of the mineral to some constituent of the bacterial electron chain that serves as the cathode. The main parameters influencing bacterial corrosion patterns may be surface defects, crystallographic orientation, and surface heterogeneity. The progress of bacterial leaching on pyrite-arsenopyrite flotation concentrates has been monitored through recording the initiation and growth of pits, grooves, holes, and jagged edges (Norman and Snyman, 1988). Arsenopyrite was destroyed before pyrite, and direct microbial attack at the point of cell contact was suggested. The pattern of bacterial distribution in a leaching tank containing arsenopyrite concentrates indicated that up to 99 percent of the *T. ferrooxidans* cells were associated with the solid phase in the pulp (Karavaiko et al., 1986). It is not known which of the bacterial cells present in a leaching reactor are more reactive—those distributed in the solution phase or those attached to the substrates. It is very important to understand the activities of attached organisms relative to unattached organisms. The rate of attachment and possible detachment, mechanisms involved in the complex interaction between

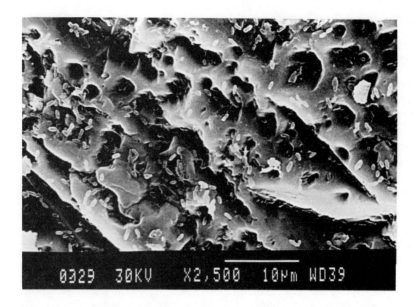

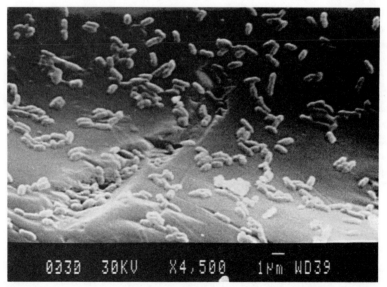

Figure 4.9 Scanning electron micrographs depicting attachment of *T. ferro-oxidans* and surface corrosion on sphalerite.

microbes and minerals, and the biochemistry of metal solubilization through bacterial attachment also need to be understood.

Increasing concentrations of heavy-metal ions such as Cu^{2+}, Ni^{2+}, Fe^{2+}, and Fe^{3+} in a bioleaching system can inhibit bacterial activity. Copper is more toxic to the bacteria than zinc is. Problems due to metal toxicity can become acute in the bioleaching of sulfide concentrates. The effect of increased copper and zinc dosages during ferrous-ion oxidation by unadapted and adapted strains of *T. ferrooxidans* has been studied by Dave et al. (1979). The toxic effect of increased copper concentration (as copper sulfate) was pronounced at 40 g/L, and bacterial activity ceased completely at 80 g of copper sulfate per liter. Similar strains, however, tolerated up to 80 g of zinc sulfate per liter. Only at about 160 g of zinc sulfate per liter was bacterial activity significantly affected. Adaptation of these strains to larger dosages of the above metal ions through successive transfers yielded organisms that were capable of optimal activity in the presence of up to 160 g of copper sulfate per liter and 640 g of zinc sulfate per liter. Bacterial strains preadapted to a chalcopyrite ore were found to be the most efficient for copper dissolution from complex lead-zinc-copper ores, and those adapted to a pyritic ore were found to be the best for sphalerite leaching from the ore (Ek and Frenay, 1983). Similarly, specially adapted copper- and nickel-tolerant strains of *T. ferrooxidans* were found to be the best suited for leaching of copper and nickel from bulk copper-nickel concentrates (Natarajan et al., 1982, 1983; Natarajan and Iwasaki, 1983).

The nature of the substrate used during bacterial growth and culturing determines the subsequent selectivity of the bacteria toward different minerals. When chalcocite was used as a substrate during growth, the bacteria oxidized the Cu^+ moiety but not the S^{2-} moiety (Nielsen and Beck, 1972). When the bacteria were grown on ferrous iron and chalcopyrite, over 90 percent of the cell population was found to be associated with the chalcopyrite substrate. The generation time and pH range of activity of the bacteria were also found to be affected by the nature of the substrate used (McGoran et al., 1969). It may be possible to develop special types of bacterial strains which can bring about selective mineral dissolution. Details of the development of specially adapted strains of *T. ferrooxidans* at the following levels of metal tolerance can be found in Karavaiko (1985):

Cu   50 g/L

Zn   120 g/L

Fe   160 g/L

Ni   72 g/L

In recent years, there has been considerable interest in the use of genetic engineering techniques for the improvement of microorganisms used in sulfide leaching. It may be possible to genetically modify strains of *T. ferrooxidans* for enhanced leaching abilities, selective attachment properties, and increased metal tolerance.

Bioleaching under Applied Potential

Through the application of an appropriate dc potential, the dissolution behavior of sulfide minerals in an acid medium can be controlled. Zinc can be selectively dissolved from a mixture of chalcopyrite and sphalerite under an applied potential in an acid electrolyte (Mahmood and Turner, 1985; Yelloji Rao and Natarajan, 1989). Electroleaching of copper concentrates as well as copper-nickel concentrates has been reported (Ammou-Chokroum et al., 1979; Natarajan et al., 1985; Pozzo et al., 1985). A previous study on the influence of direct current on the leaching of pyrite in percolators in the presence of *T. ferrooxidans* indicated no direct electrophysiological effect of the current on bacteria (Tepper and Naveke, 1977).

Recent studies have established that the biomass of *T. ferrooxidans* can be significantly increased through electrochemical means. For example, *T. ferrooxidans* was grown in an electrolytic bioreactor. When a current passed through it, bacterially generated ferric ion was reduced to the ferrous state (Yunker and Radovich, 1986). Bacteria preadapted to such electrolytic conditions were found to be more efficient, as attested to by a 3.7-fold increase in protein (cell) concentration and an increase in the ferrous oxidation rate by 1½ times. Russian workers (Kovrov et al., 1984; Grishin et al., 1985; Karavaiko, 1985) have cultivated *T. ferrooxidans* in a ferrous-ion-containing medium with continuous electrochemical reduction of ferric iron. The daily biomass output of such a system amounted to 100 g dry weight per square meter of cathode area of the reactor. The effect of different applied dc potentials on the reduction of ferric iron to the ferrous form in a bioleaching medium was reported by Jyothi et al. (1988). Electrochemical methods efficiently regenerated ferrous ions in a bacterial leaching system and also enhanced bacterial activity through a continuous supply of the energy source.

The effect of applied potentials on the dissolution of zinc, copper, and iron from sphalerite and chalcopyrite is illustrated in Fig. 4.10. The electrolytic current measured in the circuit is also indicated in the figure. The two current peaks correspond to -500 mV and $+400$ mV, respectively, during the electrochemical dissolution of sphalerite, whereas chalcopyrite has only one such dissolution peak, at $+600$ mV.

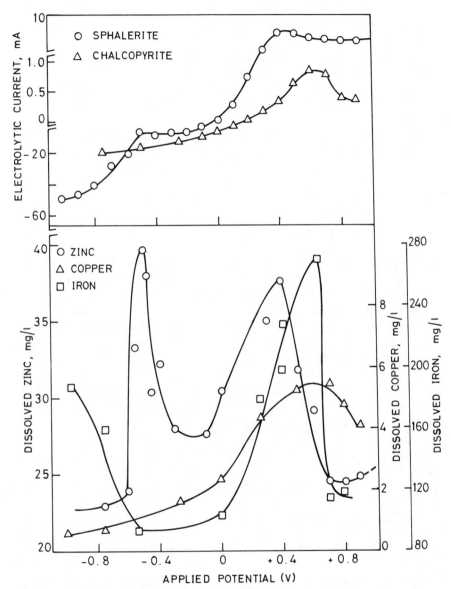

Figure 4.10 Variation of electrolytic current and dissolved metal ion concentration as a function of applied potential in a bioleaching system containing sphalerite and chalcopyrite.

Maximum zinc dissolution with negligible copper in solution was achieved at a negative potential of -500 mV.

Considerable amounts of iron were solubilized from chalcopyrite at both negative (-800 mV and above) and positive ($+600$ mV) potentials. Selective zinc dissolution in preference to copper was obtained, if a chalcopyrite-sphalerite mineral mixture was bioleached under an applied potential of -500 mV (SCE).

Results of detailed electrochemical bioleaching studies on sphalerite are illustrated in Table. 4.5. Enhanced dissolution of zinc occurred at a potential of -500 mV; the presence of T. *ferrooxidans* further accelerated the mineral dissolution. For example, zinc dissolution after 8 days of bioleaching was almost four times higher than that in the absence of bacteria at the applied potential. The application of potential led to a twofold increase in zinc dissolution compared with leaching under open-circuit conditions for 8 days.

In the sphalerite bioleaching system discussed above, efficient reconversion of ferric ion back to ferrous ion under the influence of the negative potential was observed. There was also a substantial increase in the biomass of T. *ferrooxidans*, as indicated by almost a 10-fold increase in the estimated protein concentration in the system at the end of 8 days under an applied potential of -500 mV. Variations in cell number and zinc concentration as a function of time in a bioleaching system containing sphalerite and T. *ferrooxidans* in the presence and absence of an imposed potential of -500 mV are illustrated in Fig. 4.11. It is evident that a significant increase in bacterial cell number occurred at the influence of the applied potential, which in turn enhances sphalerite dissolution. Under electrolytic growth conditions, shortening of the generation time of bacteria can also be expected. The beneficial effects of bioleaching of mixed sulfides under an applied potential are clearly evident from the above results.

Dissolution of an undesirable mineral in a complex sulfide ore can be similarly arrested through the external application of potentials. Bacterial activity can be promoted or inhibited by controlling the applied potential. Recent studies in the author's laboratory have indi-

TABLE 4.5 Leaching of Sphalerite under Open-Circuit and Applied Potential Conditions in the Presence and Absence of T. *ferrooxidans*

| | Dissolved zinc, mg/L | | | |
| | Open circuit | | Applied potential (-500 mV) | |
Duration, days	Uninoculated	Inoculated	Uninoculated	Inoculated
1	16.3	71.5	278.0	336.0
4	230.0	657.5	317.5	877.5
8	300.0	1190.0	580.0	2437.5

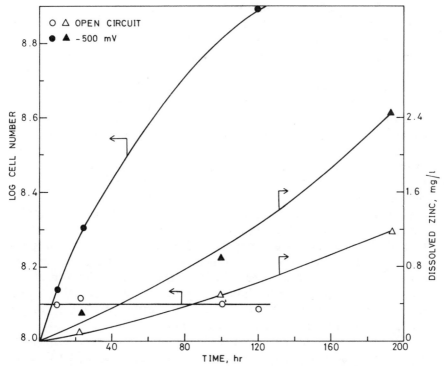

Figure 4.11 Variation of cell number and zinc concentration with time in the bioleaching of sphalerite under open-circuit and applied potential conditions.

cated that applied dc potentials in the range of +800 to +1000 mV (SCE) inhibit *T. ferrooxidans*.

The above experimental observations open up an entirely new and significant vista in the bioprocessing of minerals. The role of electrochemistry needs to be emphasized while the various biooxidation mechanisms are considered. Whether it be through galvanic interactions or as a polarization phenomenon, electrochemical principles help not only in the proper understanding of bioleaching mechanisms but also in the design and development of leaching systems.

Conclusions

Sulfide minerals behave as electrodes in an acid leaching medium and, based on their electrochemical activity, can be arranged in the form of a galvanic series. Measurement of combination potential and galvanic current in sulfide mineral combinations helps in assessing the rate of anodic oxidation. Selective dissolution of active minerals such as sphalerite can be achieved from multisulfides containing nobler sulfides, such as pyrite and chalcopyrite. The rate of such selective galvanic dissolution depends on several factors, such as relative anode-to-cathode surface area, proximity and duration of contact, conductivity of the minerals, and presence or absence of oxygen and bacteria. Laboratory tests have demonstrated that it is possible to control the selective dissolution of a mineral by adjusting the mineral-to-mineral ratio in the mixture. The contribution of galvanic interaction to the bioleaching of mixed sulfides has been demonstrated in cases involving complex sulfides such as lead-zinc-copper and copper-nickel-iron. The presence of *T. ferrooxidans* accelerates the galvanic dissolution process. An electrobiochemical model has been proposed to explain the various mechanisms involved in the bioleaching of multisulfides.

Selective dissolution of a desired mineral can also be achieved under the application of an appropriate dc potential. Under an applied potential of -500 mV (SCE), selective dissolution of sphalerite in preference to chalcopyrite can be achieved in a bioleaching system containing a mixture of the two minerals. Under the influence of the potential, ferric iron can be reduced back to ferrous, providing a continuous energy source for the bacteria. A significant increase in the bacterial biomass occurred under this condition.

Successful use of microbes in the bioleaching of complex sulfides depends also on their ability to attach to preferential mineral sites, thereby causing extensive corrosion through a direct attack mechanism. Similarly, strain improvement either through conventional adaptation or genetic engineering techniques is essential in order to develop metal-tolerant and very reactive bacteria useful in the leaching of sulfide concentrates.

References

Ahonen, L., Hiltunen, P., and Tuovinen, O. H., The role of pyrrhotite and pyrite in the bacterial leaching of chalcopyrite ores, in R. W. Lawrence, R. M. R. Branion, and H. G. Ebner (eds.), *Fundamental and Applied Biohydrometallurgy*, Elsevier, Amsterdam, 1986, pp. 13–22.

Ammou-Chokroum, M., Sen, P. K., and Fouques, F., Electrooxidation of chalcopyrite in acid chloride medium, kinetics, stoichiometry and reaction mechanism, in J. Laskowski (ed.), *Proc. Thirteenth IMPC*, Elsevier, Amsterdam, 1979, pp. 759–809.

Ashokkumar, V., Bacterial leaching of complex zinc-lead-copper ores, Master's thesis, Univ. of Mysore, 1980.

Baas Becking, L. G. M., Kaplan, I. R., and Moore, D., Limits of natural environment in terms of pH and oxidation-reduction potentials, *J. Geol.* 68:243–284 (1968).

Berry, V. K., Murr, L. E., and Hiskey, J. B., Galvanic interaction between chalcopyrite and pyrite during bacterial leaching of low grade waste, *Hydrometallurgy* 3:309–326 (1978).

Bruynesteyn, A., and Duncan, D. W., Effect of particle size on the microbiological leaching of chalcopyrite bearing ore, in F. F. Aplan, W. A. McKinney, and A. D. Pernichele (eds.), *Solution Mining Symposium*, AIME, New York, 1974, pp. 324–337.

Bryner, L. C., and Anderson, R., Microorganisms in the leaching of sulfide minerals, *Ind. Eng. Chem.* **49**:1721–1724 (1957).

Dave, S. R., Microbiological and bioleaching studies on metallurgical bacteria cultured from Indian sulfidic mine waters, Ph.D. thesis, Univ. of Mysore, 1980.

Dave, S. R., Natarajan, K. A., and Bhat, J. V., Biooxidation studies with T. ferrooxidans in the presence of copper and zinc, *Trans. Inst. Min. Metall.* **88**:C234–C237 (1979).

Dutrizac, J. E., and MacDonald, R. A. J., The effect of some impurities on the rate of chalcopyrite dissolution, *Can. Metall.* **12**:409–420 (1973).

Ek, C., and Frenay, J., Bioleaching of simple and complex sulfide ores, in G. Rossi and A. E. Torma (eds.), *Recent Progress in Biohydrometallurgy*, Associazione Mineraria Sarda, Italy, 1983, pp. 347–360.

Espejo, R. T., and Ruiz, P., Growth of free and attached *Thiobacillus ferrooxidans* in ore suspension, *Biotechnol. Bioeng.* **30**:586–592 (1987).

Gottschalk, V., and Buehler, H., Oxidation of sulfides, *Econ. Geol.* **5**:28–35 (1910).

Grishin, S. I., Skakun, T. O., Adamov, E. V., Polkin, S. I., Kovrov, B. G., Denisov, G. V., and Kovalenko, T. F., Intensification of bacterial oxidation of iron and sulfide minerals by a *T. ferrooxidans* culture at a high cell concentration, in G. I. Karavaiko and S. N. Groudev (eds.), *Proc. Intern. Sem. Modern Aspects of Microbiological Hydrometallurgy*, Projects GKNT, Moscow, 1985, pp. 259–265.

Hiskey, J. B., and Wadsworth, M. E., Galvanic conversion of chalcopyrite, *Metall. Trans. B.* **6B**:183–190 (1975).

Imai, K., On the mechanism of bacterial leaching, in L. E. Murr, A. E. Torma, and J. A. Brierley (eds.), *Metallurgical Applications of Bacterial Leaching and Related Microbiological Phenomena*, Academic Press, New York, 1978, pp. 275–295.

Jyothi, N., The role of galvanic interactions in the bioleaching of mixed sulfides, Master's thesis, Indian Institute of Science, Bangalore, 1988.

Jyothi, N., Brahmaprakash, G. P., Natarajan, K. A., and Ramananda Rao, G., Role of galvanic interactions in the bioleaching of mixed sulfides, in *Proc. First Intern. Conf. Hydrometallurgy*, Beijing, 1988.

Jyothi, N., Sudha, K. N., Brahmaprakash, G. P., Natarajan, K. A., and Ramananda Rao, G., Electrochemical aspects of bioleaching of mixed sulfides in *Symp. Biotechnol. in Mineral and Metal Processing*, SME (AIME), Las Vegas, 1988.

Karavaiko, G. I., *Microbiological Processes for the Leaching of Metals from Ores*, UNEP, Moscow, 1985.

Karavaiko, G. I., Chuchalin, L. K., Pivovarova, T. A., Yemelyanov, B. A., and Dorofeyev, A. G., Microbiological leaching of metals from arsenopyrite containing concentrates, in R. W. Lawrence, R. M. R. Branion, and H. G. Ebner (eds.), *Fundamental and Applied Biohydrometallurgy*, Elsevier, Amsterdam, 1986, pp. 115–126.

Kovrov, V. G., Denisov, G. V., and Sedelnikov, S. M., The culturing of ferrooxidizing bacteria on electric energy, in *Nauka*, Novosibirsk, U.S.S.R., 1984, 80 pp.

Mahmood, M. N., and Turner, A. K., The selective leaching of zinc from chalcopyrite-sphalerite concentrates using slurry electrodes, *Hydrometallurgy* **14**:317–329 (1985).

Malouf, E. E., and Prater, J. D., Role of bacteria in the alteration of sulfide minerals, *J. Met.* **13**:353–356 (1961).

McGoran, C. J. M., and Duncan, D. W., and Walden, C. C., Growth of *T. ferrooxidans* on various substrates, *Can. J. Microbiol.* **15**:135–138 (1969).

Mehta, A. P., and Murr, L. E., Kinetic study of sulfide leaching by galvanic interaction between chalcopyrite, pyrite and sphalerite in the presence of *T. ferrooxidans* and a thermophilic microorganism, *Biotech. Bioeng.* **24**:919–940 (1982).

Mehta, A. P., and Murr, L. E., Fundamental studies of the contribution of galvanic interaction to acid-bacterial leaching of mixed metal sulfides, *Hydrometallurgy* **9**:235–256 (1983).

Natarajan, K. A., Electrochemical studies on the growth of *T. ferrooxidans*, in *Proc. Intern. Symp. Industrial Electrochemistry*, Madras, India, 1976.

Natarajan, K. A., Microbe-mineral interactions of relevance in the hydrometallurgy of complex sulfides, in S. P. Mehrotra and T. R. Ramachandran (eds.), *Progress in Metallurgical Research: Fundamental and Applied Aspects*, Tata McGraw-Hill, New Delhi, 1985, pp. 105–112.

Natarajan, K. A., and Iwasaki, I., Role of galvanic interactions in the bioleaching of Duluth gabbro copper-nickel sulfides, *Sep. Sci. Tech.* **18**:1095–1111 (1983).

Natarajan, K. A., and Iwasaki, I., Microbe-mineral interactions in the leaching of complex sulfides, in J. A. Clum and L. A. Haas (eds.), *Microbiological Effects on Metallurgical Processes*, Met. Soc. AIME, Warrendale, PA, 1985, pp. 1–13.

Natarajan, K. A., Iwasaki, I., and Reid, K. J., Some aspects of microbe mineral-interactions of interest to Duluth gabbro copper-nickel sulfides, in G. Rossi and A. E. Torma (eds.), *Recent Progress in Biohydrometallurgy*, Associazione Mineraria Sarda, Italy, 1983, pp. 169–184.

Natarajan, K. A., Reid, K. J., and Iwasaki, I., Microbial aspects of hydrometallurgical processing and environmental control of copper/nickel bearing Duluth gabbro, in *Forty-Third Ann. Min. Symp.*, Univ. of Minnesota, Minneapolis, 1982.

Natarajan, K. A., Reimer, S. C., Iwasaki, I., and Reid, K. J., Electroleaching studies on selective nickel dissolution from a bulk copper-nickel concentrate, in *Proc. XV IMPC*, Cannes, France, 1985, pp. 413–423.

Nicol, M. J., Mechanism of aqueous reduction of chalcopyrite by copper, iron and lead, *Trans. Min. Metall.* **84:**C206–C209 (1975).

Nielsen, M. M., and Beck, J. V., Chalcocite oxidation and coupled carbon dioxide fixation by *T. ferrooxidans*, *Science* **175:**1124–1126 (1972).

Norman, P. F., and Snyman, C. P., The biological and chemical leaching of an auriferous pyrite/arsenopyrite flotation concentrate: A microscopic examination, *Geomicrobiol. J.* **6:**1–10 (1988).

Polkin, S. I., Panin, V. V., Adamov, E. V., Karavaiko, G. I., and Chernyk, A. S., Theory and practice of utilizing microorganisms in processing difficult-to-dress ores and concentrates, in *Proc. XI IMPC*, Cagliary, Italy, 1975, pp. 901–923.

Pozzo, R. L., Natarajan, K. A., Iwasaki, I., and Reid, K. J., Electrochemical processing of a complex copper-nickel sulfide concentrate, in *Proc. XV IMPC*, Cannes, France, 1985, pp. 303–313.

Rodriguez-Leiva, M., and Tributsch, H., Morphology of bacterial leaching patterns by *T. ferrooxidans* on synthetic pyrite, *Arch. Microbiol.* **149:**401–405 (1988).

Rossi, G., Discussion sections, in *Proc. XI IMPC*, Cagliary, Italy, 1975, pp. 964–969.

Rossi, G., Torma, A. E., and Trois, P., Bacteria-mediated copper recovery from a cupriferrous pyrrhotite ore: Chalcopyrite-pyrrhotite interactions, in G. Rossi and A. E. Torma (eds.), *Recent Progress in Biohydrometallurgy*, Associazione Mineraria Sarda, Italy, 1983, pp. 185–199.

Silverman, M. P., and Lundgren, D. G., Studies on the chemoautotrophic iron bacterium *Ferrobacillus ferrooxidans* I. An improved medium and a harvesting procedure for securing high cell yields, *J. Bacteriol.* **77:**642–647 (1959).

Tepper, K. P., and Naveke, R., Experiments on combined electro and bacterial leaching, in *Conf. Bacterial Leaching*, Veslay Chemie, Weinheim, 1977, pp. 211–216.

Tomashov, N. D., *Theory of Corrosion and Protection of Metals*, Macmillan, New York, 1966, pp. 228–248.

Wakao, N., Mishina, M., Sakurai, Y., and Shiota, H., Bacterial pyrite oxidation III—Adsorption of *T. ferrooxidans* cells on solid surface and its effect on iron release from pyrite, *J. Gen. Appl. Microbiol.* **30:**63–77 (1984).

Yelloji Rao, M. K., and Natarajan, K. A., Electrochemical aspects of sphalerite dissolution in the presence and absence of chalcopyrite, *Miner. Metall. Process.* **6:**29–35 (1989).

Yunker, S. B., and Radovich, J. M., Enhancement of growth and ferrous iron oxidation rates of *T. ferrooxidans* by electrochemical reduction of ferric iron, *Biotechnol. Bioeng.* **28:**1867–1875 (1986).

Bioleaching of Uranium

R. G. L. McCready
W. D. Gould

Introduction

Heap leaching of low-grade ores to recover metal values was first employed in Germany during the sixteenth century. The role of the acidophilic thiobacilli in this process was not realized until their isolation from an acidic mine drainage by Colmer and Hinkle in 1947. The commercial application of heap leaching for the recovery of uranium from low-grade ores or mine wastes in North America was first reported by Mashbir (1964). Western Nuclear Inc. chemically heap-leached uranium at their two uranium properties: the Gas Hills Mine and the Spook Mine in Wyoming.

In the early 1960s, mine operators at several of the uranium mines in the Elliot Lake area in northern Ontario noticed that the mine drainage had become very acidic and contained appreciable quantities of soluble iron and uranium (Harrison et al., 1966). Harrison and his coworkers isolated *Thiobacillus ferrooxidans* from Denison Mine water and demonstrated that the bacteria were necessary for the extraction of uranium from the ore. In the late 1960s and early 1970s all but two of the eight mines in the Elliot Lake area had shut down. Economic quantities of uranium were recovered from these mines, however, by periodically spraying the stope walls with acidic mine drainage (Fisher, 1966; MacGregor, 1969; Fletcher, 1970). In 1964 and 1965 a total of 127,000 lb of U_3O_8 were recovered by this spraying technique from Rio Algom's Milliken Mine, which had suspended mining operations earlier (Fisher, 1966).

In most ores the uranium occurs as a mixture of minerals containing the uranium in either the tetravalent or the hexavalent state. Uranium is mostly soluble in its most oxidized state, that being as the hexavalent ion (Brierley, 1978; Lundgren and Silver, 1980). Tetravalent uranium can be oxidized to the soluble form by ferric ions, but the oxidation occurs much more rapidly in the presence of the iron-oxidizing *T. ferrooxidans* (Lundgren and Silver, 1980).

$$UO_2 + Fe_2(SO_4)_3 \rightarrow UO_2SO_4 + 2FeSO_4$$

Bacterial leaching of uranium is via an indirect mechanism; the bacteria oxidize the indigenous pyrite of the ore, generating an acidic ferric-sulfate solution which carries out the chemical oxidation of the tetravalent uranium to the soluble hexavalent state. Although *T. ferrooxidans* is sensitive to high concentrations of uranium, it can be adapted to higher concentrations by selective subculturing (Tuovinen and Kelly, 1972). *T. ferrooxidans* can directly oxidize reduced compounds of uranium (uranous sulfate and UO_2) without the involvement of extraneous Fe^{3+}/Fe^{2+} couples as the chemical electron carrier (Tuovinen, 1986). The direct microbial oxidation of uranium is not significant during dump or heap leaching due to the abundance of ferric iron generated from the pyrrhotite and pyrite present within most ores.

Attempts have been made to optimize the bacterial leaching of uranium during the stope-spraying operations in the various Elliot Lake mines. Duncan and Bruynesteyn (1971) reported that adding phosphate to the wash water had no effect on the uranium leaching rate, but that stimulation was observed when ammonium or ferrous iron was added. In these studies, ammonium sulfate was added at a rate of 3 lb (1.4 kg) per stope, potassium phosphate at 0.5 lb (0.23 kg) per stope, and ferrous sulfate at 25 lb (11.4 kg) per stope as a dry powder or it was sprayed over the stope as a concentrated solution. Due to the variation in the volumes of wash water used, the size of the stopes, and the quantity and grade of the muck in the stope, valid comparisons of the uranium leaching rates under the different experimental conditions could not be accurately assessed.

Over the past two decades two Canadian mines have had the courage to attempt to increase the profitability of their operations by utilizing in-place, bacterial leaching of rubbilized ore underground.

Bacterial Leaching at the Agnew Lake Mine, Espanola, Ontario

CANMET (Canada Centre for Mineral and Energy Technology), Energy, Mines and Resources Canada had a research program on bacterial leaching throughout the 1960s which resulted in a cooperative agreement between CANMET, Kerr Addison Mines, and Agnew Lake

Mines to conduct a pilot-plant study on the bacterial leaching of the Agnew Lake uranium ore and to develop a flowsheet for the full-scale implementation of bacterial leaching at the Agnew Lake property. The study was initiated in the fall of 1969 by setting up a 2 ft by 16 ft (0.61 m by 4.88 m) leaching column containing 2 tons of Agnew Lake uranothorite ore. This ore preferentially breaks along the uranothorite mineralization, and uranothorite is readily dissolved in acid solutions which are bacterially produced from the pyrrhotite and pyrite present within the ore matrix. The leaching column was loaded with less than 8-inch run-of-mine (rom) ore and trickle-leached at a solids-to-solution ratio of 5:1. During the initial phases of leaching (16 weeks) the solution was recycled through the column every 4 h; this resulted in 50 percent extraction of the available uranium over this time period. The leach cycle was then changed to recycle the solution through the column once an hour for only 1 day per week. This pilot leach program indicated that 70 percent extraction could be achieved in 1 year and that the recovery would only increase to 82 percent extraction after 2 years of leaching.

Following the successful pilot laboratory study, an experimental program was set up to bacterially leach a stope containing 50 tons of broken ore at the 900-ft level within the Agnew Lake Mine. Two thousand gallons of inoculated 9 K medium (Silverman and Lundgren, 1959) were pumped over the surface of the muck for 4 h, pumping was discontinued for 4 h to allow the muck to "drip-dry," and the cycle was repeated. The underground solution temperature ranged from 10 to 12°C throughout the test leach period. After 30 weeks in operation 57 percent of the available uranium had been extracted from the ore, but the test had to be discontinued due to shutdown of the mining operations at that time (McCreedy et al., 1972; Parsons, 1974).

With the success of these two studies, plans were developed to recover uranium by bacterial leaching at the Agnew Lake Mine. In 1976, Kerr-Addison Inc., the major shareholder, announced plans to spend $37,000,000 to bring the mine into production of 1,000,000 lb/year of U_3O_8. The deposit is a highly fractured oligomictic quartz pebble conglomerate containing uranothorite located in two multiple-bed zones. The beds strike east-west at the surface, dip 60 to 85° south, and average 17.2 ft in thickness. The ore zone has been traced to a depth of 3100 ft and has undergone major faulting. The average ore grade is 1.0 lb/ton, and there are proven and probable reserves of 12,000,000 tons. Stope drilling for ore rubbilization was by downhole drilling of 4.5-in blastholes 100-ft long spaced 6 ft apart. Blasting was accomplished with Anfomet or Hydromex at a powder factor of 1.8 lb/ton of ore. In situ leaching was to be performed on 100- and 200-ft lifts of ore (Fig. 5.1). Because of the steeply dipping nature of the orebody, solution distribution was such that only 15 to 20 percent of the ore was wetted (Anonymous, 1976, 1978a, 1978b).

The ore was to be rubbilized to − 8 inches within the stopes, and the swell (approximately one-third of the ore) was to be transported to the surface, crushed to − 3 inches, and placed on PVC-lined heap-leaching pads. The remaining two-thirds was to be treated underground by trickle leaching and recovered underground by ion exchange followed by ammonium precipitation of the IX eluant, either underground or on the surface.

In 1977, bacterial leaching at the Agnew Lake Mine resulted in the production of only 70,000 lb of U_3O_8 due to poor solution distribution throughout the muck. To further complicate the picture, several faults were discovered during stope development, which resulted in less ore being broken than was planned and occasionally resulted in pregnant solution losses. Further, ore rubbilization was not as good as had been anticipated; often large slabs of ore were seen in the stopes. Occasionally the stopes would become flooded due to pump failures, or fines plugged the muck, resulting in complete flooding of the ore and development of anaerobic conditions which inhibited leaching.

In June 1980, Kerr Addison announced that it was closing down the Agnew Lake Mine due to declining ore grade and poor uranium recovery. Although mine development and underground stope leaching

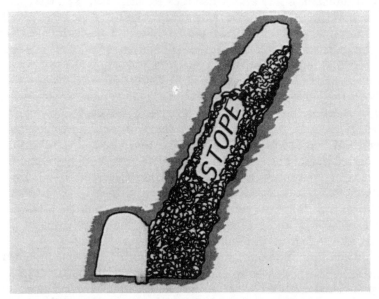

Figure 5.1   A diagram of the leaching stopes at Agnew Lake Mine.

were terminated, the surface heap leaching operations continued until 1985.

Bioleaching of Uranium at Denison Mines, Elliot Lake, Ontario

In 1984, Denison Mines, with support from CANMET, negotiated a cost-shared agreement under the National Research Council Biotechnology Industrial Research Assistance Program (IRAP) to develop a bioleaching process to commercial scale for the recovery of uranium.

The spatial geometry of the orebody in the Denison Mine is ideal for bacterial leaching. A lower reef is separated from a lower-grade upper reef by about 2 m of barren rock. Because the lower reef was mined by a room-and-pillar style of mining and the uranium has been recovered by conventional sulfuric acid leaching followed by IX and ammonium precipitation, the upper reef can be drilled and blasted into these large voids, which are ideal leaching vessels (see Fig. 5.2).

The Denison orebody is on the north limb of the Quirke Lake syncline and makes up a part of the Huronian sediments in the region. The uranium-bearing minerals, brannerite, monazite, and uraninite, are concentrated at the pebble-quartz interfaces in this quartz-pebble conglomerate. The ore has a specific gravity of 2.75 and contains in excess of 75% silica and approximately 7% pyrite-pyrrhotite. The upper reef has an average grade of about 1 lb U_3O_8 per ton (Wadden and Gallant, 1985).

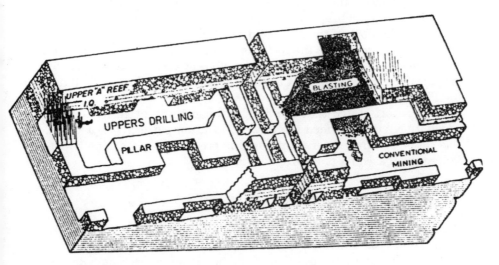

Figure 5.2   The spatial geometry of the Denison leaching stopes.

The objectives of the proposed research project were to:

Determine the drilling pattern required to provide optimal ore fragmentation

Determine whether bacterial nutrient supplementation was required for optimal leaching

Attempt to isolate, identify, and test psychrophilic strains of *T. ferrooxidans*

Compare spray-trickle leaching to flood leaching

Develop a method of protecting workers from the higher levels of radon emanations resulting from the large volumes of fragmented ore

Develop an aeration system capable of supplying sufficient air to maintain the growth of *T. ferrooxidans*

Drilling Pattern Tests

During the initial phases of the study, Denison utilized three different drilling patterns in its test program. The base patterns were a standard 3 ft by 4 ft (0.91 m by 1.22 m) spacing, which had been used by Denison for some time, 3 ft by 3 ft (0.91 m by 0.91 m), and a 2 ft by 4 ft (0.61 m by 1.22 m) spacing of drill holes. Two different types of drill rigs were used to drill the 200,000 ft of holes required for the tests: electric-hydraulic uppers jumbos and air-operated bar-and-arm drills (Fig. 5.3). Once drilling was completed, perforated air lines were installed on the floor of the stope and carefully covered with ore to prevent their collapse when the roof was blasted. Once the compressed air lines were in place and protected, the stopes were blasted and photographically analyzed by the Canadian Industries Limited (CIL) Blasting Physics Section.

Their report stated that the 3 ft by 3 ft drilling pattern produced adequate results provided the proper initiation pattern was adopted. The closer drilling patterns appeared to produce only marginally better results, based on photographic analysis.

Bacterial Nutrient Optimization Studies

While Denison Mines was conducting its research on stope development, Dalhousie University, under contract to Denison, was investigating the nutrient composition of the acidic mine drainage and isolating indigenous species of *T. ferrooxidans* from various locations within the mine. From laboratory studies on nutrient requirements,

Figure 5.3 Bar-and-arm drilling of the upper reef.

including both shake-flask leaching tests and laboratory-column leaching tests, it was concluded that sufficient nitrogen (from the explosives), magnesium, and iron (from ore constituents) were present to support growth of the thiobacilli. However, the mine water was deficient in phosphate (Table 5.1). A recommendation was made to Denison that sufficient H_3PO_4 be added to their mine water for stope leaching to produce a phosphate concentration of 15 to 20 ppm (McCready et al., 1986).

During the isolation of the thiobacilli from various locations within the mine, one very interesting observation was made. *T. ferrooxidans* were isolated from six different locations within the mine; however,

TABLE 5.1   Major Ion Contents of Denison Mine Water Samples in ppm (mg/L)

Sample	$H_2PO_4^-$	Mg^{2+}	NH_4^+	NO_2^-	NO_3^-
30D + 32N discharge	3	15	50	7	170
46076 discharge	4	10	50	7	190
46078 discharge	9	12	75	11	350
26305 sump discharge	1	8	20	2	95
31881 sump discharge	3	7	50	6	162
Total mine water	1	15	30	3	170
Average values	3.5	11.2	45.8	6	189.5

during studies on the effect of temperature, only two of the isolates would grow below 15°C. When the sump locations were checked on the mine map, it was realized that the two cold-tolerant strains had been isolated from sumps which were within 50 to 100 m of a mine ventilation intake shaft.

Similar findings were reported by Laurentian University from its Denison contract work, which involved the isolation of psychrophilic strains of *T. ferrooxidans* (Ferroni et al., 1986). Thus, during the winter months, these two sumps would be exposed to fresh air being drawn from the surface. It was recommended that one or both of these sumps be used as inocula for in situ stope leaching, because the organisms associated with these sumps were somewhat psychrophilic.

Underground Bacterial Column Leaching Studies

Six large columns were constructed and installed in the underground research laboratory at the Denison Mine. The columns were 3 m high and 0.61 m in diameter, constructed of rubberized mild steel with open tops. Each column had a capacity for just under 1.5 tonnes of broken ore (Fig. 5.4). Comparative tests were carried out to determine the effect of phosphate supplementation of the mine water and to assess the effect of inoculation with a large batch of laboratory-grown cells from Laurentian University.

When these results from tests with these two test columns were compared with the leaching data obtained from two columns leached with indigenous mine water, it was found that the large-scale inoculation resulted in a 15 percent higher uranium extraction than that observed in a control column containing ore from the same blasting pattern. The column which received phosphate-supplemented medium showed a 12 percent greater extraction than the control column. However, as the large bacterial inoculum was grown on 9 K medium, the

Figure 5.4 Denison underground bacterial leaching columns.

leach solution within this column had as much residual phosphate as the nutrient-supplemented column.

Spray Leaching Study

In the initial stage of the research project, Denison Mines set up eight research stopes within the mine; these stopes were used to compare the different blasting patterns and to compare spray leaching with flood leaching. Once the ore had been drilled and blasted, excess ore was removed from the stope and a hitch was cut into the rock around the stope opening. A reinforced concrete dam 4 ft high was then built across the stope opening to act as a solution reservoir. Because men had to climb over the muck pile to install and maintain the sprinkler system, the roof had to be bolted to prevent sloughing of ore fragments in this area.

The sprinkler system consisted of polyethylene hose and Model P3P509 Rainbird sprinklers which had a 40-ft (12-m) radius. Twenty sprinklers were required to obtain 100 percent coverage of the fragmented ore in each spray-leaching stope. With this system a percolation rate of 0.003 U.S. gal/(ton · min) [0.01 L/(ton · min)] was maintained by pressurizing the whole system to 70 lb/in^2.

Spray leaching resulted in many problems, including deterioration of all nonplastic or stainless steel parts of the sprinkler system and plugging of the spray nozzles. The spray nozzle orifices could not be increased because solution volumes would have become excessive and unmanageable. In addition, access to the surface of the broken ore was required to maintain the sprinkler system, so workers could only work in this area for limited times due to the emanation of radon gas from the broken ore.

Flood Leaching of Uranium at the Denison Mine

For the stopes selected for flood leaching, the excess ore was removed from the front of the stope after blasting. A large hitch was cut across the floor of the stope, up both sides of the entrance, and across the back (ceiling). A reinforced concrete bulkhead, capable of supporting a head pressure of up to 48 lb/in^2 or 110 ft of head, was constructed in each stope (Fig. 5.5).

After the concrete bulkheads had cured, the flood leaching cycle was begun. Acid mine drainage, having approximately pH 2.3 and a redox potential of over 450 mV, was pumped into the stope to completely cover the ore. Flooding generally required three days of continuous pumping at a rate of 25 L/s (400 gal/min). An additional 3 days were required to drain the stope. Once drained, the stope was allowed to rest for 3 weeks, and then the flooding cycle was reinitiated. After the first rest period, the pregnant liquor drained from the stope during the second flooding was pumped to the surface mill for uranium recovery by ion exchange followed by ammonium precipitation to form yellow cake (Campbell et al., 1987).

A cost comparison was made by Denison Mines between the flood leaching and spray leaching processes. Mining costs for the two techniques were equivalent, but construction costs were higher for flood leaching. However roof bolting, ventilation, and maintenance costs were much greater for spray leaching than for flood leaching. The flood leaching technique was less expensive per ton of ore treated than was spray leaching. Flood leaching was also much safer, because the broken ore was contained in a "sealed" container and stope ventilation to remove the radon gas emanating from the broken ore was more easily controlled. In contrast, the spray leaching stopes were open, required roof bolting, and needed a much higher rate of ventilation to allow the maintenance crews access to the stope to service the sprinkler systems.

Fungal Contamination of a Leaching Stope

In one of the Denison stopes, bacterial leaching had progressed very well to a point at which about 50% uranium recovery had been

Figure 5.5   Reinforced concrete bulkhead at the stope entrance.

achieved. On one of the flushes, although the E_h and the Fe^{2+}/Fe^{3+} ratio of the leach solution were ideal, little or no uranium was present in the leach solution. Denison requested assistance from Dalhousie University, and samples of the muck were collected and microbiologically analyzed. The investigators found six different species of acid-tolerant yeasts and fungi present in the stope even though the solution pH was less than 2.3. Laboratory studies indicated that two of the species of fungi isolated could readily adsorb uranium from solution. When the mine records were checked, it was found that this stope had been used as an underground garage at one time. Thus, it was postulated that the spilled hydrocarbon petroleum products that were adsorbed into the stope floor had slowly gone into solution, providing the carbon source for the growth of the aciduric fungi and yeasts.

Following these findings, Denison conducted a laboratory investigation to determine the concentration of dilute sulfuric acid required to release the uranium from the biomass. Following their laboratory study, Denison personnel flushed the contaminated stope with dilute sulfuric acid and recovered a large amount of uranium. The stope was then leached to completion without any further problems.

Results of the Two-year Development Program

During 1987, Denison Mines personnel mined into each of the initial six stopes to obtain tails samples to determine the degree of ore fragmentation and to analyze for the uranium content remaining in these residues (Tables 5.2 and 5.3). From the data in Tables 5.2 and 5.3 it is apparent that the grade of the ore was highest in the smaller particles. During ore deposition the uranium mineralization occurred on the pebble surfaces followed by sedimentation and compaction into the quartz-pebble conglomerate. The ore preferentially fractures at the pebble-gangue interface, resulting in the fine material being predominantly the uranium minerals. Except for the larger-size fractions, the leached ore fragments (Table 5.3) showed much lower grades than did the head sample (Table 5.2). Also, leaching tended to increase the percentage of finer material due to the removal of the uranium mineralization as well as the pyrite, which allowed disintegration of the particles into smaller pieces. Similar material breakdown has been observed during the bacterial leaching of pyrite from coal (McCready, 1985; McCready and Zentilli, 1985). From the test stopes it was con-

TABLE 5.2   Particle Size Analyses and Uranium Content of Fragmented Ore prior to Leaching

Size fraction	% Distribution	Grade (lb/ton)
> 12 in	18.26	0.47
> 8 < 12	14.37	0.23
> 6 < 8	9.67	0.42
> 4 < 6	13.44	0.35
> 2 < 4	21.79	0.61
> 1 < 2	11.05	0.55
> 0.5 < 1	3.83	0.32
> 0.25 < 0.5	1.94	0.81
< 0.25	5.66	1.11

TABLE 5.3   Particle Size Distribution and Grade of the Ore from the Test Stopes after Bacterial Leaching

Fraction size	Average % distribution	% Distribution range	Grade (lb/ton)
> 12 in	13.97	5.9–22.4	0.23
> 8 < 12	7.11	3.4–13.7	0.39
> 6 < 8	6.55	2.7–10.8	0.36
> 4 < 6	7.25	3.1–11.5	0.32
> 2 < 4	16.04	11.7–19.8	0.26
> 1 < 2	13.32	8.7–15.5	0.21
> 0.5 < 1	10.71	5.9–14.0	0.16
> 0.25 < 0.5	8.12	7.5–10.6	0.13
< 0.25	16.93	7.2–23.8	0.18

cluded that the overall recovery of uranium ranged from 69 to 86 percent of the available metal values.

Effect of Temperature on Bacterial Leaching

During the winter months, the acid mine drainage within the Denison Mine has an average temperature of 12°C, and the air temperature varies from 0 to 15°C. During the late summer months the air temperature underground ranges as high as 22°C, and the high humidity within the mine causes the rock faces to "sweat." Although the air temperature is elevated, the temperature of the mine drainage holds constant at around 12°C (McCready, 1988).

Mine personnel can easily determine when a particular stope is actively leaching. Early in the project it was noted that the 1.6 million gal of acid mine drainage being pumped into the stope were at 12°C, but on draining of the stope the solution temperature could be as high as 15°C.

Seasonal variation in the air temperature greatly affects the bacterial leaching process, as shown in Fig. 5.6. During the late summer, when the air temperature and humidity within the mine are high, high rates of uranium extraction have been observed (Fig. 5.6).

Production Levels at Denison

The mine plan for 1988 projected that 1,100,000 tons of ore would be maintained as an inventory for bacterial leaching, and 90 flood leaching stopes were in various stages of operation or preparation for flood leaching. During 1987 the Denison Mine produced 840,000 lb of uranium from its leaching operations. The forecast for 1988 was to increase production to 1,000,000 lb of uranium. The mine plans to in-

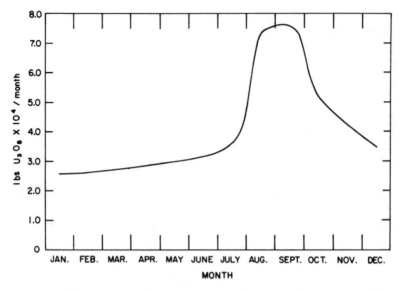

Figure 5.6   Effect of seasonal temperature variation on uranium recovery at the Denison mine.

crease production by leaching to a level where bacterial leaching will account for 25 percent of Denison's annual uranium production by the early 1990s.

Recent Modifications in the Denison Leaching Process

Originally, Denison personnel would lay perforated pipes on the floor of the stopes before blasting the ore in each stope. During the leaching process, compressed air was pumped through these pipes to provide oxygen for the microbial activity. Unfortunately, the amount of compressed air used was increasing proportionally with the number of flood leaching stopes being brought into production. This consumption of compressed air was affecting production in the conventional mining areas. Denison conducted a study in which high-pressure blowers (fans) were installed at the top of the stopes and connected through the top bulkhead to 2-in-diameter perforated plastic pipes that were laid on the stope floor before the stope was blasted. During the rest cycle these fans blew air through the muck pile and supplied oxygen for the leaching process. Overpressuring the stopes was avoided by venting through previously drilled holes in the highest part of the stope. This procedure not only provided aeration of the muck pile but removed the radon gas and airborne radon daughters from the stope, which were eventually vented to the atmosphere via an independent exhaust duct network that was installed to service the flood leaching area of the mine (Marchbank, 1987).

Labor and construction costs for the leaching process have steadily declined over the past few years at Denison. Through the experience gained in the construction of the numerous bulkheads, the crews have become more efficient and labor costs have been substantially reduced. Also, as the leaching personnel became more familiar with the process, they have varied the flood-drain cycles to improve the grade of the pregnant liquor, thereby reducing the volume of low-grade liquor being pumped to the surface.

A substantial quantity of the ore in the Denison reserves has been affected by intrusive diabase dikes which were formed after ore deposition. These dikes cause a chloritic alteration in the adjacent orebody which makes it impossible to treat the ore by conventional extraction techniques. A large sample of freshly broken chloritic ore was divided into equal parts; one part was left as mined, and the second part was crushed to −51 mm. Each sample was loaded into a 0.61 m by 3 m leaching column in the underground laboratory and flood-leached on a monthly cycle. Because chloritic ore contains apatite, no phosphate had to be added to the leach solution. Leaching of the chloritic ore produces very high iron concentrations (6 g/L) during the flush cycle; this means that the iron concentration of the interstitial water in the muck pile may be as high as 20 g/L, which may promote ferric-hydroxysulfate or jarosite precipitation within the pile. This may inhibit bacterial leaching (Fig. 5.7).

Collection and analysis of the tails of these two columns showed that 67 percent extraction was achieved with the rom sample, and 77 percent extraction was achieved with the crushed chloritic ore sample. Thus, bacterial leaching is economically feasible for the extraction of uranium from this refractory ore which cannot be treated by conventional chemical extraction processes.

Effect of Leaching on Denison Operations

The success of the bacterial leaching process has brought on stream an additional 4,000,000 tons of dike-contaminated ore which could not have been processed by conventional technologies and has enabled the in situ leaching of many millions of tons of low-grade material. Thus, tens of millions of pounds of uranium will be recovered without the production of tailings which must be stored on the surface. Because the leached ore remains underground, essentially stored in a vault, there are no environmental consequences due to the mining and recovery of uranium from this previously marginally economic ore. Al-

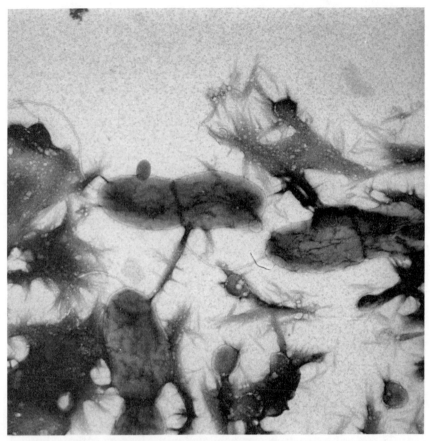

Figure 5.7 A scanning electron micrograph of *T. ferrooxidans* entrapped in an iron precipitate in liquid medium. (mag. 34,000 ×)

though exact processing cost figures are not available, Denison estimates that the cost per pound of uranium is substantially lower than its costs for the conventional sulfuric acid extraction process. In addition to recovering mineral values from the low-grade ore, the leaching process also allows Denison to recover value from the normally refractory chloritic ore which cannot be treated by conventional technologies.

Bacterial Heap-Leaching of Uranium, Rio Algom Mines

The Rio Algom Mine is located just west of the Denison Mine on the opposite shore of Quirke Lake, near Elliot Lake, Ontario. Although both mines are working the same uranium deposit, Rio Algom does not have an upper reef overlaying its major ore zone.

Due to the difference in the geology of the ore deposit at the Rio Algom mine, its activity is limited to bacterial heap leaching of low-grade and waste material left in stopes during mining. This low-grade ore is collected from stopes on a given level and piled in one specific stope. Acid mine drainage is then sprayed over the heap at the upper level, and the effluent is collected in a sump and pumped to the heap leaching stope on the next lower level and sprayed over that heap. The solution is thus passed through five heaps on five different levels within the mine; it is collected at the lowest level and pumped to the surface for uranium recovery. Using this technique, Rio Algom hopes to produce about 250,000 lb of uranium per year from below-cutoff-grade ore left in the stopes during mining.

Conclusions

The large-scale leaching studies within Canada have provided opportunities to test various engineering and microbiological concepts to optimize the leaching process. Several dominant factors which must be considered by mine management contemplating the implementation of metal recovery by bacterial leaching have been demonstrated in these large-scale studies. The most important factor in considering bacterial leaching is the spatial geometry of the orebody. If the orebody is steeply sloping, distribution of the solution throughout the fractured or rubbilized ore is extremely difficult, as was shown in the Agnew Lake underground leaching operations.

Bacterial leaching is generally a process for the recovery of metal values from low-grade ore or from below-cutoff-grade material left within the stopes during conventional mining. However, the Denison project has shown that the process can also be feasible for the recovery of metal values from high-grade contaminated ores which cannot be treated by conventional technology. A recent study by Noranda Mines has shown that a company cannot afford to develop a low-grade orebody with recovery based on bacterial leaching. The cost of mine development and ore rubbilization account for approximately 90 percent of the total cost of metal recovery ($14.74 per ton of ore developed). Because this high expenditure must be made initially and because the rate of metal recovery by bacterial leaching is relatively slow, the economics of the situation dictate that the process is nonviable (McCready and Sanmugasunderam, 1985).

Since the chemistry and mineralization of each deposit are unique, preliminary studies on the concentration and type of nutrients which will be released by dilute acid solutions must be analyzed to determine if nutrient supplementation is required for successful bacterial leaching (McCready et al., 1986).

The use of indigenous microorganisms, isolated from the specific mine environment, appears to be the most expedient approach for establishing a bacterial leaching program for a specific deposit. Indigenous organisms have been exposed to the specific soluble toxic ions associated with the particular deposit and thus have acquired a natural tolerance to their chemical and environmental conditions.

Acknowledgments

The authors wish to thank Denison Mines Ltd. for the use of its underground photographs. We particularly wish to thank Mr. A. Marchbank, bioleaching metallurgist with Denison Mines Ltd., for his informative discussions and information which he provided.

References

Anonymous, Agnew Lake plans $37M in-situ uranium leaching project, *Can. Min. J.* **97** (4):99 (1976).

Anonymous, Agnew Lake Mines encounters problems in leaching uranium ore, *Eng. Min. J.* **179** (7):43 (1978a).

Anonymous, Agnew Lake Mines: Taking giant steps in solution mining, *Eng. Min. J.* **179** (11):158–161 (1978b).

Brierley, C. L., Bacterial leaching, *CRC Crit. Rev. Microbiol.* **6**:207–262 (1978).

Campbell, M. C., Wadden, D., Marchbank, A., McCready, R. G. L., and Ferroni, G., In-place leaching of uranium at Denison Mines Limited, in *Developments of Projects for the Production of Uranium Concentrates*, International Atomic Energy Agency, IAEA-TC-453.5/11, 1987, pp. 151–165.

Colmer, A. R., and Hinkle, M. E., The role of microorganisms in acid mine drainage, *Science* **106**:253–256 (1947).

Duncan, D. W., and Bruynesteyn, A., Enhancing bacterial activity in a uranium mine, *CIM Trans.* **74**:116–120 (1971).

Ferroni, G. D., Leduc, L. G., and Todd, M., Isolation and temperature characterization of psychrotrophic strains of *Thiobacillus ferrooxidans* from the environment of a uranium mine, *J. Gen. Microbiol.* **32**:169–175 (1986).

Fisher, J. R., Bacterial leaching of Elliot Lake uranium ore, *Can. Min. Metall. Bull.* **59**:588–592 (1966).

Fletcher, A. W., Metal winning from low-grade ore by bacterial leaching, *Trans. Inst. Min. Metall.* **79**:C247–C252 (1970).

Harrison, V. F., Gow, W. A., and Ivarson, K. C., Leaching of uranium from Elliot Lake ore in the presence of bacteria, *Can. Min. J.* **87**:64–67 (1966).

Lundgren, D. G., and Silver, M., Ore leaching by bacteria, *Ann. Rev. Microbiol.* **34**:263–283 (1980).

MacGregor, R. A., Uranium dividends from bacterial leaching, *Min. Eng.* **21**:54–55 (1969).

Marchbank, A., Update on uranium leaching at Denison Mines, in *Proc. Fourth Annual BIOMINET Meeting, Sudbury, Canada*, CANMET Report SP87-10, 1987, pp. 3–18.

Mashbir, D. S., Heap leaching of low grade uranium ore, *Min. Congr. J.* **50**:50–54 (1964).

McCready, R. G. L., Microbiological studies on high-sulphur coals. Part II, Atlantic Research Laboratory Technical Report 52, NRCC 24793, National Research Council, Canada, 1985.

McCready, R. G. L., Progress in the bacterial leaching of metals in Canada, in P. R. Norris and D. P. Kelly (eds.), *Biohydrometallurgy, Proc. Intern. Symp., Warwick,* Antony Rowe, Chippenham, Wiltshire, 1988, pp. 177–195.

McCready, R. G. L., and Sanmugasunderam, V., The Noranda contract reports on the pre-feasibility study on in-place bacterial leaching: A summation, CANMET Report 85-5E, 1985.

McCready, R. G. L., Wadden, D., and Marchbank, A., Nutrient requirements for the in-place leaching of uranium by *Thiobacillus ferrooxidans, Hydrometallurgy* **17:**61–71 (1986).

McCready, R. G. L., and Zentilli, M., A feasibility study on the reclamation of coal waste dumps by bacterial leaching, *CIM Bull.* **78:**67–68 (1985).

McCreedy, H. H., Lendrum, F. C., and Gow, W. A., Stope leaching of uranium ore, in *Proc. Ann. Mtg. Can. Uranium Producers Metallurgical Committee, Ottawa, Canada,* 1972, pp. 9–33.

Parsons, H. W., Bacterial leaching of Agnew Lake Mines uranium ore, Mines Branch Report EMI 74-16, Department of Energy, Mines and Resources, Canada, 1974.

Silverman, M. P., and Lundgren, D. G., Studies on the chemoautotrophic iron bacterium *Ferrobacillus ferrooxidans.* An improved medium and a harvesting procedure for securing high cell yield, *J. Bacteriol.* **77:**642–647 (1959).

Tuovinen, O. H., Acid leaching of uranium ore materials with microbial catalysis, in H. L. Ehrlich and D. S. Holmes (eds.), *Biotechnology and Fossil Fuel Processing Industries, Biotech. Bioeng, Symp. No. 16,* Wiley, New York, 1986, pp. 65–72.

Tuovinen, O. H., and Kelly, D. P., Biology of *Thiobacillus ferrooxidans* in relation to the microbial leaching of sulphide ores, *Z. Allg. Mikrobiol.* **12:**311–346 (1972).

Wadden, D., and Gallant, A., The in-place leaching of uranium at Denison Mines, *Can. Metall. Q.* **24:**127–134 (1985).

Biotreatment
of Gold Ores

Richard W. Lawrence

Introduction

Biological leaching, which for many years has been applied to the recovery of copper and uranium, has only recently been considered for the treatment of refractory gold ores and concentrates. However, it is this application which now offers considerable commercial possibility and reward. Copper and uranium recovery operations have been essentially secondary or scavenging processes, utilizing heap leaching, dump leaching, or ore-remnant leaching. For refractory gold-bearing materials, commercial application appears imminent for large-scale processing utilizing controlled agitated reactor systems.

Refractory precious metal ores in which gold and silver are finely disseminated in sulfide minerals such as pyrite and arsenopyrite are becoming of greater interest and importance to the mining industry. This is due to a depletion of simple, free-milling ores and to a significant increase in the price of gold. The gold in many ores can be recovered in a gravity circuit or directly leached with cyanide to achieve over 90 percent extraction. However, the refractory ores contain very fine gold which cannot be separated by gravity and which respond poorly to cyanidation. Such ores require decomposition of the precious-metal-bearing sulfide minerals to liberate the precious metals for increased recoveries.

Traditional methods of pretreating refractory ores to decompose sulfides, such as roasting, are often no longer satisfactory. This can be due to economic or environmental considerations, or both. With the advent of stricter environmental regulations designed to restrict the quantities and control the quality of wastes from mining, milling, and smelting operations, some established techniques no longer have a technical and/or economic advantage. New technologies for refractory ore pretreatment are being developed to overcome these shortcomings, notably including biological oxidation systems.

This chapter discusses primarily the application of biological technology to the treatment of the sulfidic refractory gold ores and concentrates. Since by far the greatest research and development efforts have centered around the mesophilic bacterium *Thiobacillus ferrooxidans*, most of the discussion features work related to this microorganism. Other microorganisms and application of biological technology to other refractory gold ore types are briefly reviewed.

The Processing of Gold Ores

Introduction

Gold usually occurs in elemental form, or alloyed with silver, in association with other minerals. Many ore types, exhibiting a wide variation in mineralogy, can contain gold. Consequently, many processing options exist for the treatment and recovery of gold (Henley, 1975). These options vary from straightforward methods using relatively simple equipment to complex processing routes incorporating many process steps with sophisticated and expensive equipment. With the exception of placer deposits in which the gold is already liberated due to natural weathering or glaciation, gold can be considered to occur in three distinct categories of ore types: free-milling ore, base-metal ores, and refractory ores.

Free-milling ores

Free-milling ores are those from which the contained gold can be almost entirely liberated from the host rock following adequate size reduction by crushing and grinding. Subsequent recovery or concentration of the gold is achieved by using amalgamation, flotation, gravity concentration, or chemical solubilization methods, such as cyanidation. Recovery of gold from such ores is usually high, and operating costs are often relatively low.

Base-metal ores

Base-metal ores containing notably copper, zinc, lead, and nickel usually contain significant values of gold and silver. In fact, gold is often the principal value present. Gold is recovered along with the base-metal sulfides in the production of concentrates and is finally recovered from the sulfide during smelting and refining processes. The gold may be free in the ore but is often very finely disseminated within the base-metal sulfides.

Refractory ores

Of increasing importance as sources of gold are *refractory ores*—ores from which gold cannot be satisfactorily liberated by size reduction, even by very fine grinding. As previously mentioned, discussion of refractory ores will be mainly confined to those in which the precious metals are finely disseminated within sulfide minerals. Sulfide minerals such as pyrite and arsenopyrite are prominent in this regard (McPheat et al., 1969; Addison, 1980).

There are other types of refractory ores, including those containing carbonaceous material, which can, in a similar manner to sulfides, lock up gold in a way which makes recovery by gravity or cyanidation methods unacceptably low. Carbonaceous material can also provide a problem in that, even if gold may be essentially free, gold solubilized in cyanide can be removed from the solution due to adsorption onto the carbonaceous material. This effect is known as *preg-robbing*.

Recovery of gold from refractory ores requires that the ore undergo a pretreatment to liberate the gold from the host mineral. Pretreatment is usually an oxidation step and can be performed by several methods. Historically, refractory sulfide ores are roasted to remove arsenic and to break down the gold-bearing sulfides to render the ores amenable to cyanidation, such as at the Giant Yellowknife Mine in the Northwest Territories of Canada (Connel and Cross, 1981).

More recently, much effort has been directed to developing alternative methods for refractory ore treatment. Considerable emphasis has been placed on the development of hydrometallurgical processes. In this respect both chemical and biological systems have been studied. Chemical systems usually necessitate oxidation under high temperature and/or pressure. Methods include sulfate systems, such as developed by Sherritt Gordon Mines (Berezowsky and Weir, 1983) and as practiced at the McLaughlin Mine in California (Guinivire, 1984), and nitrate systems (Beattie and Raudsepp, 1985). Following oxidation in an autoclave, the gold in the residue is recovered, usually by leaching in cyanide.

Biological systems, notably utilizing iron- and sulfide-oxidizing bacteria, are usually operated at ambient temperatures and pressures. As with pressure leaching, the treated ore can be leached in cyanide to extract the gold (Lawrence and Bruynesteyn, 1983). Although no large-scale commercial plants have been built, the process has been piloted on a large scale in North America (Marchant, 1986b) and operated on a small commercial basis in South Africa (van Aswegen and Haines, 1988). Biological oxidation promises to result in lower capital and operating costs compared with the alternative technologies (Randol, 1987; Gilbert et al., 1988).

Biological Treatment of Gold Ores— Principles

Introduction

The ability of microorganisms to break down insoluble inorganic materials such as sulfide minerals is not in itself particularly remarkable. Microorganisms have been involved in many of the natural processes by which mineral deposits are formed, transformed, and degraded since the earliest times. What is extraordinary is that mineral breakdown can now be achieved at rates that can be considered as having potential for commercial mineral processing and extractive metallurgical application.

This has been achieved in the relatively very short period of time since $T.$ $ferrooxidans$ was first identified by Colmer and Hinkle (1947) as being responsible for acidic drainage in coal mines, and the same organism was found to be active in the oxidation of copper sulfide minerals in copper mine wastes (Bryner et al., 1954). Since that time, numerous research organizations have carried out research programs to gain a better understanding of the manner in which the bacteria oxidize mineral sulfides. As a result, the kinetics of the oxidation reactions have been improved substantially over those observed in nature, producing strains of microorganisms active under extreme conditions of pH and soluble metal concentrations. This has led not only to improved leaching processes for waste-rock materials, but also to extensive investigation and the development of controlled reactor processes. Most prominently, tank leach processes for the treatment of refractory gold ores, concentrates, and tailings have been developed in the late 1980s. This has been due to the need for new technology to treat ores which are not treatable by other technologies and to the high added value of the gold product.

The greatest interest in the biological treatment of gold ores is in the oxidation of sulfidic refractory ores to liberate the precious metals in agitated reactor systems. The potential use of heap leaching has also been discussed for lower-grade sulfidic ores (Bruynesteyn and Lawrence, 1983; Lawrence, 1988). Other types of refractoriness, such as silicate encapsulation, could also be approached with a biological treatment possibility, but discussion of such applications is beyond the scope of this chapter. A brief review of other aspects of gold biohydrometallurgy is given by Torma (1986).

Gold recovery using *Thiobacillus ferrooxidans*

Although several specialized types of bacteria are known to be able to derive energy from the utilization of sulfide minerals (Kelly et al., 1979; Norris et al., 1986; Wiegel and Ljungdahl, 1986), the basis of biological technology for sulfidic gold ores is the capability of the mesophilic bacterium *T. ferrooxidans* for chemolithotrophic growth in acidic environments. This bacterium, which has been the most thoroughly studied microorganism for mineral oxidation and leaching applications, obtains energy from the oxidation of reduced sulfur compounds and ferrous iron. These bacteria are nearly always found to be active wherever oxygenated waters contact exposed sulfide mineral deposits, such as at mine sites and tailings piles. Their action breaks down the sulfide minerals, releases metals into solution, and usually increases the acidity of the solution. It is therefore able to break down gold-bearing sulfides such as pyrite (FeS_2) and arsenopyrite (FeAsS) to liberate the contained precious metals for recovery by cyanidation.

Numerous papers have been published and presented in the 1980s, which show that significant enhancement in precious metal recovery can be achieved following oxidation by *T. ferrooxidans,* and that the oxidation can take place at practical rates and with favorable economics (Lawrence and Bruynesteyn, 1983; Livesey-Goldblatt et al., 1983; Renner et al., 1984; Bruynesteyn and Hackl, 1985; Lawrence, 1985; Lawrence and Gunn, 1985; Marchant, 1985; Pol'kin et al., 1985; Hackl et al., 1988; Karavaiko et al., 1986; Marchant, 1986a,b; Marchant and Lawrence, 1986; Hutchins et al., 1987; Pooley et al., 1987; Gilbert et al., 1988; Lawrence and Marchant, 1988; van Aswegen and Haines, 1988).

Details of the microbiological fundamentals of *T. ferrooxidans* and the oxidation mechanisms by which it utilizes sulfide minerals is discussed in Chap. 3.

Chemistry of gold ore oxidation

The chemistry of the bacterial oxidation of gold ores and concentrates can be represented by the following reactions, which use pyrite and arsenopyrite as examples. These minerals are the major sources of sulfide refractoriness.

The reactions of pyrite are typified by Eqs. (6.1) to (6.4). These reactions demonstrate the ability of *T. ferrooxidans* to (1) oxidize the mineral directly, and (2) oxidize the ferrous ions produced by mineral oxidation or by the chemical reduction of ferric iron, and (3) oxidize sulfur produced by ferric ion oxidation of the mineral.

$$4FeS_2 + 14O_2 + 4H_2O = 4FeSO_4 + 4H_2SO_4 \qquad (6.1)$$

$$4FeSO_4 + O_2 + 2H_2SO_4 = 2Fe_2(SO_4)_3 + 2H_2O \qquad (6.2)$$

$$FeS_2 + Fe_2(SO_4)_3 = 3FeSO_4 + 2S^0 \qquad (6.3)$$

$$2S^0 + 2H_2O + 3O_2 = 2H_2SO_4 \qquad (6.4)$$

The chemistry of the biological leaching of arsenopyrite has been interpreted in various ways by several authors (Ehrlich, 1964; Pinches, 1975; Livesey-Goldblatt et al., 1983; Panin et al., 1985; Karavaiko et al., 1986; Lawrence and Marchant, 1988), although there is general consensus on the final products of reaction. In any case, the chemistry of arsenopyrite oxidation and associated reactions is closely connected with the reactions of pyrite oxidation. Of particular significance is the ferric sulfate produced by bacterial action [Eq. (6.2)], which plays an important role in arsenopyrite oxidation reactions and is a critical factor in determining the ability to dispose of arsenic in a stable form in the environment after bioleaching.

Oxidation of arsenopyrite can be represented by the following reaction, in which pentavalent arsenic is formed:

$$4FeAsS + 13O_2 + 6H_2O = 4H_3AsO_4 + 4FeSO_4 \qquad (6.5)$$

The ferrous sulfate produced by this reaction is then oxidized according to Eq. (6.2) to produce ferric sulfate.

Trivalent arsenic may be formed as an intermediate product, followed by oxidation of the trivalent arsenic by ferric iron to the pentavalent form.

$$4FeAsS + 11O_2 + 2H_2O = 4HAsO_2 + 4FeSO_4 \qquad (6.6)$$

$$HAsO_2 + Fe_2(SO_4)_3 + 2H_2O = H_3AsO_4 + 2FeSO_4 + H_2SO_4 \qquad (6.7)$$

Oxidation of arsenopyrite by ferric sulfate may also take place.

$$2FeAsS + Fe_2(SO_4)_3 + 4H_2O + 6O_2$$
$$= 2H_3AsO_4 + 4FeSO_4 + H_2SO_4 \qquad (6.8)$$

In the presence of ferric iron in solution, ferric arsenate may be formed from pentavalent arsenic and precipitate.

$$2H_3AsO_4 + Fe_2(SO_4)_3 = 2FeAsO_4 + 3H_2SO_4 \qquad (6.9)$$

Effect of biooxidation on gold recovery

The biological oxidation of the sulfide minerals containing the gold in a refractory ore results in the liberation of the precious metal, which is then recoverable by conventional methods, such as dissolution in cyanide. In an ore containing a single refractory mineral, say pyrite, with the gold homogeneously disseminated, perhaps in solid solution, within the mineral, it might be expected that for a particular degree of oxidation of the mineral a proportional amount of the gold will be made available for cyanidation. This linear relationship has been demonstrated for a flotation concentrate containing predominantly pyrite (Lawrence and Bruynesteyn, 1983). Other test work on the same concentrate has shown that the direct relationship between oxidation and recovery is not exclusive for bioleaching. The same relationship exists for other pretreatment alternatives (Robinson, 1983).

In many refractory ores, however, the gold may be distributed in uneven proportions in more than one mineral species. Alternatively, even when there is essentially only one mineral present, the gold can be nonhomogeneous if its distribution is associated with structural features within the mineral lattice. In both of these cases the relationship between gold extraction and oxidation may not be linear.

In the first case, different sulfide minerals may be oxidized at different rates, largely determined by their electrode potentials. In a pyrite-arsenopyrite assemblage, it is the arsenopyrite which is usually more rapidly oxidized, due to its lower potential (see Chap. 4). If the gold is predominantly associated with the more reactive mineral, then the contained gold can be liberated without oxidizing the less reactive mineral. For many such ores the relative proportion of arsenopyrite to pyrite is small so that gold can be liberated at a low overall degree of sulfide oxidation.

This apparent selectivity in oxidation in a biological system also applies to the case where the gold is associated with lattice structural features. Enhanced reactivity of specific sites within the mineral crystal caused by slight variations in composition, lattice distortion, dislocations, and other mineralogical effects can result in gold being liberated for cyanide recovery without oxidizing all of the mineral.

The presence of gold within the pyrite crystal matrix has been shown by two groups (Southwood and Southwood, 1986; Hansford and Drossou, 1987) to provide preferred sites of bacterial corrosion. These workers propose a propagating pore theory which suggests that deeply penetrating pores found in pyrite particles after bioleaching result from direct bacterial oxidation of structurally disturbed regions of the mineral crystals. Since chemical (ferric sulfate) oxidation of pyrite can produce a similar pore structure in the mineral (Keller and Murr, 1982), the direct bacterial oxidation theory may not be proven by these findings. However, the selective oxidation of specific sites within a mineral and its implication on gold recovery is a significant observation.

Lawrence and Marchant (1988) provide examples of the effect of selective oxidation on gold extraction following oxidation and indicate that selectivity is often more pronounced during continuous-feed biooxidation compared with batch operation. This observation has been made by other investigators (Pol'kin et al., 1985).

Since the major operating costs of a biooxidation system are directly related to the quantity of sulfur oxidized (i.e., for the power for agitation and aeration for the oxidation itself and for the lime costs for the neutralization of the sulfate produced), the benefits of minimizing the quantity of sulfide oxidized is obvious. It is this apparent selectivity in oxidation that gives biological treatment an advantage over the pretreatment alternatives. Although the kinetics of biooxidation are slow compared with roasting and pressure leaching, it is the rapid kinetics of the alternatives which does not allow the differences in mineral reactivity to be exploited.

Factors affecting oxidation of gold ores

The many factors affecting the oxidation of sulfide minerals by *T. ferrooxidans* can be divided broadly into three groups: physicochemical, biological, and technological/process-related factors, although considerable overlap exists between all three categories. Since the effects of many of these factors on biooxidation are widely published (Tuovinen and Kelly, 1972; Brierley, 1978; Lundgren and Silver, 1980) and are discussed together with the fundamentals of bioleaching in Chap. 3, only brief consideration of the factors applicable to the treatment of gold ores is provided here.

Physicochemical factors. The physicochemical factors include pH, redox, temperature, oxygen concentration, and carbon dioxide concentration. For the biotreatment of gold ores, the effects of the physicochemical factors are largely the same as is reported in the preceding references. Specific considerations are as follows:

pH. For concentrates, the major concern usually is low pH values, since the oxidation of materials containing high levels of pyrite results in the reduction of pH well below values normally considered suitable for the growth of *T. ferrooxidans* (< 1.0). To maintain pH in a suitable range, it may be desirable to add lime (Pol'kin, 1985) or to operate a solution bleed-exchange system (Lawrence and Gunn, 1985). Ores may not contain enough sulfide to maintain the pH below appropriate values upon oxidation. In some cases, it may be necessary to add sulfuric acid to prevent an undesirable rise in pH to a range in which ferric precipitation can occur (> 2.2). Therefore for ores, the acid-base ratio (theoretical ratio between the acid production based on sulfide content to the acid consumption by alkaline ore constituents) is an important factor in determining the suitability of an ore to biotreatment.

Redox. Redox, or the oxidation-reduction potential (ORP), provides a measure of the oxidizing ability of the process environment. The oxidation of pyritic materials provides an abundance of iron in solution so that, in an active biological system, high redox values (> 750 mV SHE*) can prevail due to the maintenance of high ferric concentrations by the bacteria. This leads to the promotion of ferric oxidation reactions of pyrite and arsenopyrite as represented by Eqs. (6.3), (6.7), and (6.8).

For arsenopyrite-containing ores and concentrates, trivalent arsenic may prevail at lower oxidation potentials (< 620 mV; Lawrence and Marchant, 1988). In a highly selective leach, the discharge redox may be too low to allow elimination of arsenic as ferric arsenate, which requires pentavalent arsenic for complete reaction as in Eq. (6.9), although solutions containing trivalent arsenic and ferrous and/or ferric iron can be rapidly oxidized if necessary.

Temperature. For $T.$ $ferrooxidans,$ the optimum temperature is usually considered to be around 35°C. For practical purposes, controlling the temperature within the range of 34 to 39°C would be satisfactory. Since pyrite oxidation is exothermic, pulp heating will occur, the degree of which will depend largely on the sulfur content of the ore or concentrate, the degree of oxidation required, and the solids content of the pulp. In many cases, temperature control by cooling will be required to maintain the optimum temperature range. For some low-sulfur ores, heating may be required.

Operation of the biotreatment process a few degrees or more above the optimum for $T.$ $ferrooxidans$ could still produce active oxidation due to the presence and development of moderately thermophilic organisms (Norris et al., 1986; see also Chap. 1) which can provide a natural succession to $T.$ $ferrooxidans$ as temperatures increase. This may be more of a consideration in heap and dump leach operations

---

*SHE = standard hydrogen electrode.

which are not controlled than for a controlled tank process (Brierley and Brierley, 1978).

Oxygen and carbon dioxide concentration. Rapid biological oxidation of refractory gold ores and concentrates requires good oxygen mass transfer rates. Proper aeration of pulps must therefore be provided to maintain adequate levels of dissolved O_2. Work by Pinches et al. (1987) indicates that the critical oxygen concentration may be below 0.5 mg/L and possibly as low as 0.1 mg/L.

For concentrates requiring high oxidation rates for high sulfide contents, supplementing the air supply with CO_2 may be required to prevent the sole source of carbon to the bacteria from becoming limiting.

Biological factors. The biological factors affecting the oxidation of sulfide minerals include type of bacterial culture, degree of culture adaptation, presence of toxic-inhibiting components, unfavorable process environment, and nutrient medium composition. Reference to detailed discussion of these factors can be made to the literature referred to. Of specific interest are the following.

Bacterial culture. Although the principal focus in this chapter is on *T. ferrooxidans,* it is possible that in both natural and controlled leaching systems mineral breakdown is not exclusively due to this organism, but is effected in part by the cooperative action of physiologically distinct species, including *Leptospirillum ferrooxidans, Thiobacillus thiooxidans, Thiobacillus organoparus,* and *Thiobacillus acidophilus* (Kelly et al., 1979). The possibility of other bacterial species being involved with *T. ferrooxidans* has already been discussed in this section in connection with the effects of temperature. Similarly, Norris (1983) has shown that *L. ferrooxidans* can become dominant over *T. ferrooxidans* at pH values low enough (< 1.0) to inhibit the latter organism. Maintenance of the desired culture can be promoted through proper process control.

Culture adaptation. Adaptation of *T. ferrooxidans* and other leaching organisms to growth on or in the presence of sulfides and their oxidation products is usually readily achieved through standard culture procedures. For complex ores and concentrates, particularly those containing arsenic minerals such as orpiment and realgar, or the antimony sulfide, stibnite, adaptation to practical process conditions may require more time, especially if the culture being used is a recent isolate from a natural environment. Use of a well-adapted culture can reduce the residence time for a biotreatment process from many weeks for a new isolate to 5 days or less.

Toxic and inhibiting components.   Oxidation of gold ores and concentrates containing the sulfides pyrite and arsenopyrite produces a variety of major soluble species, including ferrous and ferric iron, trivalent and pentavalent arsenic, and high levels of sulfate. In addition, many trace elements and factors will be solubilized, depending on the specific mineralogy of the ore or concentrate. In nearly all cases, with proper culture adaptation, toxic or inhibiting effects are minimal. An exception is the effect of pH as previously noted. High concentrations of cations such as arsenic (>15 g/L) can be tolerated by *T. ferrooxidans.*

Unfavorable process environment.   Prevention of undesirable effects due to an unfavorable process environment is a matter of process control to prevent significant changes in process parameters such as pH, temperature, and dissolved oxygen. However, the leaching bacteria are hardy and can withstand the normal fluctuations and periodic losses of control experienced in plant operations. The comparatively slow reaction rates of a biological system allow ample advance warning of the onset of a changing process environment or loss of close control so that complete loss of control leading to the loss of the culture would be an extremely unlikely event.

Technological and process-related factors.   The technological and process-related factors include residence time, degree of oxidation, particle size, pulp density, method of agitation and aeration, reactor design and configuration, and other aspects of process design and equipment selection. Several of these factors will be considered in the section "Biological Treatment of Gold Ores—Applications."

Potential for thermophilic bacteria

In addition to *T. ferrooxidans,* other microorganisms are capable of oxidizing sulfide minerals. Some of these organisms are moderately or extremely thermophilic, functioning at higher temperatures than the mesophilic *T. ferrooxidans.* Several review articles and technical papers can be found on this topic (Kelly et al., 1979; Norris et al., 1986; Wiegel and Ljungdahl, 1986). One particular organism, the extremely thermophilic *Sulfolobus,* which has a temperature optimum at around 70°C, has shown particular promise as an alternative or superior organism to *T. ferrooxidans* for sulfide oxidation.

Since its isolation in 1972 (Brock et al., 1972), *Sulfolobus* has been well studied, particularly for its application in the leaching of copper (Brierley, 1980), in the oxidation of molybdenum sulfide (Brierley, 1974), and in the desulfurization of coal (Kargi and Robinson, 1985). Work carried out on refractory gold ores is limited. Some workers

(Norris and Parrott, 1986) claim higher pyrite oxidation rates by *Sulfolobus* compared with *T. ferrooxidans,* although this work used low pulp densities of 1 to 10 percent solids in the leach tests. Superior oxidation kinetics by the thermophile have also been claimed by Hutchins et al. (1987), who have presented data on the possible treatment of both sulfidic and carbonaceous refractory ores.

A detailed study carried out to investigate the potential for using *Sulfolobus* for refractory gold ore treatment (Lawrence, 1987; Lawrence and Marchant, 1987) showed better metallurgical performance for *T. ferrooxidans* for a number of ores and concentrates, particularly at practical pulp densities. This study also showed that the economic benefits of operating at the higher temperature, thereby reducing the reactor cooling requirements, are very small. Only by achieving much superior oxidation kinetics could the thermophilic system provide a significant economic advantage. However, operating pulp densities comparable to the mesophilic systems (>15 percent solids for concentrates and >25 percent solids for ores) must be demonstrated.

Operation of a semicontinuous bench-scale oxidation system using *Sulfolobus* to oxidize arsenopyrite has been described (Lindstrom and Gunneriuson, 1988). These workers found that leaching was limited by soluble arsenic concentrations above 1.5 g/L. Others (Norris and Parrott, 1986) found that the cation tolerance of *Sulfolobus* is similar to that of the mesophiles.

Biological Treatment of Gold Ores— Applications

Selection of biotreatment

The choice of process for refractory ore or concentrate treatment is based on many technical, economic, and environmental factors. For new technologies such as biotreatment, novelty is an additional factor which, even if technical and economic parity or superiority with other technologies can be demonstrated, can weigh against its selection as the process of choice for large-scale application. This latter consideration has been discussed by Curtin (1983), Spisak (1985), and Cupp (1986).

Haines (1986) has discussed the factors influencing the choice of technology for the recovery of gold from refractory ores. He lists the factors that need to be considered as mineralogy, recovery, value of product and by-product, environmental pollution, plant hygiene, operability and technological risk, and capital and operating costs.

Marchant (1985) has compared the oxidative pretreatment alternatives and shows some of the economic advantages and disadvantages of biooxidation compared with roasting and pressure leaching (Table 6.1). Others have compared biotreatment with alternative methods (Renner et al., 1984; Haines, 1986; Randol, 1987; Gilbert et al., 1988).

Whichever pretreatment process is selected, the oxidation step represents only a small proportion of the cost of the overall process design from mining of the ore to the production of gold bullion, especially for large tonnage operations. It can be anticipated, therefore, that biooxidation will initially be employed commercially on a relatively small scale where the oxidation step has a more significant impact on overall project cash flow. Larger operations will more likely employ proven processes in the near future due to lower inherent risk factors.

TABLE 6.1   Comparison of Oxidative Pretreatment Alternatives

	Treatment		
Cost factor	Biological	Pressure	Roast
1. Selective sulfide oxidation	Yes	No	No
2. Historic data and design available	No	Yes	Yes
3. Amenable to low sulfide concentrates	No	Yes	Yes
4. Conventional mineral processing equipment	Yes	No	No
5. Sophisticated design	No	Yes	No
6. Sophisticated operator knowledge	No	Yes	No
7. High-level process control	No	Yes	No
8. Special materials of construction	No	Yes	Yes
9. High maintenance/wear equipment	No	Yes	Yes
10. High-pressure vessels, lines valves, etc.	No	Yes	No
11. Sophisticated off-gas treatment	No	Yes	Yes
12. Vessel cooling/heating control	Yes	Yes	No
13. Oxygen and/or steam plant required	No	Yes	No
14. Amenable to alkaline and/or carbonaceous ores	No	Yes	Yes
15. Oxide product washing recommended	Yes	Yes	Yes
16. Saleable acid product	No	No	Yes
17. Possible difficulties in silver metallurgy	No	Yes	Yes
18. Arsenic product as As^{5+}	Yes	Yes	No

SOURCE: From Marchant (1985).

Ore or concentrate?

Biotreatment of refractory gold ores in agitated reactors can be considered either for the whole ore or for a flotation or gravity concentrate. The choice will depend on several factors, including grade and tonnage or ore, acid-base account, mineralogy of sulfides and waste rock, sulfide content, and selectivity of oxidation of the gold-bearing sulfides. These factors have been discussed by Marchant and Lawrence (1986).

Where sulfide and/or gold concentration ratios are poor (< 4:1) or where the gold recovery in the concentrate is unacceptable, it may be necessary to treat the whole ore rather than a concentrate. Successful biotreatment of ore requires the presence of sufficient oxidizable sulfide to offset the acid consumption of the ore or a reasonable acid addition requirement. Although biological treatment is most often considered for concentrates due to the economics of treating a smaller pulp volume, the additional expense of supplying a larger plant to treat ore can easily be justified even if only small gold losses occur during concentration.

Other advantages of ore treatment are the ability to operate at a higher operating density (related to the sulfur content of the bioleach feed), reduced cooling requirements, and reduced problems associated with acid or soluble metal buildup in solution. Disadvantages include more difficult liquid-solid separation and justifying the cost of fine-grinding the whole ore.

Bioleach flowsheets and design considerations

Conceptual process flowsheets for the biotreatment of refractory ores and concentrates in agitated tank systems have been proposed by a number of workers. In general terms, these flowsheets are similar and use mechanically agitated reactors to provide solid suspension and air mass transfer (Lawrence and Bruynesteyn, 1983; Renner et al., 1984; Marchant, 1986b; Marchant and Lawrence, 1986; Gilbert et al., 1988), although pachuca reactors are also considered (Pol'kin et al., 1985; Karavaiko et al., 1986).

Figure 6.1 shows a schematic flowsheet for a general biooxidation and gold extraction process for a refractory ore or concentrate. The circuit provides for the optional concentration of the refractory sulfide in the ore, usually by flotation, prior to conditioning of the ore with makeup acid, nutrients, and bioleachate recycle (if required) in an agitated tank, which may also act as a surge tank for the bioleach stage. The refractory material is oxidized in a multistage bioleach circuit comprising agitated tanks equipped with agitation, aeration, and cool-

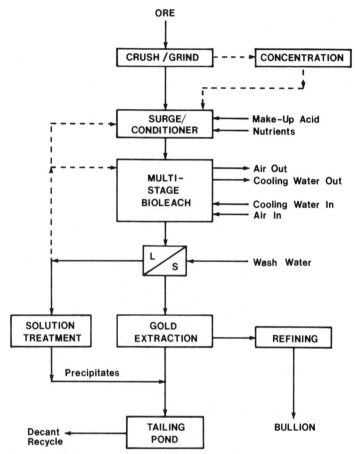

Figure 6.1 Biooxidation and gold extraction process for refractory ores and concentrates.

ing devices. Oxidized ore is separated from the bioleachate and washed as required by using countercurrent decantation or filtration. It is then treated in a gold extraction circuit. Bioleachate is neutralized to precipitate soluble cations for disposal.

Critical bioleach parameters for use in flowsheet development, feasibility studies, and plant design can be defined in continuous bioleach testing at the laboratory and pilot-plant scale. The variables listed in Table 6.2 must be defined.

The bioleach reactor configuration can be optimized based on the function relating gold extraction and sulfide oxidation and using process modeling such as described by Pinches et al. (1987). Typically, circuits will require two or more reactors in parallel in the first stage to provide biomass maintenance, followed by one or more reactors con-

TABLE 6.2   Critical Bioleach Parameters for Plant Design

Bioleach parameter	Operating range
1. Sulfur content	5–42%
2. Degree of sulfide oxidation	10–100%
3. Acid consumption of feed	Variable
4. Temperature	30–45°C
5. Pulp density	15–30% solids
6. Aeration rate	0.05–0.50 m^3 air/(min · m^3) pulp
7. Residence time	30–150 h
8. pH	1.0–1.8
9. Redox potential	450–680 mV SCE*
10. Nutrients	Variable
11. Solution exchange ratio	0–4

*SCE = saturated calomel electrode.

nected in series in subsequent stages. This and other strategies for process control have been discussed by Marchant and Lawrence (1986). Mass transfer rates to achieve oxidation targets can be calculated from laboratory data to provide aeration requirements. Agitation power is derived to satisfy mass transfer and solids suspension criteria. The net heat balance for each reactor can be optimized based on reactor configuration and aeration and agitation optimization. This allows calculation of cooling requirements taking into consideration the overall plant water balance.

Pilot-plant and commercial applications

Large-scale applications of biotreatment technology for refractory gold ores and concentrates have been few to date, although use on a significant scale is considered imminent for some operations.

The most prominent commercial application is the 10 to 12 tons per day plant treating flotation concentrate at the Fairview Mine of General Mining Union Corporation (Gencor) in South Africa (van Aswegen and Haines, 1988). This plant, which has been running successfully since October 1986, was designed to treat 40 percent of the pyrite and arsenopyrite concentrate produced at the mine, with the remainder being treated in Edwards roasters. It is understood that decisions have been made to expand the biooxidation plant to replace the roasters.

The process at Fairview is simple to operate, with the bacteria able to withstand the mishaps normally associated with industrial operations. Gold recovery in cyanidation following around 4 days of biotreatment is greater than 95 percent.

In North America application has so far been limited to pilot operations, although some of the most comprehensive information available on the practical aspects of the technology is that related to the pilot plant operated at Equity Silver Mines, British Columbia, Canada, during the year 1984–1985 (Marchant, 1985, 1986a,b). This detailed study was carried out to determine the feasibility of the biotreatment of an arsenical bulk sulfide concentrate.

The Equity pilot plant provides a good example of the effect of selective biooxidation. The pilot-plant feed was a concentrate assaying 5.5 to 6.0 g/t Au and 75 to 100 g/t Ag. At that grade, a number of hydrometallurgical and pyrometallurgical alternatives to enhance cyanidation response were shown to be uneconomical. Preferential oxidation of the gold-bearing arsenopyrite in the concentrate during bioleaching increased gold recovery from less than 20 to 75 percent following only 14 percent total sulfide oxidation.

The pilot plant was operated at a nominal 2 t/day for 9 months, including extended periods of continuous-feed operation. The study demonstrated satisfactory and predictable scale-up of the biohydrometallurgy and gold metallurgy from laboratory testing and confirmed laboratory data. The most critical effects of scale occurred in the design of pulp density, aeration rate, and residence time. These variables are highly interdependent and must be optimized on a reasonably large scale for proper plant design.

Heap leaching

Considerable quantities of refractory gold occur in low-grade ores. For such ores, the margin for viable economics is much less and is often not sufficient to support biotreatment or, much less, other oxidation pretreatments.

The recovery of gold from lower-grade oxide ores by heap leach methods is widely practiced. Similarly, the leaching of metals, principally copper and uranium, from low-grade ore heaps and waste dumps by using bacteria is a well-known and accepted technology (see Chaps. 5 and 6). The question is, can these two proven technologies be brought together to assist in the recovery of refractory gold from low-grade ores? There are some positive indicators. For one, the unwelcome production of acidic mine drainage from sulfidic wastes demonstrates that natural bacterial action in piles of rock is a powerful force. Furthermore, it can be observed that, due to bacterial oxidation, the quantity of gold recoverable with cyanide in the surface layers of waste piles of refractory ore or tailings is often significantly higher than that obtained from the material found at depth.

However, unlike oxide gold heap leaching and biological copper leaching, where the desired value is removed from the pile of rock in the solution for recovery, oxidation of refractory sulfidic material leaves the gold in the pile. Recovery of the gold must involve another leach stage. The combination of an acidic biological leach and subsequent alkaline cyanide leach is likely to present chemical and potential engineering problems. Before cyanide can be used, the heap will have to be washed and neutralized to remove acid and ferric salts. This may cause precipitation of ferric salts in the heap and lead to poor solution flow and high cyanide consumptions. Good solution management will be required during the bioleach stage to minimize these effects. Clayey ores will be difficult to treat by heap leaching due to percolation difficulties, and high acid-consuming ores may not be economical to process.

Several groups are interested in applying biological heap leach technology and field trials are in progress. Randol (1987) describes some of the attempts made to date.

A related application is the biotreatment of tailings in slime dams as demonstrated in South Africa (Livesey-Goldblatt, 1986). Tests have shown that by plowing up a layer of compacted tailings into a loose, granular mass to allow air and water penetration and inoculation of the mass with bacteria, the pyritic material is oxidized within 60 days. The oxidized material is removed for washing and cyanidation to recover the liberated gold. The next layer of the dump can be treated in the same way.

Heap leaching possibilities are discussed by Bruynesteyn and Lawrence (1983) and Lawrence (1988).

Process limitations

Biological pretreatment can generally be applied to most refractory sulfide minerals. Particularly attractive ores are those in which selective sulfide oxidation can be exploited. In addition, the technology offers significant economic advantage over roasting and pressure leaching. However, there are limitations to the technology:

1. For ores that exhibit high acid consumption and which cannot be preconcentrated to a sulfide concentrate, acid costs to maintain the process pH may become excessive.

2. Gold and silver bioleachate loading can occur, although it is not a common problem and can be overcome by suitable control.

3. Heat exchange might be a consideration in hot climates where cooling water temperature is close to the process temperature or water is in short supply.

4. Solution exchange, if required, may be a problem for poor settling or filtering material.

5. High-sulfur feeds requiring high sulfide oxidation levels may re-

quire excessive lime addition or large volumes of exchange solution for treatment and recycle.

6. Providing adequate air diffusion in finely ground slurries at high pulp densities might be difficult.

Outlook

The many years of research and development of biological processes for the mining industry have advanced laboratory investigations into engineering realities. Today, evaluation of biotreatment as a processing option is often considered at the start of a metallurgical test program as one of the alternatives alongside other pretreatment options, allowing full and proper evaluation of the technology. Although biotreatment may not be suitable for the treatment of all refractory ores, the opportunities for application are tremendous in several areas of the world, including Nevada and other western states in the United States, Canada, South America, parts of Africa, Western Australia, and Southeast Asia. It appears that significant commercial application, both in North America and worldwide, will take place within the next 2 years.

As Spisak (1985) has observed, success will not come easily. The minerals industry will have to be willing to make a real commitment to what is still a nonconventional technology. Controversial decisions will have to be made and new ground broken. It is likely that the first successful commercial plant of reasonable size will be followed in close succession by a number of plants. It is known that a number of companies with undeveloped refractory properties are watching the developments with keen interest. Like any new technology, the plant designs of today will likely be very different from those 10 or 20 years from now. At present, the developmental effort is relatively small. As plants get built, however, the number of engineers and other interdisciplinary professionals working on process development will increase, leading to accelerated advances in the technology.

References

Addison, R., Gold and silver extraction from sulfide ores, *Min. Congr. J.* **66**(10):47–54 (1980).

Beattie, M. J. V., and Raudsepp, R., The Arseno process—an update, in *Projects 88, Proc. 18th Hydrometallurgical Meeting of CIM, Edmonton, Alberta*, Conference of Metallurgists, 1988.

Berezowsky, R. M. G..S., and Weir, D. R., Pressure oxidation for treating refractory uranium and gold ores in *Proc. 22nd Conf. Metallurgists, CIM, Edmonton, Alberta*, 1983.

Brierley, C. L., Molybdenite leaching: Use of a high temperature microbe, *J. Less Common Met.* **36**:237–247 (1974).

Brierley, C. L., Bacterial leaching, *Crit. Rev. Microbiol.* **6**:207–262 (1978).

Brierley, C. L., Leaching of chalcopyrite ore using *Sulfolobus* species, *Dev. Ind.*

Microbiol. 21:435–444 (1980).

Brierley, J. A., and Brierley, C. L., Microbial leaching of copper at ambient and elevated temperatures, in L. E. Murr, A. E. Torma, and J. A. Brierley (eds.), *Metallurgical Applications of Bacterial Leaching and Related Microbiological Phenomena*, Academic Press, New York, 1978, pp. 477–490.

Brock, T. D., Brock, K. M., Belly, R. T., and Weiss, R. L., *Sulfolobus:* A new genus of sulfur-oxidizing bacteria living at low pH and high temperature, *Arch. Mikrobiol.* 84:54–68 (1972).

Bruynesteyn, A., and Hackl, R. P., The biotankleach process for the treatment of gold/silver concentrates, in J. A. Clum and L. A. Haas (eds.), *Microbiological Effect on Metallurgical Processes*, Met. Soc., AIME, Warrendale, PA, 1985, pp. 121–127.

Bruynesteyn, A., and Lawrence, R. W., A new bioleach process for sulfidic gold ores, in *Heap and Dump Leaching Practice Symp.*, Salt Lake City, 1983.

Bryner, L. C., Beck, J. V., Davis, D. B., and Wilson, D. E., Microorganisms in leaching sulfide minerals, *Ind. Eng. Chem.* 46:2578–2592 (1954).

Colmer, A. R., and Hinkle, M. E., The role of micro-organisms in acid mine drainage: A preliminary report, *Science* 106:252–256 (1947).

Connel, L., and Cross, B., Roasting process at the Giant Yellowknife (arsenic) mine, in *Proc. 20th Conf. Metallurgists, Hamilton, Ontario*, CIM, 1981.

Cupp, C. R., Biohydrometallurgy—biomythology or big business, in R. W. Lawrence, R. M. R. Branion, and H. E. Ebner (eds.), *Fundamental and Applied Biohydrometallurgy*, Elsevier, New York, 1986, pp. 3–10.

Curtin, M. E., Microbial mining and metal recovery: Corporations take the long and cautious path, *Biotechnology* **May**:229–235 (1983).

Ehrlich, H. L., Bacterial oxidation of arsenopyrite and enargite, *Econ. Geol.* 59:1306–1312 (1964).

Gilbert, S. R., Bounds, C. O., and Ice, R. R., Comparative economics of bacterial oxidation and roasting as a pretreatment step for gold recovery from an auriferous pyrite concentrate, *CIM Bull.* 81(910):89–94 (1988).

Guinivire, R., McLaughlin Project: Process, project and construction development, in *Proc. 1st Intern. Symp. on Precious Metal Recovery, Reno, Nevada*, 1984.

Hackl, R. P., Wright, F., and Bruynesteyn, A., A new biotech process for refractory gold-silver concentrates, in R. G. L. McCready (ed.), *Proc. 3rd Annual Meeting of BIOMINET*, CANMET Special Pub. SP 86-9, Supply and Services, Ottawa, Canada, 1986, pp. 71–90.

Haines, A. K., Factors influencing the choice of technology for the recovery of gold from refractory arsenical ores, in *Gold 100, Proc. Intern. Conf. on Gold* Vol. 2: *Extractive Metallurgy of Gold*, SAIMM, Johannesburg, 1986, pp. 227–233.

Hansford, G. S., and Drossou, M., A propagating pore model for the batch bioleach kinetics of refractory gold-bearing pyrite, in P. R. Norris and D. P. Kelly (eds.), *Biohydrometallurgy, Proc. Intern. Symp., Warwick*, Science and Technology Letters, Kew, U.K., 1987, pp. 525–527.

Henley, K. J., Gold ore mineralogy and its relation to metallurgical treatment, *Miner. Sci. Eng.* 7(4):289–312 (1975).

Hutchins, S. R., Brierley, J. A., and Brierley, C. L., Microbial pretreatment of carbonaceous gold ores, in *Proc. 116th Annual Meeting of AIME, Denver*, 1987.

Karavaiko, G. I., Chuchalin, L. K., Pivovarova, T. A., Yemel'yanov, B. A., and Dorofeyev, A. G., Microbiological leaching of metals from arsenopyrite containing concentrates, in R. W. Lawrence, R. M. R. Branion, and H. E. Ebner (eds.), *Fundamental and Applied Biohydrometallurgy*, Elsevier, New York, 1986, pp. 115–126.

Kargi, F., and Robinson, J. M., Biological removal of pyritic sulfur from coal by the thermophilic organism *Sulfolobus acidocaldarius*, *Biotechnol. Bioeng.* 27:41–49 (1985).

Keller, L., and Murr, L. E., Acid-bacterial and ferric sulfate leaching of pyrite single crystals, *Biotechnol. Bioeng.* 24:83–96 (1982).

Kelly, D. P., Norris, P. R., and Brierley, C. L., Microbiological methods for the extraction and recovery of metals, in A. T. Bull, D. C. Elwood, and C. Ratledge (eds.), *Microbial Technology: Current State, Future Prospects*, Cambridge University Press, Cambridge, U.K., 1979, pp. 263–308.

Lawrence, R. W., Biological preoxidation leaching for refractory gold and silver, in *Proc. 17th Canadian Mineral Processors Congress*, Energy, Mines and Resources Canada, Ottawa, Ontario, 1985, pp. 227–237.

Lawrence, R. W., The potential of thermophilic bacteria in the pretreatment of refractory gold ores, in R. G. L. McCready (ed.), *Proc. 4th Annual Meeting of BIOMINET*, CANMET Special Pub. SP 87-10, Supply and Services, Ottawa, Canada, 1987, pp. 75–102.

Lawrence, R. W., Opportunities for enhanced recovery using bacteria, in *Symp. Heap Leaching in a Canadian Environment, Timmins, Ontario*, 1988.

Lawrence, R. W., and Bruynesteyn, A., Biological preoxidation to enhance gold and silver recovery from refractory pyrite ores and concentrates, *CIM Bull.* **76**(857):107–110 (1983).

Lawrence, R. W., and Gunn, J. D., Biological preoxidation of a pyritic gold concentrate, in J. F. Spisak and G. V. Jurgensen (eds.), *Frontier Technology in Mineral Processing*, Soc. Min. Eng. AIME, New York, 1985, pp. 13–17.

Lawrence, R. W., and Marchant, P. B., Comparison of mesophilic and thermophilic oxidation for the treatment of refractory gold ores and concentrates, in P. R. Norris and D. P. Kelly (eds.), *Biohydrometallurgy, Proc. Intern. Symp., Warwick*, Science and Technology Letters, Kew, U.K., 1987, pp. 525–527.

Lawrence, R. W., and Marchant, P. B., Biochemical pretreatment in arsenical gold ore processing, in R. G. Reddy, J. L. Hendrix, and P. B. Queneau (eds.), *Arsenic Metallurgy—Fundamentals and Applications*, Met. Soc., AIME, Warrendale, PA, 1988, pp. 199–211.

Lindstrom, E. B., and Gunneriuson, L., Semi-continuous leaching of arsenopyrite with a *Sulfolobus* strain at 70°C, in P. R. Norris and D. P. Kelly (eds.), *Biohydrometallurgy, Proc. Intern. Symp.*, Warwick, Science and Technology Letters, Kew, U.K., 1988, pp. 525–527.

Livesey-Goldblatt, E., Bacterial leaching of gold, uranium, pyrite bearing mine tailing slimes, in R. W. Lawrence, R. M. R. Branion, and H. E. Ebner (eds.), *Fundamental and Applied Biohydrometallurgy*, Elsevier, New York, 1986, pp. 89–96.

Livesey-Goldblatt, E., Norman, P., and Livesey-Goldblatt, D. R., Gold recovery from arsenopyrite/pyrite ore by bacterial leaching and cyanidation, in G. Rossi and A. E. Torma (eds.), *Recent Progress in Biohydrometallurgy*, Assoc. Mineraria Sarda, Iglesias, Italy, 1983, pp. 627–641.

Lundgren, D. G., and Silver, M., Ore leaching by bacteria, *Ann. Rev. Microbiol.* **34**:263–283 (1980).

Marchant, P. B., Plant scale design and economic considerations for biooxidation of an arsenical sulfide concentrate, in *Internat. Symp. Complex Sulfides, San Diego*, CIM-AIME, 1985.

Marchant, P. B., Commercial application of biotechnology to enhance gold extraction from complex sulfides, in *Proc. 18th Canadian Mineral Processors Congress*, Energy, Mines and Resources Canada, Ottawa, Ontario, 1986a, pp. 407–437.

Marchant, P. B., Commercial piloting and the economic feasibility of plant scale continuous biological tank leaching at Equity Silver Mines Limited, in R. W. Lawrence, R. M. R. Branion, and H. E. Ebner (eds.), *Fundamental and Applied Biohydrometallurgy*, Elsevier, New York, 1986b, pp. 53–76.

Marchant, P. B., and Lawrence, R. W., Flowsheet design, process control and operating strategies in the biooxidation of refractory gold ores, in R. G. L. McCready (ed.), *Proc. 3rd Annual Meeting of BIOMINET*, CANMET Special Pub. SP 86-9, Supply and Services, Ottawa, Canada, 1986, pp. 39–51.

McPheat, I. W., Gooden, J. E. A., and Townend, R., Submicroscopic gold in a pyrite concentrate, *Proc. Aust. Inst. Min. Metall.* No. **231**:19–25 (1969).

Norris, P. R., Iron and mineral oxidation with *Leptospirillum*-like bacteria, in G. Rossi and A. E. Torma (eds.), *Recent Progress in Biohydrometallurgy*, Assoc. Mineraria Sarda, Iglesias, Italy, 1983, pp. 83–96.

Norris, P. R., and Parrott, L. M., High temperature, mineral concentrate dissolution with *Sulfolobus*, in R. W. Lawrence, R. M. R. Branion, and H. E. Ebner (eds.), *Fundamendal and Applied Biohydrometallurgy*, Elsevier, New York, 1986, pp. 355–365.

Norris, P. R., Parrott, L. M., and Marsh, R. M., Moderately thermophilic mineral-oxidizing bacteria, in H. L. Ehrlich and D. S. Holmes (eds.), *Biotechnology for the Mining, Metal-Refining, and Fossil Fuel Processing Industries, Biotechnology and Bioengineering Symp.* No. 16, Wiley, New York, 1986, pp. 253–262.

Panin, V. V., Karavaiko, G. I., and Pol'kin, S. I., Mechanism and kinetics of bacterial oxidation of sulfide minerals, in G. I. Karavaiko and S. N. Groudev (eds.), *Biogeotechnology of Metals,* United Nations Environment Program, Centre of International Projects GKNT/UNEP, Moscow, 1985, pp. 197–215.

Pinches, A., Bacterial leaching of an arsenic-bearing sulfide concentrate, in A. R. Burkin (ed.), *Leaching and Reduction in Hydrometallurgy,* Inst. Min. Met., London, 1975, pp. 28–34.

Pinches, A., Chapman, J. T., de Riele, W. A. M., and van Staden, M., The performance of bacterial leach reactors for the preoxidation of refractory gold-bearing sulfide concentrates, in P. R. Norris and D. P. Kelly (eds.), *Biohydrometallurgy, Proc. Intern. Symp.* Warwick, Science and Technology Letters, Kew, U.K., 1987, pp. 525–527.

Pol'kin, S. I., Adamov, E. V., Panin, V. V., Karavaiko, G. I., Yudina, I. N., Aslanukov, R. Y., and Grishin, S. I., Bacterial leaching of metals in tanks. Non-ferrous concentrate treatment:technology and flowsheets, in G. I. Karavaiko and S. N. Groudev (eds.), *Biogeotechnology of Metals,* United Nations Environment Program, Centre of International Projects GKNT/UNEP, Moscow, 1985, pp. 239–258.

Pooley, F. D., Shrestha, G. N., Errington, M. T., and Gibbs, H. E., The separate generator concept applied to the bacterial leaching of auriferous minerals, in G. A. Davies (ed.), *Separation Processes in Hydrometallurgy,* Wiley, New York, 1987, pp. 58–67.

Randol, *Gold and Silver Recovery Innovation Phase III,* Vol. 3, Randol International Ltd., Golden, CO, 1987, p. 1420.

Renner, C. W., Errington, M. T., King, D. J., and Pooley, F. D., Economics of bacterial leaching, in *Proc. 23rd Conf. Metallurgists, Quebec, CIM,* 1984.

Robinson, P. C., Mineralogy and treatment of refractory gold from the Porgera deposit, Papua New Guinea, *Trans. Inst. Min. Metall. Sect. C.* **92:**83–89 (1983).

Southwood, M. J., and Southwood, A. J., Mineralogical observations on the bacterial leaching of auriferous pyrite, in R. W. Lawrence, R. M. R. Branion, and H. E. Ebner (eds.), *Fundamental and Applied Biohydrometallurgy,* Elsevier, New York, 1986, pp. 98–113.

Spisak, J. F., Biotechnology and minerals, legitimate challenge or costly myth, in J. F. Spisak and G. V. Jurgensen (eds.), *Frontier Technology in Mineral Processing,* Soc. Min. Eng. AIME, New York, 1985, pp. 3–12.

Torma, A. E., Biohydrometallurgy of gold, in R. G. L. McCready (ed.), *Proc. 3rd Annual Meeting of BIOMINET,* CANMET Special Pub. SP 86-9, Supply and Services, Ottawa, Canada, 1986, pp. 3–24.

Tuovinen, O. H., and Kelly, D. P., Biology of *Thiobacillus ferrooxidans* in relation to the microbiological leaching of sulfide ores, *Z. Allg. Mikrobiol.* **12:**311–346 (1972).

Van Aswegen, P. C., and Haines, A. K., Bacteria enhance gold recovery, *Intern. Min.* **May:**19–23 (1988).

Wiegel, J., and Ljungdahl, L. G., The importance of thermophilic bacteria in biotechnology, *Crit. Rev. Microbiol.* **3:**39–108 (1986).

Other Bioleaching Processes

Giovanni Rossi
Henry L. Ehrlich

Introduction

Although up to this time, industrial-scale bioleaching has only been applied to low-grade copper and some uranium ores, and to biobeneficiation of gold ores, a potential for bioleaching or biobeneficiation of some other ores exists. In this chapter, we discuss microbial activities which exhibit such potentials.

Lead and Zinc

Lead- and zinc-bearing deposits large enough to be commercially exploited are limited in number. The minerals galena (PbS) and sphalerite (ZnS) generally occur in association with each other and are considered together for this reason. Iron frequently replaces zinc isomorphously in sphalerite, resulting in the nonstoichiometrically defined mineral marmatite (Kullerud, 1953). Arsenic, antimony, and mercury are sometimes present in large amounts in zinc sulfide minerals in addition to cadmium, germanium, gallium, and mercury. The mercury in the ore can contribute to air pollution during pyrometallurgical treatment. Galena is rarely free from associated silver, bismuth, and antimony.

Zinc and lead sulfides occur in a variety of orebodies, which may be massive or consist of sequences of lenses, veins of varying thickness, or subhorizontal beds. Massive and bedded deposits are usually located in limestone or dolomitic limestone formations. Carbonates are the major gangue minerals in this case. Vein-shaped deposits and lenses occur in country rock that is most frequently constituted of quartz with minor amounts of barite and carbonates (Bateman, 1952). The feasibility of bioleaching of such ores depends largely on the mineralogical nature of the gangue and the country rock, since limestone and calcite are acid-consuming minerals.

The overall grade of a lead-zinc ore amenable to commercial extraction ranges from 15 to 30 percent in the richest North and South American mines to 5 to 10 percent in European mines. Lead and zinc metals cannot be recovered from these run-of-the-mine (rom) ores by direct smelting: they must first undergo suitable concentration in mineral dressing plants. Sphalerite concentrates usually contain from 40 to 50 percent zinc, galena concentrates from 60 to 70 percent lead.

When the ore contains minor amounts of pyrite (FeS_2) and chalcopyrite ($CuFeS_2$), or when the sphalerite (marmatite) or galena are finely disseminated in the gangue and/or intimately intergrown with each other, the most suitable mineral dressing method is froth flotation. However, a primary prerequisite for successful flotation is the liberation of sphalerite and galena from each other and from the gangue. For finely disseminated and intergrown minerals, the fine grinding required is expensive. In current commercial practice, zinc-sulfide concentrate is subjected to oxidizing roasting, and the resulting calcine is leached with sulfuric acid. Zinc metal is recovered by electrolysis. Likewise, galena is converted to lead oxide followed by a reducing roast, the product of which is then thermally refined. When, as is true in the case of the majority of small and medium-sized deposits, no metallurgical facilities for processing concentrates exist at the mine, direct sale of the concentrate must be considered. In custom smelting, the cost of production of a metal like silver from galena or cadmium from sphalerite is very rarely recoverable when the metal is not present in sufficient concentration or is only partially recoverable. In these cases, bioleaching may represent a profitable alternative to conventional metal recovery processes.

The presence of isomorphously dissolved iron in the sphalerite lattice has a major effect on its electrode potential and, consequently, on its amenability to bioleaching. On the basis of the experimentally determined electrode potential versus pH curve, Yakhontova et al. (1980) proposed Eq. (7.1) for the chemical oxidation of sphalerite in a pH range below 5:

$$5ZnS + 8.75O_2 + 2.5H_2O \rightarrow$$

$$Zn^{2+} + 4Zn[HSO_4]^+ + HSO_4^- + 5e^- \quad (7.1)$$

In the presence of *Thiobacillus ferrooxidans* and possibly other acidophiles, the ferrous ions are readily converted to ferric ions. From the relationship

$$E_h = 0.771 + 0.059 \log \left[\frac{Fe^{3+}}{Fe^{2+}} \right] \tag{7.2}$$

the oxidizing potential of solutions containing just 10 parts of ferric ions per 10^6 parts of ferrous ions amounts to 0.476 V. Such solutions have oxidizing potentials high enough to cause sphalerite dissolution. Thus, sphalerite (marmatite) is readily bioleached (Torma et al., 1970, 1972; Bruynesteyn and Duncan, 1971; Gormely et al., 1975; Khalid and Ralph, 1977). The final product of bioleaching is a zinc-sulfate solution, and, when cadmium is present, a cadmium-sulfate solution. Zinc recoveries higher than 60 to 70 percent can only be achieved by multistage leaching with regrinding of the intermediate residues. Regrinding is needed to remove the inhibiting layer of iron-oxyhydroxide, iron-sulfate, and jarosite precipitates that form on sulfide particles during leaching (Lazaroff et al., 1982).

Galena is also readily oxidized in acid solutions. This is related to the low electrode potential of galena (< 0.3 V), although the electrode potentials of samples of galena of different origins may vary, depending on the conductivity of the mineral (Yakhontova et al., 1980; Pridmore and Shuey, 1976). For pH values below 3 (which correspond to the pH range most suitable for bioleaching with *T. ferrooxidans*), Eqs. (7.3) and (7.4) were proposed to represent the galena oxidation process (Yakhontova et al., 1980):

$$PbS + 1.75O_2 + 0.5H_2O \rightarrow Pb^{2+} + SO_4^{2-} + H^+ + e^- \tag{7.3}$$

$$\rightarrow Pb^{2+} + SO_3^{2-} + H^+ + e^- \tag{7.4}$$

The final product is lead sulfate, which is insoluble and precipitates. For this reason, the process is not applicable to in situ, dump, or heap leaching.

Bioleaching may be competitive with conventional metallurgical processes for rom lead-zinc ore below the cutoff grade for conventional mining and mineral dressing. If bioleached in a reactor, a pregnant solution containing zinc sulfate and, when present in the ore, cadmium sulfate is produced, and a sludge rich in lead sulfide and sulfate is formed (Garrels and Christ, 1967). The metal can be extracted by sintering and blast-furnace smelting.

Complex Sulfides

Finely intergrown associations of chalcopyrite ($CuFeS_2$), galena (PbS), sphalerite (ZnS), and pyrite (FeS_2) or pyrrhotite ($Fe_{1-x}S$) having a base-metal content higher than 5 percent are called *complex sulfide ores* and represent important resources of nonferrous and precious metals (Barbery et al., 1980). Tin, arsenic (in the form of arsenopyrite), and precious metals are frequently associated with complex sulfides.

Deposits of complex sulfides have been gaining importance with the progressive depletion of simple sulfide deposits and with the development of economically sound technologies for metal recovery from these ores, such as selective flotation. For some time, however, their economic potential lay in their copper values. Reserves of copper in these mixed sulfide ores alone were estimated to be 37×10^6 tons and to have an average grade of 1.35 percent (range 0.8 to 2.0 percent; Pelissonier, 1972). It has been estimated that these deposits represent about 8 percent of the known copper reserves.

The economic value of mixed sulfide ores may become appreciable. In current commercial practice, the metals in them are recovered by treating either the individual concentrates, obtained by selective flotation (Rossi, 1958, 1974; Ghiani et al., 1987), or the bulk flotation concentrates. The flowsheets for the production of the concentrates by flotation present numerous drawbacks, primarily the need for a very fine mesh-of-grind (e.g., –400 mesh Tyler—38 µm) in order to achieve an acceptable liberation of the mineral species from each other and the gangue. In addition to the higher energy, grinding, and reagent costs, the requirement for a fine mesh size results in poor selectivity and, consequently, low concentrate grades and metal recoveries. Fine grinding presents other physical constraints as well (Carta et al., 1980a).

Bulk flotation concentrates are processed in *smelter complexes,* a good example being the Roennskaer works of Boliden (Sundstroem, 1969; Petersson, 1976; Georgeaux et al., 1977). These smelter complexes consist of a series of metallurgical plants that produce pure metals but are arranged to facilitate the transfer or recycling of residues and by-products from one process stage to another for the recovery of all metal values. Each of these plants is, however, specially designed for treating bulk concentrates of ores from specific orebodies of a size that warrants the development of sophisticated flowsheets. When an orebody is not large enough to justify the construction of a smelter complex, the concentrates must be sold with all the negative financial implications such sale entails.

A serious problem of pyrometallurgical processing of sulfide concentrates that is not to be underestimated is that it is environmentally hazardous, owing to the extensive atmospheric pollution produced by roasting and smelting plants, whose gaseous effluents contain vapors of associated metals such as Sb, Bi, As, Sn, and others in addition to S and N oxides.

The unfavorable economic and environmental considerations stimulated investigations of bioleaching of complex sulfide ores as a competitive alternative to the above-mentioned commercial processes. In principle, complex metal sulfides can be microbially leached in the form of fragmented rom ore or bulk flotation concentrate. In the first case, the only recommended commercial process for ores with negligible amounts of lead appears to be in situ leaching, since the costs involved in removing the ore from the stopes ("mucking"), hoisting, and transportation to form dumps or heaps would not be compensated by the minor improvements in metal recovery attainable with operations easier to control than in situ leaching. In situ leaching is not recommended when the lead content is high enough to justify its recovery. As pointed out earlier, lead sulfate, being practically insoluble in acid leach solutions, would precipitate and encrust the fragmented ore mass; thus it is not recoverable. In situ bioleaching of complex sulfide ore is profitable only if all metal values are amenable to solubilization and can be recovered from the pregnant solution.

Yet, complex sulfide ores are often low-grade with metal values finely disseminated throughout the gangue. Bioleaching of ores containing carbonate and/or porous gangue is impractical. A carbonate gangue entails high acid consumption, a condition which is unfavorable to bioleaching unless sulfuric acid is added to the leach liquor, an expensive procedure. When porosity of the gangue is very low, as is the case with several igneous siliceous rocks whose porosity is almost zero (Goodman, 1980), access of the leach liquor to the ore particles is impeded and dissolution rates become insignificant. Since the gangues of most complex sulfide ores are characterized by such unfavorable conditions, bioleaching of fragmented rock masses may not be feasible. If these conditions exist, the gangue may first have to be separated and the sulfides collected by bulk flotation. The metal values can then be recovered from the bulk concentrates by bioleaching in stirred tanks or reactors. Bioleaching of bulk flotation concentrate merits close investigation when an ore contains metal values other than copper and zinc.

The relative abundance of the various mineral values in a complex sulfide ore seems to affect strongly the kinetic parameters of bioleaching and metal recovery (Dave et al., 1979; Carta et al., 1980b). For instance, when an ore contains the same amounts by weight of sphalerite and chalcopyrite, the dissolution rate is much higher for zinc than for copper, but this difference in rate becomes insignificant when the grade of the ore containing both metals is poor. When zinc is present in the mixture in relatively small amounts, chalcopyrite solubilization appears to be enhanced (Dutrizac and MacDonald, 1973;

Carta et al., 1980b). The relative abundance of the various mineral species may play a key role in certain instances (Rossi et al., 1983; Suzuki et al., 1986).

Although microbial leaching of bulk flotation concentrates appears to be very promising for recovering metal values from complex sulfide ores that cannot be beneficiated by conventional mineral dressing and pyrometallurgical techniques, some practical problems still need to be satisfactorily solved, such as development of faster leaching rates, attack of higher solid-to-liquid ratios of pulps in bioreactors, and prevention of precipitation of ferric oxyhydroxides, sulfates, and jarosite-like compounds during leaching. Solving the first two problems would result in considerable reduction in residence times of pulps and, therefore, substantial cuts in investments and power consumption. The solution of the third problem would prevent the formation of inhibiting encrustations on the mineral particles and would eliminate the need for stagewise bioleaching with regrinding of residues of intermediate stages.

Silver from Complex Sulfide Ores

The leaching of silver contained in mixed sulfide ores consisting, for instance, of tetrahedrite, galena, sphalerite, and pyrite, in which silver may replace some of the copper in tetrahedrite, some of the lead in galena, and some of the zinc in sphalerite, can be accelerated in 0.9 K iron medium by *T. ferrooxidans* (Ehrlich, 1986, 1988a). Although silver ion is very toxic for *T. ferrooxidans* (Hoffman and Hendrix, 1976; Norris and Kelly, 1978; Tuovinen et al., 1971), complexation diminishes this toxicity (e.g., Tuovinen et al., 1985). Ehrlich (1986, 1988a) found that 0.9 K iron medium at pH 2.0 containing 0.9 instead of 9 g of iron as $FeSO_4 \cdot 7H_2O$ per liter (Silverman and Lundgren, 1959) solubilized significant amounts of silver in ore from the Lucky Friday Mine (Hecla Mining Co., Montana) and that *T. ferrooxidans* accelerated this leaching to different degrees under different conditions. In shake culture, solubilization of silver reached a maximum in 1 week, after which the silver reprecipitated, probably as a silver jarosite (Ehrlich, 1986). When leached in columns fed with the 0.9 K medium from the bottom of the column, *T. ferrooxidans* accelerated continuous silver extraction from the ore (Ehrlich, 1986), but at a slower rate than in a stirred reactor. Little copper or zinc and minor amounts of lead were concurrently extracted (Ehrlich, 1988a). In a stirred reactor containing continuously fed 0.9 K medium at a constant volume and in which silver-containing ore remained settled on the bottom of the reactor, silver leaching was accelerated by *T. ferrooxidans* under defined conditions of pH and stirring of lixiviant (Ehrlich, 1988a). Stirring of the lixiviant was essential to move solubilized metal species from interstitial solution among the ore particles into the overlying medium.

Antimony

Of the numerous antimony minerals, stibnite (Sb_2S_3), tetrahedrite ($Cu_{12}Sb_4S_{13}$), and lead ores are the major sources as far as commercial extraction is concerned. Native antimony and oxidation products such as cervantite only account for a minor part. At present, antimony is produced by pyrometallurgical treatment of the sulfide—that is, oxidizing, roasting, and smelting of the oxide. Although antimony is used in relatively small quantities, it has many diversified and indispensable industrial uses. Its competitive price on the world market has led to the discontinuation of mining of low-grade ores by several traditional producers, Italy included. Stibnite concentrates are produced from ore by flotation or gravity concentration. The amenability of antimony-bearing complex sulfide ores and of stibnite and tetrahedrite to biooxidation is well-documented. The results reported in a pioneering paper on the solubilization of tetrahedrite by *T. ferrooxidans* (Bryner et al., 1954) were soon substantiated by further experimental evidence (Lyalikova et al., 1972). The biological oxidation of synthetic or natural antimony sulfide by *T. ferrooxidans* was investigated by Lyalikova (1966), Rossi (1971), and Torma and Gabra (1977). Lyalikova (1974) reported the discovery of specific antimony-oxidizing microorganisms, *Stibiobacter senarmontii,* which can oxidize antimony(III) oxide to antimony(V) oxide. *T. ferrooxidans* oxidizes stibnite to antimony(III) sulfate according to reaction (7.5):

$$Sb_2S_3 + 6O_2 \xrightarrow{T.\,ferrooxidans} Sb_2(SO_4)_3 \tag{7.5}$$

The antimony(III) sulfate partially hydrolyzes to produce an insoluble antimony(III) oxosulfate:

$$Sb_2(SO_4)_3 + 2H_2O \leftrightarrow (SbO)_2SO_4 + 2H_2SO_4 \tag{7.6}$$

The overall reaction can be written

$$Sb_2S_3 + 6O_2 + 2H_2O \xrightarrow{T.\,ferrooxidans} (SbO)_2SO_4 + 2H_2SO_4 \tag{7.7}$$

Owing to the very low solubility of antimony oxosulfates, the total dissolved antimony concentration remains very low, less than 2 g/dm^3. Hence, commercial bioleaching of antimony sulfide ore is not feasible by in situ, dump, or heap leaching.

Molybdenum

Of the commercially important metals, molybdenum was isolated over 200 years ago (1782), but its importance was recognized in 1927. Molybdenum is used as a very potent hardener in the production of iron and steel alloys. Molybdenum steels are widely used in mechanical construction. Cast iron strength and ductility are greatly enhanced by this metal.

The major molybdenum mineral is molybdenite (MoS_2). All molybdenum deposits are of igneous origin and range from pegmatites to contact metasomatic, disseminated replacement, fissure veins and pipes (Bateman, 1952). Molybdenum ores seldom exceed 1% molybdenite. Molybdenite is also recovered as a by-product of copper mining. Utah copper ores, for instance, contain about 0.04% molybdenite. Molybdenite ores are concentrated by selective flotation. Molybdenum is recovered from its concentrates by pyrometallurgical processes with the usual environmental hazards.

The amenability of molybdenite to biological oxidation by *T. ferrooxidans* has long been recognized (Bryner and Anderson, 1957). Encouraging results were achieved in shake flasks at 60°C by using a thermophilic isolate, *Sulfolobus* (Brierley and Brierley, 1986). Brierley and Murr (1973) found 13.3 percent of the molybdenum in 98.5 percent pure molybdenite having a particle size range of 12 to 62 μm to be oxidized at a rate of 26.5 mg/(L · day) in 30 days in the presence of 0.02% yeast extract and 1% $FeSO_4 \cdot 7H_2O$. *Sulfolobus* can tolerate concentrations of soluble molybdenum as high as 2000 ppm (Brierley and Murr, 1973) and can grow at soluble molybdenum concentrations of 750 ppm (Brierley, 1974, 1976). When submarginal copper-molybdenum ore was tested for microbial oxidation, no molybdenum was detected in solution. This negative finding was attributed to acid-consuming gangue and, possibly, to the precipitation of molybdenum as iron compounds such as ferromolybdenite. In situ, dump, and heap leaching must, therefore, be ruled out as possible biohydrometallurgical methods for molybdenum recovery. Continuous bioreactor leaching, an environmentally sound method, appears promising provided that higher solubilization rates are achieved.

The recently reported ability of *T. ferrooxidans* to reduce Mo^{6+} to molybdenum blue [$(MoO)_3(MoO_4)_2$ or $(MoO)_2MoO_4$] with elemental sulfur (Sugio et al., 1988) may be exploitable in precipitating hexavalent molybdenum from solution.

Cobalt

Cobalt is used primarily in the manufacture of special steels, magnet steel, stellite steel for metal-cutting, temperature-resistant alloys, carbide-type alloys, and corrosion-resistant steels. It is extracted pyrometallurgically or hydrometallurgically as a by-product of other ores, chiefly copper, nickel, and silver sulfides, with which it is usually associated as a minor component. The commercially important cobalt minerals are linnaeite $Co_3S_4(Fe,Cu)$, cobaltite (CoAsS), smaltite-cholanthite $(CoAs_2NiAs_2)$, and the black oxides. The amenability of several cobalt sulfides to leaching mediated by chemoautotrophic acidophilic microorganisms of the genus *Thiobacillus* was repeatedly demonstrated in the laboratory and is well-documented (Sutton and Corrick, 1961; De Cuyper, 1964; Lyalikova, 1966; Torma, 1971; Groudev, 1981). In one instance, cobalt sulfide was quickly oxidized by *T. ferrooxidans* with cobalt extraction of more than 85 percent and a dissolution rate of 490 mg/(dm^3 · h) (Torma, 1971). The reaction proceeds according to Eqs. (7.8) and (7.9):

$$CoS + H_2SO_4 + \tfrac{1}{2}O_2 \xrightarrow{\text{\textit{T. ferrooxidans}}} CoSO_4 + H_2O + S^0 \qquad (7.8)$$

$$S^0 + \tfrac{3}{2}O_2 + H_2O \xrightarrow{\text{\textit{T. ferrooxidans}}} H_2SO_4 \qquad (7.9)$$

CoS dissolution is enhanced by the presence of ferric ion in the mineral suspension. Owing to the sensitivity of *T. ferrooxidans* to Co ions during sulfur oxidation, Co^{2+} concentrations higher than about 30 g/dm in the leach liquor might compromise a commercial process. These findings indicate that cobalt recovery by bioleaching from metal sulfide ores with which it is usually associated in nature should be feasible, if the process was applied to the recovery of all the metals contained in the ore.

Nickel

Two types of nickel deposits of commercial importance are presently mined: sulfide and oxide (laterite) deposits. The primary nickel mineral in the sulfide deposits is pentlandite $(Fe,Ni)_9S_8$, which is always associated with pyrite and pyrrhotite. At least two major classes of nickel-sulfide ores have up to now proved refractory to processing with conventional techniques. One class is constituted of ores in which pentlandite occurs as exsolved fine lamellae in pyrrhotite that are extremely difficult to liberate. This intimate intergrowth poses a difficult choice between grade and recovery, since the low-grade concentrates produced to attain acceptable recoveries by conventional mineral dressing techniques cannot be economically extracted by processes such as pressure leaching (Southwood et al., 1985; Miller et al., 1986). The other class is the so-called Duluth Gabbro of northeastern Minnesota, the reserves of which are estimated at 4.6 billion tons with an average assay of 0.60% copper and 0.18% nickel. The difficulties encountered in laboratory tests aimed at producing separate copper- and nickel-rich concentrates have discouraged up to now any attempt to develop commercial operations. Low-cost leaching processes, such as microbial leaching, therefore appear particularly attractive. The only nickel minerals on which microbial leaching with *T. ferrooxidans* has been attempted are the sulfides. Since several researchers have found that in the laboratory, bacterial leaching of nickel can result in relatively high yields (Duncan and Trussell, 1964; Duncan, 1967) and that *T. ferrooxidans* can tolerate up to 72 g/dm^3 of nickel in solution (Torma, 1975), nickel extraction from refractory ores by microbial leaching may represent an economically sound process. Pentlandite biooxidation proceeds according to Eq. (7.10):

$$(Ni,Fe)_9S_8 + (141/8)O_2 + (13/4)H_2SO_4 \xrightarrow{\text{microorganisms}}$$

$$(9/2)NiSO_4 + (9/4)Fe_2(SO_4)_3 + (13/4)H_2O \quad (7.10)$$

A cyclic biohydrometallurgical batch process has been proposed for nickel (and cobalt) recovery (Torma, 1975). Encouraging results have been obtained in heap leaching of á preconcentrated South African nickeliferous ore (Southwood et al., 1985). A flowsheet has been proposed for agitated microbial leaching in airlift pachuca reactors of nickel-sulfide-bearing mixed concentrates produced from the same South African ore (Miller et al., 1986). Equally interesting is a procedure proposed for bioleaching copper-nickel sulfides in Duluth Gabbro (Natarajan et al., 1982).

Aluminosilicates

The most important source of aluminum is bauxite. Although North America and western Europe are by far the major producers of aluminum (76 percent of world production in 1981), practically all the bauxite required for this production is imported from other continents, since North American and west European bauxite deposits are practically depleted. Large occurrences of a wide range of aluminosilicates are known to exist in North America and Europe. Some of them have an aluminum content close to that presently considered economically acceptable for extraction. Of these, kyanite, spodumene, and nepheline are already used in commercial operations (Rossi, 1985). Italian occurrences of leucite [$KAl(SiO_3)_2$] amount to more than 9 billion tons of alumina (Al_2O_3) equivalent (Rossi, 1978). However, owing to several difficulties—mostly high energy requirements of the processes proposed to date—the commercial exploitation of the majority of these aluminosilicates is not competitive with the conventional production of aluminum from bauxite. Several researchers around the world have investigated the possibilities of microbial extraction of aluminum from aluminosilicates. They found that the metabolic products of several microorganisms extensively destroy the crystal lattices of aluminosilicates and silicates and thereby release aluminum. Among active organisms, *Aspergillus niger, Scopulariopsis brevicaulis,* and *Penicillium expansum* deserve mention. The latter microorganisms solubilized from 21 to 27 percent of a leucite sample in 150 days (De Grazia and Camiola, 1906; Rossi, 1978). *A. niger* forms metabolic products which are responsible for olivine, dunite, serpentine (Departement Laboratoires du B. R. G. M., 1971), muscovite (Goni et al., 1973a), feldspar (Goni et al., 1973b; Leleu et al., 1973), spodumene (Avakyan et al., 1981), kaolin (Groudev et al., 1982), and nepheline (King and Dudeney, 1987) degradation. *Penicillium simplicissimum* causes degradation of basalt in a similar way (Mehta et al., 1978, 1979). All the above-mentioned researchers attributed the degradation of the aluminosilicates by the molds to production of acids or alkalis, which then attacked the rock (by acidolysis or alkalolysis, respectively) or to complexing agents which in some way extract some constituents from the crystal lattice (complexolysis). They ruled out the possibility of direct attack. With the present state of knowledge, the only possible commercial application appears to be in situ leaching or leaching in large, open-air ponds, at temperatures compatible with the mesophilic nature of the microbes utilized.

Yet another approach has been widely investigated by researchers in Slavic countries in which they identified and isolated microbial strains that were specifically able to decompose silicates and aluminosilicates. Aleksandrov and Zak (1950) claimed to have isolated strains of "silicate" bacteria. Subsequent studies showed that the "silicate" bacteria are not a new species but strains of *Bacillus circulans* Jordan (1890) (Tesic and Todorovic, 1952, 1955). Encouraging results were reported by Groudev and Genchev (1978), who found that silica-bound aluminum was solubilized from kaolinite and a solid residue assaying 56.8% Al_2O_3 and 25.9% SiO_2. Although for the time being no commercial application of these processes can be anticipated, the possibilities offered by bacterial strains capable of solubilizing aluminosilicates are promising and warrant further investigations.

Gallium

Gallium has acquired a special importance industrially. It is important in the form of compounds or as a metal to the electronics industry for the manufacture of light-emitting, laser, Gunn-effect, and switching diode, for microwave applications, light detectors, and photoelectric materials, semiconductors in fiber optics, calculator, radio, television, and high-fidelity equipment, as a component of superconductive and other alloys, and in solid-state devices. It is also used in the manufacture of selenium rectifiers, in some dental alloys, and as sealant in glass joints in mass spectrometers (Petkof, 1985). In nature, gallium occurs in economically significant quantities in bauxite ore (up to 0.01 percent) and in sphalerite mineral deposits (up to 0.02 percent). The complex sulfide mineral germanite, whose major constituents include copper, iron, and germanium, may include up to 1.85% gallium (Petkof, 1985).

Bioleaching of gallium from bauxite has so far not been reported, but that should not be interpreted to mean that it is not feasible. Presently, gallium is leached chemically from bauxite by the application of the Bayer, Frary, or de la Breteque processes and separated from the aluminum by fractional precipitation and electrolysis.

Gallium has been leached on a laboratory scale by microbial oxidation of (1) sphalerite from the Ekaterina-Blagodatskogo and the Savinskogo deposits in the U.S.S.R. (Lyalikova and Kulikova, 1965), and (2) gallium-containing chalcopyrite concentrate (Torma, 1978). Torma (1978) also reported biooxidation of gallium sulfide (Ga_3S_2). In all instances, *T. ferrooxidans* was responsible for the leaching. Lyalikova and Kulikova (1965) performed their leaching experiments in the mineral medium of Leathen et al. (1951). Torma (1978) performed some of his experiments in a Warburg respirometer and others in shake cultures in 250-mL Erlenmeyer flasks, in each instance with the ore suspended in the iron-free mineral salts solution [modified 9 K medium of Silverman and Lundgren (1959)] at pH 1.8. He found that his strain of *T. ferrooxidans* exhibited a higher specific rate of oxygen uptake on gallium-containing chalcopyrite than on gallium sulfide in the Warburg experiments. He also found that the culture at 35°C and a rotational shake rate of 250 rev/min leached gallium fastest from gallium-containing chalcopyrite when the initial pulp density was 25 percent. He tested leaching over a pulp density range of 5 to 25 percent. He proposed that gallium was leached from chalcopyrite by both an indirect effect from oxidation by ferric iron released by the bacteria from the chalcopyrite,

$$Ga_2S_3 + 3Fe_2(SO_4)_3 \rightarrow Ga_2(SO_4)_3 + 6FeSO_4 + 3S^0 \qquad (7.11)$$

$$2CuFeS_2 + 8\frac{1}{2}O_2 + H_2SO_4 \xrightarrow{\text{bacteria}} 2CuSO_4 + Fe_2(SO_4)_3 + H_2O \qquad (7.12)$$

and by direct bacterial attack in which the bacteria appeared to oxidize the sulfide moiety of the gallium sulfide,

$$Ga_2S_3 + 6O_2 \xrightarrow{\text{bacteria}} Ga_2(SO_4)_3 \tag{7.13}$$

The ferrous sulfate and elemental sulfur produced in the chemical oxidation of gallium sulfide are microbiologically oxidized to ferric sulfate and sulfuric acid, respectively:

$$2FeSO_4 + H_2SO_4 + \tfrac{1}{2}O_2 \xrightarrow{\text{bacteria}} Fe_2(SO_4)_3 + H_2O \tag{7.14}$$

$$S + \tfrac{3}{2}O_2 + H_2O \xrightarrow{\text{bacteria}} H_2SO_4 \tag{7.15}$$

Manganese

Manganese is important industrially because it serves as a desulfurizing, deoxidizing, and/or alloying element in the manufacture of steel and cast iron. It is also a component of some nonferrous alloys. In addition, it has uses in the manufacture of carbon-zinc batteries and as a chemical reagent (Jones, 1985).

Marine ferromanganese concretions and crusts

Ferromanganese concretions and crusts on the floor of the world's oceans constitute a potentially important natural resource of base metals such as Mn, Cu, Co, and Ni. The Mn in these concretions is mainly in an insoluble, tetravalent state (Murray et al., 1984). It is found chiefly in the minerals todorokite [$(Mn^{2+},Mg,Ca)Mn_6^{4+}O_{13} \cdot 3\text{-}4H_2O$] or birnessite [$(Ca,Mg,Na_2,K_2)_xMn^{4+}Mn_2(O,OH)_2$], depending on the source of the nodules. Its concentration in nodules may be as high as 25 percent by weight or higher, and the concentration of Fe may be half that of the Mn or higher. Co, Cu, and Ni are present in variable concentrations, which can be as high as 2 percent in the case of Ni and Cu, and 1 percent in the case of Co (Mero, 1962; Cronan and Tooms, 1969). The tetravalent Mn can be enzymatically reduced to soluble, divalent Mn in a respiratory process by some bacteria native to the nodules. Glucose has been used experimentally as the electron donor in this reduction (Trimble and Ehrlich, 1968; Ehrlich et al., 1973). The Co, Cu, and Ni are solubilized as the Mn is reduced, but little of the Fe in the concretions is solubilized (Ehrlich et al., 1973). The lack of Fe solubilization from ferromanganese concretions is surprising because at least some of the Mn reducers are capable of reducing ferric iron in goethite, limonite (FeOOH), and hematite (e.g., De Castro and Ehrlich, 1970). The enzymatic reduction of the Mn(IV) in the ferromanganese is not inhibited by the presence of O (see Ehrlich, 1988b, for an explanation). Reported rates of Mn bioleaching from ferromanganese nodules in stationary batch cultures

(Trimble and Ehrlich, 1968; Ehrlich et al., 1973) are too slow to be competitive with chemical processes described in the literature (e.g., Fuerstenau and Han, 1983). However, a possibility for significant acceleration of bioleaching exists. Such acceleration may be achieved by selecting more efficient bacterial cultures, optimizing the culture medium and physical conditions for growth, and/or performing the leaching in columns or reactors. Recovery of dissolved metal species can be accomplished by solvent extraction and electrowinning.

Bacteria active in leaching Mn from ferromanganese concretions include gram-positive and gram-negative organisms (Ehrlich, 1981). The pathway of electrons from electron donor to MnO_2 as terminal electron acceptor is not the same in all these organisms (Ghiorse and Ehrlich, 1976; Ehrlich, 1980).

Terrestrial manganese ores

Terrestrial manganese ores can be divided into low-grade ores, containing less than 35% manganese by weight, and high-grade ores, which contain concentrations of manganese higher than 35 percent. Commonly, the manganese in these ores is in the form of manganous carbonate or rhodochrosite ($MnCO_3$), manganous silicate or rhodonite ($MnSiO_3$), hausmannite (Mn_3O_4), braunite [$(MnSi)_2O_3$], and pyrolusite. High-grade ores are presently preferred because the manganese can be extracted from them as ferromanganese or spiegeleisen by conventional pyrometallurgical methods or fused-salt electrolysis with a minimum of ore beneficiation. Manganese metal and manganese dioxide can be produced by electrolyzing a solution of manganous sulfate obtained by hydrometallurgical extraction of a suitable high-grade ore (Jones, 1985).

Economical exploitation of low-grade ores by hydrometallurgical methods such as the Dean-Leute ammonium carbamate process is possible (DeHuff, 1965) but is not presently favored (Jones, 1985). Yet, the alternative pyrometallurgical or electrolytic processes are energy-intensive. Biohydrometallurgical extraction, which is not energy-intensive and uses relatively inexpensive reagents, is possible (Perkins and Novielli, 1962; Agate and Deshpande, 1977; Sakhvadze et al., 1986) but has so far not been practiced on a commercial scale.

When the manganese in an ore is present as manganous carbonate or silicate, the microbial mode of leaching is via reaction with an acid which has been metabolically produced. The acids involved may be organic (acetic, lactic, pyruvic, citric, oxalic, etc.) or inorganic (sulfuric or nitric acids; e.g., Ehrlich, 1981; Babenko et al., 1987). The carbonates will react with the acids much more readily than will the silicates.

When the manganese is present as an oxide of manganese(III) or manganese(IV), enzymatic reduction or chemical reduction by metabolic products to manganese(II) is necessary to bring the manganese into solution. Occurrence of such reduction of manganese in terrestrial ores or minerals has been reported (e.g., Vavra and Frederick, 1952; Babenko et al., 1982; Holden and Madgwick, 1983; Gottfreund and Schweisfurth, 1983; Kozub and Madgwick, 1983; Ghosh and Imai, 1985a,b; Yurchenko et al., 1987). Enzymatic reduction of terrestrial manganese oxides involves a form of respiration as in the case of marine ferromanganese nodules or crusts. Chemical reduction of terrestrial manganese oxides by metabolic products may involve reactions with formate, oxalate, or pyruvate, among others (Ehrlich, 1981; Stone, 1987; Yurchenko et al., 1987).

Biohydrometallurgical extraction of low-grade manganese ores on a commercial scale will require improvement in leaching rates, the identification of inexpensive organic carbon and energy sources, and selective culture conditions for the active organism(s) which do not require asepsis. Up to now all the organisms studied (except for *T. ferrooxidans* and *T. thiooxidans*; see Ghosh and Imai, 1985a,b) that are capable of leaching manganese have been heterotrophs. Culture media which can support manganese leaching activity by these heterotrophs can easily be contaminated by other heterotrophs incapable of leaching manganese. The application of culture conditions favoring the manganese leaching microbes is essential. Attempts at improving heterotrophic manganese leaching rates using Groote Eylandt manganese ore have been made by Hart and Madgwick (1986), Holden and Madgwick (1983), and Mercz and Madgwick (1982).

Chromium

Although not a common mineral, crocoite ($PbCrO_4$), found in the oxidation zone of chromite deposits, can be leached by *Pseudomonas chromatophila* sp. nov., a facultative heterotrophic anaerobe (Lebedeva and Lyalikova, 1979). The organism is capable of anaerobically reducing chromate ion and chromate in crocoite to chromium(III), using lactate as electron donor and the chromate as electron acceptor.

Chromium, Nickel, and Vanadium Extraction from Ores or Industrial Wastes by Biologically Generated Acid

Some metals in ores, such as laterites, and in industrial wastes exist in compounds which are soluble in dilute acid. Examples are chromium, nickel, and vanadium. Some can be solubilized effectively by dilute sulfuric acid generated in the oxidation of elemental sulfur by an acidophilic autotroph such as *T. thiooxidans* [see Eq. (7.9)]. For example, 100 percent of the chromium was leached with sulfuric acid generated from sulfur by *T. thiooxidans* in 146 days from a galvanic sludge at a concentration of 20.2 percent at a pulp density of 10 g per 100 mL (Bosecker, 1986a). Under similar conditions, 100 percent of the vanadium contained in a filter-press residue at a vanadium concentration of 5.04 percent was leached at a pulp density of 11 g per 100 mL (Bosecker, 1986a). Although chemical leaching with dilute sulfuric acid was equally effective, the cost of bacterial leaching was estimated to have been lower when the costs of sulfur and sulfuric acid and their transportation costs were considered (Bosecker, 1986a). Extensive extraction of chromium from chromite ore by biogenic sulfuric acid was not successful (Ehrlich, 1983).

In another study (Bosecker, 1986b), approximately 70 percent of the nickel contained in a silicate-bearing lateritic ore (2.2% nickel concentration) at a pulp density of 5 g per 100 mL was leached in 29 days by organic acid (presumably citric acid) produced by *Penicillium* P6 from glucose. Similar leaching of limonitic nickel-containing laterite (1.5% nickel concentration) was much less effective. At best only 14 percent of the nickel was bioleached in 29 days. Chemical leaching with a citric acid solution required a higher concentration of the acid to achieve similar results.

Secondary Biological Recovery of Vanadium and Selenium

Vanadium

Vanadium values are recoverable from acidic solutions generated by *T. ferrooxidans* in the bioleaching of uranium ores containing vanadium (Goren, 1966). After the uranium has been stripped from solution by ion exchange or solvent extraction, the vanadium is recovered by oxidizing it from the +4 to the +5 oxidation state with ferric iron produced by *T. ferrooxidans* or direct oxidation by the organism. The oxidized vanadium can be recovered as iron vanadate precipitate or inert carrier material at pH 2.5 to 3. This precipitate is acid-soluble (Goren, 1966).

Selenium

Although bacteria generally cannot reduce metal salts to their metallic state, mercury being an exception, they can reduce metalloids such as selenate and tellurate to selenium and tellurium, respectively. The enzymatic reduction of selenate and selenite to selenium may be put to practical use in detoxifying polluted surface or ground water and leachates from industrial wastes containing such compounds or in recovering selenium from solution for commercial exploitation. The reduction of selenate or selenite can be viewed as a form of respiration which can be carried out by different microorganisms (bacteria, including actinomycetes, and fungi; see review by Ehrlich, 1981). The selenium may form a red deposit in the cell envelope of active cells, or it may accumulate in the culture medium. Selenate and selenite may also be reduced to volatile selenide compounds, including dimethyl selenide and dimethyl diselenide (Chau et al., 1976; Reamer and Zoller, 1980; Ehrlich, 1981). Collecting volatile selenide compounds may be a better approach to industrial recovery of selenium than reducing it to the elemental form, but this approach seems not to have been considered to date.

Conclusion

As the foregoing shows, the potential exists for industrial-scale microbial extraction of a wide variety of metals from ores. More bench-scale experimentation is needed to explore this potential further. After the most promising processes are optimized, pilot-scale experimentation will be needed to assess the economics of the chosen processes.

References

Agate, A. D., and Deshpande, H. A., Leaching of manganese ores using *Arthrobacter* species, in W. Schwartz (ed.), *Conf.—Bacterial Leaching*, Gesellschaft für Biotechnologische Forschung mbH, Braunschweig-Stoeckheim, Verlag Chemie, Weinheim, 1977, pp. 243–250.

Aleksandrov, V. G., and Zak, G. A., The bacteria which destroy aluminosilicates ("silicate" bacteria), *Mikrobiologiya* 19:97–104 (1950).

Avakyan, Z. A., Karavaiko, G. I., Melnikova, E. O., Krutsko, V. S., and Ostroushko, Yu. I., The role of microscopic fungi in the weathering of rocks and minerals from a pegmatite deposit, *Mikrobiologiya* 50:156–162 (1981).

Babenko, Yu. S., Dolgikh, L. M., and Serebryanaya, M. Z., Characteristics of the bacterial breakdown of primarily oxidized manganese ores from the Nikopol deposit, *Mikrobiologiya* 52:851–856 (1982).

Babenko, Yu. S., Serebryanaya, M. Z., Petrova, L. N., and Dolgikh, L. M., Manganese dioxide solubilization with the Krebs cycle metabolites, *Prikl. Biokhim, Mikrobiol.* 23:245–252 (1987).

Barbery, G., Fletcher, A. W., and Sirois, L. L., Exploitation of complex sulfide deposits: A review of processing options from ore to metals, in M. J. Jones (ed.), *Complex Sulfide Ores,* The Institution of Mining and Metallurgy, London, 1980, pp. 135–150.

Bateman, A. M., *Economic Mineral Deposits,* 2nd ed., Wiley, New York, 1952.

Bhappu, R. B., Reynolds, D. H., Roman, R. J., and Schwab, D. A., Hydrometallurgical recovery of molybdenum from Questa mine, State Bur. Mines Mineral Resources, Circular 81, Socorro, New Mexico, 1965.

Bosecker, K., Bacterial metal recovery and detoxification of industrial waste, in H. L. Ehrlich and D. S. Holmes (eds.), *Workshop on Biotechnology for the Mining, Metal-Refining and Fossil Fuel Processing Industries, Biotechnol. Bioeng. Symp. No. 16,* Wiley, New York, 1986a, pp. 105–120.

Bosecker, K., Leaching of lateritic nickel ores with heterotrophic microorganisms, in

R. W. Lawrence, R. M. R. Branion, and H. G. Ebner (eds.), *Fundamental and Applied Biohydrometallurgy*, Elsevier, Amsterdam, 1986b, pp. 367–382.

Brierley, C. L., Molybdenite leaching: use of a high-temperature microbe, *J. Less-Common Metals* **36**:237–247 (1974).

Brierley, C. L., Biogenic extraction of copper and molybdenum at high temperatures. Report for the U.S. Department of the Interior, Bureau of Mines, Washington, D.C., 1976.

Brierley, C. L., and Murr, L. E., Leaching: Use of a thermophilic and chemoautotrophic microbe, *Science* **179**:488–490 (1973).

Brierley, J. A., and Brierley, C. L., Microbial mining using thermophilic microorganisms, in T. D. Brock (ed.), *Thermophiles: General, Molecular, and Applied Microbiology*, Wiley, New York, 1986, pp. 279–305.

Bruynesteyn, A., and Duncan, D. W., Microbiological leaching of sulfide minerals, *Can. Metall. Q.* **10**:57–63 (1971).

Bryner, L. C., and Anderson, R., Microorganisms in leaching sulfide minerals, *Ind. Eng. Chem.* **49**:1721–1724 (1957).

Bryner, L. C., Beck, J. V., Davis, D. B., and Wilson, D. G., Microorganisms in leaching sulfide minerals, *Ind. Eng. Chem.* **46**:2587–2592 (1954).

Carta, M., Ghiani, M., and Rossi, G., Beneficiation of a complex sulfide ore by an integrated process of flotation and bioleaching, in M. J. Jones (ed.), *Complex Sulfide Ores* The Institute of Mining and Metallurgy, London, 1980a, pp. 178–185.

Carta, M., Ghiani, M., and Rossi, G., Complex sulfides concentrates bioleaching performance was related to feed composition, in *Proc. Intern, Conf. Use of Microorganisms in Hydrometallurgy*, Hungarian Academy of Sciences, Local Committee of Pecs, 1980b, pp. 203–209.

Chau, Y. K., Wong, P. T. S., Silverberg, B. A., Luxon, P. L., and Bengert, G. A., Methylation of selenium in the aquatic environment, *Science* **192**:1130–1131 (1976).

Cronan, D. S., and Tooms, J. S., The geochemistry of manganese nodules of associated pelagic deposits from the Pacific and Indian Oceans, *Deep-Sea Res.* **16**:335–359 (1969).

Dave, S. R., Natarajan, K. A., and Bhat, J. V., Bio-oxidation studies with *Thiobacillus ferrooxidans* in the presence of copper and zinc, *Trans. Inst. Min. Metall. Sect. C* **88**:C234–C237 (1979).

De Castro, A. F., and Ehrlich, H. L., Reduction of iron oxide minerals by a marine bacillus, *Antonie v. Leeuwenhoek* **36**:313–327 (1970).

De Cuyper, J. A., Bacterial leaching of low grade copper and cobalt ores, in M. Wadsworth and F. Davis (eds.), *Intern. Symp. on Unit Processes in Hydrometallurgy*, Gordon and Breach, New York, 1964, pp. 126–142.

De Grazia, S., and Camiola, G., Sull'intervento dei microorganismi nella utilizzazione della potassa leucitica del suolo da parte delle piante superiori, *Le stazioni sperimentali agrarie italiane* **39**:829–840 (1906).

DeHuff, G. L., Manganese, in *Mineral Facts and Problems*, Bureau of Mines Bulletin 630, U.S. Department of the Interior, Washington, D.C., 1965, pp. 553–572.

Departement Laboratoires du Bureau de Recherches Geologiques et Mineres, Methodes d'etudes de l'alteration supergene des mineraux et des roches, Section II, No. **5**:35–56 (1971).

Duncan, D. W., Microbiological leaching of sulfide minerals, *Aust. Min. Nov.* (1967).

Duncan, D. W., and Trussell, P. C., Advances in the microbiological leaching of sulfide ores, *Can. Metall. Q.* **3**:43–45 (1964).

Dutrizac, J. E., and MacDonald, R. J. C., The effect of some impurities on the rate of chalcopyrite dissolution, *Can. Metall. Q.* **12**:409–420 (1973).

Ehrlich, H. L., Bacterial leaching of manganese ores, in P. A. Trudinger, M. R. Walter, and B. J. Ralph (eds.), *Biogeochemistry of Ancient and Modern Environments*, Australian Academy of Science, Canberra, and Springer-Verlag, Berlin, 1980, pp. 609–614.

Ehrlich, H. L., *Geomicrobiology*, Dekker, New York, 1981.

Ehrlich, H. L., Leaching of chromite ore and sulfide matte with dilute sulfuric acid generated by *Thiobacillus thiooxidans* from sulfur, in G. Rossi and A. E. Torma (eds.), *Recent Progress in Biohydrometallurgy*, Associazione Mineraria Sarda, Iglesias, Italy, 1983, pp. 19–42.

Ehrlich, H. L., Bacterial leaching of silver from a silver-containing mixed sulfide ore by a continuous process, in R. W. Lawrence, R. M. R. Branion, and H. G. Ebner (eds.), *Fundamental and Applied Biohydrometallurgy*, Elsevier, Amsterdam, 1986, pp. 77–88.

Ehrlich, H. L., Bioleaching of silver from a mixed sulfide ore in a stirred reactor, in P. R. Norris and B. P. Kelley (eds.), *Biohydrometallurgy*, Kew, Surrey, U.K., 1988a, pp. 223–231.

Ehrlich, H. L., Manganese oxide reduction as a form of anaerobic respiration, *Geomicrobiol. J.* **5**:423–431 (1988b).

Ehrlich, H. L., Yang, S. H., and Mainwaring, J. D., Jr., Bacteriology of manganese nodules. VI. Fate of copper, nickel, cobalt, and iron during bacterial and chemical reduction of the manganese (IV), *Z. Allg. Mikrobiol.* **13**:39–48 (1973).

Fuerstenau, D. W., and Han, K. N., Metallurgy and processing of marine manganese nodules, *Miner. Process. Technol. Rev.* **1**:1–83 (1983).

Garrels, R. M., and Christ, C. L., *Equilibre des Mineraux et de Leurs Solutions Aqueuses*, Gauthier-Villars, Paris, 1967.

Georgeaux, A., Scheidt, C., Demarthe, J.-M., and Gandon, L., Traitement hydrometallurgique de minerais sulfures polymetalliques avec ou sans concentration globale prealable, *Ind. Miner.-Mineralurgie* **2-77**:86–93 (1977).

Ghiani, M., Rossi, G., and Trois, P., Use of bacteria in mineral processing, preprint No. 87-18, presented at the SME Annual Meeting, Denver, CO, 1987.

Ghiorse, W. C., and Ehrlich, H. L., Electron transport components of the MnO_2 reductase system and the location of the terminal reductase in a marine *Bacillus*, *Appl. Environ. Microbiol.* **31**:977–985 (1976).

Ghosh, J., and Imai, K., Leaching of manganese dioxide by *Thiobacillus ferrooxidans* growing on elemental sulfur, *J. Ferment. Technol.* **63**:259–262 (1985a).

Ghosh, J., and Imai, K., Leaching mechanism of manganese dioxide by *Thiobacillus ferrooxidans*, *J. Ferment. Technol.* **63**:295–298 (1985b).

Goodman, R. E., *Introduction to Rock Mechanics*, Wiley, New York, 1980.

Goni, J., Gugalski, T., and Sima, J., Solubilisation du potassium de la muscovite par voie microbienne, *Bull. B.R.G.M.*, *2eme, serie, Section IV-No. 1*, pp. 31–47 (1973a).

Goni, J., Greffard, J., Gugalski, T., and Leleu, M., La geomicrobiologie et la biomineralurgie, *Bull. Soc. Fr. Miner. Crystallogr.* **96**:252–266 (1973b).

Goren, M. B., Oxidation and recovery of vanadium values from acidic aqueous media, U.S. patent 13,252,756, 1966.

Gormely, L. S., Duncan, D. W., Branion, R. M. R., and Pinder, K. L., Continuous culture of *Thiobacillus ferrooxidans* on a zinc sulfide concentrate, *Biotechnol. Bioeng.* **17**:31–49 (1975).

Gottfreund, J., and Schweisfurth, R., Mikrobiologische Oxidation und Reduktion von Manganspecies, *Fresenius Z. Anal. Chem.* **316**:634–638 (1983).

Groudev, S. N., Leaching of cobalt from synthetic cobalt sulfide by *Thiobacillus ferrooxidans*, *C. Rend. Acad. Bulg. Sci.* **34**:217–220 (1981).

Groudev, S. N., and Genchev, F. N., Bioleaching of bauxites by wild and laboratory-bred microbial strains, in *Proc. 4th Intern. Congr. for the Study of Bauxites, Alumina, and Aluminum*, Athens, Greece, Vol. 1, 1978, pp. 271–278.

Groudev, S. N., Genchev, F. N., and Groudeva, V. I., Use of microorganisms for recovery of aluminum from aluminosilicates—achievements and prospects, *Trav. ICSOBA* **12**:203–212 (1982).

Hart, M. J., and Madgwick, J. C., Biodegradation of manganese dioxide tailings, *Aust. I.M.M. Bull. Proc.* **291**:61–64 (1986).

Hoffman, L. E., and Hendrix, J. L., Inhibition of *Thiobacillus ferrooxidans*, *Biotechnol. Bioeng.* **18**:1161–1165 (1976).

Holden, P. J., and Madgwick, J. C., Mixed culture bacterial leaching of manganese dioxide, *Proc. Aust. Inst. Min. Metall.* No. **286**:61–63 (1983).

Jones, T. S., Manganese, in *Mineral Facts and Problems*, Bureau of Mines Bulletin 675, U.S. Department of the Interior, Washington, D.C., 1985, pp. 483–497.

Khalid, A. M., and Ralph, B. J., The leaching behavior of various zinc sulfide minerals with three *Thiobacillus* species, in W. Schwartz (ed.), *Conference Bacterial Leaching*, Verlag Chemie, Weinheim, 1977, pp. 165–173.

King, A. B., and Dudeney, A. W. L., Bioleaching of nepheline, *Hydrometallurgy* **19**:69–81 (1987).

Kozub, J. M., and Madgwick, J. C., Microaerobic microbial manganese dioxide leaching, *Proc. Aust. Inst. Min. Metall.* No. 288:51–54 (1983).

Kullerud, G., The FeS-ZnS system, a geological thermometer, *Nor. Geol. Tidsskr.* **32**:61–147 (1953).

Lazaroff, N., Sigal, W., and Wasserman, A., Iron oxidation and precipitation of ferric hydroxysulfates by resting *Thiobacillus ferrooxidans* cells, *Appl. Environ. Microbiol.* **43**:924–938 (1982).

Leathen, W. W., McIntyre, L. D., and Braley, S. A., A medium for the study of the bacterial oxidation of ferrous iron, *Science* **114**:280–281 (1951).

Lebedeva, E. V., and Lyalikova, N. N., Reduction of crocoite by *Pseudomonas chromatophila* sp. nov., *Mikrobiologiya* **48**:517–522 (1979).

Leleu, M., Sarcia, C., and Goni, J., Essai d'interpretation des phenomenes naturels d'alteration de silicates: le cas de l'abilite et de l'orthose, in *Congres Intern. Geochimie Organique, Rueil*, 1973.

Lyalikova, N. N., Sulfide oxidation by *Thiobacillus ferrooxidans*, *Trans. Moscow Soc. Nat.* **24**:211–216 (1966).

Lyalikova, N. N., *Stibiobacter senarmontii* gen. nov. et sp. nov.—a new microorganism oxidizing antimony, *Microbiologiya* **43**:941–943 (1974).

Lyalikova, N. N., and Kulikova, M. F., Leaching of rare elements from sulfide ores under the influence of bacteria, *Dokl. Akad. Nauk. USSR* **164**:674–676 (1965).

Lyalikova, N. N., Shlayn, L. B., Unonova, O. G., and Anisimova, L. S., On the transformation products of compound antimony and lead sulfides under the action of the bacteria, *Izv. Akad. Nauk. USSR Ser. Biol.* **4**:564–567 (1972).

Mehta, A., Torma, A. E., and Murr, L. E., Biodegradation of aluminum-bearing rocks of *Penicillium simplicissimus*, *IRCS Med. Sci.* **6**:416 (1978).

Mehta, A. P., Torma, A. E., and Murr, L. E., Effect of environmental parameters on the efficiency of biodegradation of basalt rock by fungi, *Biotechnol. Bioeng.* **21**:875–885 (1979).

Mercz, T. I., and Madgwick, J. C., Enhancement of bacterial manganese leaching by microalgal growth products, *Proc. Aust. Inst. Min. Metall.* No. 283:43–46 (1982).

Mero, J. L., Ocean-floor manganese nodules, *Econ. Geol.* **57**:747–767 (1962).

Miller, P. C., Huberts, R., and Livesey-Goldblatt, E., The semicontinuous bacterial agitated leaching of nickel sulfide material, in R. W. Lawrence, R. M. R. Branion, and H. G. Ebner (eds.), *Fundamental and Applied Biohydrometallurgy*, Elsevier, Amsterdam, 1986, pp. 23–42.

Murray, J. W., Balistrieri, L. S., and Paul, B., The oxidation state of manganese in marine sediments and ferromanganese nodules, *Geochim. Cosmochim. Acta* **48**:1237–1247 (1984).

Natarajan, K. A., Reid, K. J., and Iwasaki, I., Microbial aspects of hydrometallurgical processing and environmental control of copper/nickel-bearing Duluth Gabbro, in *Forty-Third Annual Mining Symp., Duluth*, 1982.

Norris, P., and Kelly, D. P., Toxic metals in leaching systems, in L. E. Murr, A. E. Torma, and J. A. Brierley (eds.), *Metallurgical Applications of Bacterial Leaching and Related Microbiological Phenomena*, Academic Press, New York, 1978, pp. 83–102.

Pelissonier, H., Les dimensions des gisements de cuivre du monde: essay de metallogenie quantitative, *Mem. B.R.G.M.*, No. 57 (1972).

Perkins, E. C., and Novielli, F., Bacterial leaching of manganese ores. U.S. Department of the Interior, Bureau of Mines Report No. 6102, 1962.

Petersson, S., Norro, A., and Eriksson, S., Treatment of copper converter slags in top blowing rotary converter, in J. C. Yannopulos and J. C. Agarwal (eds.), *Extractive Metallurgy of Copper*, Vol. 1, A.I.M.C., New York, 1976, pp. 317–330.

Petkof, B., Gallium, in *Mineral Facts and Problems*, Bureau of Mines Bulletin 675, U.S. Department of the Interior, Washington, D.C., 1985, pp. 291–296.

Pridmore, F. E., and Shuey, R. T., The electrical resistivity of galena, pyrite and chalcopyrite, *Am. Miner.* **61**:248–259 (1976).

Reamer, D. C., and Zoller, W. H., Selenium biomethylation products from soil and sewage sludge, *Science* **208**:500–502 (1980).

Rossi, G., The Funtana Raminosa (Sardinia, Italy) mine and mill, Internal Report to Montecatini S.p.A., Milan, 1958.

Rossi, G., The microbiological leaching of ore minerals. II. The action on stibnite ores of microorganisms found in acid drainage waters of Italian mines, *Res. Sediment. Assoc. Min. Sarda, Iglesias* **76**:1–14 (1971).

Rossi, G., Discussion on the paper by G. M. Hughes, "Sherritt Gordon's copper-zinc concentrators: Their design and operation," in M. J. Jones (ed.), *Proc. X Intern. Min. Proc. Congr.*, Institution of Mining and Metallurgy, London, 1974, pp. 888–889.

Rossi, G., Potassium recovery through leucite bioleaching: Possibilities and limitations, in L. E. Murr, A. E. Torma, and J. A. Brierley (eds.), *Metallurgical Applications of Bacterial Leaching and Related Microbiological Phenomena*, Academic Press, New York, 1978, pp. 297–319.

Rossi, G., Aluminosilicate biodegradation: progress and perspectives, in G. I. Karavaiko and S. N. Groudev (eds.), *Biogeotechnology of Metals*, Centre of International Projects G.K.N.T., Moscow, U.S.S.R., 1985, pp. 396–408.

Rossi, G., Torma, A. E., and Trois, Bacteria-mediated copper recovery from a cupriferous pyrrhotite ore: Chalcopyrite-pyrrhotite interactions, in G. Rossi and A. E. Torma (eds.), *Recent Progress in Biohydrometallurgy*, Associazione Mineraria Sarda, Iglesias, Italy, 1983, pp. 185–200.

Sakhvadze, L. I., Papavadze, B. A., and Kerkadze, N. D., Bacterial leaching of manganese, U.S.S.R. patent SU 1,239,158, 1986.

Silverman, M. P., and Lundgren, D. G., Studies on the chemoautotrophic iron bacterium *Ferrobacillus ferrooxidans*. I. An improved medium and a harvesting procedure for securing high cell yields, *J. Bacteriol.* **77**:642–647 (1959).

Southwood, A. J., Miller, P. C., and Corrans, I. J., Parameters affecting the bacterial heap leaching of low-grade nickeliferous material, in *XV Intern. Min. Process. Congr.* Tome II: *Flottation-Hydrometallurgie*, Soc. Ind. Minerale, 1985, pp. 400–412.

Stone, A. T., Microbial metabolites and the reductive dissolution of manganese oxides: oxalate and pyruvate, *Geochim. Cosmochim. Acta* **51**:1919–1925 (1987).

Sugio, T., Tsujita, Y., Karagiri, T., Imagaki, K., and Tano, T., Reduction of Mo^{6+} with elemental sulfur by *Thiobacillus ferrooxidans*, *J. Bacteriol.* **170**:5956–5959 (1988).

Sundstroem, O. A., The slag forming plant at the Roennskaer Works of the Boliden Akteibolag, Skellefthamn, Sweden, *J. Met.* **21**:15–21 (1969).

Sutton, J. A., and Corrick, J. D., Bacteria in mining and metallurgy: Leaching selected ores and minerals; experiments with *Thiobacillus thiooxidans*, U.S. Bureau of Mines Report of Investigations No. 5839, Washington, D.C., 1961.

Suzuki, I., Lizama, H. M., and Oh, J. K., Bacterial leaching of a complex sulfide ore, in R. G. L. McCready (ed.), *Proc. 3rd Annual General Meeting of BIOMINET*, CANMET Special Publications, SP86-9 Toronto, Canada, 1986, pp. 127–135.

Tesic, Z., and Todorovic, M., A contribution to the study of "silicate bacteria," *Zemljiste i biljka I* **1**:3–18 (1952).

Tesic, Z., and Todorovic, M., On the issue of the species of "silicate bacteria," in *Atti del VI Congr. Internaz, di Microbiol. Roma*, Vol. 6, 1955, pp. 356–361.

Tesic, Z., and Todorovic, M., Contribution to knowledge of the specific properties of silicate bacteria, *Zemljiste i biljka* **8**:233–240 (1958).

Torma, A. E., Microbiological oxidation of synthetic cobalt, nickel and zinc sulfides by *Thiobacillus ferrooxidans*, *Rev. Can. Biol.* **30**:209–216 (1971).

Torma, A. E., Microbiological extraction of cobalt and nickel, Paper No. A72-7, The Metallurgical Society of A.I.M.E., New York, 1975.

Torma, A. E., Oxidation of gallium sulfides by *Thiobacillus ferrooxidans*, *Can. J. Microbiol.* **24**:888–891 (1978).

Torma, A. E., and Gabra, G. G., Oxidation of stibnite by *Thiobacillus ferrooxidans*, *Antonie v. Leeuwenhoek* **43**:1–6 (1977).

Torma, A. E., Walden, C. C., and Branion, R. M. R., Microbiological leaching of a zinc sulfide concentrate, *Biotechnol. Bioeng.* **12**:501–517 (1970).

Torma, A. E., Walden, C. C., Ducan, D. W., and Branion, R. M. R., The effect of carbon dioxide and particle surface area on the microbiological leaching of a zinc sulfide concentrate, *Biotechnol. Bioeng.* **14**:777–786 (1972).

Trimble, R. B., and Ehrlich, H. L., Bacteriology of manganese nodules. III. Reduction of

MnO_2 by two strains of nodule bacteria, *Appl. Microbiol.* **16**:695–702 (1968).

Tuovinen, O. H., Niemela, S. I., and Gyllenberg, H. G., Tolerance of *Thiobacillus ferrooxidans* to some metals, *Antonie v. Leeuwenhoek* **37**:489–496 (1971).

Tuovinen, O. H., Pukakka, J., Hiltunen, P., and Dolan, K. M., Silver toxicity to ferrous iron and pyrite oxidation and its alleviation by yeast extract in culture of *Thiobacillus ferrooxidans*, *Biotechnol. Lett.* **7**:389–394 (1985).

Vavra, J. P., and Frederick, L. R., The effect of sulfur oxidation on the availability of manganese, *Soil Sci. Soc. Am. Proc.* **16**:141–144 (1952).

Yakhontova, L. K., Nesterovich, L. G., Grudev, A. P., and Postnikova, V. P., New data on oxidation of galena and sphalerite, *Dokl. Akad. Nauk SSSR* **250**:718–721 (1980).

Yurchenko, V. A., Karavaiko, G. I., Remizov, V. I., Klyushnikova, T. M., and Tarusin, A. D., Role of the organic acids produced by *Acinetobacter calcoaceticus* in manganese leaching, *Prikl. Biokhim. Mikrobiol.* **23**:404–412 (1987).

Economics of Bioleaching

T. P. McNulty
David L. Thompson

Introduction

The economics of any new process technology must be assessed relative to those of existing technologies. In this chapter we compare biological pretreatment of refractory gold ores to pyrometallurgical and hydrometallurgical pretreatment, both standard commercial processes. This will be done by means of a hypothetical case study of a typical refractory gold ore in a typical western U.S. location.

We have selected gold ore processing for good reasons. Since it became a free-market commodity in the mid-1970s, gold has attracted continuous interest after some 30 years of dormancy. During the inflation of the late 1970s it enjoyed spectacular increases in price. During the general minerals recession of the early and mid-1980s, it remained one of the few metals whose value exceeded its usual production cost. United States gold production consequently grew from 1 million troy ounces in 1977 to nearly 5 million troy ounces in 1987 and is still increasing rapidly as dozens of new mines are opened or expanded each year. The gross annual value of domestic production presently exceeds $2 billion.

Gold Ores

Gold almost always occurs in metallic form. If the particles are coarse (larger than about 70 μm) and not cemented to other particles, they can often be recovered by simple gravity concentration. Gravity techniques take advantage of the high specific gravity of gold (19.3) compared with most gangue minerals (2.5 to 3.0). Gold concentrates recovered in this manner can be sold directly to a smelter or refiner. If the gold is present as particles too small for gravity concentration, or if it is locked with other minerals, often fine grinding followed by froth flotation can be used to produce a gold concentrate. Flotation concentrates are usually of much lower grade than gravity concentrates, and although they are occasionally sold directly to a smelter, more often they are processed by cyanidation. Sometimes the gold is so finely disseminated in the ore that it cannot be physically concentrated at all, and these ores are processed directly by cyanidation.

The vast majority of gold in this country is recovered by cyanidation, either with or without preconcentration. Cyanidation is performed with two basic techniques. The first, heap leaching, is applied to low-grade ores. The material is crushed and stacked on an impermeable pad, and cyanide solution is trickled through it. The second technique, agitation leaching, is applied to higher-grade materials. The ore is finely ground and slurried with cyanide solution in agitated tanks. Agitation leaching is more expensive than heap leaching, but usually results in higher gold recovery.

Refractory gold ores

Gold has traditionally been a "resource-limited" commodity. That means if you were lucky enough to find a gold deposit of commercial grade, you could make money working it. The extractive technologies were well established, inexpensive, and in the public domain. In recent years, however, it has become harder and harder to find easily extractable gold deposits in the United States due to the extremely intense exploration efforts previously conducted. Current exploration is revealing more and more ores that do not respond to either gravity concentration or cyanide leaching with or without preconcentration. Such ores are said to be *refractory*. The gold still occurs in these ores as the metal, with poor leaching response being caused by the host minerals.

The major categories of refractory gold ores are pyritic, arsenopyritic, carbonaceous, and mixed (a combination of two or more of the foregoing). Ores classified as pyritic typically contain up to 8 percent sulfur and less than 0.1 percent arsenic. Gold recovery by cyanidation is usually proportional to fineness of grind, but often less than 70 percent is recoverable even with the finest grinds achievable in standard milling equipment. The gold occurrence is often as very fine inclusions in the pyrite, 0.1 to 0.5 μm in diameter or even smaller.

Arsenopyritic ores typically contain up to 5 percent sulfur and up to 2 percent arsenic. They can also contain complex sulfosalts with up to 0.2 percent antimony, lead, bismuth, silver, copper, zinc, and/or mercury. As in pyritic ores, the gold may be very finely disseminated, but arsenical ores are also significant oxygen consumers during cyanide leaching. The reaction

$$4FeAsS + 4Ca(OH)_2 + 11O_2 \rightarrow 4FeSO_4 + 4CaHAsO_3 + 2H_2O$$

competes directly for oxygen used in the basic cyanidation reaction

$$4Au + 8CN^- + O_2 + 2H_2O \rightarrow 4Au(CN)_2^- + 4OH^-$$

Carbonaceous ores contain sufficient organic matter in an active form to adsorb the gold from solution as it is being leached. Graphitic carbon does not usually present a serious problem, but ores containing kerogen or humates are more troublesome. Sometimes the carbonaceous matter can be "blinded" by addition of kerosene or fuel oil, but this is not always totally effective.

There are other types of refractory gold ores such as tellurides, but they are relatively rare and will not be addressed here.

Oxidation Processes

There are three principal techniques for preoxidation of refractory gold ores: roasting, aqueous pressure oxidation, and microbial oxidation. Many others have been proposed and tested, but are not yet commercialized. Chemical oxidation, such as with chlorine, is commercially practiced but of very limited application. We consider only roasting, pressure oxidation, and microbial oxidation in this chapter.

Roasting

The objective of roasting is to convert the sulfide minerals to their oxides, thus rendering the gold available to subsequent cyanidation:

$$4FeS_2 + 11O_2 \rightarrow 2Fe_2O_3 + 8SO_2$$

Common commercial roasters are rotary kilns, multiple-hearth furnaces, and fluidized-bed reactors. Kilns are appropriate for coarse ore, multiple hearths for both coarse and fine ore, and fluidized beds for fine ore and concentrates. Temperature and atmosphere control during roasting are critical, as it is possible for some of the fine gold particles to become trapped by agglomeration or recrystallization of the gangue minerals. Roasting is not the best choice for all refractory ores, as subsequent cyanide leaching may yield as low as 80 percent recovery even with optimum roasting conditions. Air emissions from roasters are also of concern, and gas cleaning equipment often is a very significant item in both capital and operating costs.

Pressure oxidation

This aggressive pretreatment is conducted at temperatures of 180 to 210°C. It results in essentially complete oxidation of the sulfide minerals and carbon. The reactions are

$$4FeS_2 + 15O_2 + 8H_2O \rightarrow 2Fe_2O_3 + 8H_2SO_4$$

$$4FeAsS + 14O_2 + 12H_2O \rightarrow 4FeAsO_4 \cdot 2H_2O + 4H_2SO_4$$

The oxidation of both pyrite and arsenopyrite generates sulfuric acid, which must be neutralized prior to cyanidation. The other principal disadvantages of pressure oxidation are the obvious ones of operating autoclaves, high-pressure pumps, and pressure letdown equipment. An advantage is that subsequent gold recoveries by cyanidation are typically quite high (greater than 90 percent).

Bacterial oxidation

The reaction during bacterial oxidation are similar to those in pressure oxidation, except that the products include soluble iron:

$$4FeS_2 + 6O_2 + 2H_2O \rightarrow 2Fe_2(SO_4)_3 + 2H_2SO_4$$

This method has the obvious advantage of not requiring high-temperature or high-pressure equipment. The main disadvantages—continuous cooling and air injection over a long period—emanate from the long retention time requirement. Much research and development effort has been directed toward shortening this time.

Development of Economic Comparisons

A base-case flowsheet, shown in Fig. 8.1, will be used to assess the relative economics of roasting, pressure oxidation, and bacterial oxidation. The assumptions and design parameters are as follows:

1. A 1000 TPD concentrator consisting of a three-stage crushing circuit, 3 days of fine ore storage, a rod-and-ball mill circuit with cyclone closure grinding to 80 percent minus 65 mesh (work index of ore = 14), a classical single-product flotation plant with rougher, scavenger, and cleaner circuits, concentrate thickening, tailings placement, and water return systems.

2. Concentrate oxidation for 90% sulfur removal by either roasting, pressure oxidation, or bacterial oxidation. The concentrate, 100 tons per day, is assumed to contain 8% sulfur as FeS_2.

3. A typical carbon-in-leach plant with carbon stripping and bullion production.

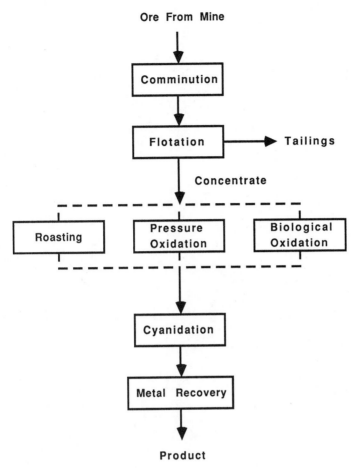

Figure 8.1   Flowsheet for base case.

Excluded from capital costs will be tailings impoundment (which is highly site-specific), primary power, primary water, and access roads beyond 5 miles. Excluded from operating costs will be depreciation, taxes, home office costs, and sales costs. The accuracy of the estimates is believed to be ±25 percent for capital costs and ±15 percent for operating costs.

All elements of the base case are held constant for comparative purposes except the concentrate oxidation section.

Roasting

The flowsheet for roasting is shown in Fig. 8.2. Major assumptions and design parameters are

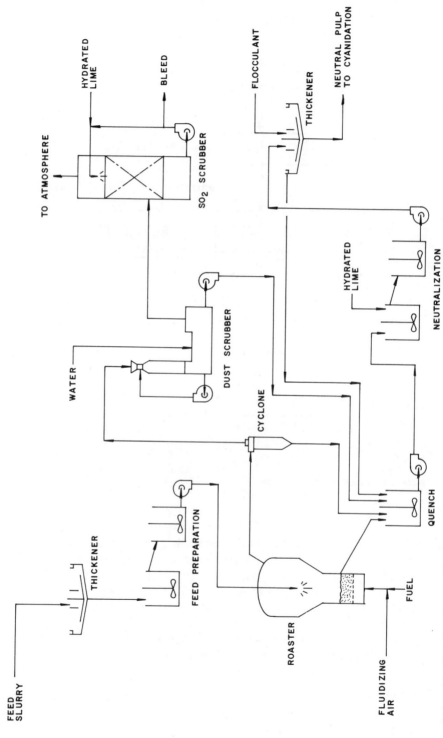

Figure 8.2 Flowsheet for roasting process.

- Fluidized-bed roasters rather than kilns or multiple hearths
- Roasting temperature 1200°F
- Twenty percent excess air for oxidation of FeS_2 to Fe_2O_3 and SO_2
- Fuel is No. 3 fuel oil, 19,900 Btu/lb

Pressure oxidation

The pressure oxidation flowsheet is shown in Fig. 8.3. Major assumptions and design parameters are

- Autoclave retention time 90 min
- Reaction temperature 375°F
- One hundred percent excess oxygen for oxidation of FeS_2 to $Fe_2(SO_4)_3$.

Biological oxidation

The biological oxidation flowsheet is shown in Fig. 8.4. Major assumptions and design parameters are

- Retention time in contactors 4 days
- Reaction temperature 95°F
- Thirty-two hundred percent excess air for oxidation of FeS_2 to $Fe_2(SO_4)_3$

Assumptions for all processes

- Flotation concentrate contains 8 percent sulfur as FeS_2.
- Ninety percent sulfur oxidation by process.
- All acid constituents neutralized with lime before cyanidation.
- Flocculant consumption 0.05 lb/ton of concentrate.

Results of Economic Comparisons

Capital costs of the three concentrate oxidation plants are shown in Table 8.1. Bacterial oxidation shows the lowest capital requirement, with pressure oxidation next and roasting the highest. Operating costs are shown in Table 8.2. Here bacterial oxidation appears to be the medium-cost alternative, with pressure oxidation lowest and roasting the most expensive.

Total capital costs for the base case, including concentration, concentrate oxidation, and carbon-in-leach (CIL), are presented in Table 8.3. Operating

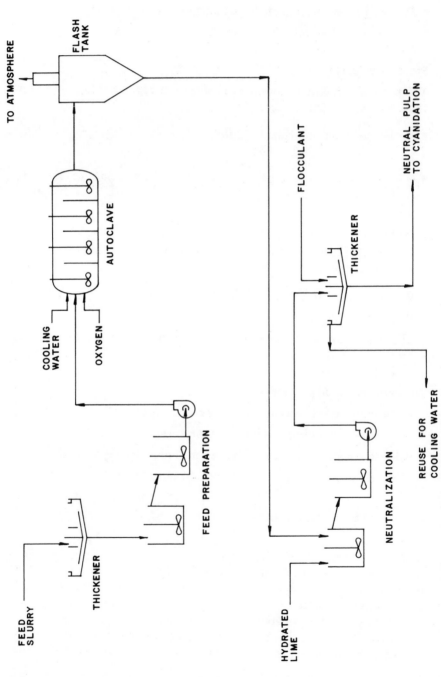

Figure 8.3 Flowsheet for pressure oxidation process.

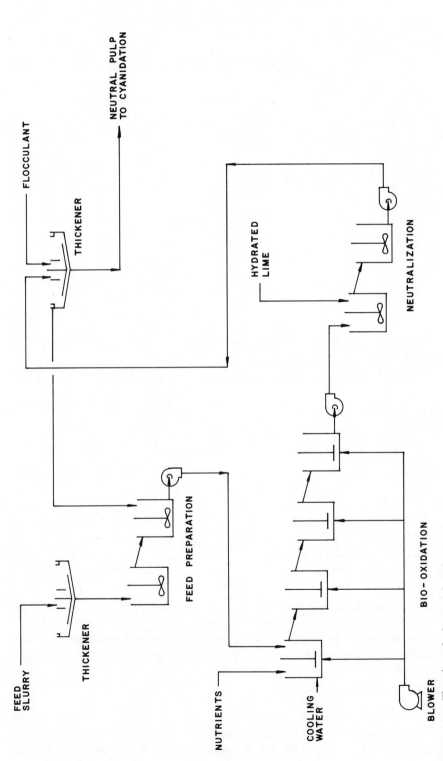

Figure 8.4 Flowsheet for bacterial oxidation process.

TABLE 8.1   Plant Cost Estimate for Refractory Gold Ore in a 100 TPD* Pretreatment Plant

(Capital costs, in $1000)

Item	Roasting	Pressure oxidation	Bacterial oxidation
Process equipment	$1,830	$1,420	$1,130
Foundations	220	180	140
Work platforms	250	110	90
Installation	550	320	230
Piping	830	740	510
Electrical	200	230	150
Instrumentation	170	110	60
Insulation	140	100	40
Painting	70	60	40
Miscellaneous	110	110	100
Total plant cost	$4,370	$3,380	$2,490

*TPD = tons per day.

TABLE 8.2   Plant Cost Estimate for Refractory Gold Ore in a 100 TPD* Pretreatment Plant

(Operating costs, in $1000/year)

Item	Roasting	Pressure oxidation	Bacterial oxidation
Labor	$ 590	$ 340	$ 320
Operating supplies	90	50	50
Maintenance supplies	50	50	20
Reagents	460	880	460
Power	100	50	630
Fuel	220		
Miscellaneous	70	70	30
Total operating cost	$1,580	$1,440	$1,510
$/ton of concentrate	45.10	41.10	43.10
$/ton of ore	4.50	4.10	4.30

*TPD = tons per day.

costs are shown in Table 8.4. Although the cost differences are within the accuracy of these estimates, they do suggest that bacterial oxidation has a capital cost advantage with the assumptions and methods used. Its operating costs are between the other two, probably too close to call.

Discussion

This study shows that microbial oxidation may be somewhat more economical than either roasting or pressure oxidation as a pretreatment

TABLE 8.3   Plant Cost Estimate for Refractory Gold Ore in a 1000 TPD* Base Case
(Capital costs, in $1000)

Item	Roasting	Pressure oxidation	Bacterial oxidation
Concentrator plant	$ 2,840	$ 2,840	$ 2,840
CIL plant	1,640	1,640	1,640
Pretreatment plant	4,370	3,380	2,490
Site facilities	3,490	3,490	3,490
Total physical plant	12,340	11,350	10,460
Design, eng. & manage.	1,240	1,160	1,050
Contingency	2,090	1,980	1,800
Working capital	1,130	1,110	1,120
Total plant cost	$16,800	$15,600	$14,430

*TPD = tons per day.

TABLE 8.4   Plant Cost Estimate for Refractory Gold Ore in a 1000 TPD* Base Case
(Operating costs, $/ton of ore)

Item	Roasting	Pressure oxidation	Bacterial oxidation
Concentrator plant	$ 5.70	$ 5.70	5.70
CIL plant	1.70	1.70	1.70
Pretreatment plant	4.50	4.10	4.30
G & A	1.70	1.70	1.70
Total operating costs	$13.60	$13.20	$13.40

*TPD = tons per day.

method for refractory gold ores. Indeed this is probably true for some, perhaps many, situations. But one of the first principles of mineral process development is that all ores are different. For purposes of our study a "typical" ore in a typical location was assumed to respond equally well to all three pretreatment methods with respect to downstream cyanidation amenability. This would not be true in the real world, and in fact extensive testing of these alternatives is necessary for any specific case. The relative economics are also quite sensitive to the assumptions used. Some refractory ores, for example, do not require 90 percent oxidation for good gold extraction. Bacterial oxidation could clearly be the process of choice in these cases if retention time is reduced.

There are also many noneconomic factors influencing process selection. Roasting is the most proven, established technology of the three, and bacterial oxidation is the least so. Many operators and financing groups wish to go with established commercial technology at all times.

The environmental permitting procedure, of late a critical item in the development of any new mineral project, varies widely from place to place and is highly dependent on the proposed process. Again, the local authorities often are more receptive to a permit application involving established processing methods with which they are familiar.

So from the mineral engineer's standpoint, it would appear at this time that bacterial oxidation of refractory gold ores is a new and valuable tool in our kit, not a panacea for all metallurgical problems. As it is further developed and refined, and as it gains commercial acceptance, it may yet become more valuable. Roasting and pressure oxidation are relatively mature technologies, and future improvements in them are likely to be incremental rather than revolutionary. Biotechnology as applied to mineral processing is, however, still in its infancy, and major breakthroughs may still be ahead of us.

Biosorption

This section deals with the scientific basis for metal binding by living and dead microbial cells. Specific examples of the use of algal, fungal, and bacterial biomass in removing metals, including radionuclides, from solution are examined, and engineering and economic considerations are discussed. This biotechnology offers promise in the treatment of some low-grade industrial waste streams and in the remediation of environmental metal pollution.

Metal-Binding Capacity of Bacterial Surfaces and Their Ability to Form Mineralized Aggregates

Robert J. C. McLean
Terry J. Beveridge

Introduction

Procaryotes (archaebacteria and eubacteria) represent the earliest and most ubiquitous form of life on this planet, having originated approximately 3.6 billion years ago (Barghoorn and Tyler, 1965; Schopf and Walker, 1983). The diversity of environments colonized by bacteria is truly impressive and encompasses underground rock formations (Balkwill and Ghiorse, 1985), hydrothermal deep sea vents (Jannasch and Taylor, 1984), areas of extreme acidity, alkalinity (Padan, 1984), and xeric cold and warm environments (Friedmann and Ocampo-Friedmann, 1984). Bacteria are also found in association with higher forms of life where they can exhibit a pathogenic, commensal, or symbiotic relationship with their host. In light of this, the successful design and extreme adaptability of bacteria become readily apparent (Beveridge, 1988).

Very few (if any) regions of the earth's surface are not affected in some way by bacteria. One need only consider the microbial cycling of elements such as N, C, S, P, Mn, Fe, and Hg (Brock et al., 1984) and the microbially assisted weathering of all mineral surfaces, including those found in such inclement environments as polar and hot deserts (Friedmann and Weed, 1987). Microfossils, the mineralized remains of bacteria in ancient organic-rich cherts and shales, attest to the antiquity of prokaryotes and emphasize their ability to effectively immobilize metals and silica from the environment into various mineral forms (Barghoorn and Tyler, 1965; Beveridge et al., 1983; Schopf and Walker, 1983; Ferris et al., 1986, 1987a, 1988). Clearly, bacterial mineralization is not only limited to rock and microfossil formation. For example, struvite ($MgNH_4PO_4$) and carbonate-apatite [$Ca_{10}(PO_4)_6 CO_3$] calculi result from bacterial mineralization processes on teeth (Driessens et al., 1985) and within the urinary tract (McLean et al., 1988) of humans.

Bacteria require diverse metals such as Na, K, Mg, Ca, Fe, Co, Cu, Mo, Mn, Ni, and Zn as inorganic nutrients (Krueger and Kolodziej, 1976; Ingraham et al., 1983; Hausinger, 1987). Because excessive quantities of these or other metals may be inhibitory or even lethal to the organisms, bacteria have developed measures to control or adapt to the metal concentrations in their local environment. One such mechanism is the ability of the cell surface to bind metals. The anionic character of the bacterial surface acts like a sponge in that it can soak up metal ions during periods of abundance. For toxic metals, this effectively stops them from entering the protoplast and killing the organism. For nutritive metals, it is possible that during periods of starvation, ions may be released to specific high-affinity transport systems for assimilation into the cell. Metals in the vicinity of bacteria are often subjected to chemical alterations. Several heavy metals which might otherwise be inhibitory to the cell can be detoxified by precipitation as insoluble salts by microbially produced sulfide or phosphate (Aiking et al., 1984). Methylation of Hg (Korthals and Winfrey, 1987) and Sn (Gilmour et al., 1987) increases the volatility of these elements, thus assisting in their removal from the vicinity of the organisms. Oxidation of Fe (Silver, 1987) and Mn (Kepkay, 1985) provide energy for some chemolithotrophic bacteria.

In this chapter we examine the metal-binding ability of bacterial surfaces. In doing so, we describe the structure of the bacterial surface, the chemistry and availability of metals in the aqueous environment of bacteria, the interaction of bacteria with metals, and the geological implications of this interaction. We conclude by establishing the significance of metal binding in nature and its potential role in biotechnology.

Bacterial Surfaces

The importance of the bacterial surface to the cell cannot be overemphasized. Surfaces offer the cell chemical and physical protection, determine cell shape, play a role in growth and division, effect diffusion of nutrients and wastes, and influence adhesion of the cell to surfaces. It is through its surface that the cell first encounters the environment.

Movement of materials into or out of the bacterial cell is controlled by a number of active (permeases, enzyme systems) or passive systems (porins, iron binding proteins) based essentially on the chemistry and structure of the surface. Bacterial surfaces are not inert chemical structures. They represent a dynamic system in which the various components are synthesized, assembled, modified, and finally broken down by autolysins and sloughed off into the environment.

For the purposes of this chapter, we shall define the bacterial surface as commencing at the plasma membrane (PM). In the case of eubacterial L-forms and the *Tenericutes* (Krieg and Holt, 1984), this represents the extent of the cell surface. All other bacteria possess a wall which is external to the PM.

One of the most striking and fundamental features of eubacteria, is that they can be divided into two broad classes, depending on their reaction to the Gram stain. Those cells which are able to retain the crystal violet iodide staining complex appear purple when viewed under the light microscope and are referred to as gram-positive. Those cells that do not retain crystal violet take on the red color of the safranin or carbolfuchsin counterstain and are referred to as gram-negative (Beveridge, 1981). In 1983, Beveridge and coworkers substantiated the conclusion that this reaction to the Gram stain was primarily determined by the thickness of the peptidoglycan (Pg) layer of these organisms (Beveridge and Davies, 1983; Davies et al., 1983). The thick Pg layer of the gram-positive cell retained the crystal violet iodide stain, whereas it diffused out of gram-negative cells due to the poor retention power of the thin Pg layer in these organisms. This division of eubacteria into two fundamental groups is very apparent upon ultrastructural examination of the cell walls (Fig. 9.1) (Beveridge, 1981; Krell and Beveridge, 1987). In addition to the eubacteria, there is a new so-called urkingdom referred to as the archaebacteria, which, among other dissimilarities to eubacteria, possess cell walls chemically and structurally unlike those of the eubacteria.

In this next section we summarize the basic structures of the gram-positive, gram-negative, and archaebacterial cell surfaces. The chemistry of these surfaces is responsible for their anionic character and their ability to bind cationic metals. An exhaustive treatment of this subject is beyond the scope of this review. The interested reader is referred to reviews in this general area by Beveridge (1981) and Krell and Beveridge (1987).

Gram-positive eubacterial cell wall

The distinguishing ultrastructural feature of gram-positive bacteria such as *Bacillus subtilis* is the thick wall seen external to the PM. In *B. subtilis* it may be 20 to 30 nm or more thick (Fig. 9.1a).

Pg is the rigid structural component of the eubacterial wall which is

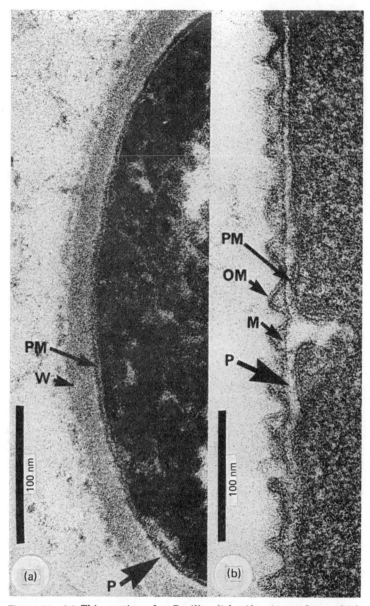

Figure 9.1 (*a*) Thin section of a *Bacillus licheniformis* envelope, which shows the thick amorphous cell wall which is representative of the gram-positive variety. (*b*) Thin section of an *Aquaspirillum serpens* envelope which is representative of the gram-negative variety. PM = plasma membrane; OM = outer membrane; M = murein or peptidoglycan layer; W = wall; P = periplasmic space. The magnification bars in each part represent 100 nm. (*Figures reproduced from Int. Rev. Cytol. 17:15–88 with permission of Academic Press.*)

primarily responsible for maintenance of cell shape and resistance to osmotic pressure. This is particularly evident in that Pg sacculi isolated and purified from bacteria retain the shape of the original cell. In addition, enzymatic removal of Pg from cells in a hypertonic solution causes a loss of cell shape and the formation of spherical, osmotically fragile protoplasts.

The [$\rightarrow$ 4 GlcNAc (β1 $\rightarrow$ 4) MurNAc 1 $\rightarrow$]$_n$ backbone of Pg was until recently thought to be of relatively constant composition among eubacteria (Schleifer and Kandler, 1972). Hayashi et al. (1973) found cell walls of *Bacillus cereus, B. subtilis,* and *B. megaterium* to contain numerous glucosamine residues which were not N-acetylated, resulting in their resistance to the Pg-degrading enzyme lysozyme. Artificial acetylation of these walls with acetic anhydride restored lysozyme sensitivity. Zipperle et al. (1984) noted similar lysozyme resistance in the walls of *Bacillus anthracis, B. cereus,* and *B. thuringiensis.* In these organisms, approximately 88 percent of the glucosamine residues and 34 percent of the muramic acid residues were not N-acetylated. The variance of N-acetylation is not restricted to gram-positive types, since Schmelzer et al. (1982) noted a partial lack of N-acetyl substitution of glucosamine in the Pg of *Rhodopseudomonas viridis,* a gram-negative eubacterium. It would thus appear that the [GlcNAc-MurNAc] Pg backbone is not as constant within eubacteria as once believed. It is also interesting that Zipperle et al. (1984) noted that newly formed septa were much more susceptible to lysozyme and mutanolysin than was the rest of the wall, implying some structural differences in Pg in these two regions of the wall.

Pg consists of a linear polymer of chains which are roughly 50 repeating units long, are oriented parallel to the cell membrane, and are generally perpendicular to the longitudinal axis of the cell (Archibald et al., 1973). Usually a short 4 to 5 amino-acid linear peptide chain is bound by its N-terminal amino acid to the carboxyl group of each MurNAc group with an amino-acid sequence consisting of alternating L and D amino acids, with L alanine at position 1 and D alanine at position 4. Amino acids at positions 2 and 3 are variable, although *meso*-diaminopimelic acid (dpm) is a common constituent at position 3 in most gram-negative eubacteria and some genera of gram-positive eubacteria, such as *Bacillus* sp. Cross-linking between individual peptidoglycan strands occurs via these peptide side chains, and, here again, gram-positive bacteria show quite a variety in the type of linking units and degree of cross-linking. The cross-linking format varies from a direct peptide bond between adjacent peptide side chains to a linking unit of variable length and composition which forms a peptide bridge. Cross-linking occurs between amino acids at either positions 3 and 4 or 2 and 4 of adjacent peptide subunits (Schleifer and Kandler, 1972).

With the presence of at least one exposed carboxyl group on the terminal amino acid of the peptide chain, there is a potential for at least one negative charge per GlcNAc-MurNAc repeating unit in an individual Pg strand. Amino acids possessing free carboxyl groups such as glutamic acid, aspartic acid, or dpm can also fill other positions, thus increasing total electronegativity in the chains. When one considers that only 10 to 50 percent of these peptide side chains are involved in cross-linking between strands (Krell and Beveridge, 1987), the anionic character of the wall contributed by the Pg fraction alone becomes readily apparent. The degree and nature of Pg cross-linking also influence the rigidity and permeability of the wall.

Pg accounts for approximately 50 percent of the *B. subtilis* wall (Beveridge, 1981). Other polymers are also incorporated and can have a profound effect on wall structure and function. Teichoic acids (TA) and teichuronic acids (TUA) are incorporated into the walls of gram-positive bacteria, where they become linked to a variable number of MurNAc residues (Ward, 1981) upon completion of Pg cross-linking (Fischer and Tomasz, 1984). This linkage may either be a direct phosphodiester bond between TA and MurNAc, as evidenced in *Staphylococcus lactis* (Button et al., 1966), or a linkage unit, such as the disaccharide in *B. subtilis* or *B. licheniformis* (Kaya et al., 1984). The importance of these polymers to the cell can be demonstrated in that *B. subtilis* mutants, which have lost the ability to produce TA, become misshapen and form irregular spheres due to the loss of autolysin function (Boylan et al., 1972; Doyle and Koch, 1987; Vitkovic, 1987). Mg^{2+} and Na^+ binding capacity of these walls is also greatly diminished. TA of *B. licheniformis* was also shown to be involved in Mg^{2+} binding and the maintenance of the correct ionic environment for membrane enzymes (Hughes et al., 1973). Purified TA from *B. subtilis* 168 possesses a "rigid-rod" conformation when suspended in distilled H_2O, whereas in the presence of Na^+, TA forms a random coil due to a change in secondary structure. Mg^{2+} had little effect on TA structure, but TA bound with Mg showed a higher sedimentation velocity than did TA bound with Na^+, indicating a possible ionic bridging of TA molecules in the presence of Mg. Doyle et al. (1974) proposed that this alteration in TA structure in the presence of high (> 0.3 M) NaCl concentration may be partially responsible for the lowered *B. subtilis* wall permeability under these conditions.

TA can constitute from 30 to 50 percent of the gram-positive wall (Lambert et al., 1977). TA are ester-linked ribitol or glycerol phosphate chains. Histochemical localization of the polyglucosyl, glycerol TA in *B. subtilis* with the lectin concanavalan A, showed most of the TA to be exposed on the outer surface where it extends a short distance from the cell (Birdsell et al., 1975). TA could also be detected in this manner within the wall fabric upon prior partial digestion of the Pg strands with lysozyme (Doyle et al., 1975). TA, associated with the cell membrane as well as the wall, is referred to as lipoteichoic acid (LTA) (Wicken and Knox, 1975). In some cases, LTA traverses the wall and can be exposed as antigens on the cell surface (Van Driel et al., 1973; Wicken and Knox, 1975). As is the case with TA (Heptinstall et al., 1970; Lambert et al., 1977; Ward, 1981), the composition of LTA varies among bacterial strains (Iwasaki et al., 1986).

Growth of gram-positive cells under conditions of phosphate limita- tion results in the repression of TA synthesis. Teichuronic acids are produced instead (Hussey et al., 1978; Kruyssen et al., 1980, 1981; Ward, 1981). In this case the anionic character of TUA is contributed by uronic acid moieties such as the glucuronic acid present in the TUA of B. licheniformis ATCC 9945 (Lifely et al., 1980). Restoration of an adequate phosphorous supply derepresses TA synthesis in B. subtilis, and it then replaces TUA as a new wall is generated. Hussey and co- workers (1978) found that while TA was preferentially incorporated into the wall fabric when phosphate was present, TUA production was maintained to some extent regardless of the nutritional state of the cell. When chemostat grown B. subtilis 168 cells were microscopically examined under varying conditions of phosphate limitation using TA- specific phage markers, Lang et al. (1982) found that regulation of TA and TUA synthesis was expressed equally among all members of the cell population and that individual cells did not make one type of poly- mer exclusively. B. licheniformis differs from B. subtilis in that its walls always contain significant amounts of TUA even when phos- phate is abundant (Beveridge, 1981). Regardless of nutritional status, a continual turnover of wall material is maintained. de Boer et al. (1981) noted that anionic wall polymers were continually being shed into the supernatant from chemostat grown B. subtilis subsp. niger cells. In the case of Mg-limited cells, TA was secreted, whereas in the case of phosphate limitation, TUA would be secreted.

Most studies to date have stressed the importance of PO_4^{3-} avail- ability as the prime regulating factor in TA or TUA synthesis. Other factors, such as glucose or Mg limitation, play insignificant roles in the synthesis of these polymers. However, in Streptococcus faecium, inhibition of protein synthesis with chloramphenicol or deprivation of valine, an essential amino acid, results in an increased carbohydrate substitution of LTA, thus elevating its mass-to-charge ratio (Kessler et al., 1983).

In addition to Pg, TA, TUA, and LTA, several other minor wall polymers have been documented. These include a neutral poly- saccharide containing N-acetyl glucosamine, N-acetylmannosamine, N-acetylgalactosamine, and glucose from B. cereus, an acidic glucosamine-galactose-rhamnose-glycerol-phosphate polysaccharide also from *B. cereus* (Kojima et al., 1985), and an *N*-acetylgalactosamine-phosphate-glucose polymer of *B. subtilis* 168 (Shibaev et al., 1973). In the latter two cases, the presence of phosphate would give these polymers an anionic nature. Neutral wall constituents also exist and, depending on the wall chemotype, can provide a substantial proportion of the wall. For example, mycolic acids and corynemycolic acids (long-chain β-hydroxyl, α-branched fatty acids) can make up to 50 percent of the dry weight mass of *Mycobacterium* and *Corynebacterium* walls (Beveridge, 1981). But since these neutral polymers are unwettable and therefore less accessible to aqueous metalloions, they will not be discussed further.

Variations in wall composition can also affect charge distribution within the gram-positive wall. Using cationized ferritin as an ultrastructural probe, Sonnenfeld et al. (1985) were able to show an asymmetrical anionic charge distribution in *B. subtilis* walls. *B. licheniformis* walls with their higher proportion of TUA demonstrated a more even distribution.

In summary, the gram-positive cell envelope is characterized by the presence of a thick (20 to 30 nm) wall containing Pg, TA, LTA, and/or TUA. Carboxyl groups and phosphate groups in these compounds give the wall an anionic character. The availability of wall anionic groups to cationic metals is determined by several factors, including the growth state of the cell which influences the accessibility of these groups. This includes the degree and nature of Pg cross-linking, the number, type, conformation, and location of secondary wall polymers, and the presence of internal ionic bonds between anionic and cationic sites resident within the wall fabric.

Gram-negative eubacterial cell wall

The major differences in structure between gram-positive and gram-negative walls become readily apparent upon ultrastructural examination (Fig. 9.1). In comparison to the gram-positive wall, the most obvious change is the lack of a thick Pg layer and the presence of a second membrane within the envelope (Fig. 9.1*b*). This second or outer membrane (OM) is separated from the inner or plasma membrane by a region known as the periplasm. The main structural component of the gram-positive wall, Pg, is still present, but is located in the periplasm (Murray et al., 1965), where it presumably forms a monolayer (Beveridge, 1981). Except in unusual cases, TA, TUA, and LTA are not found (Beveridge, 1988). Instead, some gram-negative bacteria such as *Salmonella, Serratia,* and *Escherichia coli* possess a lipoprotein which is bound on average to every tenth repeating

GlcNAc-MurNAc unit of peptidoglycan (Braun et al., 1970). This lipoprotein is thought to play a role in binding the Pg and OM together (Braun, 1975). The fatty-acid acyl chains of the lipoprotein are cemented into the hydrophobic domain of the membrane, whereas the N-terminus of the protein is covalently bonded into the Pg. Only one-third of the resident copies are attached to the Pg. Other gram-negative bacteria, such as *Proteus mirabilis* and some aquaspirilla, do not possess this lipoprotein (Braun et al., 1970; Gmeiner, 1979); others, such as *Pseudomonas aeruginosa,* possess them but do not covalently bond them to Pg. It is unclear what binds the Pg and OM in these cells, but, presumably, ionic bonding and salt bridging (mediated by Mg^{2+}) play a significant role.

The region between the OM and PM of gram-negative bacteria contains a thin layer of Pg. This Pg differs from that of gram-positive organisms in that it is almost exclusively the A1γ chemotype (Schleifer and Kandler, 1972). Pg in gram-negative bacteria is generally much less abundant and, depending on its conformation, may be present only as a monolayer. It serves the same structural function as in gram-positive bacteria. Anionic groups in gram-negative Pg do bind metal (Hoyle and Beveridge, 1983), but the overall quantity of metal bound is considerably less due to the reduced abundance of Pg.

The presence of two membranes in gram-negative cell envelopes is confirmed by freeze-fracture studies (Bayer and Remsen, 1970). However the greater tendency of the PM to be cleaved by this technique (Beveridge, 1981) suggests differences in membrane structure. This is borne out by biochemical analysis. The PM of gram-negative bacteria bears a close resemblance to the PM of gram-positive organisms. It is a typical phospholipid-containing bilayer. The polar head groups in the PM of *E. coli* and *Salmonella typhimurium* are separated by 5.4 to 5.5 or 6.4 to 6.5 nm, depending on whether the membrane is above or below its phase transition temperature (Cronan et al., 1987). Major PM phospholipids in *E. coli* are phosphatidylethanolamine (70 to 80 percent), phosphatidylglycerol (5 to 15 percent), and cardiolipin (5 to 15 percent). The exact composition of the membrane fatty-acid chains in *E. coli* is influenced by the environment and the organism's state of growth (Cronan and Vagelos, 1972). As in gram-positive bacteria, many enzymes involved in respiration and biosynthesis, as well as transport proteins, are localized here.

The OM of gram-negative bacteria is considerably different from the PM (Nikaido and Vaara, 1987). Phospholipids constitute a much smaller fraction of the OM, contain larger quantities of phosphatidylethanolamine and saturated fatty acids than do those of the PM (Lugtenberg and Peters, 1976), and are placed on its inner leaflet. Lipopolysaccharide (LPS), a compound unique to gram-negative eubacteria, is a major constituent of the OM (Nikaido and Vaara, 1987). This molecule consists of three distinct chemical regions: lipid A, the core polysaccharide, and the O side chain. Lipid A is hydrophobic and is therefore anchored in the OM. It consists of a phosphorylated glucosamine dissacharide which is esterified with fatty acid. Attached to the lipid A is the core oligosaccharide. Two sugars commonly associated with this core region include heptose (L-glycero-D-mannoheptose) and KDO (3-deoxy-D-mannooctulosonic acid). The outermost constituent of LPS is a polysaccharide chain, called the O-specific side chain. In *Pseudomonas aeruginosa* PAO1, Lam et al. (1987) showed the O-side chains to extend 20 to 40 nm away from the cell. While the lipid A and core polysaccharide regions are quite similar among enteric gram-negative bacteria, the sugar composition of the O-side chain is generally very species- and strain-specific. Serological strain typing of many enteric bacteria such as *Salmonella* sp. is done with antisera raised against this O-specific side chain (Krieg and Holt, 1984). Agar colonies of gram-negative bacteria with intact LPS are referred to as smooth strains due to their smooth, glistening appearance. Mutants lacking some or all of the O-side chain or the O-side chain and core polysaccharide region are referred to as rough and deep rough strains, respectively. To our knowledge, the complete lack of LPS (i.e., a "deep–deep rough mutation") represents a lethal mutation, because no isolates of this phenotype have been described.

Electron microscopy of plasmolyzed gram-negative cells shows zones of adhesion between the outer and plasma membranes (Bayer, 1968). These "Bayer adhesion zones" are thought to represent regions where new OM growth and LPS incorporation occur at points where the molecule diffuses rapidly throughout the OM (Mühlradt et al., 1973; Bayer et al., 1982).

Average spacing between LPS molecules in the OM is 3 to 4 nm (Mühlradt et al., 1974). LPS is asymmetrically located in the outer leaflet of the OM (Mühlradt and Golecki, 1975; Funahara and Nikaido, 1980). This asymmetrical distribution of LPS is maintained by the Pg layer, since in its absence LPS occurs on both OM faces (Mühlradt and Golecki, 1975).

LPS molecules are amphilytes due to the hydrophobic fatty acids and hydrophilic phosphate groups in the lipid A portion. LPS interacts with other molecular constituents within the OM by means of a variety of noncovalent bonds, but Mg^{2+} and Ca^{2+} are essential for its proper conformation (Shands and Chun, 1980). The bonding of these metals is affected by chelating agents such as EDTA (Hancock, 1984) or the presence of competing cations such as aminoglycoside antibiotics (Martin and Beveridge, 1986; Walker and Beveridge, 1988). In either case the integrity of the OM is disrupted and LPS is the major component released.

Iron deprivation of gram-negative organisms such as *Ps. aeruginosa* induces the formation of high-molecular-weight proteins in the outer membrane as well as low-molecular-weight siderophores (Neilands, 1981; Brown and Williams, 1985). These siderophores possess a voracious appetite for iron [particularly Fe(III)] being able to remove it from human iron-chelating glycoproteins such as transferrin. This is quite significant in that the free-iron concentration in the body of humans is about 10^{-18} M (M. R. W. Brown, personal communication; Griffiths, 1983). The iron-binding proteins, which are derepressed under conditions of iron limitation, generally act as receptors for iron-siderophore complexes; they also transport this iron into the cell (Neilands, 1981; Griffiths, 1983). In order for bacteria to survive in such an iron-poor environment, the need for an efficient iron-scavenging system becomes quite apparent. While we acknowledge the great importance of siderophore production and OM iron-binding proteins, these compounds are considered, at present, to be primarily of medical importance and will not be dealt with further. The interested reader is referred to the review by Brown and Williams (1985). Numerous other proteins have also been described in the OM. For a recent survey of these proteins and their roles in structure and membrane transport, see the review by Nikaido and Vaara (1987).

Surface arrays

In nature, the walls of many bacteria possess a layer of regularly arrayed proteins on their surfaces (Fig. 9.2). These proteins are synthesized within the cell and exported to the surface, where they self-assemble into a two-dimensional paracrystalline layer (S-layer). In some cells, the S-layer can represent up to 10 percent of the total cellular protein. Although its importance has not been established in all bacteria, the S-layer has been implicated in the virulence of the fish pathogen *Aeromonas salmonicida* (Kay et al., 1981). Here it presumably protects the organism by lowering the permeability of the wall to host serum killing factors (Stewart et al., 1986). In other bacteria, the S-layer has also been implicated in resistance to predation and in facilitating adhesion to surfaces (Sleytr and Messner, 1983).

S-layers in at least some bacteria require metals such as Ca^{2+}, Sr^{2+} (Beveridge and Muray, 1976; Buckmire and Murray, 1976), or Mg^{2+} (Beveridge, 1979). Presumably these divalent cations are involved in either salt bridging or protein conformational changes which affect protein-protein or protein-wall interactions. Other than the role of metals in S-layer stabilization, little work has been done to study the

Figure 9.2 Negative stain of the S-layer of *Aquaspirillum putridiconchylium*; bar = 50 nm.

influence of S-layers on cell surface metal binding. One study showed that *Sporosarcina ureae* was capable of growth in the presence of heavy metals because they could be immobilized by aggregating with S-layer protein subunits (Beveridge, 1984). With the widespread occurrence of S-layers among natural bacterial isolates (Sleytr and Messner, 1983), more work needs to be done on the interactions of these proteins with metals.

Extracellular polymers

In many bacteria, capsules or slime represent the outermost layer of the cell surface. It is with this region that these cells first make contact with their environment. Capsules and slime are differentiated from one another in that the former are extracellular polymers tightly associated with the cell surface and can exclude a negative stain such as India ink when viewed by light microscopy. Slime is more loosely associated with the cell surface, often sloughs off the cell into the external milieu, and does not exclude India ink. Recently all extracellular bacterial polymers (including capsules and slime but excluding LPS, TA, LTA, and TUA) have been broadly grouped together and called the glycocalyx (Costerton et al., 1981). Since not all bacterial capsules are made entirely of carbohydrates, we will continue to use the older term capsule.

Capsules differ widely in composition among bacterial species and strains. They consist of linear polymers of polysaccharide or amino-acid repeating units (Troy, 1973; Sutherland, 1977). The polysac-charide varieties may or may not contain monosaccharide side chains. The individual repeating units generally consist of one or more sugars or acid derivatives (Sutherland, 1985). In contrast to the rigid cell wall, the bacterial capsule is a loosely arranged structure containing over 90% H_2O. Its true magnitude in relation to the cell is only apparent by electron microscopy when special preparative techniques, such as antibody stabilization (Bayer et al., 1985) or freeze substitution (Graham, Harris, and Beveridge, unpublished observations; Fig. 9.3), are used to minimize dehydration artifacts. Because capsules may contain anionic groups such as carboxyl groups and, occasionally, phosphate and sulfate groups (Smiley and Wilkinson, 1983; Sutherland, 1985; Altman et al., 1986), it is not surprising that they are capable of binding metals (Table 9.1; McLean and Beveridge, 1988). One might expect differences between capsule and wall metal-binding characteristics in that the anionic groups within the wall are generally much more rigidly oriented than are those of the capsule.

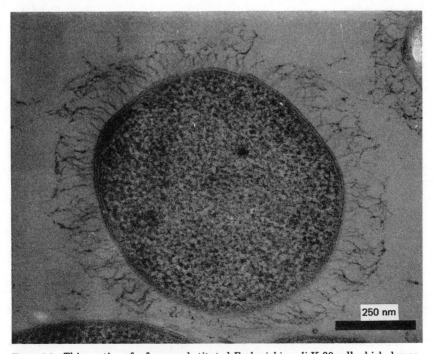

Figure 9.3 Thin section of a freeze-substituted *Escherichia coli* K-30 cell which demon-strates the capsule on its surface; bar = 250 nm.

TABLE 9.1   Metal Binding by Bacterial Capsules*

Metal	*Bacillus licheniformis*	*Klebsiella pneumoniae* K-20
Na^+	$0.259^\dagger$	0.018
K^+	0.223	0.001
Mg^{2+}	0.073	0.246
Ca^{2+}	1.044	3.500
Cr^{3+}	0.940	0.083
Mn^{2+}	0.071	0.001
Fe^{3+}	1.340	2.340
Co^{2+}	0.100	0.001
Ni^{2+}	0.080	0.001
Cu^{2+}	0.890	0.003
Zn^{2+}	0.149	0.011

*Metal binding was measured by suspending lyophilized capsule (1 mg/mL) in a trace metal solution containing 1 mM of the chloride salts of the above metals.
$\dagger$Expressed as μmol metal bound/mg (dry wt) capsule.
SOURCE: McLean and Beveridge (1988).

Metal binding by bacterial capsules will only be dealt with briefly here since it is addressed in much greater detail in Chap. 10.

Archaebacterial cell walls

The walls of the archaebacteria are chemically and structurally quite different from the eubacteria. Aside from their dissimilarity to eubacteria, there is a very great diversity in the structure and composition among these organisms, as shown in Table 9.2 (Woese et al., 1978; Kandler, 1982).

TABLE 9.2   Cell Wall and Envelope Structures in Archaebacteria

Genus	Wall type	Polymer
Methanobacteriales	Rigid wall	Pseudomurein
Methanosarcina	Rigid wall	Heteropolysaccharide
Methanococcus	S-layer	Protein
Methanomicrobium	S-layer	Protein
Methanogenium	S-layer	Glycoprotein
Methanothrix	Protein sheath	Protein
Methanospirillum	Protein sheath	Protein
Halobacterium	S-layer	Glycoprotein
Halococcus	Rigid wall	Sulfated heteropolysaccharide
Sulfolobus	S-layer	Glycoprotein
Thermoproteus	S-layer	Glycoprotein
Desulfurococcus	S-layer	Glycoprotein
Thermoplasma	None	Glycoproteins extending above the membrane surface

SOURCE: From Kandler (1982).

Membranes of archaebacteria contain phospholipids, phospho-glycolipids, and glycolipids (Woese et al., 1978; Kushwaha et al., 1981; Sprott et al., 1983); however, the fatty-acid residues are joined to the glycerol backbone via ether linkages rather than the ester linkages characteristic of eubacteria (Woese et al., 1978). The PM of *Methanospirillum hungatei* was shown by Sprott et al. (1983) to contain approximately 35 to 37 percent lipids (mainly as biphytanyl-diglycerol tetraether glycolipid), 45 to 50 percent protein, and 10 to 12 percent carbohydrates. The tetraetherglycolipid spans the bilayer and effectively covalently bonds the two membrane faces. Kushwaha et al. (1981) suggested that this imparted stability to the membranes of these methanogens. It is of interest that Ni accounts for 0.16 percent of the dry weight of *Msp. hungatei* membranes (Sprott et al., 1983). The membranes of *Thermoplasma acidophilum* contain a 152 kdal glycoprotein, which confers some rigidity to the membrane in the absence of a cell wall and is also responsible for cation binding (particularly Ca and Al; Yang and Haug, 1979). The anionic character of membrane phosphate groups would presumably be responsible for most of the metal binding to this membrane in other archaebacteria.

Archaebacteria possessing rigid walls include the order *Methano-bacteriales*. The main structural polysaccharide in these methanogens contains GlcNAc and *N*-acetyl-L-talosaminuronic acid (NAcTalNU) in a B1 → 3 linkage (Kandler, 1982). Short peptide stems arise from each (NAcTalNU) residue and are involved in cross-linking between strands. The common Pg constituents, dpm and D amino acids, are absent. In spite of the chemical differences to eubacterial Pg, Kandler and König (1978) proposed the name *pseudomurein,* since this polymer serves a very similar function in the archaebacterial cell wall. One would therefore expect any free carboxyl groups present in the peptide stems of pseudomurein to contribute to the anionic character of archaebacterial walls.

Other wall components have also been described. These include the wall heteropolysaccharides found in *Methanosarcina* and *Halococcus* (Kandler, 1982). In the case of *Halococcus,* these polysaccharides are sulfated (Steber and Schleifer, 1975); accordingly, carboxyl and sulfate groups would be expected to contribute to the anionic character of these polymers.

Msp. hungatei and *Methanothrix concilii* possess a very unusual regularly arrayed, proteinaceous sheath (Patel et al., 1986). Unlike most protein arrays, these sheaths are very resilient to physical and chemical perturbants; they can withstand shear forces, high temperatures, strong acids, and, in the case of *Mtx. concilii,* 5 *M* NaOH (Beveridge et al., 1986; Patel et al., 1986). Electron diffraction patterns of these sheaths revealed a periodic repeating-surface structure with a unit cell of 5.6 nm by 2.8 nm (Stewart et al., 1985). Most recent work indicates that COO^- groups are regularly positioned between surface subunits (Beveridge et al., 1988). One might anticipate that bacterial surfaces with naturally occurring periodic anionic groups would conceivably make ideal templates for metal binding and the formation of crystalline minerals (Mann, 1988). *Msp. hungatii* sheaths contain metals, notably Ca and Mg, whereas Zn is the predominant metal in the sheath of *Mtx. concilii*. Both sheaths have a strong affinity for exogenously added Ni and Co (G. D. Sprott, G. Patel, G. Southam, and T. J. Beveridge, unpublished observations; Fig. 9.4).

In several archaebacteria, the wall entirely comprises a paracrystalline S-layer composed of protein or glycoprotein subunits (Kandler, 1982). In the case of *Halobacterium halobium* these protein subunits are highly anionic because of the presence of covalently bonded sulfate groups (Wieland et al., 1980). The high Na^+ concentrations present in the environment of these organisms are necessary to neutralize the anionic nature of adjacent proteins and to promote otherwise weak intermolecular hydrophobic forces (Lanyi, 1974). Low ionic conditions dissolve the wall and lyse these cells. The wall protein

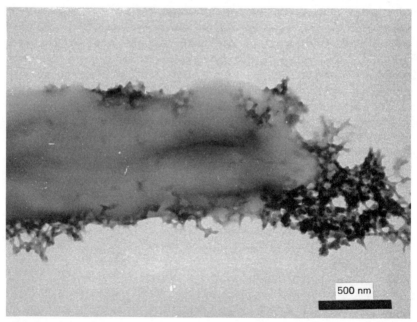

500 nm

Figure 9.4 The cell surface of *Methanothrix concilii* after incubation in a dilute solution of $CoCl_2$. No stain has been used, and the electron density is due to the Co mineral which is formed; bar = 500 nm. (*Collaborators in this work are G. Southam, University of Guelph, and G. Patel and G. D. Sprott, National Research Council of Canada, Ottawa.*)

subunits in *Methanococcus vannielii* are also held together by hydrophobic forces, but, unlike the halobacteria, high salt concentrations are not required (Jones et al., 1977). The highly anionic glycoprotein subunits of the acidophilic thermophile *Sulfolobus acidocaldarius* are stabilized by membrane interactions (Weiss, 1974). In all these cases, metals play a vital role in neutralizing the anionic charge of these subunits and in promoting otherwise weak hydrophobic interactions between adjacent proteins.

Some Aqueous Metal Characteristics Influencing Bacterial Metal Binding

The chemical reaction between an anionic bacterial surface (S^-) and a cationic metal ion (M^+) should ideally represent a simple electrostatic attraction: $M^+ + S^- \rightarrow S\text{---}M$. In reality, the chemistry of this situation is much more complicated. In aqueous solution, metal ions are usually hydrated. In natural water systems, metal ions can be attracted to a number of dissolved, colloidal, or solid organic and inorganic substances. In order for binding to a bacterial surface to occur, it may be necessary for the hydration shell or the complexing ligand to be displaced from M^+. When the effects of pH, E_h, interfacial nature, and chemical attributes of the metal (heat of hydration, charge density, electronic shell, etc.), and the presence of competing ions and other biological surfaces are also taken into account, we can begin to grasp the true complexity of bacterial metal binding.

In this section we introduce some of the chemical factors influencing bacterial metal binding. For reasons of space, it is impractical to address totally all aspects of aqueous metal chemistry. We refer the interested reader to books and reviews by Sherbert (1978), Förstner (1979), Babich and Stotzky (1980), Snoeyink and Jenkins (1980), Faust and Aly (1981), Florence (1982), Moore and Ramamoorthy (1984), and Salomons and Förstner (1984).

Dissolved substances in aqueous environments tend to concentrate on surfaces. In a general way this is termed the *interfacial effect*. Not all of these concentrating or bonding factors are understood. Metal ions are certainly no exception to this in that they are especially adsorbed onto hydrophilic anionic surfaces, such as sulfide minerals (Jean and Bancroft, 1986), aluminum hydroxides (Gessa et al., 1984), iron oxides (Laxen, 1985; Schoer, 1985), clays (El-Sayed et al., 1970; Marshall, 1975), and soils (Petruzzelli et al., 1985). Adsorption of metals to inorganic surfaces is very pH-dependent. Various metals, among them Cu, Fe, Mn, and Al, form insoluble oxides or hydroxides at neutral or alkaline pH. As a result, their free ions are not available for binding. At acid pH, metal binding is often limited by the increasing tendency of protons to compete with metal ions for anionic binding sites. Typically, metal ions will not bind to surfaces below a certain "threshold" pH. The exact value of this pH threshold varies with the affinity of the particular metal to the surface in question (Jean and Bancroft, 1986; McIlroy et al., 1986).

As mentioned previously, several metals form insoluble oxides and/or hydroxides at alkaline pH. While this precipitation effectively removes the particular metalloion from solution, coprecipitation of other metals may also occur at this time. Frimmel and Geywitz (1987) showed that Pb, Cu, Zn, and Cd would coprecipitate with ferric hydroxide at pH 7.5. Lowering the pH to 4.5 caused the dissolution of freshly formed ferric-hydroxide flocs and a remobilization of iron and the other coprecipitated metals. The reversible nature of this reaction was not seen if these precipitates were first allowed to mature for several days, suggesting that a molecular rearrangement of the $Fe(OH)_3$ precipitate to a more stable form had occurred. Field studies by Schoer (1985) have shown that iron precipitates exist in four different forms in river sediments. These include amorphous Fe(III) hydroxide, crystalline Fe(III) oxide, and Fe(III) and Fe(II) attached to silicate lattices. Schoer (1985) also demonstrated that a statistically significant immobilization of heavy metals occurred when they became coated with iron compounds. Although direct bonding of trace metals to iron oxides does occur (Hayes et al., 1987), the presence of organic matter, such as humic acids, will often enhance this adsorption (Laxen, 1985) by forming a bridging ligand between the metal oxide and the trace metal (Davis and Leckie, 1978).

Most natural systems contain large quantities of dissolved and suspended organic matter. Many of these poorly biodegradable organic compounds of microbial, plant, and animal origin are referred to as *humic substances*. Humic substances are a loosely defined group of "amorphous brown or black, hydrophilic, acidic, polydisperse substances of molecular weights ranging from several hundreds to tens of thousands" (Snoeyink and Jenkins, 1980). They represent the bulk of the soluble, nonliving organic content of soils and water. These substances are arbitrarily divided into two diverse groups based on their solubility in weak acid and base. Fulvic acids are soluble in both dilute acid and base, whereas humic acids are soluble only in dilute base. Depending on pH, anionic carboxyl, phenolic, alcoholic-OH, carbonyl, ether, ester, and quinone groups bind metals readily (Snoeyink and Jenkins, 1980; Gamble, 1986). Indeed in aquatic environments with a high organic matter content, such as swamps and sediments, virtually all the soluble metals present are complexed to some extent with organic matter, especially humic acids (Gardiner, 1974; Raspor et al., 1984a,b; Salomons and Förstner, 1984; Huljev,

1986). Conversely, in environments with a low organic matter content, such as the open ocean, dissolved metals tend to be present as free ions (Raspor et al., 1984b). The importance of metal speciation can be shown in that heavy metals when present as free ions are often most toxic to organisms such as eucaryotic marine life-forms (Bernhard and George, 1986), cyanobacteria (Jardim and Pearson, 1985), and methanogens (Jarrell et al., 1987). Complexation of metal ions with organic compounds often reduces metal toxicity.

As stated previously, under oxic conditions, iron usually resides in two particulate inorganic forms, either as amorphous or crystalline minerals. The recent investigations of Lovley have demonstrated that iron is more easily assimilated and reduced by bacteria if present as amorphous iron hydroxide rather than when it is in a more highly ordered crystalline form (Lovley and Phillips 1986a,b; Lovley, 1987). We have recently observed that metal binding of Fe to purified *B. licheniformis* capsule is strongly affected by the oxidation state of the metal (McLean and Beveridge, 1988). Relatively little work has been done to investigate the importance of metal speciation to bacteria (Olson, 1986). In light of this, it would appear that more work needs to be done in this area.

Several experimental approaches can be used to determine the nature of metal speciation. The electrochemical charge of some metal ions is altered or even eliminated by complexation with organic compounds. These changes can be detected with a high degree of sensitivity by a variety of voltametric techniques (Nürnberg, 1982; Richardson, 1985; Van den Berg, 1986; Buffle et al., 1987; Van Leeuwen, 1987), although the usefulness of this technique is limited to electrochemically active metal species (Gorman and Skogerboe, 1986). Several free metal ions, such as Cu^{2+}, Ca^{2+}, and Na^+, can be detected by ion-selective electrodes (Van Loon, 1982; Fish and Morel, 1985), and many metal species can be separated by ion chromatography (IC) or high-performance liquid chromatography (HPLC). When coupled to an element-specific detector, such as an atomic absorption spectrophotometer, an inductively coupled plasma emission spectrophotometer, or a mass spectrometer, IC and HPLC represent very powerful analytical tools (Van Loon, 1982; Chau, 1986; Krull, 1986; Smith, 1988). All of these techniques are quite sensitive and can distinguish metal concentrations in the parts per million to parts per billion range.

Bacterial Metal Binding

Examples of bacterial metal binding are quite widespread in nature. A common example is the slimy, rust-colored coating in water pipes and on rocks in streams. This is due to the binding of iron by the microbial biofilms which grow on these surfaces (Costerton et al., 1987). If the bacteria within these films are examined unstained by transmission electron microscopy (TEM), their walls and capsules can be seen due to the electron scattering power of their bound metals (Fig. 9.5). Indeed the anionic nature of most bacterial surfaces allows their staining by the cationic heavy-metal stains routinely used in TEM (Beveridge, 1978).

Early microelectrophoresis experiments showed that the electronegative surface property of bacteria could be altered by the addition of cationic surface-active agents (Dyar and Ordal, 1946; McQuillen, 1950; Harden and Harris, 1953), and soluble polyvalent metals (Schott and Young, 1953). Later work (Sherbert, 1978) has shown that anionic groups present in the Pg and wall acids of gram-positive bacteria and the Pg, phospholipids, and LPS of gram-negative bacteria to

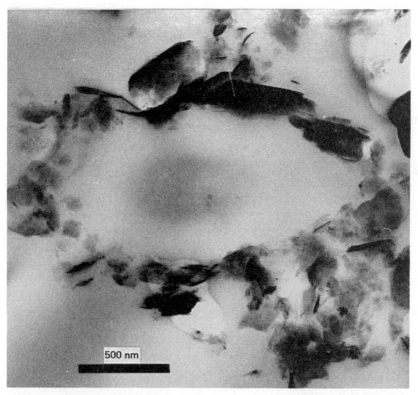

Figure 9.5 Thin section of a bacterial cell which is embedded in a series of platy minerals. This sample is representative of a biofilm scraping taken from a freshwater stream in Southern Ontario by the authors; bar = 500 nm.

be chiefly responsible for the anionic character and metal-binding ability of the wall.

The gram-positive wall, which is an amalgam of Pg, TA, and TUA, generally has a high affinity for metal ions. One would expect that the relatively rigid conformation in which the wall constituents are placed would result in the gram-positive wall having a specificity for certain metallic ions. If we examine the metal-binding character of *B. subtilis* walls, we do encounter preferential binding of several metals (Table 9.3) (Beveridge and Murray, 1976b). Extraction of TA and TUA and chemical modification of the carboxyl groups of Pg in this organism showed that Pg is chiefly responsible for metal binding in this organism (Matthews et al., 1979; Beveridge and Murray, 1980). Chemical modification of the various charged wall groups in *B. subtilis* results in an alteration of the metal-binding character of the wall. Using this approach, Beveridge and Murray (1980) showed that modification of amine groups did not reduce the ability of *B. subtilis* walls to bind metal cations. In contrast, neutralization or reversal of the anionic charge of the carboxyl groups severely limited cationic metal binding (Table 9.4). TA removal by mild alkali treatment did reduce cationic metal binding due to the removal of phosphate groups, but not to the same extent as alteration of carboxyl groups. In addition to altering the metal-binding ability of *B. subtilis* walls, carboxyl group alteration resulted in a redistribution of the residual metal-binding sites and the formation of internal crystal formation when these walls were exposed to gold chloride. Doyle et al. (1980) confirmed these observations and also showed that chemical alteration of wall amine groups partially increased the ability of *B. subtilis* walls to bind cationic metals due to the removal of competing counterions (most

TABLE 9.3   Metal Binding by *Bacillus subtilis* and *Bacillus licheniformis* Walls*

Metal	*Bacillus subtilis*	*Bacillus licheniformis*
Na^+	2.697	0.910
K^+	1.944	0.560
Mg^{2+}	8.226	0.400
Ca^{2+}	0.399	0.590
Mn^{2+}	0.801	0.662
Fe^{3+}	3.581	0.760
Ni^{2+}	0.107	0.520
Cu^{2+}	2.990	0.490
Au^{3+}	0.363	0.031

*Expressed as μmol metal bound/mg (dry wt) wall fragments. Refer to references for experimental detail.

SOURCE: From Beveridge and Murray (1976b) and Beveridge et al. (1982).

TABLE 9.4 Metal Binding by *B. subtilis* Walls with Chemically Modified Amine and Carboxyl Groups*

Metal	Modified amine groups		Modified carboxyl groups		
	A	B	C	D	E
Na^+	2.680	2.206	0	0	0
K^+	1.600	2.000	0	0	0
Mg^{2+}	7.664	8.806	0.520	0.300	0.160
Ca^{2+}	0.851	1.082	0.380	0.360	0.300
Mn^{2+}	0.820	0.880	0.732	0.680	0.100
Fe^{3+}	2.680	2.860	2.260	0.240	0.240
Ni^{2+}	0.320	0.341	0.024	0.004	0.004
Cu^{2+}	0.760	0.860	0.993	0.506	0.260
Au^{3+}	0.391	0.427	0.214	0.103	0.018

*Expressed as μmol metal bound/mg (dry wt) wall fragments. Refer to Beveridge and Murray (1980) for experimental detail.
A—amine groups of wall modified with s-acetylmercaptosuccinic anhydride
B—amine and hydroxyl wall groups modified with sodium iodoacetate
C—wall carboxyl groups neutralized with glycine ethyl ester
D—wall carboxyl groups made slightly electropositive with glycinamide
E—wall carboxyl groups made electropositive with ethylenediamine
SOURCE: From Beveridge and Murray (1980).

probably diaminopimellic acid) from the vicinity of carboxyl groups. Competition experiments between metals for binding sites suggested that select metal cations bind to identical wall sites. At low pH, protons compete effectively with metal cations for anionic metal-binding sites.

In contrast to *B. subtilis*, the major sites for metal binding in *B. licheniformis* walls are contributed by the TA and TUA fractions, since these polymers account for the major mass of this wall. Removal of TA and TUA components from *B. licheniformis* walls by mild alkali and trichloroacetic acid treatments, respectively, reduced their metal-binding capacity by about 90 percent (Beveridge et al., 1982). TUA from *B. licheniformis* was shown by sedimentation velocity studies to be a particularly potent metal chelator, possessing a high degree of affinity for Mg^{2+} and Na^+, which were the only metals used in the experiment. The walls of *B. subtilis* and *B. licheniformis* grown in the presence of phosphate differ in that the walls of the latter organism contain TUA as well as TA. As can be seen from Table 9.3, *B. licheniformis* walls bind less metal and demonstrate a different affinity profile for metal cations than do *B. subtilis* walls.

Metal binding by *B. subtilis* walls is considered to be a two-stage process. Initially, metal ions are bound to anionic sites in the Pg. These bound metals then act as nucleation sites whereby additional metals might be bound (often forming aggregates visible by TEM; Beveridge, 1978; Beveridge and Jack, 1982). In contrast, most metal binding by *B. licheniformis* walls appears to be a single-step process which is associated with TA and TUA in that metal aggregates are not usually detected by TEM (Beveridge et al., 1982). Thus it would appear that the wall chemistry of bacteria not only affects its metal-binding profile, but also influences the nucleation of metallic aggregates and crystals.

Gram-negative bacterial envelopes have a lower overall charge density and are chemically and structurally very different from gram-positive walls (Beveridge, 1981). In light of this, one would expect their metal-binding capacity and characteristics to differ from those of gram-positive organisms. Beveridge and Koval (1981) examined the metal binding of purified *E. coli* K-12 envelopes (Table 9.5). In most cases, TEM of metal-loaded envelopes suggested that the metal ions complexed with the polar head groups of the membrane, producing a conventional bilayer appearance (Fig. 9.6). The binding of some transition metals (Hf, Zr, Pr, and Sm) was asymmetric in that the outer face of the membrane was preferentially stained. One could speculate that the instability of these metal ions in solution would cause their precipitation and removal on the first surface (outer surface of the membrane vesicle) that they encountered before they had an opportunity to cross the membrane. The possibility of chemical differences in binding affinity between the outer and inner faces of these membranes cannot be discounted; however, the native lipid asymmetry of the OM is usually lost during the preparation of these vesicles.

As is the case in gram-positive bacteria such as *B. subtilis,* purified Pg from gram-negative *E. coli* binds metals quite well (Table 9.5; Hoyle and Beveridge, 1984). The overall contribution of Pg to metal binding in gram-negative organisms is quite small due to the limited

TABLE 9.5 Metal Binding by *Escherichia coli* AB264 Wall Envelopes and Purified Pg

Metal	Wall envelopes*	Purified Pg*
Na^+	0.042	0.290
K^+	0.082	0.058
Mg^{2+}	0.256	0.035
Ca^{2+}	0.035	0.038
Mn^{2+}	0.140	0.052
Fe^{3+}	0.200	0.100
Ni^{2+}	0.002	0.019
Cu^{2+}	0.090	Not done
Au^{3+}	0.056	Not done

*Expressed as μmol metal bound/mg (dry wt) envelope or Pg. Refer to references for experimental detail.

SOURCE: From Beveridge and Koval (1981) and Hoyle and Beveridge (1983).

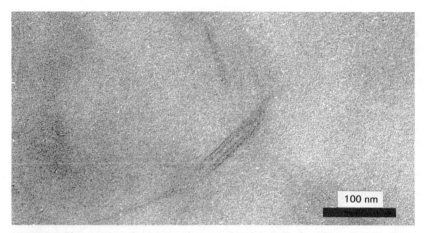

Figure 9.6 Thin section of outer membrane isolated from *Escherichia coli* K-12 which was suspended in a 5-mM solution of $CuCl_2$. No stain has been added to the sample, and the electron density is derived from the sorbed copper (for more information see Beveridge and Koval, 1981); bar = 100 nm.

quantity of this substance in the cell envelope and its shielding from the environment by the overlying OM.

^{31}P NMR studies show that phosphate groups in the phospholipids and LPS of *E. coli* OM readily bind metals such as Ca^{2+}, Cd^{2+}, Yb^{3+}, La^{3+} (Strain et al., 1983), and Eu^{3+} (Ferris and Beveridge, 1984). Other anionic carboxyl groups such as those' present in KDO would also be expected to represent potential metal-binding sites. Ferris and Beveridge (1986) showed that of the three COO^- groups present on KDO, two were shielded by their proximity to $—NH_3^+$ groups and were therefore unavailable for metal binding.

Other bacterial surface components which have been implicated in metal binding include S-layers and capsules. The importance of divalent metal cations in maintaining S-layer structure and assembly was addressed earlier. Anionic groups of bacterial capsules differ from those present in walls in that they are not as rigidly locked into a particular orientation by cross-links or complexation with neighboring molecules. In light of this, capsules might be expected to have very little metal-binding selectivity. Interestingly, this turns out to not be so. Mittelman and Geesey (1985) found purified capsules from a freshwater bacterium to have a high degree of affinity for Cu. We have found that exposure of the polyglutamic-acid capsule of *B. licheniformis* to a solution containing 1-mM concentrations of 12 elements results in a preferential binding of some transition elements (Table 9.1), most notably Fe and Cu. In contrast, under identical conditions, the man-$(gal)_2$-glcA capsule of *Klebsiella pneumoniae* K-20 preferentially binds Fe and Ca (McLean and Beveridge, 1988). Metal binding by bacterial capsules is therefore more complex than first believed. This will be further addressed in Chap. 10.

Metal binding by bacterial surfaces is largely a passive phenomenon in that the whole process represents an electrostatic interaction between cationic metals and anionic cell surface groups. Consequently, it is not necessary that the organisms be viable, only that their surfaces remain intact. Indeed, several studies have shown that dead microorganisms will often bind greater quantities of metals than do living cells (Kurek et al., 1982; DiSpirito et al., 1983). This may be due in part to an increase in the number or exposure of metal-binding sites following cell death due to autolysin action.

Chemical Alteration of Bound Metals and Mineralized Aggregate Formation

The external environment and internal metabolism of living bacteria often exert a profound influence on the chemistry of bound metals. These influences include changes in oxidation state, formation of organometallic compounds, and formation of precipitates due to detoxification or energy-yielding mechanisms of the bacterial protoplast. Alternatively, the metals may be affected indirectly by the production of metabolic end products such as SO_4^{2-} and S^{2-}, or an alteration in the local pH and/or E_h. The results of microbial activity may ultimately lead to metal immobilization, remobilization, and/or the formation of metal aggregates.

Some heavy metals such as Hg and Sn are potentially toxic to microorganisms due to their inhibitory activity on many enzymes. The soluble Hg and Sn cations of these metals are the most toxic form to the cell. Consequently, bacteria living in environments containing appreciable quantities of heavy metals have often adapted to their environment by evolving resistance mechanisms (Barkay, 1987), which are often found on plasmids and transposons. Resistance strategies generally involve a decrease in susceptibility to heavy metals by an alteration in structure and/or quantity of the susceptible enzyme(s), a decrease in heavy-metal membrane permeability, active transport (efflux) of heavy metals from the cell, or a variety of detoxification mechanisms which are often plasmid-mediated (Silver and Misra, 1988).

Metal sulfide compounds and many metal phosphate compounds are very insoluble. Thus the production of these two anions represent simple microbial detoxification mechanisms, because the resultant insoluble heavy-metal precipitates are relatively nontoxic (Aiking et al., 1984, 1985; Erardi et al., 1987). These precipitates are quite often retained in the vicinity of the cell by being bound to the capsule or the walls.

An alternative detoxification strategy employed by some bacteria involves converting the offensive heavy-metal species to a more volatile form, which then evaporates from the cellular vicinity. Mercury is often detoxified by complexation with organic ligands or by being reduced to its elemental form. Sn organometallic complexes resulting from detoxification have also been described (Compeau and Bartha, 1984, 1985; Summers, 1986; Foster, 1987; Gilmour et al., 1987; Korthals and Winfrey, 1987). These changes in oxidation state or speciation can often result in an altered binding affinity of the metal to the bacterial surface. We have noted that Fe^{2+} binds to a lesser extent to purified B. licheniformis capsule than does Fe^{3+}. In addition, reduction of bound Fe^{3+} results in a remobilization of this metal (McLean and Beveridge, 1988).

Reduction or oxidation of metals can also occur as a metabolic function of bacteria. In the case of lithotrophic autotrophs (Brock et al., 1984), oxidation of inorganic compounds, including metals, provides energy which can then be used by the bacterium to fix carbon. Fe^0, Fe^{2+}, Mn^{2+}, and U^{4+} are examples of metals which can be oxidized by lithotrophic autotrophic bacteria (DiSpirito and Tuovinen, 1982a,b; Kepkay, 1985; Daniels et al., 1987). Alternatively, metals and metal ions can act as electron acceptors during anaerobic bacterial growth. Iron and manganese are commonly reduced in this fashion (Sørensen, 1982; Jones et al., 1984; Burdige and Nealson, 1985; Sakata, 1985; Lovley, 1987; Lovley and Phillips, 1987; Silver, 1987). Tugel et al. (1986) found that attachment of bacteria to iron particles was a prerequisite for iron reduction since most oxidized forms of iron are insoluble in nature (and would otherwise be unavailable to the organisms). Iron reduction will result in the formation of a variety of minerals such as hematite and maghematite (different crystalline forms of Fe_2O_3), siderite ($FeCO_3$), magnetite (Fe_3O_4), geothite [$FeO(OH)$], vivianite [$Fe_3(PO_4)_2 \cdot 8H_2O$), pyrite (FeS_2), mackinawite (FeS), and greigite (Fe_3S_4). Mineral formation is influenced largely by the cellular environment present (pH, E_h, and other elements which are present) at the time of iron reduction (Bell et al., 1987; Lovley, 1987). Under conditions of low pH and E_h, free Fe^{2+} ions can also be present. Further changes to these minerals, such as the conversion of iron oxides to sulfides, can occur due to bacterial action (Karlin and Levi, 1983).

Numerous bacteria in aquatic sediments encounter and bind a wide variety of metals in their environment. As the sediments accumulate, they and their bacterial components become subject to geological forces that eventually result in rock formation. During this time, chemical and physical changes occur within the sediments as diagenesis progresses. The surfaces of bacteria present in these sediments make ideal biological templates for the concentration of metals and the nucleation of crystals (Beveridge et al., 1983; Mann, 1988) and can often greatly influence the initial mineralization processes.

Bacteria and their remains in metal-rich waters (Ferris et al., 1987a) and geothermal sediments (Ferris et al., 1986, 1987b) have been shown by TEM, analytical electron microscopy, and electron diffraction to act as nucleation sites for authigenic minerals. Iron-silica crystalling species were associated with bacteria in geothermal sediments (Ferris et al., 1986), whereas the bacteria in metal-rich waters were associated with Fe-Al silicate (clay) polymorphs, millerite (NiS), and mackinawite (FeS_{1-x}). In the latter environment, bacteria are thought to initially bind and concentrate Fe and Ni. These concentrated metals then react with bacterially produced sulfide from the sediments or soluble silica to produce the corresponding mineral. Similarly, todokorite (MnO_6) (Ferris et al., 1987b) and Mg calcite (marine peloids) (Chafetz, 1986) deposits have also been shown to be nucleated by bacteria.

Bacterial mineral authigenesis can be simulated in the laboratory. Lovley et al. (1987) recently demonstrated the production of magnetite (Fe_3O_4) from amorphous iron hydroxide by the growth of dissimilatory iron-reducing bacteria. Beveridge et al. (1983) modeled low-temperature geological processes by suspending metal-loaded *B. subtilis* cells in artificial sediments at 100°C for 1 to 200 days. They found that a variety of crystalline mineral phosphates, metal sulfides, and polymeric metal-complexed organic residues formed. The bacterial surface provided phosphate to the minerals, and in the absence of elemental S (S^0) authigenesis of phosphate minerals occurred. When S^0 was present, metal sulfides evolved; however, if the pH was elevated, metal sulfide formation was blocked and phosphates were promoted. Initially, metal deposition occurred as small localized microcrystals. Eventually these crystals increased in size and distribution quite noticeably (Fig. 9.7).

One possible long-term consequence of bacterial mineralization is the formation of microbial fossils. Laboratory experiments by Ferris et al. (1988) simulated this process by suspending *B. subtilis* cells in a dilute aqueous solution of silica. They showed that in vitro "fossilization" of these organisms resulted from the incorporation and binding of silica to the cell wall. The success of this process was enhanced if the cells were loaded with iron prior to exposure to the silica; this was likely due to the inhibitory effect of iron on cell autolysin action.

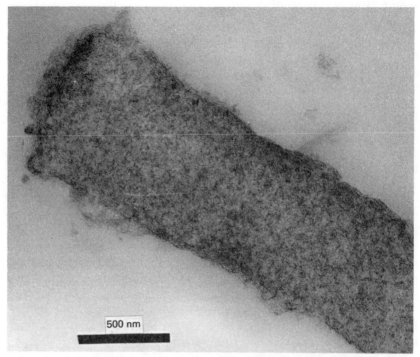

Figure 9.7 Thin section of a *Bacillus subtilis* 168 cell which was loaded with uranium and mineralized in a matrix of quartz and calcite during a low-temperature diagenesis simulation. Tiny crystals of uranium phosphate abound, and the cell is now essentially a solid mineral (for more information see Beveridge et al., 1983); bar = 500 nm.

Summary and Conclusions

Bacterial cell surfaces are able to interact with and bind metals in their environment due to their anionic nature. While this feature often benefits the organisms themselves, it is very important from a terrestrial perspective. The localized high concentrations of metals which may develop on bacterial surfaces are subject to mineral formation and diagenesis, which in some cases may be modified by bacterial metabolism.

This feature of bacteria has many potential industrial uses for the biological concentration and recovery of metals. We have shown that in our chemically defined bacterial surface models (*B. subtilis* walls, *E. coli* envelopes, and *B. licheniformis* and *K. pneumoniae* K-20 capsules) that a certain selectivity or preference of metals occurs. It would be expected that dissimilar bacterial surface chemistries would result in different metal preferences. The possibility exists of selecting a particular organism or strain to obtain a desired metal or mineral. Metal selectivity could be even more enhanced if anionic groups of the surface were chemically tailored to suit the desired metal.

To be suitable for industry, the metal sorbing agent must be cheap and easily recycled. Bacteria can utilize and be grown on a number of industrial organic wastes, and several studies have shown that chemical treatment of the metal-loaded bacteria can reclaim metal (Hutchins et al., 1986). For example, most bacterial surface-bound metals can be released upon acidification, which allows their recovery. Norberg and Persson (1984) were able to selectively release bound Cu, Cd, and UO_2 ions from *Zoogloea ramigera* biomass by simply lowering pH.

There are several advantages to using bacteria. The rapid growth of these organisms and their adaptability to an assortment of substrates and environments are some useful attributes. It is not necessary to use living organisms, since cell remnants will often adsorb more metal than their living counterparts. Following chemical treatment to desorb bound metals, microbial biomass can often be reused for metal reclamation. In some cases microbial biomass can be recycled through a number of metal-loading-desorption cycles. For full utilization of the natural metal-binding nature of bacteria, more work needs to be done to survey microorganisms for their metal-binding characteristics and their capability to accomplish this on an industrial scale.

Acknowledgments

Our work has been supported by the Natural Sciences and Engineering Research Council of Canada (NSERC) and Medical Research Council of Canada operating grants to TJB. Analytical electron microscopy was performed at the Guelph Regional STEM facility, which is supported by the University of Guelph, NSERC, and user fees. RJCM was supported by a postdoctoral fellowship from the Alberta Heritage Foundation for Medical Research and is currently supported by a Career Scientist Fellowship from the Ontario Ministry of Health.

We thank F. G. Ferris, C. A. Flemming, W. S. Fyfe, R. J. Harris, B. Hoyle, R. Humphrey, S. F. Koval, N. L. Martin, R. G. E. Murray, G. Southam, and S. G. Walker for their support and collaboration through the years.

References

Aiking, H., Govers, H., and van't Riet, J., Detoxification of mercury, cadmium, and lead in *Klebsiella aerogenes* NCTC 418 growing in continuous culture, *Appl. Environ. Microbiol.* **50**:1262–1267 (1985).

Aiking, H., Stijnman, A., van Garderen, C., van Heerikhuizen, H., and van't Riet, J., Inorganic phosphate accumulation and cadmium detoxification in *Klebsiella aerogenes* NCTC 418 growing in continuous culture, *Appl. Environ. Microbiol.* **47:**374–377 (1984).

Altman, E., Brisson, J.-R., and Perry, M. B., Structural studies of the capsular polysaccharide from *Haemophilus pleuropneumoniae* serotype, *Biochem. Cell Biol.* **64:**707–716 (1986).

Archibald, A. R., Baddiley, J., and Heckels, J. E., Molecular arrangement of teichoic acid in the cell wall of *Staphylococcus lactis, Nature New Biol.* **241:**29–31 (1973).

Babich, H., and Stotzky, G., Environmental factors that influence the toxicity of heavy metal and gaseous pollutants to microorganisms, *Crit. Rev. Microbiol.* **7:**99–145 (1980).

Balkwill, D. L., and Ghiorse, W. C., Characterization of subsurface bacteria associated with two shallow aquifers in Oklahoma, *Appl. Environ. Microbiol.* **50:**580–588 (1985).

Barghoorn, E. S., and Tyler, S. A., Microorganisms from the Gunflint chert, *Science (USA)* **147:**563–577 (1965).

Barkay, T., Adaptation of aquatic microbial communities to Hg^{2+} stress, *Appl. Environ. Microbiol.* **53:**2725–2732 (1987).

Bayer, M. E., Areas of adhesion between wall and membrane of *Escherichia coli, J. Gen. Microbiol.* **53:**395–404 (1968).

Bayer, M. E., Carlemalm, E., and Kellenberger, E., Capsule of *Escherichia coli* K29: Ultrastructural preservation and immunoelectron microscopy, *J. Bacteriol.* **162:**985–991 (1985).

Bayer, M. H., Costello, G. P., and Bayer, M. E., Isolation and partial characterization of membrane vesicles carrying markers of the membrane adhesion sites, *J. Bacteriol.* **149:**758–767 (1982).

Bayer, M. E., and Remsen, C. C., Structure of *Escherichia coli* after freeze-etching, *J. Bacteriol.* **101:**304–313 (1970)

Bell, P. E., Mills, A. L., and Herman, J. S., Biogeochemical conditions favoring magnetite formation during anaerobic iron reduction, *Appl. Environ. Microbiol.* **53:**2610–2616 (1987).

Bernhard, M., and George, S. G., Importance of chemical species in uptake, loss and toxicity of elements for marine organisms, in M. Bernhard, R. E. Brinckman, and P. J. Sadler (eds.), *The Importance of Chemical "Speciation" in Environmental Processes,* Springer-Verlag, Berlin, 1986, pp. 385–422.

Beveridge, T. J., The response of cell walls of *Bacillus subtilis* to metals and to electron-microscopic strains, *Can. J. Microbiol.* **24:**89–104 (1978).

Beveridge, T. J., Surface arrays on the wall of *Sporosarcina ureae, J. Bacteriol.* **139:**1039–1048 (1979).

Beveridge, T. J., Ultrastructure, chemistry, and function of the bacterial wall, *Int. Rev. Cytol.* **72:**229–317 (1981).

Beveridge, T. J., Mechanisms of the binding of metallic ions to bacterial walls and the possible impact on microbial ecology, in M. J. Klug and C. A. Reddy (eds.), *Current Perspectives in Microbial Ecology,* ASM Press, Washington, 1984, pp. 601–607.

Beveridge, T. J., The bacterial surface: General considerations towards design and function, *Can. J. Microbiol.* **34:**363–372 (1988).

Beveridge, T. J., Wall ultrastructure: How little we know, in P. Actor, L. Daneo-Moore, M. L. Higgins, M. R. J. Salton, and G. D. Shockman (eds.), *Antibiotic Inhibition of the Bacterial Cell Surface Assembly and Function,* American Society for Microbiology, Washington, 1988b, chap. 1.

Beveridge, T. J., and Davies, J. A., Cellular responses of *Bacillus subtilis* and *Escherichia coli* to the Gram stain, *J. Bacteriol.* **156:**846–858 (1983).

Beveridge, T. J., Forsberg, C. W., and Doyle, R. J., Major sites of metal binding in *Bacillus licheniformis* walls, *J. Bacteriol.* **150:**1438–1448 (1982).

Beveridge, T. J., Harris, B. J., Patel, G. B., and Sprott, G. D., Cell division and filament splitting in *Methanothrix concilii, Can. J. Microbiol.* **32:**779–786 (1986).

Beveridge, T. J., and Jack, T., Binding of an inert, cationic, osmium probe to walls of *Bacillus subtilis, J. Bacteriol.* **149:**1120–1123 (1982).

Beveridge, T. J., and Koval, S. F., Binding of metals to cell envelopes of *Escherichia coli* K-12, *Appl. Environ. Microbiol.* **42:**325–335 (1981).

Beveridge, T. J., Meloche, J. D., Fyfe, W. S., and Murray, R. G. E., Diagenesis of metals chemically complexed to bacteria: Laboratory formation of metal phosphates, sulfides, and organic condensates in artificial sediments, *Appl. Environ. Microbiol.* **45**:1094–1108 (1983).

Beveridge, T. J., and Murray, R. G. E., Dependence of the superficial layers of *Spirillum putridiconchylium* on Ca^{2+} or Sr^{2+}, *Can. J. Microbiol.* **22**:1233–1244 (1976a).

Beveridge, T. J., and Murray, R. G. E., Uptake and retention of metals by cell walls of *Bacillus subtilis, J. Bacteriol.* **127**:1502–1518 (1976b).

Beveridge, T. J., and Murray, R. G. E., Sites of metal deposition in the cell wall of *Bacillus subtilis, J. Bacteriol.* **141**:876–887 (1980).

Beveridge, T. J., Southam, G., Sara, M., Pum, D., and Sleytr, U., Surface charge of the sheath of *Methanospirillum hungatei* GPI, *Abstr. 88th Annu. Meet. Am. Soc. Microbiol.* J3, 1988, p. 205.

Birdsell, D. C., Doyle, R. J., and Morgenstern, M., Organization of teichoic acid in the cell wall of *Bacillus subtilis, J. Bacteriol.* **121**:726–734 (1975).

Boylan, R. J., Mendelson, N. H., Brooks, D., and Young, F. E., Regulation of the bacterial cell wall: Analysis of a mutant of *Bacillus subtilis* defective in biosynthesis of teichoic acid, *J. Bacteriol.* **110**:281–290 (1972).

Braun, V., Convalent lipoprotein from the outer membrane of *Escherichia coli, Biochim. Biophys. Acta* **415**:335–377 (1975).

Braun, V., Rehn, K., and Wolff, H., Supra-molecular structure of the rigid layer of the cell wall of *Salmonella, Serratia, Proteus,* and *Pseudomonas fluorescens.* Number of lipoprotein molecules in a membrane layer, *Biochem.* **9**:5041–5049 (1970).

Brock, T. D., Smith, D. W., and Madigan, M. T., *Biology of Microorganisms,* 4th ed., Prentice-Hall, Englewood Cliffs, NJ, 1984, pp. 204–208.

Brown, M. R. W., and Williams, P., The influence of environment on envelope properties affecting survival of bacteria in infections, *Ann. Rev. Microbiol.* **39**:527–556 (1985).

Buckmire, F. L. A., and Murray, R. G. E., Substructure and in vitro assembly of the outer, structured layer of *Spirillum serpens, J. Bacteriol.* **125**:290–299 (1976).

Buffle, J., Vuilleumier, J. J., Tercier, M. L., and Parthasarathy, N., Voltammetric study of humic and fulvic substances. V. Interpretation of metal ion complexation measured by anodic stripping voltammetric methods, *Sci. Total Environ.* **60**:75–96 (1987).

Burdige, D. J., and Nealson, K. H., Microbial manganese reduction by enrichment cultures form coastal marine sediments, *Appl. Environ. Microbiol.* **50**:491–497 (1985).

Button, D., Archibald, A. R., and Baddiley, J., The linkage between teichoic acid and glycosaminopeptide in the walls of a strain of *Staphylococcus lactis, Biochem. J.* **99**:11c–14c (1966).

Chafetz, H. S., Marine peloids: A product of bacterially induced precipitation of calcite, *J. Sediment. Petrol.* **56**:812–817 (1986).

Chau, Y. K., Analytical aspects of organometallic species determination in fresh water systems, in M. Bernhard, R. E. Brinckman, and P. J. Sadler (eds.), *The Importance of Chemical "Speciation" in Environmental Processes,* Springer-Verlag, Berlin, 1986, pp. 149–167.

Compeau, G., and Bartha, R., Methylation and demethylation of mercury under controlled redox, pH, and salinity conditions, *Appl. Environ. Microbiol.* **48**:1203–1207 (1984).

Compeau, G. C., and Bartha, R., Sulfate-reducing bacteria: Principal methylators of mercury in anoxic estuarine sediment, *Appl. Environ. Microbiol.* **50**:498–502 (1985).

Costerton, J. W., Cheng, K.-J., Geesey, G. G., Ladd, T. I., Nickel, J. C., Dasgupta, M., and Marrie, T. J., Bacterial biofilms in nature and disease, *Ann. Rev. Microbiol.* **41**:435–464 (1987).

Costerton, J. W., Irvin, R. T., and Cheng, K.-J., The bacterial glycocalyx in nature and disease, *Ann. Rev. Microbiol.* **35**:299–324 (1981).

Cronan, J. E., Jr., Gennis, R. B., and Maloy, S. R., Cytoplasmic membrane, in R. C. Neidhardt (ed.), *Escherichia coli* and *Salmonella typhimurium Cellular and Molecular Biology,* American Society for Microbiology, Washington, 1987, pp. 31–55.

Cronan, J. E., Jr., and Vagelos, P. R., Metabolism and function of the membrane phospholipids of *Escherichia coli, Biochim. Biophys. Acta* **265**:25–60 (1972).

Daniels, L., Belay, N., Rajagopal, B. S., and Weimer, P. J., Bacterial methanogenesis and growth from CO_2 with elemental iron as the sole source of electrons, *Science (USA)* 237:509–511 (1987).

Davies, J. A., Anderson, G. K., Beveridge, T. J., and Clark, H. C., Chemical mechanism of the Gram stain and synthesis of a new electron-opaque marker for electron microscopy which replaces the iodine mordant of the stain, *J. Bacteriol.* 156:837–845 (1983).

Davis, J. A., and Leckie, J. O., Effect of adsorbed complexing ligands on trace metal uptake by hydrous oxides, *Environ. Sci. Technol.* 12:1309–1315 (1978).

DiSpirito, A. A., Talnagi, J. W., Jr., and Tuovinen, O. H., Accumulation and cellular distribution of uranium in *Thiobacillus ferrooxidans, Arch. Microbiol.* 135:250–253 (1983).

DiSpirito, A. A., and Tuovinen, O. H., Kinetics of uranous ion and ferrous iron oxidation by *Thiobacillus ferrooxidans, Arch. Microbiol.* 133:33–37 (1982a).

DiSpirito, A. A., and Tuovinen, O. H., Uranous ion oxidation and carbon dioxide fixation by *Thiobacillus ferrooxidans, Arch. Microbiol.* 133:28–32 (1982b).

Doyle, R. J., and Koch, A. L., The functions of autolysins in the growth and division of *Bacillus subtilis, Crit. Rev. Microbiol.* 15:169–222 (1987).

Doyle, R. J., Matthews, T. H., and Streips, U. N., Chemical basis for selectivity of metal ions by the *Bacillus subtilis* cell wall, *J. Bacteriol.* 143:471–480 (1980).

Doyle, R. J., McDannel, M. L., Streips, U. N., Birdsell, D. C., and Young, F. E., Polyelectrolyte nature of bacterial teichoic acids, *J. Bacteriol.* 118:606–615 (1974).

Doyle, R. J., McDannel, M. L., Helman, J. R., and Streips, U. N., Distribution of teichoic acid in the cell wall of *Bacillus subtilis, J. Bacteriol.* 122:152–158 (1975).

Driessens, F. C., Borggreven, J. M. P. M., Verbeeck, R. M. H., van Dijk, J. W. E., and Feugin, F. F., On the physicochemistry of plaque calcification and the phase composition of dental calculus, *J. Periodont. Res.* 20:329–336 (1985).

Dyar, M. T., and Ordal, E. J., Electro-kinetic studies on bacterial surfaces. The effects of surface-active agents on the electrophoretic mobilities of bacteria, *J. Bacteriol.* 51:149–167 (1946).

El-Sayed, M. H., Burau, R. G., and Babcock, K. L., Thermodynamics of copper(II)-calcium exchange on bentonite clay, *Soil Sci. Soc. Am. Proc.* 34:397–400 (1970).

Erardi, F. X., Failla, M. L., and Falkinham, J. O. III, Plasmid-encoded copper resistance and precipitation by *Mycobacterium scrofulaceum, Appl. Environ. Microbiol.* 53:1951–1954 (1987).

Faust, S. D., and Aly, O. M., *Chemistry of Natural Waters,* Ann Arbor Science, Ann Arbor, MI, 1981.

Ferris, F. G., and Beveridge, T. J., Binding of a paramagnetic metal cation to *Escherichia coli* K-12 outer membrane vesicles, *FEMS Microbiol. Lett.* 24:43–46 (1984).

Ferris, F. G., and Beveridge, T. J., Site specificity of metallic ion binding in *Escherichia coli* K-12 lipopolysaccharide, *Can. J. Microbiol.* 32:52–55 (1986).

Ferris, F. G., Beveridge, T. J., and Fyfe, W. S., Iron-silica crystallite nucleation by bacteria in a geothermal sediment, *Nature (London)* 320:609–611 (1986).

Ferris, F. G., Fyfe, W. S., and Beveridge, T. J., Bacteria as nucleation sites for authigenic minerals in a metal-contaminated lake sediment, *Chem. Geol.* 63:225–232 (1987a).

Ferris, F. G., Fyfe, W. S., and Beveridge, T. J., Manganese oxide deposition in a hot spring microbial mat, *Geomicrobiol. J.* 5:33–42 (1987b).

Ferris, F. G., Fyfe, W. S., and Beveridge, T. J., Metallic ion binding by *Bacillus subtilis:* Implications for the fossilization of microorganisms, *Geology* 16:149–152 (1988).

Fischer, H., and Tomasz, A., Production and release of peptidoglycan and wall teichoic acid polymers in Pneumococci treated with beta-lactam antibiotics, *J. Bacteriol.* 157:507–513 (1984).

Fish, W., and Morel, F. M. M., Propagation of error in fulvic acid titration data: A comparison of three analytical methods, *Can. J. Chem.* 63:1185–1193 (1985).

Florence, T. M., The speciation of trace elements in water, *Talanta* 29:345–364 (1982).

Förstner, V., *Metal Pollution in the Aquatic Environment,* Springer-Verlag, Berlin, 1979.

Foster, T. J., The genetics and biochemistry of mercury resistance, *Crit. Rev. Microbiol.* 15:117–140 (1987).

Friedmann, E. I., and Ocampo-Friedmann, R., Endolithic microorganisms in extreme dry environments: Analysis of a lithobiontic microbial habitat, in M. J. Klug and C. A. Reddy (eds.), *Current Perspectives in Microbial Ecology*, American Society for Microbiology, Washington, D.C., 1984, pp. 177–185.

Friedmann, E. I., and Weed, R., Microbial trace-fossil formation, biogenous, and abiotic weathering in the Antarctic cold desert, *Science (USA)* **236**:703–705 (1987).

Frimmel, F. H., and Geywitz, J., Direct polarographic recording of metal elimination from aquatic samples by coprecipitation with ferric hydroxide, *Sci. Total Environ.* **60**:57–65 (1987).

Funahara, Y., and Nikaido, H., Asymmetric localization of lipopolysaccharides on the outer membrane of *Salmonella typhimurium*, *J. Bacteriol.* **141**:1463–1465 (1980).

Gamble, D. S., Interactions between natural organic polymers and metals in soil and freshwater systems: Equilibria, in M. Bernhard, F. E. Brinkman, and P. J. Sadler (eds.), *The Importance of Chemical "Speciation" in Environmental Processes*, Springer-Verlag, Berlin, 1986, pp. 217–236.

Gardiner, J., The chemistry of cadmium in natural water. II. The adsorption of cadmium on river muds and naturally occurring solids, *Water Res.* **8**:157–164 (1974).

Gessa, C., De Cherchi, M. L., Melis, P., Micera, G., and Erre, L. S., Anion-induced metal binding in amorphous aluminum hydroxide, *Colloids Surf.* **11**:109–117 (1984).

Gilmour, C. C., Tuttle, J. H., and Means, J. C., Anaerobic microbial methylation of inorganic tin in estuarine sediment slurries, *Microb. Ecol.* **14**:233–242 (1987).

Gmeiner, J., Covalent linkage of lipoprotein to peptidoglycan is not essential for outer membrane stability in *Proteus mirabilis*, *Arch. Microbiol.* **121**:177–180 (1979).

Gorman, W. C., and Skogerboe, R. K., Effects of metal complexation on the accuracy of anodic stripping voltammetry, *Anal. Chim. Acta* **187**:325–331 (1986).

Griffiths, E., Availability of iron and survival of bacteria in infection, in C. S. F. Easmon, J. Jelijaszewicz, M. R. W. Brown, and P. A. Lambert (eds.), *Medical Microbiology 3, Role of the Envelope in the Survival of Bacteria in Infection*, Academic Press, London, 1983, pp. 153–177.

Hancock, R. E. W., Alterations in outer membrane permeability, *Ann. Rev. Microbiol.* **38**:237–264 (1984).

Harden, V. P., and Harris, J. O., Isoelectric point of bacterial cells, *J. Bacteriol.* **65**:198–202 (1953).

Hausinger, R. P., Nickel utilization by microorganisms, *Microbiol. Rev.* **51**:22–42 (1987).

Hayashi, H., Araki, Y., and Ito, F., Occurrence of glucosamine residues with free amino groups in cell wall peptidoglycan from Bacilli as a factor responsible for resistance to lysozyme, *J. Bacteriol.* **113**:592–598 (1973).

Hayes, K. F., Roe, A. L., Brown, G. E., Jr., Hodgson, K. O., Leckie, J. O., and Parks, G. A., In situ x-ray absorption study of surface complexes: Selenium oxyanions on αFeOOH, *Science (USA)* **238**:783–786 (1987).

Heptinstall, S., Archibald, A. R., and Baddiley, J., Teichoic acids and membrane function in bacteria, *Nature (London)* **225**:519–521 (1970).

Hoyle, B., and Beveridge, T. J., Binding of metallic ions to the outer membrane of *Escherichia coli*, *Appl. Environ. Microbiol.* **46**:749–752 (1983).

Hoyle, B. D., and Beveridge, T. J., Metal binding by the peptidoglycan sacculus of *Escherichia coli* K-12, *Can. J. Microbiol.* **30**:204–211 (1984).

Hughes, A. H., Hancock, I. C., and Baddiley, J., The function of teichoic acids in cation control in bacterial membranes, *Biochem. J.* **132**:83–93 (1973).

Huljev, D. J., Interaction of some metals between marine-origin humic acids and aqueous solutions, *Environ. Res.* **40**:470–478 (1986).

Hussey, H., Sueda, S., Cheah, S.-C., and Baddiley, J., Control of teichoic acid synthesis in *Bacillus licheniformis* ATCC 9945, *Eur. J. Biochem.* **82**:169–174 (1978).

Hutchins, S. R., Davidson, M. S., Brierley, J. A., and Brierley, C. L., Microorganisms in reclamation of metals, *Ann. Rev. Microbiol.* **40**:311–336 (1986).

Ingraham, J. L., Maaløe, O., and Neidhardt, F. C., *Growth of the Bacterial Cell*, Sinauer, Sunderland, MA, 1983.

Iwasaki, H., Shimada, A., and Ito, E., Comparative studies of lipoteichoic acids from several *Bacillus* strains, *J. Bacteriol.* **167**:508–516 (1986).

Jannasch, H. W., and Taylor, C. D., Deep-sea microbiology, *Ann. Rev. Microbiol.* **38**:487–514 (1984).

Jardim, W. F., and Pearson, H. W., Copper toxicity to cyanobacteria and its dependence on extracellular ligand concentration and degradation, *Microb. Ecol.* **11**:139–148 (1985).

Jarrell, K. F., Saulnier, M., Ley, A., Inhibition of methanogenesis in pure cultures by ammonia, fatty acids, and heavy metals, and protection against heavy metal toxicity by sewage sludge, *Can. J. Microbiol.* **33**:551–554 (1987).

Jean, G. E., and Bancroft, G. M., Heavy metal adsorption by sulphide mineral surfaces, *Geochim. Cosmochim. Acta* **50**:1455–1463 (1986).

Jones, J. B., Bowers, B., and Stadtman, T. C., *Methanococcus vanniellii:* Ultrastructure and sensitivity to detergents and antibiotics, *J. Bacteriol.* **130**:1357–1363 (1977).

Jones, J. G., Gardner, S., and Simon, B. M., Reduction of ferric iron by heterotrophic bacteria in lake sediments, *J. Gen. Microbiol.* **130**:45–51 (1984).

Kandler, O., Cell wall structures and their phylogenetic implications, *Zbl. Bakt. Hyg. I Abt. Orig. C* **3**:149–160 (1982).

Kandler, O., and König, H., Chemical composition of the peptidoglycan-free cell walls of methanogenic bacteria, *Arch. Microbiol.* **118**:141–152 (1978).

Karlin, R., and Levi, S., Diagenesis of magnetic minerals in recent haemipelagic sediments, *Nature (London)* **303**:327–330 (1983).

Kay, W. W., Buckley, J. T., Ishiguro, E. E., Phipps, B. M., Monette, J. P. L., and Trust, T. J., Purification and disposition of a surface protein associated with virulence of *Aeromonas salmonicida, J. Bacteriol.* **147**:1077–1084 (1981).

Kaya, S., Yokoyama, K., Araki, Y., and Ito, E., N-Acetylmannosaminyl (1-4) N-acetyl-glucosamine a linkage unit between glycerol teichoic acid and peptidoglycan in cell walls of several *Bacillus* strains, *J. Bacteriol.* **158**:990–996 (1984).

Kepkay, P. E., Kinetics of microbial manganese oxidation and trace metal binding in sediments: Results from an in situ dialysis technique, *Limnol. Oceanogr.* **30**:713–726 (1985).

Kessler, R. E., Wicken, A. J., and Shockman, G. D., Increased carbohydrate substitution of lipoteichoic acid during inhibition of protein synthesis, *J. Bacteriol.* **155**:138–144 (1983).

Kojima, N., Araki, Y., and Ito, E., Structural studies on the acidic polysaccharide of *Bacillus cereus* AHV 1356 cell walls, *Eur. J. Biochem.* **148**:479–484 (1985).

Korthals, E. T., and Winfrey, M. R., Seasonal and spatial variations in mercury methylation and demethylation in an oligotrophic lake, *Appl. Environ. Microbiol.* **53**:2397–2404 (1987).

Krell, P. J., and Beveridge, T. J., The structure of bacteria and molecular biology of viruses, *Int. Rev. Cytol.* **17**:15–88 (1987).

Krieg, N. R., and Holt, J. G., *Bergey's Manual of Systematic Bacteriology,* vol. 1, 9th ed., Williams and Wilkins, Baltimore, 1984.

Krueger, W. B., and Kolodziej, B. J., Measurement of cellular copper levels in *Bacillus megaterium* during exponential growth and sporulation, *Microbios.* **17**:141–147 (1976).

Krull, I. S., Analysis of inorganic species by ion chromatography and liquid chromatography, in M. Bernhard, F. E. Brinckman, and P. J. Sadler (eds.), *The Importance of Chemical "Speciation" in Environmental Processes,* Springer-Verlag, Berlin, 1986, pp. 579–611.

Kruyssen, F. J., de Boer, W. R., and Wouters, J. T. M., Effects of carbon source and growth rate on cell wall composition of *Bacillus subtilis* subsp. niger, *J. Bacteriol.* **144**:238–246 (1980).

Kruyssen, F. J., de Boer, W. R., and Wouters, J. T. M., Cell wall metabolism in *Bacillus subtilis* subsp. niger: Effects of changes in phosphate supply to the culture, *J. Bacteriol.* **146**:867–876 (1981).

Kurek, E., Czaban, J., and Bollag, J. M., Sorption of cadmium by microorganisms in competition with other soil constituents, *Appl. Environ. Microbiol.* **43**:1011–1015 (1982).

Kushwaha, S. C., Kates, M., Sprott, G. D., and Smith, I. C. P., Novel complex polar lipids from the methanogenic archaebacterium *Methanospirillum hungatei, Science (USA)* **211**:1163–1164 (1981).

Lam, J. S., Lam, M. Y. C., MacDonald, L. A., and Hancock, R. E. W., Visualization of *Pseudomonas aeruginosa* O antigens by using a protein A-dextran-colloidal gold conjugate with both immunoglobulin G and immunoglobin M monoclonal antibodies, *J. Bacteriol.* **169**:3531–3538 (1987).

Lambert, P. A., Hancock, I. C., and Baddiley, J., Occurrence and function of membrane teichoic acids, *Biochim. Biophys. Acta* **472**:1–12 (1977).

Lang, W. K., Glassey, K., and Archibald, A. R., Influence of phosphate supply on teichoic acid and teichuronic acid content of *Bacillus subtilis* cell walls, *J. Bacteriol.* **151**:367–375 (1982).

Lanyi, J. K., Salt-dependent properties of protein from extremely halophilic bacteria, *Bacteriol. Rev.* **38**:272–290 (1974).

Laxen, D. P. H., Trace metal adsorption/coprecipitation on hydrous ferric oxide under realistic conditions, *Water Res.* **19**:1229–1236 (1985).

Lifely, M. R., Tarelli, E., and Baddiley, J., The teichuronic acid from the walls of *Bacillus licheniformis* ATCC 9945, *Biochem. J.* **191**:305–318 (1980).

Lovley, D. R., Organic matter mineralization with the reduction of ferric iron: A review, *Geomicrobiol. J.* **5**:375–399 (1987).

Lovley, D. R., and Phillips, E. J. P., Availability of ferric iron for microbial reduction in bottom sediments of the freshwater tidal Potomac river, *Appl. Environ. Microbiol.* **52**:751–757 (1986a).

Lovley, D. R., and Phillips, E. J. P., Organic matter mineralization with reduction of ferric iron in anaerobic sediments, *Appl. Environ. Microbiol.* **51**:683–689 (1986b).

Lovley, D. R., and Phillips, E. J. P., Competitive mechanisms for inhibition of sulfate reduction and methane production in the zone of ferric iron reduction in sediments, *Appl. Environ. Microbiol.* **53**:2636–2641 (1987).

Lovley, D. R., Stolz, J. F., Nord, G. L., Jr., and Phillips, E. J. P., Anaerobic production of magnetite by a dissimilatory iron-reducing microorganism, *Nature (London)* **330**:252–254 (1987).

Lugtenberg, E. J. J., and Peters, R., Distribution of lipids in cytoplasmic and outer membranes of *Escherichia coli* K-12, *Biochim. Biophys. Acta.* **441**:38–47 (1976).

Mann, S., Molecular recognition in biomineralization, *Nature (London)* **332**:119–124 (1988).

Marshall, K. C., Clay minerology in relation to survival of soil bacteria, *Ann. Rev. Phytopath.* **13**:357–373 (1975).

Martin, N. L., and Beveridge, T. J., Gentamicin interaction with *Pseudomonas aeruginosa* cell envelope, *Antimicrob. Agents Chemother.* **29**:1079–1087 (1986).

Matthews, T. H., Doyle, R. J., and Streips, U. N., Contribution of peptidoglycan to the binding of metal ions by the cell wall of *Bacillus subtilis*, *Curr. Microbiol.* **3**:51–53 (1979).

McIlroy, L. N., DePinto, J. V., Young, T. C., and Martin, S. C., Partitioning of heavy metals to suspended solids of the Flint River, Michigan, *Environ. Toxicol. Chem.* **5**:609–623 (1986).

McLean, R. J. C., and Beveridge, T. J., Influence of metal ion charge on their binding capacity to bacterial capsules, *Abstr. 88th Ann. Meet. Am. Soc. Microbiol.* Q146, 1988, p. 307.

McLean, R. J. C., Nickel, J. C., Cheng, K.-J., and Costerton, J. W., The ecology and pathogenicity of urease-producing bacteria in the urinary tract, *Crit. Rev. Microbiol.* **16**:37–79 (1988).

McQuillen, J., The bacterial surface. I. Effect of cetyltrimethylammonium bromide on the electrophoretic mobility of certain gram-positive bacteria, *Biochim. Biophys. Acta* **5**:463–471 (1950).

Mittelman, M. W., and Geesey, G. G., Copper-binding characteristics of exopolymers from a freshwater sediment bacterium, *Appl. Environ. Microbiol.* **49**:846–851 (1985).

Moore, J. W., and Ramamoorthy, S., *Heavy Metals in Natural Waters*, Springer-Verlag, New York, 1984.

Mühlradt, P. F., and Golecki, J. R., Asymmetrical distribution and artifactual reorientation of lipopolysaccharide in the outer membrane bilayer of *Salmonella typhimurium*, *Eur. J. Biochem.* **51**:343–352 (1975).

Mühlradt, P. F., Menzel, J., Golecki, J. R., and Speth, V., Outer membrane of *Salmo-*

nella. Sites of export of newly synthesized lipopolysaccharide on the bacterial surface, *Eur. J. Biochem.* **35:**471–481 (1973).

Mühlradt, P. F., Menzel, J., Golecki, J. R., and Speth, V., Lateral mobility and surface density of lipopolysaccharide in the outer membrane of *Salmonella typhimurium,* *Eur. J. Biochem.* **43:**533–539 (1974).

Murray, R. G. E., Steed, P., and Elson, H. E., The location of the mucopeptide in sections of the cell wall of *Escherichia coli* and other gram-negative bacteria, *Can. J. Microbiol.* **11:**547–560 (1965).

Neilands, J. B., Microbial iron compounds, *Ann. Rev. Biochem.* **50:**715–731 (1981).

Nikaido, H., and Vaara, M., Outer membrane, in F. C. Neidhardt (ed.), *Escherichia coli and Salmonella typhimurium Cellular and Molecular Biology,* vol. 1, American Society for Microbiology, Washington, D.C., 1987, pp. 7–22.

Norberg, A. B., and Persson, H., Accumulation of heavy-metal ions by *Zoogloea ramigera, Biotechnol. Bioeng.* **26:**239–246 (1984).

Nürnberg, H. W., Voltametric tract analysis in ecological chemistry of toxic metals, *Pure Appl. Chem.* **54:**853–878 (1982).

Olson, G. J., Microbial intervention in trace element-containing industrial process streams and waste products, in M. Bernhard, F. E. Brinckman, and P. J. Sadler (eds.), *The Importance of Chemical "Speciation" in Environmental Processes,* Springer-Verlag, Berlin, 1986, pp. 493–512.

Padan, E., Adaptation of bacteria to external pH, in M. J. Klug and C. A. Reddy (eds.), *Current Perspectives in Microbial Ecology,* American Society for Microbiology, Washington, D.C., 1984, pp. 49–55.

Patel, G. B., Sprott, G. D., Humphrey, R. W., and Beveridge, T. J., Comparative analyses of the sheath structures of *Methanothrix concilii* GP6 and *Methanospirillum hungatei* GPI and JFI, *Can. J. Microbiol.* **32:**623–631 (1986).

Petruzzelli, G., Guidi, G., and Lubrano, L., Ionic strength effect on heavy metal adsorption by soil, *Commun. Soil. Sci. Plant. Anal.* **16:**971–986 (1985).

Raspor, B., Nurnberg, H. W., Valenta, P., and Branica, M., Studies in seawater and lake water on interactions of trace metals with humic substances isolated from marine and estuarine sediments. I. Characterization of humic substances, *Mar. Chem.* **15:**217–230 (1984a).

Raspor, B., Nurnberg, H. W., Valenta, P., and Branica, M., Studies in seawater and lake water on interactions of trace metals with humic substances isolated from marine and estuarine sediments. II. Voltammetric investigations on trace metal complex formatin in the dissolved phase, *Mar. Chem.* **15:**231–249 (1984b).

Richardson, D. H. S., Applications of voltammetry in environmental science, *Environ. Pollut. Ser. B.* **10:**261–276 (1985).

Sakata, M., Diagenetic remobilization of manganese, iron, copper and lead in anoxic sediment of a freshwater pond, *Water Res.* **19:**1022–1038 (1985).

Salomons, W., and Förstner, V., *Metals in the Hydrocycle,* Springer-Verlag, Berlin, 1984.

Schleifer, K. H., and Kandler, O., Peptidoglycan types of bacterial cell walls and their taxonomic implications, *Bacteriol. Rev.* **36:**407–477 (1972).

Schmelzer, E., Weckesser, J., Warth, R., and Mayer, H., Peptidoglycan of *Rhodopseudomonas viridis:* Partial lack of N-acetyl substitution of glucosamine, *J. Bacteriol.* **149:**151–155 (1982).

Schoer, J., Iron-oxo-hydroxides and their significance to the behaviour of heavy metals in estuaries, *Environ. Technol. Lett.* **6:**189–202 (1985).

Schopf, J. W., and Walker, M. R., Archean microfossils: New evidence of ancient microbes, in J. W. Schopf (ed.), *Earth's Earliest Biosphere: Its Origin and Evolution,* Princeton University Press, Princeton, N.J., 1983, pp. 214–239.

Schott, H., and Young, C. Y., Electrokinetic studies of bacteria. III. Effect of polyvalent metal ions on electrophoretic mobility and growth of *Streptococcus faecalis, J. Pharm. Sci.* **62:**1797–1801 (1953).

Shands, J. W., Jr., and Chun, P. W., The dispersion of gram-negative lipopolysaccharide by deoxycholate, *J. Biol. Chem.* **225:**1221–1226 (1980).

Sherbert, G. V., *The Biophysical Characterization of the Cell Surface,* Academic Press, London, 1978, pp. 78–92.

Shibaev, V. N., Duckworth, M., Archibald, A. R., and Baddiley, J., The structure of a polymer containing galactosamine from walls of *Bacillus subtilis* 168, *Biochem. J.*, **135**:383–384 (1973).

Silver, M., Distribution of iron-oxidizing bacteria in the nordic uranium tailing deposit, Elliot Lake, Ontario, Canada, *Appl. Environ. Microbiol.* **53**:846–852 (1987).

Silver, S., and Misra, T. K., Plasmid-mediated heavy metal resistances, *Ann. Rev. Microbiol.* **42**:717–743 (1988).

Sleytr, U. B., and Messner, P., Crystalline surface layers on bacteria, *Ann. Rev. Microbiol.* **37**:311–339 (1983).

Smiley, D. W., and Wilkinson, B. J., Survey of taurine uptake and metabolism in *Staphylococcus aureus*, *J. Gen. Microbiol.* **129**:2421–2428 (1983).

Smith, R. E., *Ion Chromatography Applications*, CRC Press, Boca Raton, FL, 1988.

Snoeyink, V. L., and Jenkins, D., *Water Chemistry*, Wiley, New York, 1980.

Sonnenfeld, E. M., Beveridge, T. J., Koch, A. L., and Doyle, R. J., Asymmetric distribution of charge on the cell wall of *Bacillus subtilis*, *J. Bacteriol.* **163**:1167–1171 (1985).

Sørensen, J., Reduction of ferric iron in anaerobic, marine sediment and interaction with reduction of nitrate and sulfate, *Appl. Environ. Microbiol.* **43**:319–324 (1982).

Sprott, G. D., Shaw, K. M., and Jarrell, K. F., Isolation and chemical composition of the cytoplasmic membrane of the archaebacterium *Methanospirillum hungatei*, *J. Biol. Chem.* **258**:4026–4031 (1983).

Steber, J., and Schleifer, K. H., *Halococcus morrhuae*: A sulfated heteropolysaccharide as the structural component of the bacterial cell wall, *Arch. Microbiol.* **105**:173–177 (1975).

Stewart, M., Beveridge, T. J., and Sprott, G. D., Crystalline order to high resolution in the sheath of *Methanospirillum hungatei*: A cross-beta structure, *J. Mol. Biol.* **183**:509–515 (1985).

Stewart, M., Beveridge, T. J., and Trust, T. J., Two patterns in the *Aeromonas salmonicida* A-layer may reflect a structural transformation that alters permeability, *J. Bacteriol.* **166**:120–127 (1986).

Strain, S. M., Fesik, S. W., and Armitage, I. M., Structure and metal-binding properties of lipopolysaccharides from heptoseless mutants of *Escherichia coli* studied by ^{13}C and ^{31}P nuclear magnetic resonance, *J. Biol. Chem.* **258**:13,466–13,477 (1983).

Summers, A. O., Organization, expression, and evolution of genes for mercury resistance, *Ann. Rev. Microbiol.* **40**:607–634 (1986).

Sutherland, I. W., Bacterial exopolysaccharides—their nature and production, in I. W. Sutherland (ed.), *Surface Carbohydrates of the Prokaryotic Cell*, Academic Press, London, 1977, pp. 27–96.

Sutherland, I. W., Biosynthesis and composition of gram-negative bacterial extracellular and wall polysaccharides, *Ann. Rev. Microbiol.* **39**:243–270 (1985).

Troy, F. A., Chemistry and biosynthesis of the poly (γ-D-glutamyl) capsule in *Bacillus licheniformis*. II. Characterization and structural properties of the enzymatically synthesized polymer, *J. Biol. Chem.* **248**:316–324 (1973).

Tugel, J. B., Hines, M. E., and Jones, G. E., Microbial iron reduction by enrichment cultures isolated from estuarine sediments, *Appl. Environ. Microbiol.* **52**:1167–1172 (1986).

Van den Berg, C. M. G., The determination of trace metals in sea-water using cathodic stripping voltammetry, *Sci. Total Environ.* **49**:89–99 (1986).

Van Driel, D., Wicken, A. J., Dickson, M. R., and Knox, K. W., Cellular location of the lipoteichoic acids of *Lactobacillus fermenti* NCTC 6991 and *Lactobacillus casei*, *J. Ultrastruct. Res.* **43**:483–497 (1973).

Van Leeuwen, H. P., Voltammetric titrations involving metal complexes: effect of kinetics and diffusion coefficients, *Sci. Total Environ.* **60**:45–55 (1987).

Van Loon, J. C., *Chemical Analysis of Inorganic Constituents of Water*, CRC Press, Boca Raton, FL, 1982.

Vitkovic, L., Wall turnover deficiency of *Bacillus subtilis* Ni 15 is due to a decrease in teichoic acid, *Can. J. Microbiol.* **33**:566–568 (1987).

Walker, S. G., and Beveridge, T. J., Amikacin disrupts the cell envelope of *Pseudomonas aeruginosa* ATCC 9027, *Can. J. Microbiol.* **34**:12–18 (1988).

Ward, J. B., Teichoic and teichuronic acids: Biosynthesis, assembly, and location, *Microbiol. Rev.* **45**:211–243 (1981).

Weiss, R. L., Subunit cell wall of *Sulfolobus acidocaldarius*, *J. Bacteriol.* **118:**275–284 (1974).

Wicken, A. J., and Knox, K. W., Lipoteichoic acids: A new class of bacterial antigen, *Science (USA)* **187:**1161–1167 (1975).

Wieland, F., Dompert, W., Bernhardt, G., and Sumper, M., Halobacterial glycoprotein saccharides contain covalently linked sulfate, *FEBS Lett.* **120:**110–114 (1980).

Woese, C. R., Magrum, L. J., and Fox, G. E., Archaebacteria, *J. Mol. Evol.* **11:**245–252 (1978).

Yang, L. L., and Haug, A., Purification and partial characterization of a procaryotic glycoprotein from the plasma membrane of *Thermoplasma acidophilum*, *Biochim. Biophys. Acta* **556:**265–277 (1979).

Zipperle, G. F., Jr., Ezzell, J. W., Jr., and Doyle, R. J., Glucosamine substitution and muramidase susceptibility in *Bacillus anthracis*, *Can. J. Microbiol.* **30:**553–559 (1984).

Extracellular Polymers
for Metal Binding

Gill Geesey

Larry Jang

Introduction

Capsular and slime exopolymers are elaborated by a wide variety of microorganisms in nature. These exopolymers serve as a buffer between the cell wall and the external environment. Under some circumstances the exopolymers serve as a barrier to harmful or toxic substances such as bacteriophage, antibiotics, and biocides. The polymers form a matrix that physically excludes large molecules, thus minimizing their contact with the cell wall and underlying cell components. Many exopolymers also carry a charge which promotes ionic and electrostatic binding of counterions. These interactions may, on the one hand, prevent excess quantities of charged molecules such as heavy metals from approaching the cell surface and yet facilitate the concentration of growth-promoting nutrients present at low concentrations in the surrounding environment for subsequent uptake by the cell.

The influence of metal ions on microorganisms is unique in that over a rather narrow concentration range, their status may change from that of an essential growth-promoting element to one that is toxic. The exopolymers therefore serve a critical function in regulating the exposure levels that the bacterial cells encounter in environments with widely fluctuating metal-ion concentrations. As one of the two main mechanisms of metal concentration by microbial cells, capsular polymers have evolved features which suggest they act as effective modulators of metal-ion concentration at the cell surface.

Tolerance of microorganisms to metals in important industrial processes has long been known to involve extracellular polymers. Early investigations on the interactions of metal ions with bacterial exopolymers arose from studies on the effects of metal wastes on sewage treatment microorganisms. The floc-forming strain 115 of *Zoogloea ramigera* was found to remove significantly more dissolved Co, Cu, Fe, and Ni from solution than a strain that did not produce a gelatinous polymeric matrix around the cells (Friedman and Dugan, 1968). Young cultures of the floc-forming strain exhibited a lower affinity for Zn than cultures which contained more exopolymer. Later, zinc binding by isolated exopolymer was confirmed.

Recent evidence suggests that interactions between capsular and slime exopolymers and certain metal ions contribute to corrosion reactions which have severe economic consequences (Geesey et al., 1986). Interestingly, the metal-recovery industry is interested in utilizing these biologically catalyzed reactions to improve metal-recovery efficiencies from mining operations and other metal-laden effluents. Thus, interactions between metals and these microbial exopolymers are of practical importance.

Ecological and geochemical processes are also influenced by the complexation reactions between metals and surface polymers of microbial cells. The importance of microorganisms in metal mobilization and diagenesis may well involve binding to exopolymers and other microbial metabolites. These processes, while far from being understood, are at least beginning to receive the attention they deserve.

In this chapter, biochemical and physicochemical features which promote interactions of microbial exopolymers with various metal ions in solution will be described. These interactions will be considered in terms of their influence on the exopolymers, the bacteria which produce the exopolymers, and the surrounding environment.

Nature of Bacterial Capsule and Slime

In this chapter, *exopolymer* refers to a macromolecule excreted outside the cell wall proper. In some cases, however, the exopolymer may be an extension of a cell wall component, such as the *O*-antigen portion of the lipopolysaccharide in the outer membrane of gram-negative bacteria. The microbial exopolymers with which most metal studies have been conducted are those which are commonly associated with capsules or slime. Capsules are invariably observed anchoring sessile microorganisms to each other, to other types of living cells, and to inanimate objects. They appear as the outermost envelope layer surrounding the cell. Capsules are elaborated by both gram-negative and gram-positive eubacteria, cyanobacteria, archaebacteria, and unicellular algae. Capsular exopolymers are firmly retained by the cell and can withstand large shear forces (Pedersen et al., 1986). This association enables the cell to establish a stable orientation with respect to its neighbors and environment. Capsular exopolymers may extend from 0.1 to 10 μm or more from the cell surface into the surrounding environment, creating a buffer zone between the outer cell wall and the external environment. The widespread appearance of capsules among natural populations of microorganisms indicates that they are essential to survival.

In laboratory cultures, capsular polymers maintain a more transient association with the cell. In some cases, the exopolymers take the form of what is frequently referred to as "slime." Slime polymers slough from the cell into the surrounding medium. The extent to which this sloughing occurs in the natural environment remains to be determined. There is some evidence that overproduction of capsular polymers leads to sloughing. Production of slime may be controlled by environmental conditions. Different strains of the same microorganism may elaborate exopolymers that maintain different associations with the cell. In aquatic and industrial fields, the term *slime* is often used to refer to the biofilm which develops on submerged surfaces. The biofilm contains cells and their exopolymers and any adsorbed material from the bulk aqueous phase. The properties of slime will therefore differ, depending on how it is defined. In this chapter, the term *slime* will refer to the assemblage of polymeric material that is excreted from microbial cells, including dissolved material adsorbed from the culture solution but excluding the cells and their wall components.

Although most capsule and slime exopolymers described to date are composed of polysaccharide, considerable amounts of protein are frequently recovered in crude exopolymer preparations. Whether the proteins perform any function in this exocellular location remains to be determined. Other cell products such as DNA and RNA are also recovered with the capsule and slime exopolymers. It is likely that some or all of this material represents excretory or lysis products that have become trapped by the polysaccharide molecules. The extent to which they contribute to the overall interactions between the capsular polysaccharide component and other substances in the environment is a question that has not been widely studied.

The exopolysaccharide component of microbial exopolymers usually possesses a repeating sequence of from two to six sugar subunits. Some polymers such as alginic acid, however, possess blocks of repeating sugars in one segment of the polymer that are different from those in other segments. Many bacterial and algal capsular polymers are composed of acidic polysaccharides. The acidic properties are contributed primarily by free carboxyl groups of uronic acid subunits or pyruvylated sugars. In some instances, the exopolysaccharide exhibits branching where two to five sugar subunits extend away from the polymer backbone. When both capsule and slime exopolymers are produced by the same bacterial culture, the chemical composition of their polysaccharides is usually similar (Cohen and Johnstone, 1964).

The subunit composition and structure determine the physical properties of the exopolymer. Most microbial exopolysaccharides are extremely hygroscopic. Greater than 99 percent of the volume of a hydrated exopolymer is contributed by water. Most capsules and slimes form a colloid or gel phase, depending on the nature of the surrounding environment. Aggregation of polymers occurs in the presence of divalent cations. Visible flocs may form under certain conditions. These flocs have been associated with "bulking" of activated sludge in the sewage treatment process. Some slime exopolymers possess unusual rheological properties that have been exploited for various industrial applications. The high viscosity of xanthan, a slime from the plant pathogen *Xanthomonas campestris* and alginic acid, a slime from several different species of bacteria and algae, has made these polymers useful as stabilizing agents in foods. In addition, metal cross-linked xanthan has been used as a fracturing gel in enhanced oil recovery (Menjivar, 1984). These characteristics make capsular and slime exopolymers unique among other naturally occurring biopolymers.

Properties of Microbial Exopolymers Relevant to Metal Binding

Many capsular and slime exopolymers of aquatic microorganisms act as polyanions under natural conditions. The capsules of most bacteria examined to date contain from 5 to 25 percent uronic acids (Sutherland, 1980). Some sugar subunits of capsules and slime possess ketal-linked pyruvate groups which generally contribute around 5 percent of the polymer weight (Sandford et al., 1977; Boyle and Reade, 1983). Several genera of enteric bacteria possess capsules containing sialic acids, which are derivatives of neuraminic acid (Dewitt and Rowe, 1961; Liu et al., 1971). The anionic character of polysaccharides containing any of the above components is conferred by the carboxylic acid moiety present. The presence of the carboxylic acid groups cause the polysaccharides to exhibit pK_a values between 3.5 and 5.0. At neutral to alkaline pH, the partially ionized carboxyl groups are available to interact with positively charged metal ions.

Polysaccharides also contain an abundance of hydroxyl groups which tend to interact with metal ions. The electronegative oxygen atoms of hydroxyl groups are likely to participate in metal interactions with anionic or neutral polysaccharides.

Some polysaccharides contain amino sugars or sugars with amide-linked functional groups. These nitrogen-containing functional groups are capable of reacting with some metals. To date, only a few exopolysaccharides have been found to contain amide or amine substitutions. Consequently, their importance in metal interactions with exopolymers remains to be determined. It is quite likely, however, that the metal-binding capacity contributed by the protein component of exopolymeric material involves these functional groups. Kihn et al. (1987) showed that Cu^{2+} was chelated by peptides and proteins extracted from walls of *Saccharomyces cervisiae*. The binding sites were formed by an amide and a strongly complexing, amine-like ligand. In slightly acidic conditions, Cu^{2+} was bound by oxygen of the amide, whereas at basic pH, NHCO became deprotonated and the negatively charged nitrogen bound the metal.

Mechanisms of Metal Binding to Bacterial Exopolysaccharides

Several classes of polymeric molecules appear to participate in exopolymer-metal interactions. Protein and polysaccharide components from a firmly bound capsular polymer of a freshwater sediment bacterium exhibited similar binding capacities toward cupric ions (Mittelman and Geesey, 1985). Only the mechanisms of interaction between polysaccharides and metal ions, however, will be considered here.

The general interaction between a metal ion and an organic molecule may be described as an acid-base reaction:

$$M^{n+} + LH \rightarrow M^{n+} - L + H^+$$

where M^{n+}, the metal ion, or H^+, the proton, represents the acid, and L, the organic molecule or ligand, represents the base. The release of protons by acidic polysaccharides during exposure to Cu ions has been reported in several instances (Zunino and Martin, 1977; Mittelman and Geesey, 1985). Binding of Cu ions resulted in a shift in the pK_a of a capsular polysaccharide from 4.90 to 4.05 (Mittelman and Geesey, 1985). These data suggest that competition exists between Cu ions and protons at the metal-binding site of acidic exopolymers.

Metal ions tend to form bonds with functional groups containing electron-donating atoms. Four types of interactions between metal ions and exopolymers have been described (Steiner et al., 1976). The two most important types were those that involved salt bridges with carboxyl groups on acidic polymers and those that involved weak electrostatic bonds with hydroxyl groups on neutral polymers. In polysaccharides containing uronic acids or pyruvate groups, lone-pair electrons on oxygen atoms of carboxyl groups have a strong tendency to interact with charge-compensating metal ions. Oxygen atoms in the ether bonds and hydroxyl groups of sugar subunits act as weak electron donors in both acidic and neutral polysaccharides (Martell, 1971). The "S-type" isotherm produced by complexes between metal ions and the capsule of Z. ramigera 115 was proposed to involve primarily hydroxyl groups of the glucose subunits even though the polysaccharide contains free carboxyl groups on the ketal-linked pyruvate residue (Brown and Lester, 1979). This type of metal binding is believed to be the most important mechanism of metal removal in activated sludge.

Manzini et al. (1984) emphasized the importance of carboxyl groups in cupric ion binding. They suggested that an intimate interaction exists between Cu ions and carboxyl groups on acidic polysaccharides. Cupric ion binding to a carboxyl group is electrovalent. In the case of polyuronates, the mode and extent of the interaction are believed to depend on several factors: the nature of the component sugars and their relative distribution in the chain, the magnitude of the overall electrostatic field, and the ratios of Cu to polymer and of Cu to simple supporting electrolyte.

The modes of interaction change as the ratio, R, of total metal ions to COO^- groups varies. As the ratio decreases from a maximum of 0.50, the average number of uronate groups in alginic acid that interact with each Cu ion increases to more than two. Under optimum conditions 85 percent of the COO^- groups can interact with Cu^{2+}. The remaining 15 percent are inaccessible for complex formation. This is due to some extent to competition between Cu ions and counterions such as Na^+ for the charged site on the polymer molecule.

Linear charge density on acidic polymers is believed to influence metal interactions. The stability of the interaction has been shown to increase with increasing charge density (Mathews, 1960). A polysaccharide exhibits a stronger interaction with metal ions when it exists in a gel state (high charge density) than when it exists in solution (lower charge density). Ionic bonds between a single polyvalent cation and two anionic groups on separate polymer chains are believed to contribute to the stability of the gels. This idea is supported by evidence that Cu^{2+} complexes with two uronate groups in the case of alginic acid (Manzini et al., 1984). Charge transfer occurs from the negative charge on the oxygen atom of the COO^- to the Cu^{2+} ion.

Charge transfer is very sensitive to the stereochemical environment. Often, the ionic radius of a metal determines whether complexation will occur. For uncharged polysaccharides, affinity for a metal usually decreases with increasing ionic radius of the metal. In contrast, carboxylated polysaccharides often exhibit preferential binding to cations with large ionic radii. Affinity of alginic acid for the transition metals decreases in the order Cu > Ni, Co > Zn > Mn which roughly follows the Irving-Williams order of decreasing complex stability.

With few exceptions, carboxylated polyanions exhibit higher selectivity for transition metals than do the alkali- and alkaline-earth metals. In some instances, however, the affinity of Mg fluctuates relative to other cations in different anionic polysaccharides. In addition, selectivity coefficients may be different, depending on whether the polymer exists in solution or in a gel state.

Volume change on dissociation of ionizable groups along alginic-acid chains is simply a linear function of the degree of ionization (Paoletti et al., 1981). Dissociation occurs without any interference among the hydration spheres of the ionizable (COO^-) groups. The ionization behavior of alginate does not depend on whether the counterion is Na^+ or K^+. Volume change is related to a change in the extent of hydration of the cation. A net volume increase occurs during the binding of Cu^{2+} to carboxylate groups on sodium alginate.

Cu ions in solution displace Ca ions bound to alginic acid. A high selectivity for Cu over Ca is a characteristic feature of carboxylated polyanions, in general. The binding mechanism that is responsible for the higher affinity for Cu than Ca appears to be independent of free hydroxyl groups in the polymer and of the steric arrangement of the carboxyl groups (Haug and Smidsrod, 1970).

Differences in degree of ion hydration and hence the distance of closest approach between the cation and the site of interaction on the ligand may influence cation selectivity. Energy required for ion dehydration must also be taken into account. Metal ions in solution are always fully coordinated or solvated by water molecules. When a metal ion becomes complexed by an organic ligand, there is a displacement and reordering of water molecules: the number of water molecules displaced depends on the dimension and charge of the metal ion and on the size of the coordinated electron-donating group of the ligand. It was proposed that Cu ions have a larger ordering effect on the water molecules than do Ca ions, and that the displacement of Ca by Cu on the polymer should be favored by the relatively large positive entropic change (Haug and Smidsrod, 1970). The reaction has a positive enthalpy of formation and is entirely entropy-driven. In alginic acid and hyaluronic acid, cations and most water molecules have well-defined positions and may determine gross configuration of the polymer chain.

Paoletti et al. (1981) suggest that the interaction between COO^- groups on alginic acid and Cu^{2+} ions loosens a relatively large number of water molecules from the solvation shells of both ionic species, thus producing a net gain of molecular degrees of freedom in the system.

This occurs when the ratio of moles of Cu^{2+} to equivalents of alginate in solution is less than 0.4. At higher ratios, gel formation occurs.

Order within the liquid subphase may also depend on the chemical properties of the capsule. In the case of M41 capsular polysaccharide of *Escherichia coli*, only the polyanions are ordered and the cations and water molecules behave like a liquid filling the spaces between polymer molecules (Moorhouse et al., 1977).

Differences in selectivity can be due to variations in stereochemistry of the sugar subunits comprising the polysaccharide. For the exchange reactions Ca-Mg, Ca-Sr, Sr-Mg, and Co-Ca, the mannuronic acid residues in alginic acid (gel) have selectivity coefficients of unity (Smidsrod and Haug, 1968). However, selectivity coefficients for exchange reactions Sr-Mg and Co-Ca for guluronic acid residues were 150 and 0.17, respectively. The high selectivity of guluronic acid residues in the alkaline-earth exchange reactions appears to depend on the existence of alginate in an insoluble gel state. Both residues demonstrate a selectivity of Cu over Ca.

The stability and selectivity of complexes between alkaline-earth ions and alginates are believed to be influenced by hydroxyl groups (Cozzi et al., 1969). Differences in affinity among these cations for polymannuronic, polygalacturonic, and polyguluronic acid were attributed to variations in the positioning of hydroxyl groups (Haug and Smidsrod, 1970). It appears that the spacing of coordinating oxygen atoms in the polyuronate molecule determine which cations will bind as well as their stability. The availability of two hydroxyl groups on the uronate subunit is more stabilizing than one. In general, the more coordination bonds that are formed between the metal and the ligand, the greater is the stability of the complex. The formation of stable complexes between metal ions and anionic polysaccharides usually involves both electrovalent (nonionic) and coordinate covalent bonding.

A series of cavities that offer 4-oxygen coordination with Ca ions exists in segments of alginic acid which contain repeating sequences of guluronic acid (Rees, 1972; Kohn, 1975). Cooperative binding of the guluronic-acid segments with functional groups on other adjacent polymer chains is necessary for cavity formation. Fewer cavities exist in segments which contain repeating sequences of galacturonic. These cavities offer only three oxygen atoms for coordination. Only shallow cavities are formed with polymannuronic acid. The depth of the cavity and the number of available coordinating oxygen atoms are believed to determine the degree of selectivity and stability for Ca and Sr ions. Steric fit therefore seems to be important and supports the evidence that ionic radius of the metal frequently determines whether complexation will occur.

Using the linear charge model for the enthalpy of dilution of polyelectrolyte solutions, a study was performed by Pass and Hales (1981) to investigate the effects of the cation on the enthalpy of dilution of alkali-metal salts of alginate. They proposed that the enthalpy change that occurs during the interaction between polyelectrolytes and counterions may be considered in terms of at least three possible effects: enthalpy changes occurring on dilution, enthalpy changes arising from interactions between the ion atmosphere of the polyelectrolyte and the counterion, and enthalpy changes involving the site-bound counterions.

The enthalpy of dilution depends on the charge density on the polyelectrolyte. The enthalpy is increasingly exothermic as the charge density decreases and as the size of the counterion decreases (K < Na < Li). The enthalpy of dilution of carboxylated polyanions shows a dependence on the counterion: Li > Na > K.

The enthalpy of interaction of a polysaccharide with a cation is influenced by the concentration of the cation. The enthalpy of mixing the salt form of a carboxylated polysaccharide with increasing concentrations of a chloride salt of the same cation becomes more endothermic as the concentration of the salt increases (Pass and Hales, 1981). The enthalpy change depends on the nature of the cation.

Mixing alginate of one cation salt with the salt of another cation alters the concentration of the counterions in the ion atmosphere and at the binding site (Pass and Hales, 1981). Enthalpy changes become more endothermic when an alkali-metal alginate is mixed with an alkali-metal chloride containing a lighter counterion. The replacement of a site-bound cation by a heavier cation is an exothermic process.

The equilibrium reactions between the alkali-metal cations and carboxylated polyanions exhibit affinities in which Li > Na > K. These observations, along with the enthalpy data, suggest that the endothermic replacement of a site-bound cation with a lighter cation is accompanied by an increase in total entropy. This would then be consistent with the observed volume changes which are related to changes in the extent of hydration of the counterion. Exchange of a cation such as Li from solution with a site-bound cation such as Na results in a positive enthalpy change and an increase in entropy when there is a net loss of bound water molecules. Metal-ion complex formation with carboxylated ligands has a positive enthalpy of formation and is also an entropy-driven process.

The bond strength between the site-bound counterion and the carboxylated polyanion is not controlled by the nature of the counterion (Paoletti et al., 1981). Binding will occur as long as there is a greater release of water molecules. Thus, binding affinity is related more to entropy changes than to bond strength, which is related to enthalpy changes.

For interactions between counterions and sulfated polyanions, less dehydration of the counterions occurs than that observed with carboxylated polyanions. The sequence of volume changes is reversed, so Li < Na < K. The enthalpy change and the equilibrium, free-energy change follow the same sequence. The sequence of enthalpy changes indicates that the bond strength between a site-bound alkali metal and a sulfated polyanion such as dextran sulfate or carageenan increases with increasing cation size.

Mo binding by extracellular slimes of rhizosphere bacteria was investigated by Tan and Loutit (1976). Greater than 90 percent of the Mo associated with cells of a *Pseudomonas* sp. was bound to the slime. Mo was mainly bound by a passive process to the uronic acid-containing subunits of the slime layer. The slime from *Pseudomonas aeruginosa* was reported to bind Mo through the glucuronic-acid subunit (Stojkovski et al., 1986). Cu was also bound by this subunit. Glucuronic acid contributed 32 percent of the total slime weight. Complexation occurred through the two oxygen atoms of the carboxyl group and the oxygen atom of the hydroxyl groups of C-3 and C-4 of the uronic-acid molecule. As much as 97 percent of the total Mo concentrated by the cells was bound to the slime.

Enzymes were used by Flatau et al. (1985) to identify sites of Cd binding by cell surface polymers of a marine pseudomonad. Of the enzymes tested, pectinase was found to liberate the greatest amount of Cd-binding material from cell envelopes, suggesting that galacturonic-acid residues were the most important complexing agent. Release of lesser amounts of Cd was achieved by treatment of the envelopes with cellulase. It was suggested that neutral sugars also contributed to the binding reaction.

A large number of metals have been reported to successfully cross-link polysaccharides. Carboxyl groups are active cross-linking sites on acidic polysaccharides while for neutral polysaccharides, any number of hydroxyl groups should be able to serve as cross-linking sites. There are several processes which compete with the cross-linking reaction. One process is the hydrolysis of the metal ion. In aqueous solutions, hydrated metal ions behave as acids and may hydrolyze by releasing one or more protons from coordinated water molecules (Menjivar, 1984). The extent of hydrolysis depends on the polarizing properties of the metal ion and the pH of the solution. At low pH, protonation of the ligand competes with the cross-linking reaction. The optimum pH range for the gelation of a polymer with a metal will depend on the acid-base properties of the ligands on the polymer and on the hydrolysis properties of the metal.

Purified preparations of the floc polysaccharide isolated from *Zoogloea* spp. bound different amounts of alkaline-earth and transition-metal ions. The relative affinity of the cell-floc material for the transition-metal ions was determined to be Fe > Cu > Co > Ni (Dugan and Pickrum, 1972). Approximately 25 percent of the floc weight was contributed by bound metal ions. The combined polysaccharide species purified from crude floc produced by cells of *Z. ramigera* 115 was reported to bind 0.25 μmol Fe^{3+} per milligram polysaccharide (Ikeda et al., 1982). Dugan proposed that metal binding occurred through an exchange reaction with polymer-bound water molecules and that the removal of water resulted in the observed flocculation that occurred in the presence of metals.

Flocculation of solutions containing either Cu or Cd with floc-producing strains of *Z. ramigera* 115 reduced the dissolved metal concentration to less than 0.1 and 0.0005 g/L, respectively. The complexed metals could be released from the flocs at pH 3 to 4. These results suggest that the metals were complexed by carboxyl groups on the polymers via ionic bonding.

Metal uptake was compared between capsule-producing and noncapsulated strains of *Klebsiella* (*Enterobacter*) *aerogenes* (Rudd et al., 1983). Of the metals tested (Cu, Cd, Ni, Mn, Co), all but Ni were accumulated to a greater extent by the encapsulated strain. Six- to 10-fold more Cu and Cd were concentrated by the capsular polymers than was taken up by the cells. A shift from cellular uptake to extracellular binding of these metals was observed in chemostat cultures as the dilution rate was decreased. The extent of metal removal from solution by the capsulated strain followed the order Cu > Cd > Co > Mn > Ni. This sequence of affinity is similar to that displayed by *Z. ramigera* 115 flocs and to laboratory-scale activated sludge biomass.

The relative affinity of isolated extracellular polymer from *E. aerogenes* for metal ions differed slightly from that exhibited by encapsulated cells. Freundlich isotherms indicated that the relative affinities of metals for sites on the exopolymer material followed the sequence Cd > Co > Ni > Mn (Brown and Lester, 1982). At 0.1 mg/L free Co, Cd, or Ni, the polymer was close to saturation at approximately 1 mg metal-bound per gram polymer.

The sheath material of a *Microcystis* species of cyanobacteria concentrated Fe and Mn by factors of 10^4 and 10^5, respectively (Amemiya and Nakayama, 1984). The sheath material contained fivefold higher concentrations of metals than whole cells. Based on Freundlich isotherms, the affinity of the sheath for metals followed the series Fe > Zn > Mn > Cu, Ni. The sheath contained 35 to 47 percent carbohydrate, 12 to 15 percent uronic acids, and 18 to 24 percent protein.

On the basis of these and other studies conducted to date, it may be concluded that bacterial exopolymers exhibit selectivity for metal ions. Oxygen atoms of carboxyl and hydroxyl groups participated in metal-ion binding by bacterial exopolysaccharides. The selectivity for one cation over another is determined by a variety of factors involving the properties of both the cation and the polysaccharide. Some features of polysaccharides which promote interactions with metal ions are clearly distinct from those exhibited by classical polyelectrolytes.

Binding Constants for Exopolymer-Metal Complexes

The type of quantitative information needed to make valid comparisons of affinities of different exopolymers for metal ions is often not available. As a result, it is difficult to compare binding strength, competition between different metals, and binding site density for those exopolymers that have been shown to interact with metal ions. Stabilities of complexes between metal ions and ligand-binding sites are based on the equilibrium constants for their formation. To date, only a few stability constants have been determined for bacterial exopolymer-metal complexes.

Rudd et al. (1984) obtained log K_i values for Cu, Cd, Co, and Ni complexes with capsular material from *E. aerogenes* Type 64. The constants were determined at one pH (6.3), one polymer concentration, and at metal concentrations ranging from 0.2 μM to 0.1 mM. An indirect correlation was observed between the stability constant and the number of binding sites on the polymer molecule. The number of binding sites per molecule (1.7 $\times$ 10^6 MW) ranged from 51 for Cd to 7 for Cu. The complexation capacity (mol/g dry weight) was 4 $\times$ 10^{-6} and 3 $\times$ 10^{-5} for Cu and Cd, respectively. Only one type of binding site was identified for each metal. The binding sites exhibited some specificity for the different metal ions. Only Co was reduced in overall binding when the polymer was exposed to combinations of metals. These observations were contradicted, however, by evidence indicating that the polymer possessed only two distinct types of binding sites. The capsule contained neutral hexose sugars, uronic acids, and sugars substituted with acetate and pyruvate.

The conditional stability constants and maximum binding activity (MBA) of Cu bindings sites on protein and polysaccharide components of the capsule from a freshwater sediment bacterium (FRI) were compared by Mittelman and Geesey (1985). As the protein(s) component was removed, binding sites were lost, but the stability constant remained relatively constant. A highly purified polysaccharide fraction exhibited an MBA of 253 nmol Cu per milligram carbohydrate. On a dry-weight basis, humic acid possessed an MBA 20 times higher than

that of the crude exopolymer. Their stability constants were comparable, however. Infrared spectroscopic analysis indicated that Cu binding involved a carboxyl residue on the polymer molecule (Geesey et
al., 1988). Experiments conducted at Cu concentrations, pH, and temperatures of sediments from which the bacterium was isolated indicated that the exopolymer was capable of binding the metal under
natural conditions. In view of these results, it is likely that bacterial
exopolymers compete with other metal complexing agents for metal
ions in nature.

Intrinsic Stability Constants

Although thermodynamic binding constants for metal-ligand complexes are used to compare affinities of different metals for a particular ligand and for different ligands, the values determined are conditional in nature when they are based on Langmuir's model and
Scatchard's plotting procedure. Stability constants obtained by these
methods vary with environmental conditions such as pH, ionic
strength, polymer concentration, and temperature. The determination
of intrinsic stability constants and binding-site densities, which are
independent of reaction conditions, requires an accurate estimation of
the activities of interacting chemical species at the site of complex formation. Such information has been difficult to obtain in the case of
anionic polysaccharides.

The electrostatic field component created by the negatively charged
reaction sites of anionic polysaccharides produces a higher concentration of counterions in the aqueous region immediately adjacent to the
polymer chain (polymer subphase) than that found in the bulk aqueous phase. The activities of the counterions in the polymer subphase
can be estimated, however, from the measured activities in the bulk
solution by applying Donnan equilibrium theory (Geesey and Jang,
1989). Base titration of the polymer at different ionic strengths provides a partition coefficient $10^{n\Delta pK}$ (where n = valence of the
counterion and $\Delta pK = pK_{HA}^{app} - pK_{HA}^{int}$, where superscripts app and
int refer to the apparent and intrinsic acid dissociation constants, respectively), which describes the distribution of the counterion between
the polymer subphase and the bulk aqueous phase. A numerical procedure based on this multiphase model, the colligative properties of
the polymer solution, and Donnan equilibrium theory may then be
used to estimate the volume of the polymer. Dividing the moles of ionizable ligands by the volume of the polymer subphase (instead of the
volume of the bulk liquid) provided an accurate description of the concentration of ligands that were available for Cu binding by alginic
acid (Jang et al., 1989).

Since polyvalent metal ions appear to bind more than one type of ligand on the polysaccharide molecule, unique stability constants exist for each type of interaction. This has been shown to be the case for Cu binding by alginic acid (Geesey and Jang, 1989). If titration of the polymer in the presence of metal yields data that do not provide the same stability constant for the reaction

$$M^{2+} + 2HA = MA^+ + 2H^+ + A^-; \qquad D_1 \text{ (monodentate complex)}$$

at different ionic strengths, fractions of ligands ionized, and polymer concentrations, then it is likely that a portion of the bound metal ions formed a bidentate complex:

$$M^{2+} + 2HA = MA_2 + 2H^+; \qquad D_2 \text{ (bidentate complex)}$$

D_1 and D_2 are obtained from the slope and Y-intercept of plots of D'_1 versus A^-/V_p and D'_2 versus V_p/A^-, where

$$D'_1 = \frac{(H^+)^2 [M_{sb}][A^-]}{(M^{2+})[HA]^2}$$

In this equation, M_{sb} ($MA^+ + MA_2$) is the moles of site-bound metal in the polymer subphase and replaces MA^+ when the apparent distribution coefficient D'_1 is defined; V_p is the polymer subphase volume, and the other symbols have their usual meanings. The parentheses refer to activity, and the brackets refer to a grouping of mathematical terms. Similarly, if MA_2 is replaced by M_{sb}, another distribution coefficient, D'_2, may be defined as

$$D'_2 = \frac{(H^+)^2 [M_{sb}]V_p}{(M^{2+})[HA]^2}$$

D_1 and D_2 are related to D'_1 by the expression

$$D'_1 = D_1 + \left[\frac{D_2 A^-}{V_p} \right]$$

and to D'_2 by the expression

$$D'_2 = D_2 + \left[\frac{D_1 V_p}{A^-} \right]$$

The intrinsic stability constants $\beta_{MA^+}{}^{int}$ and $\beta_{MA_2}{}^{int}$ of monodentate and bidentate complexes, respectively, formed between the metal and ligands on the polymer are obtained by dividing D_1 or D_2 by the square of the inverse of the acid dissociation constant (K_{HA}). By this approach, stability constants for the monodentate complex and bidentate complex between alginic acid and Cu^{2+} ions were found to be 832 and 67, respectively (Jang et al., 1989).

The proportion of metal bound to the polymer via monodentate and bidentate complexes can be calculated from the expression

$$\frac{\text{Moles monodentate complex}}{\text{Moles bidentate complex}} = \frac{D_1^{\text{int}}}{D_2^{\text{int}}} \times \frac{A^-}{V_p} = \frac{\beta_1^{\text{int}}}{\beta_2^{\text{int}}} \times \frac{A^-}{V_p}$$

From these determinations it was possible to compare the values obtained for the intrinsic stability constant $\beta_{\text{MA}}^{\text{int}}$ and the conditional stability constant $\beta_{\text{MA}}^{\text{app}}$. For the monodentate complex constant the value for $\beta_{\text{MA}}^{\text{app}}$ is higher than the $\beta_{\text{MA}}^{\text{int}}$ by the factor $10^{2\Delta pK}$, and for the bidentate complex $\beta_{\text{MA}}^{\text{app}}$ is higher than $\beta_{\text{MA}}^{\text{int}}$ by the factor $10^{2\Delta pK}V_s/V_p$, where subscripts s and p refer to the volumes of the bulk liquid phase and polymer subphase, respectively. Both the partition coefficient $10^{2\Delta pK}$ and the volume of the polymer subphase vary with environmental conditions as well as the degree of polymer ionization and extent of metal binding to the polymer.

Influence of Environmental Conditions on Properties of Exopolysaccharides Relevant to Metal Interactions

Bacterial exopolymers are highly influenced by growth conditions in the environment. Metal adsorption by crude floc from *Z. ramigera* 115 was influenced by pH. A maximum of 3 mmol Cu was adsorbed per gram dry weight at pH 6 to 7 and a maximum of 1.8 mmol Cd per gram dry weight at pH 8.0 (Norberg and Persson, 1984). The binding capacity of the floc was not influenced by changes in ionic strength. The capacity of the floc to bind iron was enhanced by the presence of phosphate (Dugan and Pickrum, 1972).

Tait et al. (1986) showed that exopolysaccharide synthesis by *X. campestris* decreased as the growth rate decreased. Exopolymer isolated from chemostat cultures at high dilution rates had higher acetyl content and lower pyruvate content than that recovered from cultures at low dilution rates. Uhlinger and White (1983) provided evidence that exopolymer formation by *Pseudomonas atlantica* was enhanced by an increase in available surface area. In estuarine sediments, the exopolymer contributed a carbon content at least as great as the cells themselves. The greatest accumulation of exopolymer occurred in the stationary phase when cells were stressed. The composition of the *P. atlantica* exopolymer changed during the growth cycle. The proportion and amount of uronic acids increased relative to neutral hexoses as the rate of exopolymer synthesis increased. In other studies, the quantity and composition of slime polysaccharide by *P. atlantica* was found to vary with dilution rate in chemostat cultures (Gordon et al., 1988). At low dilution rates (0.005 h^{-1}) the uronic acid content was lower than at high dilution rates (0.01 to 0.04 h^{-1}). Galacturonic-acid content increased with increasing dilution rate. Glucuronic acid appeared only at the highest dilution rate. These results indicate that the properties of bacterial exopolysaccharides are quite variable and appear to be determined by conditions of the surrounding environment. It is therefore likely that the metal-binding characteristics of the exopolymers also depend on the environmental conditions.

Differences in the intrinsic stability constants for Cu binding were observed for slime polymers from chemostat cultures of *P. atlantica* grown at different dilution rates (unpublished results). The highest stability constants for both mono- and bidentate complexes occurred at a dilution rate of 0.015 h^{-1}. At this dilution rate, the exopolymer exhibited the highest carbohydrate content (400 µg/mg dry-weight polymer). These results suggest that the carbohydrate component of the exopolymer exhibits the greatest stability with respect to cupric ions.

Influence of Metals on Properties of Microbial Exopolysaccharides

Metals are known to influence the production of capsule and slime polysaccharides. Metal and metalloid ions serve as cofactors in polysaccharide synthesis. Exopolysaccharide synthesis by *E. aerogenes* is stimulated by Mg, K, and Ca ions (Wilkinson and Stark, 1956). Polysaccharide production in *Chromobacterium violaceum* is enhanced in the presence of Fe and Ca ions (Corpe, 1964). The yield and form of exopolymer produced by a coryneform bacterium isolated from Cr-polluted sediments were influenced by the amount of Cr in the culture medium (Bremer and Loutit, 1986a). Exopolysaccharide production increased as more Cr(III) was added to the medium.

The presence of metals in the culture medium has been shown to influence the properties of the exopolysaccharide synthesized during bacterial growth. The morphology of the capsule of *Azotobacter chroococcum* changes as a result of metal availability in the growth medium (Ferala et al., 1986). Under iron- or molybdenum-sufficient conditions, the capsule is condensed, whereas under deficient conditions the capsule appears more diffuse and extensive. It is not yet known whether these morphological differences reflect direct interaction between the metal and the polymers or whether the elevated metal concentrations affect polymer subunit chemistry at the level of synthesis.

Interactions with certain metal ions cause many capsular and slime exopolymers to flocculate. Metals that exist as free cations or as hydroxides promote flocculation and rapid settlement of bacterial exopolymers in solution (Brierley and Lanza, 1985). At very high metal concentrations (0.2 M), extracellular polymers from *P. atlantica* formed a water-insoluble precipitate with Fe, Cu, and Pb (Corpe, 1975). Binding of metal ions causes some slime polymers to condense into gels. Carboxylated polymers often form gels in the presence of Cu ions (Brown and Lester, 1979). Controlled gel formation of alginates by Ca ions has created new applications for this polysaccharide in a variety of commercial enterprises.

Coprecipitation is frequently observed when bacterial exopolysaccharides come in contact with metal ions. Purified polysaccharide from *Zoogloea* sp. flocculated in the presence of several metals (Ikeda et al., 1982). Maximum polymer precipitation occurred in the presence of 0.5-mM ferric-iron solution. Ferrous iron was about 50 percent as effective in this regard as ferric iron. Since binding of metal ions to this exopolymer promoted flocculation, the reaction was proposed as a means of removing metal ions in a concentrated form from acid mine water (Dugan, 1970).

The capacity of capsular polymers to bind metals depends on the form in which the polymer exists. Rudd et al. (1984) compared the metal-binding capacity of colloidal and soluble forms of capsular material recovered from *Klebsiella* (*Enterobacter*) *aerogenes*. They determined that approximately five times more Ni complexed with the soluble form than with the colloidal form. Cu, Cd, Mn, and Co were also complexed to a greater extent by the soluble form of the polymer. However, estimation of the metal-binding capacity of polymer in these studies did not employ the rigorous approach required for polymers that form a colloid or gel phase. Consequently, these results should be viewed with caution.

Removal of Ca from culture medium leads to an increase in the relative concentration of mannuronic acid in alginic acid synthesized by *Azotobacter vinelandii* (Couperwhite and McCallum, 1974; Annison and Couperwhite, 1984). Appanna and Preston (1987) showed that exopolysaccharide production by an arctic *Rhizobium* was affected by amendments of Mn to the culture medium. Additions of 400 μM Mn^{2+} as $MnCl_2 \cdot H_2O$ resulted in a 14 percent increase in carbohydrate as glucose, a 20 percent decrease in uronic acid, and a reduction in viscosity in the exopolymer produced by the cells. These results demonstrate that the metal-ion makeup of the surrounding environment clearly affects the properties of bacterial exopolysaccharides.

Ramifications of Exopolymer-Metal Interactions

Metal tolerance

Under laboratory conditions, bacteria that elaborate exopolymers are often more tolerant to heavy metals than mutants that are unable to produce protective capsules and slime. Survival of capsule-producing strains of E. aerogenes in the presence of 10 μg/mL Cu^{2+} or Cd^{2+} was significantly greater than in mutants which lost their capsule-producing ability (Bitton and Friehofer, 1978). Furthermore, capsular material isolated from the wild-type strain enhanced the survival of the mutant strain when added to a metal-containing suspension of the latter. The capsular polymer was shown to bind both Cu and Cd ions. The capsular polymer was also demonstrated to provide protection against Cd (Bauda and Block, 1985).

Nonmucoid variants of a coryneform bacterium isolated from Cr-contaminated sediments were less tolerant to Cr(III) than a mucoid wild-type strain (Aislabie and Loutit, 1986). The mucoid strain accumulated more Cr(III) than the nonmucoid strain. Greater than 80 percent of the metal was complexed to the exopolymer. These and other studies conducted to date relate the tolerance to a reduction in free metal concentration at the cell surface.

Bioaccumulation

Complexation of metals by microbial biofilms has been suggested to play an important role in controlling toxic metal concentrations in the bulk water of aquatic environments. Bottom sediments contain high densities of sessile bacteria that are attached to particle surfaces via their capsular exopolymers. The polymers extend over the surface of and between particles to form a biofilm and are optimally oriented to interact with metal ions in solution that approach the particles.

Since sessile bacteria contribute a significant portion of the total living biomass in sediments and, in addition, represent a source of high nutritive value, they are a preferred food source for many benthic invertebrates and vertebrates. Bacterial exopolymers were suggested to be a source of nutrition for the deposit feeding clam Macoma balthica (Harvey and Luoma, 1984). Sediment-bound Cd, Zn, and Ag were taken up by the clam during ingestion of the bacterial exopolymer material (Harvey and Luoma, 1985). Patrick and Loutit (1978) demonstrated that metals, initially concentrated by bacteria and passed on through oligochaete worms, eventually accumulated in fish. In another study, it was shown that the Cr content of the snail Amphibola, which was fed Cr-complexed bacterial polysaccharide, contained threefold higher levels of the metal than snails fed polysaccharide containing very low levels of Cr (Bremer and Loutit, 1986b). Examination of body parts demonstrated that the bulk of the Cr was associated with internal tissues with lesser amounts adsorbed to the shell of the snail. It was further shown that polysaccharide with bound Cr was degraded more slowly than Cr-free polysaccharide by the heterotrophic sediment bacteria. These studies suggest that detritivores are capable of accumulating exopolymer-bound metals during their ingestion of the bacteria. The capsular and slime exopolymers excreted by sessile bacteria therefore provide a means of metal entry into the aquatic food chain.

Capsular polymers are also used by epiphytic bacteria to attach to plant material. Montgomery and Price (1979) indicated that the largest net uptake of metals from a source of sewage pollution occurred in the epilithic microbial community of a model turtle grass–mangrove ecosystem. In another study, bacterial epiphytes were shown to be a major factor in contributing to the total Cr, Cu, Fe, Pb, and Zn of aquatic plants (Patrick and Loutit, 1977). Binding of Mo by extracellular polysaccharides produced by the epiphytic bacteria resulted in a reduction in the amount of metal that entered the plant (Lee and Loutit, 1977). Thus, the capsular polymers of bacteria can either enhance or inhibit metal uptake by higher life forms.

Metal deposition/oxidation

The acidic polysaccharides and protein material which are present in the sheaths of the *Sphaerotilus-Leptothrix* group have been suggested to serve as the matrix for ferromanganese-oxide deposition (Ghiorse, 1986). The sheath appears as an extensive fibrillar network of fine threads not unlike capsular polymers of other bacteria when viewed by transmission electron microscopy. Although the chemical composition of bacterial sheaths has not been well-characterized, Romano and Peloquin (1963) found that the sheath of *Sphaerotilus natans* contained carbohydrate (36 percent), amino sugars (11 percent), protein (27 percent), and lipid (5 percent). Reaction of the sheath with ruthenium red suggests that the sheath of *Leptothrix discophora* possesses acidic groups (Ghiorse, 1986). A variant of a strain of *L. discophora,* which produced no sheath, still oxidized Mn^{2+} (Adams and Ghiorse, 1986). Thus, it appears that the Mn-precipitating action of the sheath is not necessarily related to Mn oxidation by this bacterium.

Macroparticles suspended in seawater below depths of 100 m contain exopolymer-producing bacteria that precipitate Mn and Fe (Cowen, 1983). Small variable amounts of Ca, Ba, Cu, and Zn were also bound by the halo of exopolymer material. The capsular metal deposits appear to contribute a major portion of the weakly bound fraction of the particulate Fe flux (Cowen and Silver, 1984). It was postulated that the Fe- and Mn-depositing bacteria associated with suspended macroparticles constitute a locally significant influence on Fe and Mn geochemistry.

Corrosion of metals

Recent studies have implicated bacterial exopolymers and other acidic polysaccharides in the corrosion of metals. White et al. (1986) showed that the corrosion of 304 stainless steel in seawater was accelerated by slime formation and biofilm development over the metal surface. The corrosion induced by two *Vibrio* strains was reversed by removal of their exopolymer from the surface. Nonhomogeneous distribution of microbes and their exopolymers on the metal surface was believed to create sites of different cathodic activity which accelerated the observed corrosion. For this reason, most microbially influenced corrosion results in the formation of irregularly spaced pits.

Iwaoka et al. (1986) showed that gum arabic, a uronic-acid-containing polysaccharide from the acacia plant, was capable of adsorbing to metallic Cu films. Approximately 50 percent of the polymer remained adsorbed to the film after gentle rinsing with water. Subsequent to polymer adsorption, a decrease occurred in the thickness of the metal film as determined by attenuated total reflectance with Fourier-transform infrared spectrometry. Since no change in film thickness occurred in the presence of water alone, it was proposed that the adsorbed polysaccharide promoted ionization of the metal. Subsequent studies demonstrated that the ionized metal was bound to the polysaccharide (Jolley et al., 1989).

Different polysaccharides, including the capsule from a sediment bacterial isolate, caused different rates of metallic Cu dissolution (Geesey et al., 1987). The Cu was ionized in either cupric or cuprous form, depending on the polysaccharide that adsorbed to the metal surface. In the presence of gum arabic, the metal was oxidized to Cu^{2+}, whereas in the presence of culture menstruum containing slime excreted by a marine bacterial isolate referred to as LST, the Cu was oxidized to Cu^+ (Jolley et al., 1989). No dissolution or ionization of the Cu film was observed, however, when it was exposed to an aqueous suspension of a slime polysaccharide from *P. atlantica* over the 40-h test period (Geesey et al., 1988). All of the polysaccharides tested, including that from *P. atlantica,* were capable of binding Cu ions (unpublished results). These results demonstrate that the exopolysaccharides elaborated by bacteria have the capacity to adsorb to metal surfaces and bind metal ions to the extent that metal ionization is enhanced. This reaction may involve the liberation of protons and a reduction in pH at the metal surface (Geesey et al., 1986). Thus, the interactions between Cu ions and bacterial exopolymers provide a mechanism whereby bacterial biofilms can promote corrosion of Cu and perhaps other metal surfaces under aerobic conditions without intervention of the bacterial cell and its associated activities. Whether such a mechanism contributes to the commonly observed pitting of Cu and Cu alloy tubing used in industrial heat exchanges and water distribution systems remains to be determined.

Application of Bacterial Exopolymers in Metal-Scavenging Processes

Numerous industrial and mining activities produce liquid wastes containing recoverable metal ions. Biomass and biopolymers have been used to complex the dissolved metal ions to a solid phase, thereby reducing their concentrations in the bulk liquid. In recent years, reactors such as the AMT-BioClaim process, marketed by Advanced Mineral Technologies, Inc., Golden, Colorado, represent successful attempts to utilize biosorption in a technologically feasible way.

Reactor design plays a crucial role in determining the efficiency of metal recovery. The nature of the solid support will generally determine many of the reactor parameters. Since bacterial exopolymers can exist as gels, colloids, and dissolved species, they exhibit a flexibility that is somewhat unique among other biopolymers and synthetic polymers. The ability of bacterial exopolymers to exist in these different phases should maximize exposure of the binding sites to the metals dissolved in the bulk liquid phase. The flexibility of the colloid and gel states should also provide advantages over other solid supports for ease of removal from the reactor.

One of the primary considerations of the solid support is its specificity or selectivity for certain metals. There is presently a great need to identify biopolymers that demonstrate high selectivity toward metals such as Cu, Co, Ni, Zn, Cr, and Ag and minimal reactivity with cations such as Na, Ca, K, and Mg. Manipulation of the culture conditions and the genetic engineering of exopolymer-producing microorganisms should provide a useful approach to achieve this goal. Specific stereochemistry features of microbial exopolymers should soon be identified with specific genes and their products so that the desired conformations can be selected. Presumably, other parameters such as binding-site density or binding capacity could be modulated at the gene level as well.

Identification of exopolymers that exhibit very high stability constants for metals will increase the value of these bioproducts in metal removal from aqueous media. There is an increasing demand for water used in processes in the electronics industry to contain less than 1 ppb of certain metals. "Polishing" industrial water and effluents is still a major challenge to industry. The high reactivity of polysaccharides to various reagents should facilitate their chemical modification to achieve stability constants that are orders of magnitude greater than their naturally occurring analogues.

One parameter that appears to be particularly intriguing in the case of bacterial exopolymers is the possibility of creating high diffusivity of metal ions within the polymer phase by manipulating the charge properties of the various functional groups on the polymer. Addition of certain anions to metal-containing wastes also appears to accelerate the rate of metal diffusion through the polymer matrix (Jang, unpublished results).

The multiple phases in which bacterial exopolymers exist offer opportunities to create more favorable flow patterns within a reactor. New flow designs should yield lower energy requirements while increasing mass transfer efficiency for metal diffusion from the bulk liquid to the surface of the biopolymer matrix (Jang, unpublished results).

The molecular volume changes that occur in exopolymers as a result of metal-ion binding in aqueous solution should facilitate separation of metal-saturated polymer from unsaturated polymer. As a polymer complexes more metal, the increased polymer density should enable the complex to migrate to a different position in a reactor than uncomplexed polymer. This characteristic as well as the plasticity of the polymer may allow for the construction of reactors that operate continuously without any downtime to replace the spent complexing agent.

It is evident from the above discussion that new and exciting prospects exist for bacterial exopolymers in metal-recovery operations. Although our current knowledge of interactions between bacterial exopolymers and metal ions is somewhat limited, the information that is available suggests that the polysaccharide fraction has the potential to form unique associations with metals that are important ecologically and economically. Presumably, research in these areas will help to advance our understanding of the role that bacterial exopolymers have in controlling the distribution of metal ions in nature.

References

Adams, L. F., and Ghiorse, W. C., Physiology and ultrastructure of *Leptothrix discophora* SS-1, *Arch. Microbiol.* **145**:126 (1986).

Aislabie, J., and Loutit, M. W., Accumulation of Cr(III) by bacteria isolated from polluted sediments, *Mar. Environ. Res.* **20**:221 (1986).

Amemiya, Y., and Nakayama, O., The chemical composition and metal adsorption capacity of the sheath materials isolated from *Microcystis*, cyanobacteria, *Jap. J. Limnol.* **45**:187 (1984).

Annison, G., and Couperwhite, I., Consequences of the association of calcium with alginate during batch culture of *Azotobacter vinlandii*, *Appl. Microbiol. Biotechnol.* **19**:321 (1984).

Appanna, V. D., and Preston, C. M., Manganese elicits the synthesis of a novel exopolysaccharide in an arctic *Rhizobium*, *FEBS Lett.* **215**:79 (1987).

Bauda, P., and Block, J. C., Cadmium biosorption and toxicity to laboratory-grown bacteria, *Environ. Technol. Lett.* **6**:445 (1985).

Bitton, G., and Friehofer, V., Influence of extracellular polysaccharide on the toxicity of the copper and cadmium toward *Klebsiella aerogenes*, *Microb. Ecol.* **4**:119 (1978).

Boyle, C. D., and Reade, A. E., Characterization of two extracellular polysaccharides from marine bacteria, *Appl. Environ. Microbiol.* **46**:392 (1983).

Bremer, P. J., and Loutit, M. W., The effect of Cr(III) on the form and degradability of a polysaccharide produced by a bacterium isolated from a marine sediment, *Mar. Environ. Res.* **20**:249 (1986a).

Bremer, P. J., and Loutit, M. W., Bacterial polysaccharide as a vehicle for entry of Cr(III) to a food chain, *Mar. Res.* **20**:235 (1986b).

Brierley, C. L., and Lanza, G. R., Microbial technology for aggregation and dewatering of phosphate clay slimes: Implications on resource recovery, in R. L. Tate and D. A. Klein (eds.), *Soil Reclamation Processes*, Dekker, New York, 1985, pp. 243–277.

Brown, M. J., and Lester, J. N., Metal removal in activated sludge: The role of bacterial extracellular polymers, *Water Res.* **13**:817 (1979).

Brown, M. J., and Lester, J. N., Role of bacterial extracellular polymers in metal uptake in pure bacterial cultures and activated sludge. I. Effects of metal concentration, *Water Res.* **16**:1539 (1982).

Cohen, G. H., and Johnstone, D. B., Extracellular polysaccharides of *Azotobacter vinlandii, J. Bacteriol.* **88**:329 (1964).

Corpe, W., Factors influencing growth and polysaccharide formation by strains of *Chromobacterium violaceum, J. Bacteriol.* **88**:1433 (1964).

Corpe, W. A., Metal-binding properties of surface materials from marine bacteria, *Dev. Ind. Microbiol.* **16**:249 (1975).

Couperwhite, I., and McCallum, M. F., The influence of EDTA on the composition of alginate synthesized by *Azotobacter vinlandii, Arch. Microbiol.* **97**:73 (1974).

Cowen, J. P., Fe and Mn depositing bacteria in marine suspended macro-particles, in P. Westbroek and E. W. de Jong (eds.), *Biomineralization and Biological Metal Accumulation*, Reidel, Boston, 1983, pp. 489–493.

Cowen, J. P., and Silver, M. W., The association of iron and manganese with bacteria on marine macroparticulate material, *Science* **224**:1340 (1984).

Cozzi, D., Desideri, P. G., and Lepri, L., The mechanisms of ion exchange with alginic acid, *J. Chromatogr.* **40**:130 (1969).

Dewitt, C. W., and Rowe, J. A., Sialic acids (*N*,7-*O*-diacetylneuraminic acid and *N*-acetylneuraminic acid) in *Escherichia coli, J. Bacteriol.* **82**:838 (1961).

Dugan, P. R., Removal of mine water ions by microbial polymers, in *Proc. 3rd Symp. Coal Mine Drainage Research*, Carnegie Mellon Institute, Pittsburgh, PA, 1970, pp. 279–283.

Dugan, P. R., and Pickrum, H. M., Removal of mineral ions from water by microbially produced polymers, in *Proc. 27th Industrial Waste Conf.*, Purdue Univ., Lafayette, IN, 1972, pp. 1019–1038.

Ferala, N. F., Champlin, A. K., and Fekete, F. A., Morphological differences in the capsular polysaccharide of nitrogen fixing *Azotobacter chroococcum* B-8 as a function of iron and molybdenum starvation, *FEMS Microbiol. Lett.* **33**:137 (1986).

Flatau, G. N., Clement, R. L., and Gauthier, M. J., Cadmium binding sites of cells of a marine pseudomonad, *Chemosphere* **14**:1409 (1985).

Friedman, B. A., and Dugan, P. R., Concentration and accumulation of metallic ions by the bacterium *Zoogloea, Dev. Ind. Microbiol.* **9**:381 (1968).

Geesey, G. G., Iwaoka, T., and Griffiths, P. R., Characterization of interfacial phenomena occurring during exposure of a thin copper film to an aqueous suspension of an acidic polysaccharide, *J. Colloid Interface Sci.* **120**:370 (1987).

Geesey, G. G., and Jang, L., Interactions between metal ions and capsular polymers, in T. Beveridge and R. J. Doyle (eds.), *Bacterial Interactions with Metallic Ions*, Wiley, New York, 1989, pp. 325–357.

Geesey, G. G., Jang, L., Jolley, J. G., Hankins, M. R., Iwaoka, T., and Griffiths, P. R., Binding of metal ions by extracellular polymers of biofilm bacteria, *Water Sci. Tech.* **20**:161 (1988).

Geesey, G. G., Mittelman, M. W., Iwaoka, T., and Griffiths, P. R., Role of bacterial exopolymers in the deterioration of metallic copper surfaces, *Mater. Performance* **25**:37 (1986).

Ghiorse, W. C., Applicability of ferromanganese-depositing microorganisms to industrial metal recovery processes, in H. L. Ehrlich and D. S. Holmes (eds.) *Biotechnology for the Mining, Metal-refining, and Fossil Fuel Processing Industries*, Wiley, New York, 1986, pp. 141–148.

Gordon, G., Quintero, E., and Geesey, G., Effect of dilution rates on *Pseudomonas atlantica* exopolymer, *Bacteriol. Abstracts*, p. 83 (1988).

Harvey, R. W., and Luoma, S. N., The role of bacterial exopolymers and suspended bacteria in the nutrition of the deposit feeding chain, *Macoma balthica, J. Mar. Res.* **42**:957 (1984).

Harvey, R. W., and Luoma, S. N., Effects of adherent bacteria and bacterial extracellular polymers upon assimilation by *Macoma balthica* of sediment bound Cd, Zn, and Ag, *Mar. Ecol. Prog. Ser.* **22**:281 (1985).

Haug, A., and Smidsrod, O., Selectivity of some anionic polymers for divalent metal ions, *Acta. Chem. Scand.* **24**:843 (1970).

Ikeda, F., Shuto, H., Saito, T., Fukui, T., and Tomita, K., An extracellular polysaccharide produced by *Zoogloea ramigera* 115, *Eur. J. Biochem.* **123**:437 (1982).

Iwaoka, T., Griffiths, P. R., Kitasko, J. T., and Geesey, G. G., Copper-coated cylindrical internal reflection elements for investigating interfacial phenomena, *Appl. Spectrosc.* **40**:1062 (1986).

Jang, L. K., Quintero, E. J., Gordon, G., Rohricht, M., and Geesey, G., The osmotic co-efficient of the sodium form of some polymers of biological origin, *Biopolymers* **28**:1485 (1989).

Jolley, J. G., Geesey, G. G., Hankins, M. R., Wright, R. B., and Wichlacz, P. L., In situ, real-time FI-IR/CIR/ATR study of the biocorrosion of copper by gum arabic, alginic acid, bacterial culture supernatant and *Pseudomonas atlantica* exopolymer, *Appl. Spectros.* **43**:1062 (1989).

Kihn, J. C., Mestdagh, M. M., and Rouxhet, P. G., ESR study of copper (II) and protoplasts of *Saccharomyces cervisiae, Can. J. Microbiol.* **33**:777 (1987).

Kohn, R., Ion binding on polyuronides-alginate and pectin, *Pure Appl. Chem.* **42**:371 (1975).

Lee, T. E., and Loutit, M. W., Effect of extracellular polysaccharides of rhizosphere bacteria on the concentration of molybdenum in plants, *Soil Biol. Biochem.* **9**:411 (1977).

Liu, T.-Y., Gotschlich, E. C., Dunne, F. T., and Jonnsen, E. K., Studies on the Meningococcal polysaccharides, *J. Biol. Chem.* **246**:4703 (1971).

Manzini, G., Cesaro, A., Delben, F., Paoletti, S., and Reisenhofer, E., Copper (II) binding by natural ionic polysaccharides Part 1. Potentiometric and spectroscopic data, *Bioelectrochem. Bioenerg.* **12**:443 (1984).

Martell, A. E., Principles of complex formation, in S. J. Faust and J. V. Hunter (eds.), *Organic Compounds in Aquatic Environments*, Dekker, New York, 1971, pp. 239–263.

Mathews, M. B., Trivalent cation binding of acid mucopolysaccharides, *Biochim. Biophys. Acta.* **37**:288 (1960).

Menjivar, J. A., The use of water-soluble polymers in oil field applications: Hydraulic fracturing, in R. R. Colwell (ed.), *Biotechnology of Marine Polysaccharides*, Hemisphere, Washington, D.C., 1984, pp. 250–280.

Mittelman, M. W., and Geesey, G. G., Copper-binding characteristics of exopolymers from a freshwater sediment bacterium, *Appl. Environ. Microbiol.* **49**:846 (1985).

Montgomery, J. R., and Price, M. T., Release of trace metals by sewage sludge and the consequent uptake by members of a turtle grass mangrove ecosystem, *Environ. Sci. Technol.* **13**:546 (1979).

Moorhouse, R., Winter, W. T., Arnold, S., and Bayer, M. E., Conformation and molecular organization in fibers of the capsular polysaccharide from *Escherichia coli* M41 mutant, *J. Mol. Biol.* **109**:373 (1977).

Norberg, A. B., and Persson, H., Accumulation of heavy-metal ions by *Zoogloea ramigera, Biotechnol. Bioeng.* **26**:239 (1984).

Paoletti, S., Cesaro, A., Ciana, A., Delben, F., Manzini, G., and Crescenzi, V., Thermodynamics of protonation and of Cu(II) binding in aqueous alginate solutions, in *Solution Properties of Polysaccharides*, Am. Chem. Soc. Symp. Ser. 150, Washington, D.C., 1981, pp. 379–386.

Pass, G., and Hales, P. W., Interaction between metal cations and anionic polysaccharides, in D. A. Bryant (ed.), *Solution Properties of Polysaccharides*, Am. Chem. Soc. Ser. 150, Washington, D.C., 1981, pp. 349–365.

Patrick, F. M., and Loutit, M. W., The uptake of heavy metals by epiphytic bacteria on

Alisma plantago-aquatica, Water Res. **11**:699 (1977).

Patrick, F. M., and Loutit, M. W., Passage of metals to freshwater fish from their food, *Water Res.* **12**:395 (1978).

Pedersen, K., Holmstrom, C., Olsson, A.-K., and Pedersen, A., Statistical evaluation of the influence of species variation, culture conditions, surface wettability and fluid shear on attachment and biofilm development of marine bacteria. *Arch. Microbiol.* **145**:1 (1986).

Rees, D. A., Polysaccharide gels, *Chem. Ind.* **19**:630 (1972).

Romano, A. H., and Peloquin, J. P., Composition of the sheath of *Sphaerotilus natans, J. Bacteriol.* **86**:252 (1963).

Rudd, T., Sterritt, R. M., and Lester, J. N., Mass balance of heavy metal uptake by encapsulated cultures of *Klebsiella aerogenes, Microb. Ecol.* **9**:261 (1983).

Rudd, T., Sterritt, R. M., and Lester, J. N., Formation and conditional stability constants of complexes formed between heavy metals and bacterial extracellular polymers, *Water Res.* **18**:379 (1984).

Sandford, P. A., Pittsley, J. E., Knutson, C. C., Watson, P. R., Cadmus, M. C., and Jeanes, A., Variations in *Xanthomonas campestris* NRRL B 1459: Characterization of xanthan products of differing pyruvic acid content, in P. A. Sandford and A. Laskin (eds.), *Extracellular Microbial Polysaccharides,* Am. Chem. Soc. Symp. Ser. 45, Washington, D.C., 1977, pp. 192–210.

Smidsrod, O., and Haug, A., Dependence upon uronic acid composition of some ion-exchange properties of alginates, *Acta Chem. Scand.* **22**:1989 (1968).

Steiner, I., McLaren, D. A., and Forster, C. F., The nature of activated sludge: The role of bacterial extracellular polymers, *Water Res.* **13**:817 (1976).

Stojkovski, S., Magee, R. A., and Leisegang, J., Molybdenum binding by *Pseudomonas aeruginosa, Aust. J. Chem.* **39**:1205 (1986).

Sutherland, I. W., Polysaccharides in the adhesion of marine and freshwater bacteria, in R. C. W. Berkeley, J. M. Lynch, J. Melling, and B. Vincent (eds.), *Microbial Adhesion to Surfaces,* Ellis Horwood, Chichester, U.K., 1980, pp. 329–338.

Tait, M. I., Sutherland, I. W., and Clarke-Sturman, A. J., Effect of growth conditions on the production, composition and viscosity of *Xanthomonas campestris* exopolysaccharide, *J. Gen. Microbiol.* **132**:1483 (1986).

Tan, E. L., and Loutit, M. W., Concentration of molybdenum by extracellular material produced by rhizosphere bacteria, *Soil Biol. Biochem.* **8**:461 (1976).

Uhlinger, D. J., and White, D. C., Relationship between physiological status and formation of extracellular polysaccharide glycocalyx in *Pseudomonas atlantica, Appl. Environ. Microbiol.* **45**:64 (1983).

White, D. C., Nivens, D. E., Nichols, P. D., Kerger, B. D., Henson, J. M., Geesey, G. G., and Clarke, C. K., Role of aerobic bacteria and their extracellular polymers in the facilitation of corrosion: Use of FTIR spectroscopy and "signature" phospholipid fatty acid analysis, in S. C. Dexter (ed.), *Biologically Induced Corrosion,* National Association of Corrosion Engineers, Houston, 1986, pp. 233–243.

Wilkinson, J. F., and Stark, G. H., The synthesis of polysaccharide by washed suspensions of *Klebsiella aerogenes, Proc. R. Phys. Soc., Edinburgh* **25**:35 (1956).

Zunino, H., and Martin, J. P., Metal-binding organic macromolecules in soil. II. Characterization of the maximum binding ability of the macromolecules, *Soil Sci.* **123**:188 (1977).

Fungi and Yeasts
for Metal Accumulation

Geoffrey Michael Gadd

Introduction

Fungi and yeasts, in common with other microbial groups such as bacteria, cyanobacteria, and algae, are capable of accumulating heavy metals and radionuclides even from dilute external solutions (Table 11.1; Zajic and Chiu, 1972; Tuovinen and Kelly, 1974; Gadd and Griffiths, 1978a; Shumate and Strandberg, 1985; Gadd, 1986a,b; Trevors et al., 1986). A variety of physical, chemical, and biological mechanisms may be involved. Such processes are of current biotechnological interest, not only because the removal of potentially toxic heavy metals and/or radionuclides from solution can lead to the detoxification of industrial effluents but also because subsequent recovery of accumulated elements is possible after destructive or nondestructive treatments (Kelly et al., 1979; Brierley and Brierley, 1983; Brierley et al., 1985, 1986; Shumate and Strandberg, 1985; Gadd, 1986b; Hutchins et al., 1986). Metal recovery may be especially desirable for valuable metals like gold and silver, in which case destructive recovery may be economically viable. It may also be feasible to recover low-value elements from biomass by nondestructive processes in order to regenerate the biomass for subsequent uptake and to concentrate a potentially toxic element providing low-volume containable waste.

Although some of the basic concepts and features of metal accumulation are common to most microbial groups, fungi possess some unique attributes which, in many ways, reflect their morphological and physiological diversity. The majority of fungi exhibit a filamen-

TABLE 11.1   Metal Accumulation by Fungi and Yeasts[a]

Organism	Element and incubation details	Element concentration in biomass (% dry weight)	Reference
Saccharomyces cerevisiae	Cd, 5–500 μM, 1 h	0.24–3.12	Gadd and Mowll (1983)
Saccharomyces cerevisiae	Cd, 200 μM, 1 h	0.89	Norris and Kelly (1977)
Candida utilis		1.46	Norris and Kelly (1979)
Rhodotorula mucilaginosa		0.37	
Aspergillus niger	Cd, 89 μM, 24 h	0.37	Kurek et al. (1982)
Mucor racemosus		0.62	
Penicillium chrysogenum		0.29	
Trichoderma viride		0.22	
Pythium sp.	Cd, 45 μM, 5 d	0.022	Duddridge and Wainwright (1980)
Scytalidium lignicola		0.025	
Dictyuchus sterile		0.044	
Penicillium spinulosum			
nongrowing	Cd, 22 μM, 2 h	0.15	Ross and Townsley (1986)
growing, midlinear phase	Cd, 22 μM	0.04	Townsley et al. (1986a)
Aureobasidium pullulans			
yeastlike cells	Cu, 80 μM, 3 h	0.09	Gadd and Mowll (1985)
mycelium		0.08	
chlamydospores		0.27	
Rhizopus arrhizus	Cu, 300 μM[b], 1 h	1.71	de Rome and Gadd (1987)
Cladosporium resinae		0.79	
Penicillium italicum		0.95	
Candida utilis	Cu, 1 mM, 2 h	0.17	Khovrychev (1973)
Penicillium spinulosum			
nongrowing	Cu, 40 μM, 2 h	0.24	Townsley and Ross (1985)
growing, lag period	Cu, 40 μM, 24 h	0.36	
growing, midlinear phase	Cu, 40 μM, 50 h	0.05	
growing, stationary phase	Cu, 40 μM, > 72 h	0.03	
Aspergillus niger			
nongrowing	Cu, 40 μM, 2 h	0.19	Townsley and Ross (1986)
growing, lag period	Cu, 40 μM, 12 h	0.51	
growing, midlinear phase	Cu, 40 μM, 72 h	0.02	
growing, stationary phase	Cu, 40 μM, 108 h	0.02	

NOTE: See p. 252 for footnotes.

TABLE 11.1   Metal Accumulation by Fungi and Yeasts[a] (*Continued*)

Organism	Element and incubation details	Element concentration in biomass (% dry weight)	Reference
Trichoderma viride			
nongrowing	Cu, 40 μM, 2 h	0.14	Townsley et al. (1986b)
growing, lag period	Cu, 40 μM, 12 h	0.35	
growing, midlinear phase	Cu, 40 μM, 36 h	0.03	
growing, stationary phase	Cu, 40 μM, 72 h	0.03	
Aureobasidium pullulans			
growing, lag period	Cu, 4 mM, 24 h	0.64	Gadd and Griffiths (1978b, 1980)
growing, midexponential phase	Cu, 4 mM, 48 h	0.24	
growing, stationary phase	Cu, 4 mM, 90 h	0.13	
Rhizopus arrhizus[c]	Cr	3.1	Tobin et al. (1984)
	Mn	1.2	
	Cu	1.6	
	Zn	2.0	
	Cd	3.0	
	Hg	5.8	
	Pb	10.4	
	U	19.5	
	Ag	5.4	
Penicillium spinulosum			
nongrowing	Zn, 38 μM, 2 h	0.13	Ross and Townsley (1986)
growing, midlinear phase	Zn, 38 μM	0.02	Townsley et al. (1986a)
Penicillium notatum[d]	Cu	8.0	Siegel et al. (1983)
	Zn	2.3	
	Cd	0.5	
	Al	5.3	
	Sn	6.1	
	Pb	0.5	
Penicillium sp.	Pb, 240 μM, 2 h	0.61	Siegel et al. (1986)
Penicillium Cl			
5-d-old mycelium	U, 420 μM, 2 h	17.0	Zajic and Chiu (1972)
15-d-old mycelium		8.0	
Aspergillus terreus	U, 126 μM[b]	0.1	Tsezos and Volesky (1981)

TABLE 11.1   Metal Accumulation by Fungi and Yeasts[a] (*Continued*)

Organism	Element and incubation details	Element concentration in biomass (% dry weight)	Reference
Aspergillus niger		1.3	
Penicillium chrysogenum		7.0	
Rhizopus arrhizus		14.0	
Candida albicans	U, 42 μM, 1 h	1.57	Horikoshi et al. (1981)
Candida robusta		2.79	
Hansenula anomala		4.04	
Rhodotorula glutinis		3.65	
Saccharomyces cerevisiae		3.38	
Aspergillus niger	U, 42 μM, 1 h	8.07	Horikoshi et al. (1981)
Aspergillus oryzae		6.34	
Chaetomium globosum		5.42	
Penicillium janthinellum		5.27	
Rhizopus oryzae		6.05	
Trichoderma viride		4.25	
Saccharomyces cerevisiae	U, 420 μM, > 2 h	10–15	Strandberg et al. (1981)
Penicillium sp.	U, 420 μM, 18 h	0.14	Galun et al. (1983)
Rhizopus arrhizus	U, 925 μM, airlift fermenter	4.23	McCready and Lakshmanan (1986)
Penicillium sp.		2.05	
Saccharomyces cerevisiae		5.08	
Aspergillus niger	U, 300 μM, 4 h	21.5	Yakubu and Dudeney (1986)
Aspergillus terreus	Th, 129 μM[b]	0.6	Tsezos and Volesky (1981)
Aspergillus niger		1.7	
Penicillium chrysogenum		14.2	
Rhizopus arrhizus		18.5	
Saccharomyces cerevisiae	Th, 3 mM[b] in 1 M HNO$_3$, 30 min	11.9	Gadd et al. (1988)
Aspergillus niger		13.9	
Rhizopus arrhizus		9.7	

[a]Metal concentrations refer to initial concentrations unless indicated otherwise. Readers are advised to consult the original references for full experimental details.
[b]Equilibrium or residual metal concentration.
[c]Maximum uptake capacities shown as determined from Langmuir adsorption isotherms.
[d]Accumulated metals measured after 30 days in contact with metal surfaces.

tous or hyphal growth form which enables efficient colonization of substrates. Hyphae exhibit apical growth and can produce lateral branches. There may be further structural and biochemical differentiation which may be relevant to mineral recovery. Different cell forms may have different uptake capacities, and living cells may have different sensitivities toward potentially toxic heavy metals (Gadd, 1986a). Yeasts are a particular group of fungi that predominantly exhibit a unicellular mode of growth, vegetatively reproducing by budding or fission. Some species can also produce mycelium.

Fungi and yeasts are, in general, easy to grow, produce high yields of biomass, and may be amenable to genetical and morphological manipulation. Their use in a variety of major industrial processes also ensures that fungal and yeast biomass may be available as a waste product for which the biotechnology of metal removal and recovery may provide a good use, since both living and dead biomass are capable of metal accumulation. *Aspergillus niger* biomass arises in substantial quantities from citric-acid production, while *Saccharomyces cerevisiae* is readily available from food and beverage industries. Other kinds of biomass arising as waste products from certain industrial processes and fermentations include *Rhizopus arrhizus, Aspergillus terreus,* and *Penicillium chrysogenum* (Tsezos and Volesky, 1981).

There may be marked differences in uptake mechanisms between living and dead cells and, therefore, a wide variety of potential applications in various industrial contexts. The use of dead biomass for metal binding circumvents toxicity problems that can occur with living cells, eliminates the economic component of nutrient supply, and subsequent metal and biomass recovery is often easier. However, living cells can accumulate metals intracellularly, often to higher levels than dead cells, and may also precipitate metals in and around cell walls or in the external medium by various products of metabolism. In addition, many fungi can be extremely tolerant of heavy metals in comparison with other microbial groups (Gadd et al., 1984a; Gadd, 1986a). Tolerance may depend on intrinsic properties of the organism, but may also arise as a result of physiological and/or genetical adaptations. Tolerance or resistance is relevant to metal accumulation since it is often associated with reduced intracellular uptake (Gadd, 1986a).

Mechanisms of Heavy-Metal and Radionuclide Accumulation by Fungi and Yeasts

A variety of mechanisms occur in fungal cells for the removal of heavy metals and radionuclides from solution. These mechanisms range from physicochemical interactions, such as adsorption, to processes that depend on cell metabolism, such as intracellular metal accumulation or extracellular precipitation of metals, as a result of excreted metabolites. It therefore follows that there may be considerable differences in the mechanisms expressed, depending on whether organisms are living or dead and, if living, whether growth takes place.

Metal uptake by living fungi can often be divided into two main phases of uptake. The first phase, which can also occur with dead cells, is metabolism-independent metal binding to cell walls and other external surfaces; the second phase is energy-dependent intracellular metal uptake across the cell membrane. In some cases, particularly where toxicity is manifest, intracellular uptake may not be linked with metabolism but may be a consequence of increased membrane permeability and the resultant exposure of further metal-binding sites within the cell. Such phases are often more readily distinguished in short-term experiments where cell suspensions are supplied with an energy source (e.g., glucose), but significant growth is precluded (Brierley et al., 1985; Gadd, 1986a). These phases of uptake may not be clearly seen in all organisms. For radionuclides, it appears that most of the accumulation may be a result of physicochemical interactions with cell walls with little or no intracellular uptake. In growing cells, either or both phases of uptake may be obscured by additional aspects of metabolism (e.g., extracellular products which may complex or precipitate metals outside the cells). For clarity, the above mechanisms of metal and/or radionuclide accumulation will be dealt with separately.

Metabolism-independent accumulation

Adsorption is frequently used to describe metabolism-independent binding of heavy metals and/or radionuclides to fungal biomass, although with increasing awareness of the multiplicity of physicochemical interactions that can occur it is clear that this term may be rather simplistic. *Adsorption* involves the accumulation or concentration of substances at a surface or interface; the material being adsorbed is called the *adsorbate,* and the adsorbing phase is called the *adsorbent. Absorption* is another term often used to describe metabolism-independent binding of metals to biomass. However, absorption occurs when the atoms or molecules of one phase almost uniformly penetrate into those of another phase to form a "solution" with the second phase (Weber, 1972).

The three main types of adsorption involve electrical attractions, van der Waals attractions, or chemical attractions of the solute to the adsorbent. The first type is related to ion exchange and is often called *exchange* adsorption (Weber, 1972). This type can occur widely in fungal biomass, and, indeed, adsorption has been defined as the attraction of positively charged ions to negatively charged ligands in cell material (Brierley and Brierley, 1983). Adsorption as a result of van der Waals forces is often termed *physical* or *ideal* adsorption and is where the adsorbed molecule can have translational movement within the interface. Chemical attractions occurring between the adsorbate and the adsorbent are called *chemical* or *activated* adsorption (Weber, 1972). It is often difficult to distinguish between physical and chemical adsorption, and most adsorption phenomena are combinations of the three forms of adsorption described (Weber, 1972). Thus, metabolism-dependent binding of metal ions or radionuclides to fungal biomass may be quite complex and include some or all of the above phenomena.

To circumvent this problem of process definition, researchers are using the term *biosorption* more frequently. The term *sorption* may be used to include both adsorption and absorption, and refers to a process where a component moves from one phase to be accumulated in another, preferably solid, phase (Weber, 1972). Thus, biosorption describes the nondirected physical-chemical reactions that occur between metal and radionuclide species and cellular components (Shumate and Strandberg, 1985). It should not be used to include those aspects of heavy-metal and/or radionuclide uptake that depend on metabolism (e.g., active transport).

Metabolism-independent biosorption of heavy metals and radionuclides to fungal biomass is generally a rapid process under common experimental conditions, often being complete within a few minutes. Amounts bound may be large (Gadd, 1986b; de Rome and Gadd, 1987). Varieties of ligands are involved in metal binding, including carboxyl, amine, hydroxyl, phosphate, and sulfhydryl groups, although the relative importance of each is often difficult to resolve (Strandberg et al., 1981). Many metals and radionuclides have complex solution chemistries, and it is not always possible to tell which species are present (Tobin et al., 1984). Differences in affinities between different elements and their ionic species may exist for the various ligands encountered in biological systems (Nieboer and Richardson, 1980). Since cell walls of fungi and yeasts can vary considerably in their overall composition, it is not surprising that differences in adsorption capacities exist between different species, between cells of different ages, and even between different cell forms of the same organism. *Penicillium italicum* spore walls take up more copper than spore walls of *Neurospora crassa* (Somers, 1963). Differences in cell wall composition could explain greater Zn^{2+} adsorption by the red-pigmented yeast *Sporobolomyces roseus* than by *S. cerevisiae* (Mowll and Gadd, 1983). Melanin-pigmented chlamydospores of *Aureobasidium pullulans* bind more metal ions than do hyaline yeastlike cells and mycelium, with the bound metal being entirely located in the thick cell walls which act as a permeability barrier in preventing intracellular uptake (Gadd, 1984; Mowll and Gadd, 1984; Gadd and Mowll, 1985).

Biosorption can therefore be affected by the composition of the biomass and by other physical and chemical factors. In $R.$ $arrhizus,$ adsorption appeared to be related to the ionic radius of La^{3+}, Mn^{2+}, Cu^{2+}, Zn^{2+}, Cd^{2+}, Ba^{2+}, Hg^{2+}, Pb^{2+}, UO_2^{2+}, and Ag^+, but not for Cr^{3+} or the alkali-metal cations Na^+, K^+, Rb^+, and Cs^+, which were not adsorbed (Tobin et al., 1984). In general, over modest ranges of temperature (e.g., 4 to 30°C), biosorption of heavy metals is relatively unaffected, in contrast to metabolism-dependent intracellular uptake (Norris and Kelly, 1977; de Rome and Gadd, 1987). Low external pH often decreases the rate and extent of adsorption of a variety of metal ions (e.g., Cu^{2+}, Cd^{2+}, Zn^{2+}, Mn^{2+}, Co^{2+}, Zn^{2+}) (Rothstein and Hayes, 1956; Fuhrmann and Rothstein, 1968; Paton and Budd, 1972; Roomans et al., 1979; Gadd and Mowll, 1985; de Rome and Gadd, 1987). The presence of other anions and cations can also affect adsorption either by precipitation (e.g., phosphates, hydroxides) or by competition for adsorption sites. For example, Mg^{2+}, Ca^{2+}, and K^+ competed with Cu^{2+} in fungal spores (Somers, 1963); Mg^{2+} and Pb^{2+} affected Cu^{2+} adsorption in $Candida$ $utilis$ (Khovrychev, 1973); Hg^{2+}, Co^{2+}, Mg^{2+}, Mn^{2+}, and Cu^{2+} inhibited Cd^{2+} binding by yeastlike cells of $A.$ $pullulans$ (Mowll and Gadd, 1984), and Cu^{2+} binding to $Penicillium$ $spinulosum$ was reduced by Mg^{2+}, Mn^{2+}, Co^{2+}, and Zn^{2+} (Townsley and Ross, 1985). It was found that Cr^{3+} was not accumulated by $Penicillium$ mycelium unless the mycelium was first incubated in $Pb(NO_3)_2$ (Siegel et al., 1986).

The biomass concentration can also have a significant effect on adsorption. At a given equilibrium concentration, yeast cells adsorb more metal ions at low cell densities than at high cell densities. The binding of Hg^{2+}, Ag^+, Cd^{2+}, Al^{3+}, Ni^{2+}, Cu^{2+}, and Pb^{2+} all showed strong dependence on cell concentration (Itoh et al., 1975). Similar findings were obtained in a study of Cu^{2+} adsorption by $R.$ $arrhizus,$ $Cladosporium$ $resinae,$ and $P.$ $italicum$ (de Rome and Gadd, 1987). Itoh et al. (1975) suggested that electrostatic interactions between cells may be a significant factor in the biomass concentration dependency of metal adsorption, with a larger quantity of cation being adsorbed when the distance between cells is great.

Both living and dead fungal and yeast biomass are capable of the adsorption of metal ions and radionuclides, which has obvious implications for biotechnological exploitation. The uptake capacity of dead cells may be greater, equivalent to, or less than the capacity of living cells, and this uptake capacity may depend on the particular killing treatment used and any subsequent alterations in wall structure (Somers, 1963; Duddridge and Wainwright, 1980). After heat treatment of *Penicillium* sp., uptake of Pb^{2+} was more rapid but the capacity was the same (Siegel et al., 1986). For industrial applications, dead biomass has the advantages that it is immune from toxicity or other adverse external conditions (e.g., extremes of pH and temperature) and that it may be available from industrial sources as a waste product. The use of dead fungal biomass is analogous to the use of commercially available ion-exchange resins, although current indications are that adsorption capacities of fungal biomass can be much greater than that of ion-exchange resins (Tsezos and Volesky, 1981; Brierley and Brierley, 1983; Tsezos and Keller, 1983). Use of dead biomass as an adsorption system also means that it can be used in established conventions, theories, and formulas already in routine use for such adsorption systems as ion exchange.

Adsorption can be expressed by a variety of isotherms, which can be useful for illustrating differences between, and the capacities of, different types of biomass. Among the variety of treatments that can be used are Freundlich, Scatchard, and Langmuir isotherms that refer to single-layer adsorption (Langmuir, 1918; Freundlich, 1926; Scatchard, 1949). The Brunauer-Emmett-Teller (BET) isotherm reflects multilayer adsorption but reduces to the Langmuir model when the limit of adsorption is a monolayer (Brunauer et al., 1938). Copper adsorption by *C. resinae* and *P. italicum* obeyed the Langmuir and Freundlich isotherms for single-layer adsorption, whereas *R. arrhizus* followed the BET isotherm, indicating that adsorption was more complex in this organism (de Rome and Gadd, 1987). Adsorption isotherms have been applied to a wide range of fungi, yeasts, metal ions, and radionuclides and seem to be an established component in uptake studies (Tsezos and Volesky, 1981; Gadd, 1986b; Ross and Townsley, 1986). However, adsorption isotherms, which were originally devised for more defined systems (such as gas adsorption to surfaces), are undoubtedly quite simplistic when applied to complex biological systems and may shed no light on mechanistic aspects of adsorption. As mentioned, the relative importance of particular ligands is difficult to resolve, and it is also unlikely that true monolayer adsorption ever occurs in such systems or that there is no movement of molecules in the surface. The adsorption characteristics of yeast for a range of metal ions differed from the Langmuir or Freundlich isotherms and were affected by cell density (Itoh et al., 1975). For *S. cerevisiae* and Hg^{2+} the initial adsorption phase did approximate a monolayer and was describable by the Langmuir equation, but this was followed by a transition region and then a multilayer or penetration phase (Brown et al., 1974). For *S. cerevisiae* and UO_2^{2+}, observed data were fitted to an adsorption model that was analogous to enzyme-substrate kinetics (Weidemann et al., 1981). The formation of obvious multilayers or "crystalline" deposition has especially been noted for radionuclides, which will be dealt with subsequently, and for Cu^{2+} and *N. crassa* (Subramanyam et al., 1983). In the latter, the mycelium turned blue in the presence of high concentrations of Cu^{2+} in a medium containing nitrate and phosphate. The walls were ultimately capable of containing 12% Cu^{2+}. Phosphate precipitation may be involved in this case. In spite of the disadvantages of adsorption isotherms, which may apply equally to more defined systems of adsorption (e.g., activated carbon; Weber, 1972), they do have some value in providing information on uptake capacities, describing equilibrium conditions for adsorption by different biomass types, and illustrating differences between species and morphological types (Paton and Budd, 1972; Duddridge and Wainwright, 1980; Tsezos and Volesky, 1981; Mowll and Gadd, 1983, 1984; Gadd and Mowll, 1985; de Rome and Gadd, 1987).

A considerable amount of work has been carried out on the uptake of radionuclides, such as U and Th, by fungal biomass. Although most work has been carried out with dead biomass, it appears that metabolism-independent biosorption is the main mechanism of uptake even in living cells. Uptake capacities can be large and, as with heavy metals, can be affected by external variables such as pH, temperature, and the presence of other chemical species. Initial rates of U uptake by *S. cerevisiae* were increased as the pH was raised from 2.5 to 5.5, with optimum equilibrium conditions at pH 3 to 4 (Rothstein and Hayes, 1956; Strandberg et al., 1981; Tsezos and Volesky, 1981). Maximal uptake of Ra by *P. chrysogenum* occurred between pH 7 and 10 with little or no uptake occurring at pH 2 (Tsezos and Keller, 1983). Effluents in which Th is found are often highly acidic because of high concentrations of HNO_3. However, significant Th uptake by fungal biomass is still possible under such conditions, and the biomass appeared relatively unaffected by the acidic conditions (Gadd et al., 1988). The solubility of Th decreases with increasing pH. At pH < 2, solubility is high, with Th^{4+} the main species. As the pH increases, various hydrolysis products appear, including $Th(OH)^{3+}$, $Th(OH)_2^{2+}$, $Th_2(OH)_2^{6+}$, $Th(OH)_3^+$, $Th_6(OH)_{15}^{9+}$, and colloidal $Th(OH)_4$ (aq). It is likely that dominant hydrolyzed species (e.g., $Th(OH)_2^{2+}$) are taken up more efficiently than Th^{4+} (Tsezos and Volesky, 1981). Increasing the temperature to 50°C enhanced U uptake by *S. cerevisiae* (Strandberg et al., 1981). Similar findings were reported for other fungal systems (Shumate et al., 1978; Tsezos and Volesky, 1981). Metal cations such as Fe^{2+}, Fe^{3+}, Zn^{2+}, Cu^{2+}, Mn^{2+}, Mg^{2+}, Ca^{2+}, and Ba^{2+} have all been shown to interfere with U uptake (Rothstein and Hayes, 1956; Tsezos and Volesky, 1982; Tsezos, 1983; Galun et al., 1984). In *R. arrhizus*, U biosorption involves at least three processes. Coordination of U to the amine N of chitin and adsorption in the cell wall chitin structure occur simultaneously and rapidly, followed by precipitation of $UO_2(OH)_2$ within the chitin microcrystalline cell wall structure at a slower rate (Tsezos and Volesky, 1982). A free radical on the chitin molecule, possibly assigned to a hydroxyl group, appears to be involved in the UO_2^{2+} ion coordination to N (Tsezos, 1983). Low pH reduced U uptake, probably because of increased H_3O^+, resulting in competition between H_3O^+ and U species for chitin complexation sites and effects on the ionic species of U present. At pH <2.5, U predominantly exists in solution as UO_2^{2+}, but at pH >2.5 it hydrolyzes and the accompanying reduction in solubility favors adsorption (Tsezos and Volesky, 1982). At pH 3.0 to 4.5, hydrolysis products can include $(UO_2)_2(OH)_2^{2+}$, $UO_2(OH)^+$, and $(UO_2)_3(OH)_5^+$ as well as UO_2^{2+} (Tsezos and Volesky, 1981). Complexation with carbonates may also occur (Strandberg et al., 1981). In unbuffered solutions, the pH of the external suspending medium increases during U adsorption by *S. cerevisiae,* indicating release of $—OH^-$ ions from the biomass and confirming that UO_2^{2+} was the main form of U bound (Shumate et al., 1978; Strandberg et al., 1981). Early work on binding of UO_2^{2+} suggested the involvement of anionic groups, particularly phosphate and carboxyl groups (Rothstein and Hayes, 1956). More recent work with *S. cerevisiae* has shown that U was deposited as needlelike fibrils in a layer approximately 0.2 μm thick on cell walls with little or no U being located within cells (Strandberg et al., 1981). On average, cell-bound U comprised 10 to 15 percent of the dry weight, but only about 32 percent of the cells had measurable amounts of U associated with them. This meant that the U concentration of that fraction approached 50 percent of the dry weight, and such a large relative amount cannot be accounted for by simple adsorption to negatively charged sites (e.g., $R—COO^-$ or $R—PO_4^{2-}$). It was suggested that additional U "crystallizes" on already bound molecules (Strandberg et al., 1981), perhaps in a way analogous to the *R. arrhizus* model (Tsezos, 1983). This feature of radionuclide adsorption by fungal cells may explain why it can be a slow process, taking approximately 1 h to reach equilibrium for U and *S. cerevisiae* (Strandberg et al., 1981). Similar results were found for U and Sr uptake by *R. arrhizus* and *S. cerevisiae,* where at 10 and 100 μM, complete equilibriation was not reached with either organism in up to 3 h incubation (Gadd et al., 1988). Again, this may be a reflection of multiple uptake mechanisms since, in the absence of metabolism, intracellular uptake (unless nonspecific) was not likely to occur. Sr removal from solution by *R. arrhizus* was decreased at low pH and in the presence of equimolar concentrations of U and Mn (Gadd et al., 1988).

As described previously for heavy metals, the biomass concentration is another variable that can affect uptake and, therefore, removal from solution. A decrease in the specific uptake of Th by *S. cerevisiae* occurred with increasing biomass concentrations, and this was also found for *A. niger* at high biomass densities. However, no simple relationship occurred with *R. arrhizus* (Gadd et al., 1988). Of course, specific uptake may be lower at high biomass concentrations, but the total removal of metal and/or radionuclide from solution is higher than at lower biomass concentrations.

Various treatments may be employed to increase the capacity of biomass for metal and/or radionuclide adsorption. As mentioned, certain killing treatments may increase the adsorption capacity, and powdering of dried biomass will also expose additional binding sites (Tobin et al., 1984). Detergent treatment has also been used to increase the capacity of fungal biomass for adsorption, since this can result in disruption of cell components such as membranes, increase cell permeability, and expose a wider variety of potential binding sites (Brierley et al., 1987). Detergent treatment of *P. spinulosum* greatly improved Cu^{2+} removal from solution, especially at high Cu^{2+} concentrations. For example, at an initial Cu^{2+} concentration of 63.5 μg/mL at pH 5.5, control biomass removed 21 percent of the available Cu^{2+}, but after detergent treatment 83 percent was removed (Ross and Townsley, 1986). Maximal metal removal by treated biomass occurred very rapidly, being complete within 5 min (Ross and Townsley, 1986). Detergent treatment was also found to increase Th uptake by *S. cerevisiae* and *R. arrhizus,* although the effect was not as significant as in the former example (Gadd et al., 1988). However, the conditions for Th uptake were markedly different, taking place in 1 M HNO_3. It was likely that the biomass was already substantially permeabilized prior to detergent treatment. In *S. cerevisiae,* U uptake was increased by HCHO and $HgCl_2$ (Strandberg et al., 1981), again probably because of increased cell permeabilization.

In some circumstances, fungi can adsorb insoluble metal compounds (e.g., sulfides), and this may represent another area of potential biotechnological application. *A. niger* oxidized Cu, Pb, and Zn sulfides to sulfates, the sulfide particles in the medium being adsorbed onto mycelial surfaces (Wainwright and Grayston, 1986). *Mucor flavus* could adsorb PbS, Zn dust, and $Fe(OH)_3$ ("ocher") from acid mine drainage. *A. niger, Fusarium solani,* and *P. notatum* could also remove ocher from solution but not as efficiently as *M. flavus* (Wainwright et al., 1986). Intact, fresh mycelium was best for adsorption at 25°C; the presence of a carbon source was unnecessary. When *M. flavus* was grown with PbS for 2 to 5 days, it converted small particles of the compound to a fine, even suspension which was then completely adsorbed after 7 days (Wainwright et al., 1986).

Uptake by isolated wall constituents

Compounds derived from fungal biomass may act as efficient biosorption agents. A wide variety of such materials are found in fungal walls, including mannans, glucans, phosphomannans, melanins, chitin, and chitosan. Up to 30 percent of *A. niger* biomass comprises an association of chitin with glucan (Muzzarelli and Tanfani, 1982). The role of fungal chitin in radionuclide uptake has been discussed previously. It is worth repeating that waste fungal biomass may provide a relatively inexpensive source of such compounds. However, it should be borne in mind that the extraction of a cellular component from biomass may sometimes affect its sequestration ability. This was found for chitin, where a decrease in U biosorption was observed after the chitin was extracted from the biomass and used alone (Tsezos, 1986).

Chitin phosphate and chitosan phosphate were found to adsorb greater amounts of U than Cu, Cd, Mn, Zn, Co, Ni, Mg, and Ca. Adsorption of U was rapid and subject to similar pH and temperature effects as those described previously (Sakaguchi and Nakajima, 1982). For chitin phosphate, the order of adsorption efficiency was $UO_2^{2+} \gg Cu^{2+} > Cd^{2+} > Mn^{2+} > Zn^{2+} > Mg^{2+} > Co^{2+} > Ni^{2+} > Ca^{2+}$, whereas for chitosan phosphate it was $UO_2^{2+} \gg Cu^{2+} > Zn^{2+} > Mn^{2+} > Co^{2+} > Ni^{2+} > Mg^{2+} > Ca^{2+}$. UO_2^{2+} and Co^{2+} could easily be separated from each other by using chitin phosphate. A good U desorbent was Na_2CO_3. Nonphosphorylated chitin and chitosan were not as efficient adsorbents as their phosphorylated derivatives (Sakaguchi and Nakajima, 1982). Alkali treatment of *A. niger* was used to obtain insoluble chitosan-glucan complexes which were found to be efficient chelating agents for a range of metal ions (Muzzarelli and Tanfani, 1982). Glucans carrying amino-acid or sugar-acid groups exhibit enhanced chelating capacities and fast binding of transition-metal ions. Such compounds include a derivative of dehydroascorbic acid, which is efficient for chelating U, and a cross-linked glycine glucan which can remove ^{60}Co from brines (Muzzarelli et al., 1986).

Fungal phenolic polymers and melanins contain, among other components, phenolic units, peptides, carbohydrates, aliphatic hydrocarbons, and fatty acids with major oxygen-containing groups being carboxyl and phenolic and alcoholic hydroxyl, carbonyl, and methoxyl groups. It was suggested that such groups are involved in transition-metal bonding to form metal-organic complexes, for example, Fe and Cu-fungal phenolic polymer complexes (Saiz-Jimenez and Shafizadeh, 1984). The maximum binding ability of fungal phenolic polymers and melanins followed the sequence Cu > Ca > Mg > Zn (Zunino and Martin, 1977). Hydrolyzed polymers for *Epicoccum purpurascens* removed 72 percent of Cu^{2+} from a $0.5 \times 10^{-4}\ M$ solution (Saiz-Jimenez and Shafizadeh, 1984).

Desorption from loaded biomass

Technical applications of metal and/or radionuclide accumulation by fungal biomass may depend on the ease of element recovery either for subsequent reclamation or further containment of toxic and radioactive substances. Recovery may also be desirable in order to regenerate the biomass for reuse in multiple adsorption-desorption cycles (Tsezos, 1984). For maximum benefit, desorption techniques should be highly efficient, economical, and result in minimal damage to the biomass so that subsequent readsorption is unimpaired.

As described previously, acidic conditions can repress metabolism-independent accumulation (and also intracellular uptake), and acids can be effective desorption agents. Dilute (0.1 M) mineral acids (e.g., HCl, HNO_3, and H_2SO_4) were effective in removing Cu^{2+} from $C.$ $utilis$ (Khovrychev, 1973), $R.$ $arrhizus$, $C.$ $resinae$, and $P.$ $italicum$ (de Rome and Gadd, 1987), and U from $R.$ $arrhizus$ (94 percent elution efficiency; Tsezos, 1984) and $S.$ $cerevisiae$ (59 percent elution efficiency; Strandberg et al., 1981). However, in the latter example, subsequent uptake by the treated biomass was reduced (Strandberg et al., 1981), and at higher concentrations of mineral acids (≥ 1 M) there may be substantial structural damage to the biomass (Tsezos, 1984).

Organic complexing agents may be efficient desorption agents and apparently do not have adverse effects on the biomass. Ethylenediaminetetraacetic acid (EDTA, 0.1 M) was effective in desorbing U from $S.$ $cerevisiae$ (72.3 percent efficient) and $Penicillium$ $digitatum$ (Galun et al., 1983). In the latter case, the adsorption capacity of the biomass was increased after EDTA treatment (Galun et al., 1983). In a study of Cd desorption from loaded $A.$ $niger$, $Mucor$ $racemosus$, $P.$ $chrysogenum$, and $Trichoderma$ $viride$, it was found that the efficiency of 0.1 M NaOH as an extractant increased with increasing amounts of biomass, whereas the extraction efficiency of 0.1 M diethylenetriaminepentaacetic acid (DTPA) decreased. It appeared that the NaOH and DTPA extracted different chemical forms of Cd (Kurek et al., 1982). Decuprified cell walls of $N.$ $crassa$ were obtained by treatment with 8-hydroxyquinoline; they were able to readsorb Cu^{2+} (Subramanyam et al., 1983).

Carbonates are also efficient desorption agents with perhaps the most commercial potential. Sodium carbonate (Na_2CO_3) is a very efficient eluant ($\sim$ 90 percent), but the equilibrium elution pH may be highly alkaline (pH 11 to 12) and may result in some damage to the biomass (Tsezos, 1984). Ammonium carbonate ($(NH_4)_2CO_3$, was also highly efficient in removing U from $Penicillium$ $digitatum$ (97 percent; Galun et al., 1983) and $S.$ $cerevisiae$ (84 percent; Strandberg et al., 1981). The adsorption capacity of the biomass was increased after treatment in both bases. Of several elution systems examined for U desorption from $R.$ $arrhizus$, sodium bicarbonate ($NaHCO_3$) appeared the most promising because of almost complete U recovery ($\geq$ 90 percent) and high U concentration factors (Tsezos, 1984). The operating pH range was 7.5 to 8.5, and there was little damage to the biomass, which enabled the uptake capacity of the biomass, after repeated adsorption-desorption cycles, to remain near 90 percent of the original value (Tsezos, 1984). Similar efficiencies were noted for U desorption from *P. digitatum* (Galun et al., 1983). The solid-to-liquid ratios that can be used for bicarbonate elution systems can exceed 120:1 (mg:mL) for a 1 M NaHCO$_3$ solution with almost 100% U recovery and U concentrations in the eluate of at least 1.98×10^4 mg/L (Tsezos, 1984). The presence of other ionic species may affect desorption. For example, So$_4^{2+}$ limited the elution of U from *R. arrhizus* by altering the structure of the cell wall chitin, which resulted in confinement of some of the adsorbed U inside the chitin network (Tsezos, 1984).

Metabolism-dependent intracellular accumulation

Metabolism-dependent intracellular uptake, whereby metal ions are transported into cells across the cell membrane, may be a slower process than adsorption and is inhibited by low temperatures, the absence of an energy source (e.g., glucose), and by glucose analogues, metabolic inhibitors, and uncouplers (Norris and Kelly, 1977, 1979; Borst-Pauwels, 1981; Gadd, 1986a,b). In certain fungi, especially yeasts, much greater amounts of metal may be accumulated by such a process than by metabolism-independent accumulation (Norris and Kelly, 1977, 1979). Many metals are essential for growth and metabolism (e.g., Cu, Fe, Zn, Co, Mn), and all organisms therefore have the ability to accumulate these intracellularly from low external concentrations so that physiological needs are satisfied. However, at the relatively high concentrations of metals frequently used in uptake studies, energy-dependent intracellular uptake may be difficult to visualize and may not be as significant a component of total uptake as general biosorption. This is particularly true for filamentous fungi, where high values of metabolism-independent biosorption mask low rates of intracellular influx (Gadd and White, 1985; Gadd et al., 1987). For such organisms, the majority of published work suggests that general metabolism-independent processes such as adsorption and complexation make up the majority of total metal uptake by the biomass (Duddridge and Wainwright, 1980; Ross and Townsley, 1986). For radionuclides like U and Th it appears that little or none may enter the cells (Shumate and Strandberg, 1985), although little detailed work on living cells has been carried out. However, yeasts often exhibit high levels of intracellular uptake. These organisms, as well as other fungi, may be able to precipitate metals around the cells as a result of metabolic processes, and are also capable of synthesizing intracellular metal-binding proteins. Many fungi have the capacity to tolerate and grow in high concentrations of heavy metals.

Energy-dependent uptake of several metals by fungi has been demonstrated, including Cu^{2+}, Cd^{2+}, Ni^{2+}, Zn^{2+}, Co^{2+}, Mn^{2+}, Sr^{2+}, Mg^{2+}, and Ca^{2+} (Borst-Pauwels, 1981; Gadd, 1986a). Influx often appears to obey saturation kinetics, although wide variations in kinetics parameters have been reported (Borst-Pauwels, 1981). Reasons for this include strain differences, complexation or adsorption effects, toxicity, the effects of surface potential, and the existence of more than one transport mechanism (Norris and Kelly, 1977; Borst-Pauwels, 1981; Parkin and Ross, 1986a,b; White and Gadd, 1987a). The mechanism of intracellular uptake is complex and still not fully understood. The membrane potential may be a prime determinant of influx, and uptake may be reduced or prevented if the membrane potential is abolished or membrane depolarization is effected (Roomans et al., 1979; Lichko et al., 1980). Conversely, influx may be increased on hyperpolarization by ionophore treatment (Gadd and Mowll, 1985).

The metabolic state of cells often determines rates of intracellular uptake. A requirement for cells to have sufficient amounts of K^+ or phosphate may exist (Fuhrmann and Rothstein, 1968; Roomans et al., 1979; Gadd and Mowll, 1985). Exponential phase cells often exhibit greater rates of intracellular uptake than cells in other growth phases (Khovrychev, 1973; Failla and Weinberg, 1977; Okorokov et al., 1979), although the situation is often unclear in instances where changes in the medium, metabolite production, and morphology can obscure metal uptake (Gadd, 1986a). Zn was hyperaccumulated by C. utilis, previously grown under Zn-limiting conditions in a chemostat (Lawford et al., 1980), although this was not the case for Cu (Parkin and Ross, 1986b). In some instances, intracellular uptake may not be linked with metabolism; for example, Pb is believed to accumulate intracellularly by diffusion in S. cerevisiae (Heldwein et al., 1977). In addition, in cases of toxicity and/or increased permeabilization of cell membranes, increased uptake may occur because of further binding to exposed intracellular sites (Borst-Pauwels, 1981; Gadd and Mowll, 1985).

After intracellular metal uptake, electroneutrality may be maintained by K^+ efflux. A relationship of $1M^{2+}$ (in) to $2K^+$ (out) has often been observed (Fuhrmann and Rothstein, 1968; Norris and Kelly, 1977; Gadd and Mowll, 1985). As mentioned, good rates of influx can depend on adequate levels of intracellular K^+; conversely, increasing external K^+ may inhibit intracellular uptake (Okorokov et al., 1979; Lichko et al., 1980; Gadd and Mowll, 1985). However, K^+ release may not be detected in all instances (Norris and Kelly, 1977; Mowll and Gadd, 1983). Its relationship with intracellular uptake is often obscured because of toxic symptoms. Heavy metals may disrupt cell membranes, and observed K^+ release may be wholly or partially representative of cell death (Gadd and Mowll, 1983; Theuvenet et al., 1987; White and Gadd, 1987b).

External factors like pH, the presence of other anions and cations, and organic materials can affect intracellular uptake. In general, rates of intracellular uptake are decreased at low pH values, and an optimum pH for influx is often observed (Gadd, 1986a). At pH values around neutrality, metal-ion availability may be reduced because of hydroxide precipitation (Gadd and Griffiths, 1978a). Other cations can repress intracellular uptake; Mn^{2+}, Fe^{2+}, Ca^{2+}, and Mg^{2+} have been shown to do this in a variety of fungi (Norris and Kelly, 1977, 1979; Mowll and Gadd, 1984; Gadd and Mowll, 1985).

In growing fungal cultures, phases of adsorption and intracellular uptake may be obscured by changes in the physiology and morphology of the fungus (see later) and the physical and chemical properties of the growth medium (Gadd, 1986a). A frequently observed phenomenon is that metal uptake by growing batch cultures is maximal during the lag period or early stages of growth and declines as the culture reaches the stationary phase. This has been shown for Cu and *A. pullulans* (Gadd and Griffiths, 1978b, 1980), *Debaryomyces hansenii* (Wakatsuki et al., 1979), and *P. spinulosum, A. niger,* and *T. viride* (Townsley and Ross, 1985, 1986; Townsley et al., 1986a). Changes in the kinetics of Cu uptake by *A. pullulans* were shown to be related to the fall in medium pH which occurred during growth and the alleviation of Cu toxicity at low pH (Gadd and Griffiths, 1980). However, for several filamentous fungi, pH reduction of the medium was not the only factor responsible for reduced Cu uptake. A reduction could occur before any significant drop in the pH and in media where the pH was maintained at 5.5 (Townsley and Ross, 1985, 1986). Explanations for this include alterations in cell wall composition during growth and/or the release of metabolites (e.g., citric acid) that can control metal availability (Townsley and Ross, 1985, 1986). Complexation may be involved in uptake and may partly explain the frequent marked differences in uptake capacities between nongrowing and growing fungal cultures (Townsley and Ross, 1986). *C. utilis* exhibits cyclical Zn accumulation in batch culture with highest uptake during the lag and late exponential phases; it was suggested that there was regulation of Zn uptake in this organism (Failla and Weinberg, 1977). At low external Mn^{2+} (10 n*M*) concentrations, intracellular Mn^{2+} remained relatively constant in batch cultures of *C. utilis,* and there appeared to be a specific, high-affinity Mn^{2+} transport system (Parkin and Ross, 1986a).

As mentioned previously, many fungi have the ability to tolerate high concentrations of potentially toxic heavy metals, which may be useful if any recovery system employing living cells is envisaged. However, tolerance may be related to decreased intracellular uptake and/or impermeability of the cells toward metal ions (Gadd, 1986a). A close connection between intracellular uptake and toxicity has been shown for Cu^{2+}, Cd^{2+}, and Zn^{2+} in $S.$ $cerevisiae$ (Ross, 1977; Ross and Walsh, 1981; Gadd and Mowll, 1983; Mowll and Gadd, 1983; Joho et al., 1983; Gadd et al., 1984b; White and Gadd, 1986). No detectable Co^{2+} influx was noted in a resistant strain of $N.$ $crassa$ (Venkateswerlu and Sastry, 1979). Melanized chlamydospores of $A.$ $pullulans$ are impermeable to heavy metals and more tolerant than hyaline cell types (Gadd, 1981; Mowll and Gadd, 1984). In contrast to the above examples, a Mn-resistant mutant of $S.$ $cerevisiae$ accumulated more Mn^{2+} than the sensitive parental strain, but this was assumed to be the result of an internal sequestering system of greater efficiency in the resistant strain (Bianchi et al., 1981).

Where a reduction in intracellular uptake occurs at low external pH, a corresponding reduction in toxicity may also occur. $Scytalidium$ sp. and $P.$ $ochro\text{-}chloron$ can grow in saturated $CuSO_4$ at pH 0.3 to 2.0, but they are sensitive to no more than 4×10^{-5} M near neutrality (Gadd and White, 1985). For $P.$ $ochro\text{-}chloron,$ a constant level of Cu uptake occurs above approximately 16 mM, but, at pH 6 and above, uptake markedly increases and is concomitant with toxicity (Gadd and White, 1985).

Fate of intracellularly accumulated metals

Once inside the cell, metal ions may be compartmentalized and/or converted to more innocuous forms by precipitation or binding. The majority of Co^{2+}, Zn^{2+}, Mg^{2+}, Mn^{2+}, and K^+ taken in by $S.$ $cerevisiae$ is located in the vacuole (Okorokov et al., 1980; Lichko et al., 1982; White and Gadd, 1986, 1987a), where it may be bound to low-molecular-weight polyphosphates (Okorokov et al., 1980; Lichko et al., 1982). The vacuolar membrane possesses a transport system for internal transfer of metal ions, and evidence suggests that it is a proton antiport (White and Gadd, 1987a). In cells lacking vacuoles, localization of Sr^{2+}, Mg^{2+}, and Ca^{2+} in polyphosphate granules has been demonstrated (Roomans, 1980). Deposition of Ag within vacuoles (and around cell walls) has been observed in $Cryptococcus$ $albidus$ (Brown and Smith, 1976). Cell wall deposits were assumed to be elemental Ag, whereas AgCl was presumed to occur in the vacuoles (Brown and Smith, 1976). Intracellular electron-dense bodies resulted after exposure of $Chrysosporium$ $pannorum$ to Hg^{2+} (Williams and Pugh, 1975) and $Neocosmospora$ $vasinfecta$ to Zn^{2+} (Paton and Budd, 1972). Some yeasts "oxidatively detoxify" Tl to TlO_2 within mitochondria which may be subsequently excreted from cells (Lindegren and Lindegren, 1973).

A common response to metal exposure is the induction of intracellular low-molecular-weight cysteine-rich, metal-binding proteins, called *metallothioneins,* which have functions in detoxification and the storage and regulation of intracellular metal ions. The Cu-induced metallothionein of *S. cerevisiae* has received the most attention, and there is a recent review dealing with the structure of the *CUP 1* locus, mechanisms of gene amplification, and relationship with resistance (Butt and Ecker, 1987). This protein of *S. cerevisiae* is induced only by Cu and not Cd or Zn (Butt and Ecker, 1987). Cu-inducible metallothionein has also been documented in *N. crassa* (Munger et al., 1985), whereas in *Dactylium dendroides* two Cu-binding proteins were induced but only one was rich in cysteine (Shatzman and Kosman, 1979). In *Schizosaccharomyces pombe,* Cd-induced synthesis of peptides that were different in structure from Cu metallothionein of *S. cerevisiae* and *N. crassa* (Murasugi et al., 1985). Subcellular distribution of Cd has also been examined in resistant and sensitive strains of *S. cerevisiae* through differential extraction procedures (Joho et al., 1986; White and Gadd, 1986). In a sensitive strain, most intracellular Cd was bound to the insoluble cytosolic material, whereas low-molecular-weight ($< 30,000$) Cd-binding proteins were detected in the cytosol of resistant strains (Joho et al., 1986). In *C. utilis,* evidence has been presented for synthesis of an inducible Zn-binding protein (Failla et al., 1976). Clearly, metal-binding proteins may have applications in metal recovery, either inside or outside active cells or in purified forms, especially since such proteins seem able to bind valuable elements like Au and Ag as well as Cu, Cd, and Zn. Future work may lead to the "engineering" of different metallothioneins with specific affinities for different metals (Butt and Ecker, 1987).

Extracellular precipitation and complexation

Many extracellular fungal products can complex or precipitate heavy metals, although the possible relevance of such processes to metal removal and/or recovery has received scant attention.

Citric acid can be an efficient metal-ion chelator, and oxalic acid can interact with metal ions to form insoluble oxalate crystals around cell walls and in the external medium (Murphy and Levy, 1983; Sutter et al., 1983). The production of H_2S by yeasts can result in extensive precipitation of metals as insoluble sulfides predominately in and around cell walls (Ashida et al., 1963; Minney and Quirk, 1985). For strains of *S. cerevisiae* growing on a Cu-containing medium, resultant colonies may appear dark in color because of CuS formation (Ashida et al., 1963). Two *Rhodotorula* strains and a *Trichosporon* sp., isolated from acid mine waters, were able to precipitate Cu^{2+} with H_2S formed from the reduction of elemental S (Ehrlich and Fox, 1967). Some postulated examples of metal recovery and/or removal systems that use living microorganisms rely on sulfide precipitation of heavy metals (Brierley et al., 1985).

Fe is of fundamental importance to living cells, and many fungi and yeasts release high-affinity Fe-binding molecules called *siderophores*. The externally formed Fe^{3+} chelates are subsequently taken up into the cell (Raymond et al., 1984; Adjimani and Emery, 1987). In several fungi and yeasts, the excretion of such Fe-binding molecules is markedly stimulated by Fe deficiency, and such compounds may also bind Ga (Adjimani and Emery, 1987). *D. hansenii* produced riboflavin, or a related compound, when grown in Fe-deficient media or in the presence of metals such as Cu, Co, and Zn. Polarographic analysis showed that the pigment was markedly capable of Fe^{3+} binding (Gadd and Edwards, 1986).

Biotechnological Aspects of Fungal Metal Accumulation

As described, both living and dead fungi can accumulate heavy metals and radionuclides by a variety of mechanisms, and some have practical potential. The current perspective appears to be that dead biomass has more advantages than living material in that it may be obtained inexpensively as waste, it is not subject to metal toxicity or adverse operating conditions, nutrient supply is unnecessary, and recovery of metals may be by relatively simple nondestructive treatments that can lead to regeneration of the biomass. Living cells, particularly of yeasts, may have higher uptake capacities than dead cells because of intracellular uptake, precipitation, and so on, but internalized or precipitated metals are difficult to extract from cells and destructive treatment (e.g., incineration) may be necessary (Kelly et al., 1979; Brierley et al., 1985). In addition, toxicity and susceptibility to adverse operating conditions may be a problem with living cells, and although it is easy to obtain tolerant strains, these may exhibit reduced uptake or impermeability. It seems that living cells may have possibilities in some circumstances, but so far these possibilities have been largely unexplored. Metallothioneins, particulate metal accumulation, extracellular precipitation, and complexation are all worthy of continued investigation, and several postulated industrial methods for biological effluent detoxification involve some of these processes (Brierley et al., 1985).

An important consideration, whether living or dead cells are used, is the form and kind of biomass to be used. Laboratory experimentation can provide much useful information, but for industrial applications biomass has some disadvantages. It is generally of small particle size and low strength and density, which can limit the choice of suitable reactors and make biomass or effluent separation difficult (Tsezos, 1986). For use in packed-bed or fluidized-bed reactors, immobilized or pelleted biomass is of greater potential (see Chap. 15). Immobilized biomass, whether within or on an inert matrix, has advantages in that high flow rates can be achieved, clogging is minimized, particle size can be controlled, and high biomass loadings are possible (Shumate et al., 1980; Tsezos, 1986; Yakubu and Dudeney, 1986). Such procedures have been used for U removal using particles of dead biomass that were immobilized using polymeric membranes. The particles can be made to any size and contain only 10 percent inert material. The multiple use of such particles for U adsorption has been confirmed experimentally (Tsezos, 1986). Commercial processes already exist for operation of such systems, particularly for the recovery of valuable metals (Brierley et al., 1986). Other examples of immobilized systems include *Aspergillus oryzae* on reticulated foam particles for Cd removal (Kiff and Little, 1986) and *Trichoderma viride* packed in molochite and used for Cu removal from simulated effluents (Townsley et al., 1986b).

Filamentous fungi often grow in pellet form in culture, and such pellets may have advantages similar to immobilized particles. Pellets (4 mm diameter) of *A. niger* have been used in a fluidized bed reactor for U removal. A simple ion-exchange process predominated, where UO_2^{2+} ions replace protons on amino-acid groups of proteins and glycoproteins within the cell wall. It was found that such a system was 14 times more efficient than the commercial ion-exchange resin IRA-400 (Yakubu and Dudeney, 1986). However, fungal pellets may break up and cause increased resistance to liquid flow, and the similarity in specific gravity between cells and the liquid medium was found to be disadvantageous in continuous process applications (Yakubu and Dudeney, 1986).

Industrial Considerations and Concluding Remarks

Living and dead fungal biomass and derived products can be effective accumulators of metals and radionuclides, but for industrial application any biomass-related process must be economically competitive with existing technologies. The economics of U removal by *R. arrhizus* grown in an airlift fermenter were unfavorable in comparison with ion exchange and reverse osmosis (McCready and Lakshmanan, 1986). The current consensus is that, for improved commercial use, immobilized or pelleted preparations should be used with recovery involving an inexpensive stripping agent which can be recycled (Brierley et al., 1986; McCready and Lakshmanan, 1986; Tsezos, 1986). It is clear that in relation to other parameters of industrial relevance, fungi and fungal products can be highly efficient removal agents and have high uptake capacities. Fuller discussion of economics will be found in other chapters and also in Brierley et al. (1986), where costs and application of a successful proprietary granulated biomass system are discussed. Such systems may be used in conjunction with systems for the removal of impurities other than metals and radionuclides (Brierley et al., 1986). Biomass is highly efficient at dilute external concentrations and therefore may not necessarily replace existing technologies, but it may serve as a "polisher" after an existing treatment that is not completely efficient (Brierley et al., 1986). With nonprecious metals and radionuclides, where recovery may not be desired, the economics are not as clear and there is the problem of contaminated biomass to consider. Any process should therefore be devised to produce low-volume easily containable waste.

It is clear that some examples of metal recovery are competitive with existing treatments, but further work is needed in several areas in order to realize the full potential of fungal systems. It is hoped that administrative authorities and those industries responsible for releasing large amounts of potentially hazardous metals and radionuclides into the environment take a prominent lead in the development and exploitation of such technologies.

References

Adjimani, J. P., and Emery, T., Iron uptake in *Mycelia sterilia* EP-76, *J. Bacteriol.* **169**:3664–3668 (1987).

Ashida, J., Higashi, N., and Kikuchi, T., An electron microscopic study on copper precipitation by copper-resistant yeast cells, *Protoplasma* **57**:27–32 (1963).

Bianchi, M. E., Carbone, M. L., Lucchini, G., and Magni, G. E., Mutants resistant to manganese in *Saccharomyces cerevisiae*, *Curr. Genet.* **4**:215–220 (1981).

Borst-Pauwels, G. W. F. H., Ion transport in yeast, *Biochim. Biophys. Acta* **650**:88–127 (1981).

Brierley, C. L., Kelly, D. P., Seal, K. J., and Best, D. J., Materials and biotechnology, in I. J. Higgins, D. J. Best, and J. Jones (eds.), *Biotechnology*, Blackwell, Oxford, 1985, pp. 163–212.

Brierley, J. A., and Brierley, C. L., Biological accumulation of some heavy metals—biotechnological applications, in P. Westbroek and E. W. de Jong (eds.), *Biomineralization and Biological Metal Accumulation*, Reidel, Dordecht, 1983, pp. 499–509.

Brierley, J. A., Goyak, G. M., and Brierley, C. L., Considerations for commercial use of natural products for metals recovery, in H. Eccles and S. Hunt (eds.), *Immobilisation of Ions by Bio-sorption*, Ellis Horwood, Chichester, 1986, pp. 105–117.

Brierley, J. A., Brierley, C. L., Decker, R. F., and Goyak, G. M., Treatment of microorganisms with alkaline solution to enhance metal uptake properties, U.S. Patent 4,690,984, 1987.

Brown, R. B., Vairo, M. L. R., and Borzani, W., Quantitative study of labelled Hg^{2+} adsorption by live yeast cells. Evaluation of the number of glucose penetration sites, *J. Ferment. Technol.* **52**:536–541 (1974).

Brown, T. A., and Smith, D. G., The effects of silver nitrate on the growth and ultrastructure of the yeast *Cryptococcus albidus, Microbios Lett.* **3**:155–162 (1976).

Brunauer, S., Emmett, P. H., and Teller, E., Adsorption of gases in multimolecular layers, *J. Am. Chem. Soc.* **60**:309–319 (1938).

Butt, T. R., and Ecker, D. J., Yeast metallothionein and applications in biotechnology, *Microbiol. Rev.* **51**:351–364 (1987).

de Rome, L., and Gadd, G. M., Copper adsorption by *Rhizopus arrhizus, Cladosporium resinae* and *Penicillium italicum, Appl. Microbiol. Biotechnol.* **26**:84–90 (1987).

Duddridge, J. E., and Wainwright, M., Heavy metal accumulation by aquatic fungi and reduction in viability of *Gammarus pulex* fed Cd^{2+} contaminated mycelium, *Water Res.* **14**:1605–1611 (1980).

Ehrlich, H. L., and Fox, S. I., Copper sulphide precipitation by yeasts from acid minewaters, *Appl. Microbiol.* **15**:135–139 (1967).

Failla, M. L., Benedict, C. D., and Weinberg, E. D., Accumulation and storage of Zn^{2+} by *Candida utilis, J. Gen. Microbiol.* **94**:23–36 (1976).

Failla, M. L., and Weinberg, E. D., Cyclic accumulation of zinc by *Candida utilis* during growth in batch culture, *J. Gen. Microbiol.* **99**:85–97 (1977).

Freundlich, H., *Colloid and Capillary Chemistry,* Methuen, London, 1926.

Fuhrmann, G. F., and Rothstein, A., The transport of Zn^{2+}, Co^{2+} and Ni^{2+} into yeast cells, *Biochim. Biophys. Acta* **163**:325–330 (1968).

Gadd, G. M., Mechanisms implicated in the ecological success of polymorphic fungi in metal-polluted habitats, *Environ. Technol. Lett.* **2**:531–536 (1981).

Gadd, G. M., Effect of copper on *Aureobasidium pullulans* in solid medium: Adaptation not necessary for tolerant behaviour, *Trans. Br. Mycol. Soc.* **82**:546–549 (1984).

Gadd, G. M., Fungal responses towards heavy metals, in R. A. Herbert and G. A. Codd (eds.), *Microbes in Extreme Environments,* Academic Press, London, 1986a, pp. 83–110.

Gadd, G. M., The uptake of heavy metals by fungi and yeasts: the chemistry and physiology of the process and applications for biotechnology, in H. Eccles and S. Hunt (eds.), *Immobilisation of Ions by Bio-sorption,* Ellis Horwood, Chichester, 1986b, pp. 135–147.

Gadd, G. M., Chudek, J. A., Foster, R., and Reed, R. H., The osmotic responses of *Penicillium ochro-chloron:* Changes in internal solute levels in response to copper and salt stress, *J. Gen. Microbiol.* **130**:1969–1975 (1984a).

Gadd, G. M., and Edwards, S. W., Heavy metal-induced flavin production by *Debaromyces hansenii* and possible connexions with iron metabolism, *Trans. Br. Mycol. Soc.* **87**:533–542 (1986).

Gadd, G. M., and Griffiths, A. J., Microorganisms and heavy metal toxicity, *Microb. Ecol.* **4**:303–317 (1978a).

Gadd, G. M., and Griffiths, A. J., Copper tolerance of *Aureobasidium pullulans, Microbios Lett.* **6**:117–124 (1978b).

Gadd, G. M., and Griffiths, A. J., Influence of pH on toxicity and uptake of copper in *Aureobasidium pullulans, Trans. Br. Mycol. Soc.* **75**:91–96 (1980).

Gadd, G. M., and Mowll, J. L., The relationship between cadmium uptake, potassium release and viability in *Saccharomyces cerevisiae, FEMS Microbiol. Lett.* **16**:45–48 (1983).

Gadd, G. M., and Mowll, J. L., Copper uptake by yeast-like cells, hyphae and chlamydospores of *Aureobasidium pullulans, Exp. Mycol.* **9**:230–240 (1985).

Gadd, G. M., Stewart, A., White, C., and Mowll, J. L., Copper uptake by whole cells and protoplasts of a wild-type and copper-resistant strain of *Saccharomyces cerevisiae, FEMS Microbiol. Lett.* **24**:231–234 (1984b).

Gadd, G. M., and White, C., Copper uptake by *Penicillium ochro-chloron:* influence of pH on toxicity and demonstration of energy-dependent copper influx using protoplasts, *J. Gen. Microbiol.* **131**:1875–1879 (1985).

Gadd, G. M., White, C., and de Rome, L., Heavy metal and radionuclide uptake by fungi and yeasts, in P. R. Norris and D. P. Kelly (eds.), *Biohydrometallurgy,* Science and Technology Letters, Kew, Surrey, U.K., 1988, pp. 421–435.

Gadd, G. M., White, C., and Mowll, J. L., Heavy metal uptake by intact cells and protoplasts of *Aureobasidium pullulans, FEMS Microbiol. Ecol.* **45**:261–267 (1987).

Galun, M., Keller, P., Feldstein, H., Galun, E., Siegel, S., and Siegel, B., Recovery of uranium (VI) from solution using fungi. II. Release from uranium-loaded *Penicillium biomass, Water Air Soil Pollut.* **20:**277–285 (1983).

Galun, M., Keller, P., Malki, D., Feldstein, H., Galun, E., Siegel, S., and Siegel, B., Removal of uranium (VI) from solution by fungal biomass: Inhibition by iron, *Water Air Soil Pollut.* **21:**411–414 (1984).

Heldwein, R., Tromballa, H. W., and Broda, E., Aufnahme von Cobalt, Blei und Cadmium durch Bäckerhafe, *Z. Allg. Mikrobiol.* **17:**299–308 (1977).

Horikoshi, T., Nakajima, A., and Sakaguchi, T., Studies on the accumulation of heavy metal elements in biological systems. XIX. Accumulation of uranium by microorganisms, *Eur. J. Appl. Microbiol. Biotechnol.* **12:**90–96 (1981).

Hutchins, S. R., Davidson, M. S., Brierley, J. A., and Brierley, C. L., Microorganisms in reclamation of metals, *Ann. Rev. Microbiol.* **40:**311–336 (1986).

Itoh, M., Yuasa, M., and Kobayashi, T., Adsorption of metal ions on yeast cells at varied cell concentrations, *Plant Cell Physiol.* **16:**1167–1169 (1975).

Joho, M., Sukenobu, Y., Egashira, E., and Murayama, T., The correlation between Cd^{2+} sensitivity and Cd^{2+} uptake in the strains of *Saccharomyces cerevisiae, Plant Cell Physiol.* **24:**389–394 (1983).

Joho, M., Yamanaka, C., and Murayama, T., Cd^{2+} accommodation by *Saccharomyces cerevisiae, Microbios* **45:**169–179 (1986).

Kelly, D. P., Norris, P. R., and Brierley, C. L., Microbiological methods for the extraction and recovery of metals, in A. T. Bull, D. C. Ellwood, and C. Ratledge (eds.), *Microbial Technology: Current State, Future Prospects,* Cambridge University Press, Cambridge, 1979, pp. 263–308.

Khovrychev, M. P., Absorption of copper ions by cells of *Candida utilis, Microbiology* **42:**745–749 (1973).

Kiff, R. J., and Little, D. R., Biosorption of heavy metals by immobilized fungal biomass, in H. Eccles and S. Hunt (eds.), *Immobilisation of Ions by Bio-sorption,* Ellis Horwood, Chichester, U.K., 1986, pp. 71–80.

Kurek, E., Czaban, J., and Bollag, J.-M., Sorption of cadmium by microorganisms in competition with other soil constituents, *Appl. Environ. Microbiol.* **43:**1011–1015 (1982).

Langmuir, I., The adsorption of gases on plane surfaces of glass, mica, and platinum, *J. Am. Chem. Soc.* **40:**1361–1403 (1918).

Lawford, H. G., Pik, J. R., Lawford, G. R., Williams, T., and Kligerman, A., Hyperaccumulation of zinc by zinc-depleted *Candida utilis* grown in chemostat culture, *Can. J. Microbiol.* **26:**71–76 (1980).

Lichko, L. P., Okorokov, L. A., and Kulaev, I. S., Role of vacuolar ion pool in *Saccharomyces carlsbergensis:* potassium efflux from vacuoles is coupled with manganese or magnesium influx, *J. Bacteriol.* **144:**666–671 (1980).

Lichko, L. P., Okorokov, L. A., and Kulaev, I. S., Participation of vacuoles in regulation of levels of K^+, Mg^{2+} and orthophosphate ions in cytoplasm of the yeast *Saccharomyces carlsbergensis, Arch. Microbiol.* **132:**289–293 (1982).

Lindegren, C. C., and Lindegren, G., Oxidative detoxification of thallium in the yeast mitochondria, *Antonie van Leeuwenhoek* **39:**351–353 (1973).

McCready, R. C. L., and Lakshmanan, V. I., Review of bioadsorption research to recover uranium from leach solutions in Canada, in H. Eccles and S. Hunt (eds.), *Immobilisation of Ions by Bio-sorption,* Ellis Horwood, Chichester, U.K., 1986, pp. 219–226.

Minney, S. F., and Quirk, A. V., Growth and adaptation of *Saccharomyces cerevisiae* at different cadmium concentrations, *Microbios* **42:**37–44 (1985).

Mowll, J. L., and Gadd, G. M., Zinc uptake and toxicity in the yeasts *Sporobolomyces roseus* and *Saccharomyces cerevisiae, J. Gen. Microbiol.* **129:**3421–3425 (1983).

Mowll, J. L., and Gadd, G. M., Cadmium uptake by *Aureobasidium pullulans, J. Gen. Microbiol.* **130:**279–284 (1984).

Munger, K., Germann, N. A., and Lerch, K., Isolation and structural organisation of the *Neurospora* copper metallothionein gene, *EMBO J.* **4:**1459–1462 (1985).

Murasugi, A., Wada-Nagawa, C., and Hayashi, Y., Formation of cadmium-binding peptide allomorphs in fission yeast, *J. Biochem. (Tokyo)* **96:**1375–1379 (1985).

Murphy, R. J., and Levy, J. F., Production of copper oxalate by some copper tolerant fungi, *Trans. Br. Mycol. Soc.* **81**:165–168 (1983).

Muzzarelli, R. A. A., Bregani, F., and Sigon, F., Chelating capacities of amino acid glucans and sugar acid glucans derived from chitosan, in H. Eccles and S. Hunt (eds.), *Immobilisation of Ions by Bio-sorption*, Ellis Horwood, Chichester, U.K., 1986, pp. 173–182.

Muzzarelli, R. A. A., and Tanfani, F., The chelating ability of chitinous materials from *Aspergillus niger, Streptomyces, Mucor rouxii, Phycomyces blakeseanus* and *Choanephora curcurbitum*, in S. Mirano and S. Tokura (eds.), *Chitin and Chitosan*, The Japanese Society of Chitin and Chitosan, Tottori, Japan, 1982, pp. 183–186.

Nieboer, E., and Richardson, D. H. S., The replacement of the nondescript term "heavy metals" by a biologically and chemically significant classification of metal ions, *Environ. Pollut.* **1**:3–26 (1980).

Norris, P. R., and Kelly, D. P., Accumulation of cadmium and cobalt by *Saccharomyces cerevisiae, J. Gen. Microbiol.* **99**:317–324 (1977).

Norris, P. R., and Kelly, D. P., Accumulation of metals by bacteria and yeasts, *Dev. Ind. Microbiol.* **20**:299–308 (1979).

Okorokov, L. A., Kadomtseva, V. M., and Titovskii, B. I., Transport of manganese into *Saccharomyces cerevisiae, Folia. Microbiol.* **24**:240–246 (1979).

Okorokov, L. A., Lichko, L. P., and Kulaev, I. S., Vacuoles: main compartments of potassium, magnesium and phosphate ions in *Saccharomyces carlsbergensis* cells, *J. Bacteriol.* **144**:661–665 (1980).

Parkin, M. J., and Ross, I. S., The specific uptake of manganese in the yeast *Candida utilis, J. Gen. Microbiol.* **132**:2155–2160 (1986a).

Parkin, M. J., and Ross, I. S., The regulation of Mn^{2+} and Cu^{2+} uptake in cells of the yeast *Candida utilis* grown in continuous culture, *FEMS Microbiol. Lett.* **37**:59–62 (1986b).

Paton, W. H. N., and Budd, K., Zinc uptake in *Neocosmospora vasinfecta, J. Gen. Microbiol.* **72**:173–184 (1972).

Raymond, K. N., Muller, G., and Matzanke, B. F., Complexation of iron by siderophores. A review of their solution and structural chemistry and biological function, *Topics Curr. Chem.* **123**:49–102 (1984).

Roomans, G. M., Localization of divalent cations in phosphate-rich cytoplasmic granules in yeast, *Physiol. Plant* **48**:47–50 (1980).

Roomans, G. M., Theuvenet, A. P. R., Van den Berg, T. P. R., and Borst-Pauwels, G. W. F. H., Kinetics of Ca^{2+} and Sr^{2+} uptake by yeast. Effect of pH, divalent cations and phosphate, *Biochim. Biophys. Acta* **551**:187–196 (1979).

Ross, I. S., Effects of glucose on copper uptake and toxicity in *Saccharomyces cerevisiae, Trans. Br. Mycol. Soc.* **69**:77–81 (1977).

Ross, I. S., and Townsley, C. C., The uptake of heavy metals by filamentous fungi, in H. Eccles and S. Hunt (eds.), *Immobilisation of Ions by Bio-sorption*, Ellis Horwood, Chichester, U.K., 1986, pp. 49–58.

Ross, I. S., and Walsh, A. L., Resistance to copper in *Saccharomyces cerevisiae, Trans. Br. Mycol. Soc.* **77**:27–32 (1981).

Rothstein, A., and Hayes, A. D., The relationship of the cell surface to metabolism. XIII. The cation-binding properties of the yeast cell surface, *Arch. Biochem. Biophys.* **63**:87–99 (1956).

Saiz-Jimenez, C., and Shafizadeh, F., Iron and copper binding by fungal phenolic polymers: An electron spin resonance study, *Curr. Microbiol.* **10**:281–286 (1984).

Sakaguchi, T., and Nakajima, A., Recovery of uranium by chitin phosphate and chitosan phosphate, in S. Mirano and S. Tokura (eds.), *Chitin and Chitosan*, The Japanese Society of Chitin and Chitosan, Tottori, Japan, 1982, pp. 177–182.

Scatchard, G., The attractions of proteins for small molecules and ions, *Ann. N. Y. Acad. Sci.* **51**:660–672 (1949).

Shatzman, A. R., and Kosman, D. J., Characterisation of two copper-binding components of the fungus *Dactylium dendroides, Arch. Biochem. Biophys.* **194**:226–235 (1979).

Shumate, S. E., and Strandberg, G. W., Accumulation of metals by microbial cells, in M. Moo-Young, C. N. Robinson, and J. A. Howell (eds.), vol. 4, *Comprehensive*

Biotechnology, Pergamon Press, New York, 1985, pp. 235–247.

Shumate, S. E., Strandberg, G. W., McWhirter, D. A., Parrott, J. R., Bogacki, G. M., and Locke, B. R., Separation of heavy metals from aqueous solutions using "biosorbents"—development of contacting devices for uranium removal, *Biotechnol. Bioeng. Symp.* **10**:27–34 (1980).

Shumate, S. E., Strandberg, G. W., and Parrott, J. R., Biological removal of metal ions from aqueous process streams, *Biotechnol. Bioeng. Symp.* **8**:13–20 (1978).

Siegel, S., Keller, P., Galun, M., Lehr, H., Siegel, B., and Galun, E., Biosorption of lead and chromium by *Penicillium* preparations, *Water Air Soil Pollut.* **27**:69–75 (1986).

Siegel, S. M., Siegel, B. Z., and Clark, K. E., Bio-corrosion: Solubilization and accumulation of metals by fungi, *Water Air Soil Pollut.* **19**:229–236 (1983).

Somers, E., The uptake of copper by fungal cells, *Ann. Appl. Biol.* **51**:425–437 (1963).

Strandberg, G. W., Shumate, S. E., and Parrott, J. R., Microbial cells as biosorbents for heavy metals: Accumulation of uranium by *Saccharomyces cerevisiae* and *Pseudomonas aeruginosa*, *Appl. Environ. Microbiol.* **41**:237–245 (1981).

Subramanyam, C., Venkateswerlu, G., and Rao, S. L. N., Cell wall composition of *Neurospora crassa* under conditions of copper toxicity, *Appl. Environ. Microbiol.* **46**:585–590 (1983).

Sutter, H. P., Jones, E. B. G., and Walchli, O., The mechanism of copper tolerance in *Poria placenta* (Fr.) Cke and *Poria vaillantii* (Pers.). Fr., *Mater. Org.* **18**:243–263 (1983).

Theuvenet, A. P. R., Kessels, B. G. F., Blankensteijn, W. M., and Borst-Pauwels, G. W. F. H., A comparative study of K^+-loss from a cadmium-sensitive and a cadmium-resistant strain of *Saccharomyces cerevisiae*, *FEMS Microbiol. Lett.* **43**:147–153 (1987).

Tobin, J. M., Cooper, D. G., and Neufeld, R. J., Uptake of metal ions by *Rhizopus arrhizus* biomass, *Appl. Environ. Microbiol.* **47**:821–824 (1984).

Townsley, C. C., and Ross, I. S., Copper uptake by *Penicillium spinulosum*, *Microbios* **44**:125–132 (1985).

Townsley, C. C., and Ross, I. S., Copper uptake in *Aspergillus niger* during batch growth and in non-growing mycelial suspensions, *Exp. Mycol.* **10**:281–288 (1986).

Townsley, C. C., Ross, I. S., and Atkins, A. S., Biorecovery of metallic residues from various industrial effluents using filamentous fungi, in R. W. Lawrence, R. M. R., Branion, and H. G. Ebner (eds.), *Fundamental and Applied Biohydrometallurgy*, Elsevier, Amsterdam, 1986a, pp. 279–289.

Townsley, C. C., Ross, I. S., and Atkins, A. S., Copper removal from a simulated leach effluent using the filamentous fungus *Trichoderma viride*, in H. Eccles and S. Hunt (eds.), *Immobilisation of Ions by Bio-sorption*, Ellis Horwood, Chichester, U.K., 1986b, pp. 159–170.

Trevors, J. T., Stratton, G. W., and Gadd, G. M., Cadmium transport, resistance, and toxicity in bacteria, algae, and fungi, *Can. J. Microbiol.* **32**:447–464 (1986).

Tsezos, M., The role of chitin in uranium adsorption by *Rhizopus arrhizus, Biotechnol. Bioeng.* **25**:2025–2040 (1983).

Tsezos, M., Recovery of uranium from biological adsorbents—desorption equilibrium, *Biotechnol. Bioeng.* **26**:973–981 (1984).

Tsezos, M., Adsorption by microbial biomass as a process for removal of ions from process or waste solutions, in H. Eccles and S. Hunt (eds.), *Immobilisation of Ions by Bio-sorption*, Ellis Horwood, Chichester, U.K., 1986, pp. 201–218.

Tsezos, M., and Keller, D. M., Adsorption of radium-226 by biological origin absorbents, *Biotechnol. Bioeng.* **25**:201–215 (1983).

Tsezos, M., and Volesky, B., Biosorption of uranium and thorium, *Biotechnol. Bioeng.* **23**:583–604 (1981).

Tsezos, M., and Volesky, B., The mechanism of uranium biosorption by *Rhizopus arrhizus, Biotechnol. Bioeng.* **24**:385–401 (1982).

Tuovinen, O. H., and Kelly, D. P., Use of microorganisms for the recovery of metals, *Int. Metall. Rev.* **19**:21–31 (1974).

Venkateswerlu, G., and Sastry, K. S., Cobalt transport in a cobalt-resistant strain of *Neurospora crassa, J. Biosci.* **1**:433–439 (1979).

Wainwright, M., and Grayston, S. J., Oxidation of heavy metal sulphides by *Aspergillus niger* and *Trichoderma harzianum*, *Trans. Br. Mycol. Soc.* **86**:269–272 (1986).

Wainwright, M., Grayston, S. J., and de Jong, P., Adsorption of insoluble compounds by mycelium of the fungus *Mucor flavus*, *Enzyme Microb. Technol.* **8**:597–600 (1986).

Wakatsuki, T., Imahara, H., Kitamura, T., and Tanaka, H., On the absorption of copper into yeast cell, *Agric. Biol. Chem.* **43**:1687–1692 (1979).

Weber, W. J., Adsorption, in W. J. Weber (ed.), *Physico-Chemical Processes for Water Quality Control*, Wiley, New York, 1972, pp. 199–259.

Weidemann, D. P., Tanner, R. D., Strandberg, G. W., and Shumate, S. E., Modelling the rate of transfer of uranyl ions onto microbial cells, *Enzyme Microb. Technol.* **3**:33–40 (1981).

White, C., and Gadd, G. M., Uptake and cellular distribution of copper, cobalt and cadmium in strains of *Saccharomyces cerevisiae* cultured on elevated concentrations of these metals, *FEMS Microbiol. Ecol.* **38**:277–283 (1986).

White, C., and Gadd, G. M., The uptake and cellular distribution of zinc in *Saccharomyces cerevisiae*, *J. Gen. Microbiol.* **133**:727–737 (1987a).

White, C., and Gadd, G. M., Inhibition of H^+ efflux and K^+ uptake, and induction of K^+ efflux in yeast by heavy metals, *Toxicol. Assess.* **2**:437–447 (1987b).

Williams, J. I., and Pugh, G. J. F., Resistance of *Chrysosporium pannorum* to an organomercury fungicide, *Trans. Br. Mycol. Soc.* **64**:255–263 (1975).

Yakubu, N. A., and Dudeney, A. W. L., Biosorption of uranium with *Aspergillus niger*, in H. Eccles and S. Hunt (eds.), *Immobilisation of Ions by Bio-sorption*, Ellis Horwood, Chichester, U.K., 1986, pp. 183–200.

Zajic, J. E., and Chiu, Y. S., Recovery of heavy metals by microbes, *Dev. Ind. Microbiol.* **13**:91–100 (1972).

Zunino, H., and Martin, J. P., Metal binding organic macromolecules in soil. I. Hypothesis interpreting the role of soil organic matter in the translocation of metal ions from rocks to biological systems, *Soil Sci.* **123**:65–76 (1977).

Microbial Oxygenic Photoautotrophs (Cyanobacteria and Algae) for Metal-Ion Binding

Benjamin Greene

Dennis W. Darnall

Introduction

Chemical composition of cyanobacteria and algal cell walls

While we realize that both procaryotes and eucaryotes are microbial oxygenic photoautotrophs, for ease of discussion we will use the older terminology *algae* to describe both.

Algae are photosynthetic microorganisms comprising thousands of genera and species, many of which have novel metal-binding properties. Current systems of algal classification recognize between four and nine divisions of the algal kingdom, which have been organized according to cell wall composition, differences in pigmentation, starch and lipid composition, or presence of flagellae and other morphological features (Bold et al., 1980). Biologically active (living) or inactive (nonliving) algal cells can reversibly bind significant quantities of metal ions from aqueous solutions, and in many ways algal biomass functions like specialty ion-exchange resins in metal-ion recovery processes (Ferguson and Bubela, 1974; Crist et al., 1981; Wood and Wang, 1983; Greene et al., 1987a,b; Krambeer, 1987; Volesky, 1987; Darnall et al., 1988; Gee and Dudeney, 1988; Kuyucak and Volesky, 1988). Potential metal-cation-binding sites in algal cell components include carboxylate, amine, imidazole, phosphate, sulfhydryl, sulfate, hydroxyl, and other chemical functional groups contained in the cell proteins and sugars (Siegel and Siegel, 1973; Crist et al., 1981; Watkins II et al., 1987). A summary of potential metal-binding sites and their location in biological polymers is shown in Table 12.1. Table 12.2 displays the approximate protein and carbohydrate content of algal cell walls. Certain functional groups, such as amines and imidazoles, are positively charged when protonated and may electrostatically bind negatively charged metal complexes. Thus, algal cells contain many polyfunctional metal-binding sites for both cationic and anionic metal complexes.

The total cell wall structure for most algae has not been completely elucidated. However, information has been gathered regarding the structure of the cell walls of algae in general. As indicated above, algae are divided into two principal groups based on cell structure: the procaryotic cyanobacteria (blue-green algae), which do not have a membrane-bound nucleus or organelles, and the eucaryotes, whose cells have both a membrane-bound nucleus and organelles and which constitute the rest of the algae.

Cyanobacteria—technically photosynthetic bacteria—have cell walls in which the major component is murein (peptidoglycans or

TABLE 12.1   Ionizable Groups in Biological Polymers Capable of Participating in Metal Binding*

Group	Location	pK_a
Carboxyl	Protein C-terminal	3.5–4
Carboxyl	Beta aspartic	4–5
Carboxyl	Gamma glutamic	4–5
Carboxyl	Uronic acid	3–4.4
Carboxyl	N-Acetylneuraminic	2.6
Carboxyl	Lactate	3.8
Sulfonic acid	Cysteic acid	1.3
Phosphate	Serine as ester	6.8,2.0[†]
Phosphate	Polyol mono ester	0.9–2.1
Phosphate	Polysaccharide diester	1.5,6.0
Hydroxyl	Tyrosine-phenolic	9.5–10.5
Hydroxyl	Saccharide-alcoholic	12–13.0
Sulfydryl	Cysteine	8.3[†]
Amino	Protein N-terminal	7.5–8.0
Amino	Cytidine (pyrimidine)	4.11
Amino	Adenosine (purine)	3.45
Amino	Lysine	8.9,10.5[†]
Imidazole	Histidine	6–7
Imidazole	Guanosine (purine)	2.3
Imino	Peptide	13

*From Hunt (1986).
†From Segel (1976).

TABLE 12.2   Approximate Protein and Carbohydrate Content of Algal Cell Walls (as % Dry Weight)

Alga	Protein	Carbohydrate content (%)	Reference
Chlorella pyrenoidosa	27–43	6–7	Gotelli and Cleland (1968), Northcote et al. (1958)
Rhodymenia palmata*	37–50	12.9	Siegel and Siegel (1973), Whyte (1971)
Eisenia bicyclis*	10–25	60.0	Gotelli and Cleland (1968), Bird and Haas (1931)
Cyanidium caldarium	50–55	15–17	Bailey and Staehelin (1968)

*Compositions of the cell wall of the cited alga were not available; instead, average reported compositions of algae in the same algal group were used.

mucopeptides associated with as many as eight amino acids) containing diaminopimelic acid, muramic acid and N-acetyl glucosamine (Bold et al., 1980). For those cyanobacteria having sheaths external to the cell wall, the sheath is composed of a matrix of pectic acids and mucopolysaccharides. Small amounts of various proteins, particularly transport proteins, are also likely to be associated with the cell walls of the cyanobacteria.

For most eucaryotic algae the cell wall structure is considerably different from that of the cyanobacteria. Furthermore, there is wide variation among eucaryotic algal species. The major algal divisions according to Bold et al. (1980) are Chlorophycophyta (green algae), Charophyta (stoneworts), Euglenophycophyta (euglenids), Phaeophycophyta (brown algae), Chrysophycophyta (golden algae, including diatoms), Pyrrophycophyta including (dinoflagellates), and Rhodophycophyta (red algae). Except for the Euglenophycophyta and some members of the Chlorophycophyta and Pyrrophycophyta that have no cell walls, the only cell wall component common to all the eucaryotic algal division is cellulose (poly-β-1:4-glucopyranose). However, one exception is the alga *Porphyra* of the division Rhodophycophyta, which lacks cellulose and contains mannans and xylans as its principal cell wall component.

In addition to cellulose, the principal cell wall components for the eucaryotic algal division Chlorophycophyta are β-1,4-D-mannopyranose (mannan), β-1,3-D-xylose (xylan), hydroxyproline, glycosides, and in some groups, ester sulfates of highly branched, water-soluble polysaccharides composed of D-glucuronic acid, D-xylose, and L-rhamnose, or sometimes D-galactose, L-arabinose, D-mannose, and D-xylose; for Phaeophycophyta—alginic acid and mucopolysaccharide sulfates; for Chrysophycophyta—silicates, arabino-galactans or mucopolysaccharides, sulfates (mucilaginous substances), and some N-acetylglucosamine (chitin); for Pyrrophycophyta—arabino-galactans or sulfated mucopolysaccharides, and so on (mucilaginous substances); and for Rhodophycophyta—xylans, pectins, and hydrocolloids (Siegel and Siegel, 1973; Bold et al., 1980).

As a result of the wide structural and chemical variation in algal cell walls depending upon the algal division, genera, species, and variety, it is expected that metal-binding properties by algal genera would vary. Methods for screening the metal-binding properties of different algal species have been reported (Jennett et al., 1979), and several investigators have observed species-dependent variations in how metal ions bind to algae. We report on these here.

Metal complexation by algae

Detailed studies of metal complexation by biopolymers appear elsewhere (Hunt, 1986). However, it is relevant here to describe some chemical characteristics of metal-ligand interactions and electrostatic interactions of metal ions that may be operational in mechanisms of metal binding to algae.

Binding or accumulation of heavy-metal ions by biologically active algal cells may occur by several mechanisms, such as surface binding or precipitation, or by intracellular transport and chelation (Wood and Wang, 1983). Binding of metal ions to biologically inactive algal cells also occurs by different mechanisms, including covalent or electrostatic binding to cell surfaces or even by more complex mechanisms involving chemical redox interactions between the metal ion and the algal cell. Metal binding to biologically inactive microorganisms has been referred to elsewhere as *biosorption* (Tsezos and Volesky, 1981).

Methods for use of algae for metal recovery: batch or column reactors

Various engineering designs are used to incorporate algae or other microbes into water treatment schemes (Krambeer, 1987; Volesky, 1987). Processes for metal-ion sequestering by algae usually involve either a *batch* or *column* configuration. The theory of batch versus column methods for separation processes is described in detail in Karger et al. (1973). For similar reasons as in ion-exchange equilibria, algal reactors employing a column configuration offer greater metal-binding capacity and higher efficiency (i.e., high-purity effluents) and are more readily adopted to automation and continuous flow than are batch reactors. In a batch configuration, free or immobilized algal cells are mixed with the metal-ion solution and, following equilibration, are removed by settling, centrifugation, or filtration. In a column configuration, the metal-ion solution is pumped through a column packed with immobilized algal cells. The column configuration is not suited for packing with free algal cells, since the cells tend to clump and excessive hydrostatic pressure is required to generate a suitable flow rate. Since algal cells are inherently fragile, attrition of the biomass may occur under high pressure. This problem has been alleviated by the use of algae immobilized in a suitable porous matrix. Polyacrylamide, calcium alginate, and silica have been used to immobilize algal cells and used in packed columns for metal-ion recovery (Nakajima et al., 1982; Darnall et al., 1986b; Greene et al., 1987b; Gee and Dudeney, 1988). Reviews on various immobilization methods for algae and other microorganisms appear elsewhere (Kennedy and Cabral, 1983; Robinson et al., 1986; Greene and Bedell, in press).

Metal Binding to Living Algae

There have been a number of reports on the feasibility of using algal cultures contained in lagoons or meander systems for heavy-metal recovery from polluted waters. However, special strains of algae and suitable environments are usually required to maintain productive cultures in toxic metal environments. Khummongkol et al. (1982) developed a model for the accumulation of heavy-metal ions by living algal cultures, and evaluated this model by using cadmium(II) and *Chlorella vulgaris*, a green alga. The model assumed that the mechanism of cadmium(II) accumulation by the alga was entirely due to adsorption of the metal ion on the outer surface of the algal cell wall. It was predicted that adsorption would be rapid and that equilibrium should be reached after about 10 min. The workers found, however, that under growth conditions there was an additional contribution from increased surface area due to the expanding biomass. Furthermore, a greater amount of cadmium(II) was accumulated by the alga than was predicted on the assumption of surface accumulation alone, which implied that another binding mechanism was involved. The researchers suggested that an improved model for heavy-metal accumulation by living algae might also include a contribution by intracellular uptake. Thus, the kinetics of metal binding by living algal cultures may be affected by the viability and growth of the biomass.

Gale and Wixon (1979) and Jennett et al. (1979) reported that benthic algal mats contained in a meander system could be used to remove heavy-metal ions from zinc and lead mining and milling effluents. The contaminated waters were circulated through a series of shallow canals that contained the algae, and within 3 h the concentration of lead(II) in the water decreased from 3 ppm to 50 ppb. Fine particulate matter, which also contained heavy-metal ions, was removed by settling into the dense algal mat. The greatest advantage of the meander system is that there are no recurring costs after the initial expenditure for construction, and cleaner water was produced as long as the algae bloomed. The disadvantage of the meander system, however, is that the metal ions are not removed from the environment but are simply immobilized in the algal mats.

Dissanayake and Kritsotakis (1984) found that naturally occurring benthic algal mats in seawater offshore of Sri Lanka contained relatively high concentrations of gold and platinum. However, no evidence was shown on whether those metals had been bound to the algae as ions or whether the dense algal growth had filtered out or trapped colloidal particles.

A laboratory-scale system for the removal of heavy-metal ions from wastewater lagoons containing live algal cultures was evaluated by Filip et al. (1979). Following sand filtration of the metal-saturated algae, it was predicted that the effluent water would contain lower concentrations of heavy-metal ions, as well as reduced amounts of solids and organic contaminants. Using mixed cultures of algae, it was found that the toxicity of heavy-metal ions was the major limiting factor in the survival of certain algal species. In the case of chromium(VI), only *Oscillatoria* was resistant to relatively high concentrations of the metal ion and could be used. The workers also pointed out that the quantity of metal ions removed by algae, and therefore the quality of the effluent, depended largely on the biomass concentration, which was highly dependent on the local environmental and climatic conditions.

A report by Brierley and Brierley (1981) indicated that U(VI) and Mo(VI) were not significantly accumulated by living cultures of the filamentous algae *Spyrogyra* and *Oscillatoria* and the benthic alga *Chara* from waters derived from a U mine water treatment facility. The pH of the water was between 7 and 8, and there was an appreciable concentration of bicarbonate ion. The workers concluded that large quantities of biomass and large pond areas would be required to achieve the desired levels of metal-ion removal.

The feasibility of using living *Chlorella pyrenoidosa* to remove Cd(II) from electroplating wastewaters was investigated by Jennett et al. (1979). The process design involved the rapid mixing of a living algal suspension (from a nearby pond culture) with pH-adjusted wastewater, followed by sand filtration of the metal-saturated alga. This process was a variation of the existing methods, since only the short-term adsorptive properties of the alga were exploited rather than a long-term biological exposure to metal-containing solution, which would likely be detrimental to cell growth.

Becker (1983) concluded that the inhibition of nitrogenase activity caused by mercury and cadmium limited the use of living cyanobacteria to remove metal ions and other contaminants from sewage waters.

Perhaps the greatest difficulties encountered with the use of living algal culture in heavy-metal-recovery operations arise from the control and maintenance of algal growth in polluted waters. A further limitation of methods based on living algal cultures is the inability to recover the metal ions from metal-saturated algal cells while maintaining their viability. The limitation exists because substantial pH adjustments or the addition of specific complexing agents (which may be toxic) are necessary to strip metal ions from algal cells, and many algae require a narrow pH range for optimum growth. A method employing only the adsorptive properties of precultured algae avoids the toxic effects of metals on cell growth.

Metal Binding to Nonliving Algae

Metals which are biosorbed

There are numerous reports in the literature on binding of different metal ions to nonliving algae. Ferguson and Bubela (1974) reported that frozen or freeze-dried preparations of *Chlorella vulgaris, Ulothix,* and *Chlamydomonas* accumulated significant quantities of Cu(II), Pb(II), and Zn(II) from laboratory solutions. Nakajima et al. (1981) studied the binding of various heavy-metal ions to freeze-dried *Chlorella regularis.* They observed relatively selective accumulation of metal ions from a solution containing an equivalent concentration (1.0 mM) of each ion. The order of selectivity of metal binding to *Chlorella* decreased in the following order: UO_2^{2+} > Cu^{2+} > Zn^{2+} > Ba^{2+} = Mn^{2+} = Cd^{2+} > Ni^{2+} = Sr^{2+}.

Darnall et al. (1986a,b) and Greene et al. (1986a,b; 1987a,b) examined the binding of a number of metal ions to freeze-dried *Chlorella vulgaris,* and to spray-dried or sun-dried *Chlorella pyrenoidosa* and *Spirulina platensis.* Under specified conditions these algae biosorbed significant quantities of Al^{3+}, Be^{2+}, Cu^{2+}, Pb^{2+}, Ni^{2+}, Zn^{2+}, Cr^{3+}, Cr^{6+}, Co^{2+}, Fe^{2+}, Fe^{3+}, UO_2^{2+}, $AuCl_4^-$, $Au(CN)_2^-$, $AuBr_4^-$, Ag^+, Hg^{2+}, $PtCl_4^{2-}$, SeO_4^{2-}, MoO_4^{2-}, and Mn^{2+}. Metal binding was found to depend on pH, temperature, and presence of competing metal ions. Specific examples are described below. The workers also developed some metal-ion separation schemes based on the pH dependence of metal-ion binding to these alga. The major advantage of the use of nonliving algae to recover metal ions was that binding and recovery of the metal ions can occur under conditions that were normally toxic to living cells.

pH dependence of metal binding: different metals. Darnall et al. (1986a,b) and Greene et al. (1987a,b) have determined the effects of pH on binding of a number of metal ions to dried biologically inactive preparations of *Chlorella vulgaris*. Most of the metal ions tested could be divided into three major categories, depending on how binding to the alga was affected by pH. Figure 12.1 displays representative pH profiles for metal binding to *Chlorella*.

One group of metal ions, including $Hg(II)$, $Au(III)$ as $AuCl_4^-$, $Ag(I)$, $Pd(II)$, and $Au(I)$ thiomalate, bound to *Chlorella vulgaris* rather independently of pH values between 2 and 7. This behavior is consistent with the general coordination chemistry of the metal ions. These metal ions are all classified as "soft" according to Pearson (1973). Soft metal ions undergo covalent binding to softer ligands, such as sulfhydryl and amine groups, and those binding interactions are generally minimally affected by ionic interactions and pH.

A second group of metal ions, which were found to bind more strongly to *Chlorella vulgaris* as pH was increased from 2 to 5, consisted of borderline soft and hard metal cations, including $Cu(II)$, $Ni(II)$, $Zn(II)$, $Co(II)$, $Pb(II)$, $Cr(III)$, $Cd(II)$, $U(VI)$, $Co(II)$, $Be(II)$, and $Al(III)$. The pH-dependent behavior resembled the binding of these metal ions to cation-exchange resins containing carboxylate or amine groups. A pH dependence of metal-cation binding generally occurs when the active metal-binding sites (such as carboxylate or amine groups) can also bind protons. Thus, metal ions and protons compete for the same binding sites.

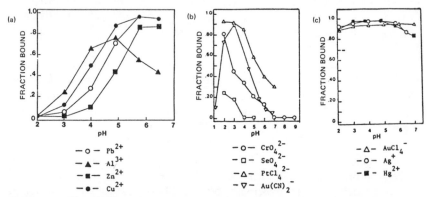

Figure 12.1 The pH dependence of metal-ion sorption by *Chlorella vulgaris*. (*From Darnall et al., Recovery of heavy metals by immobilized algae, in R. Thompson (ed.), Trace Metal Removal from Aqueous Solution, The Royal Society of Chemistry, Burlington House, London, 1986b, pp. 1–24.*)

A third group of metal ions was found to be more strongly bound to *Chlorella vulgaris* at pH 2 than at pH 5. These ions included mainly the oxoanions MoO_4^{2-}, SeO_4^{2-}, and CrO_4^{2-} and other anionic metal complexes, including $PtCl_4^{2-}$ and $Au(CN)_2^-$. The increased binding of these metal ions at low pH was consistent with electrostatic binding to ligands such as amines or imidazoles, which would be positively charged due to protonation at low pH values. Also, since the isoelectric point for many algal species lies between 3 and 4 (Crist et al., 1981), the overall net charge on the cell wall promotes easier access of anions to positively charged binding sites as the pH is decreased below the isoelectric pH.

Metal binding to different algal species.   Greene et al. (1987b) and Gardea-Torresdey (1988) have determined that different algal species exhibit different metal-binding characteristics at a given pH. This presumably occurs because of differences in the cell wall compositions of the different organisms, which result in unique strong binding sites in certain algae (Table 12.2). For example, *Eisenia bicyclis* was more effective in binding aluminum(III) at pH 2.0 than were *Cyanidium caldarium, Spirulina platensis,* and *Chlorella pyrenoidosa. Cyanidium caldarium* bound more copper(II) at pH 2.0 than did *Eisenia bicyclis, Spirulina platensis,* and *Chlorella pyrenoidosa.* The capacity for tetrachloroaurate(III) accumulation at pH 2.0 decreased in the order *Chlorella pyrenoidosa > Cyanidium caldarium > Eisenia bicyclis.* These differences in the pH dependence of metal binding to different algal species were used for metal-ion separation schemes (Greene et al., 1987a; Darnall et al., 1988; Gardea-Torresdey, 1988).

Darnall et al. (1988) and Gardea-Torresdey (1988) studied the binding of tetrachloroaurate(III) to *Rhodymenia palmata, Macrocystis pyrifera, Phophyra yezoensis, Laminaria japonica,* and other algal species and observed substantial differences in the pH dependence of gold binding (Fig. 12.2). While the mechanism of gold binding to algae is quite complex (Greene et al., 1986b), there may also be a species-dependent variation in the mechanism. Kuyucak and Volesky (1988) reported that various marine algae biosorbed significant quantities of $AuCl_4^-$, and demonstrated that the seaweed *Sargassum natans* could accumulate nearly 50 percent of its dry weight in gold. Thus screening of different algal species for their gold-binding abilities may reveal organisms that are particularly useful in methods of gold recovery.

Spirulina platensis and *Cyanidium caldarium* and other algal species exhibited increased binding of dicyanoaurate(I) as the pH was lowered from 7 to 2 (Fig. 12.3). However, *Cyanidium* had a greater binding capacity for dicyanoaurate(I) than did *Spirulina.* This suggests the presence of more unique strong gold-binding sites on *Cyanidium* than on *Spirulina.*

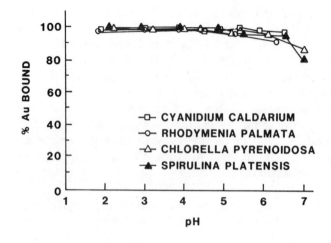

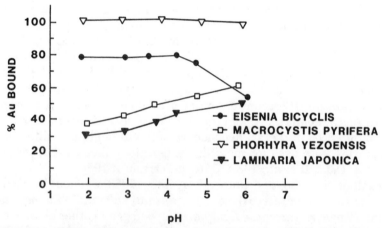

Figure 12.2   Effect of pH on the binding of tetrachloroaurate(III) to algae. Alga (5 mg/mL) was agitated for 1 h with 1×10^{-4} M HAuCl$_4$ at different pH values. Following centrifugation, the free concentration of Au and the final pH values in the supernatant solution were determined. (*From Darnall et al., Gold binding to algae, in P. R. Norris and D. P. Kelly (eds.), Biohydrometallurgy, Science and Technology Letters, Kew, Surrey, U.K., 1988, pp. 487–498.*)

Separation of metals

Since not all metal ions exhibit identical properties in binding to algae, several procedures have been successful for separation or selective binding of metal ions from solutions containing several metal ions. The control of pH and addition of competing ligands or complexing agents are major factors in these procedures. Thus, separation of metal ions may be accomplished either by adjusting condi-

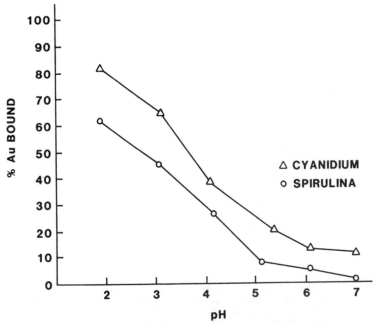

Figure 12.3 Binding of dicyanoaurate(I) to *Cyanidium caldarium* and *Spirulina platensis*. Alga (5 mg/mL) was agitated for 1 h with 1×10^{-4} M $KAu(CN)_2$ at different pH values. Following centrifugation, the free Au concentrations and the final pH values were determined. (*From Darnall et al., Gold binding to algae, in P. R. Norris and D. P. Kelly (eds.), Biohydrometallurgy, Science and Technology Letters, Kew, Surrey, U.K., 1988, pp. 487–498.*)

tions so that there is selective binding to the algae or by simultaneous binding of a number of metal ions present in solution followed by selective elution of each metal ion bound to algae. These methods have direct application for metal recovery in wastewater treatment or mining processes (Krambeer, 1987).

Control of pH is a major factor affecting the separation of metal ions from solutions containing several metal ions. Figure 12.4 shows a pH profile for the binding to *Chlorella vulgaris* of nine different metal ions present in equimolar concentrations (0.1 mM). With the notable exception of Ag(I), the metal ions Cu(II), Ni(II), Co(II), Cd(II), Cr(III), Zn(II), Pb(II), and Al(III) exhibited minimal binding near pH 2. As a result of the alga's selectivity, only Ag(I) was effectively removed from the solution at pH 2.0. Furthermore, within the group of ions whose binding was pH-dependent, there were significant differences in behavior. As pH was increased from 2 to 5, the fraction of each metal bound, or relative selectivity in binding, differed in the order Al(III) > Cu(II) > Pb(II) > Cr(III) > Cd(II) > Ni(II) > Zn(II) > Co(II).

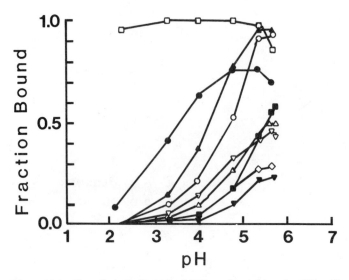

Figure 12.4 The effect of pH on the binding of metal ions by *Chlorella vulgaris* from solutions containing several metal ions. *Chlorella* (5 mg/mL) was suspended in solutions containing 0.1 mM each of the following metal ions in 0.05 M sodium acetate adjusted to different pHs. The metal-ion salts were aluminum(III) nitrate (●), cadmium(II) nitrate (△), chromium(III) nitrate (▽), cobalt(II) nitrate (▼), copper(II) nitrate (▲), lead(II) acetate (○), nickel(II) nitrate (◇), silver(I) nitrate (□), or zinc(II) acetate (■). The reaction mixtures were shaken for 2 h, centrifuged, decanted, and the final pH's recorded. Metal-ion analyses were performed on the supernatants, and the fractions of metal ions bound to the algae were calculated.

This selectivity forms a basis for which pH-dependent resolution of metal ions on a chromatographic column containing an algal packing can be made. Examples of these experiments are described below.

Schemes based on selective binding at certain pH values have been devised for separation of metal ions, using algae in batch and in column configurations (Darnall et al., 1986b; Greene et al., 1987a,b; Gardea-Torresdey, 1988). For example, tetrachloroplatinate(II) was selectively separated from Cu(II) at pH 2 in a batch system using *Chlorella* as the biosorbent. Likewise with *Chlorella vulgaris*, tetrachloroaurate(III) (0.01 mM) was effectively removed from a large excess of Cu(II) (9.9 mM) at pH 2.0, and Cr(VI) was selectively removed from electroplating wastewaters also containing Cu(II), Zn(II), and Ni(II) near pH 2.0. Novel separation of Cr(VI) from Cr(III) was performed by binding the Cr(VI) at pH 1.5, thereby leaving Cr(III) in solution. Since it was possible to recover algal-bound Cr(VI) by elution at higher pH values, this methods may have direct applications in the treatment of electroplating wastewaters (Krambeer, 1987).

The separation of metal ions through selective binding to algae occurred in experiments where a competing ligand or complexing agent was added to the solution. Although tetrachloroaurate(III) and Hg(II) showed relatively pH-independent binding to *Chlorella vulgaris*, the presence of high chloride concentrations inhibited Hg(II) binding, whereas tetrachloroaurate(III) binding was essentially unaffected (Greene et al., 1987a). Thus, selective gold binding to *Chlorella vulgaris* occurred in a solution containing 0.013 mM chloride ion and 0.1 mM concentrations of $AuCl_4^-$ and Hg(II) (Greene et al., 1987a). Other experiments confirmed that tetrachloroaurate(III) and Hg(II) in separate solutions were each strongly bound to *Chlorella vulgaris* when no excess chloride ion was present. However, when the two ions were mixed together with *Chlorella*, chloride ions liberated by bound gold caused an interference with Hg(II) binding (Greene et al., 1987a). A further experiment showed that mercury bound strongly to *Chlorella vulgaris* that was initially treated with tetrachloroaurate(III) and later washed free of excess chloride. This result suggests that interference of mercury binding was a consequence of chloride complexation by mercury rather than preferential binding of gold in the presence of mercury.

Stripping of metals from algae

Studies on effects of pH and competing ligands or complexing agents on metal-ion binding to algae have demonstrated novel methods for stripping and recovering the bound metal ions. Representative examples of stripping methods are described below. For metal ions that show a marked pH dependence in binding to algae, stripping of bound metal ions can be accomplished upon pH adjustment. For example, binding of metal cations such as Cu(II), Cr(III), Ni(II), Pb(II), Zn(II), Cd(II), and Co(II) to *Chlorella vulgaris* at pH 5.0 is quantitatively reversed by lowering the pH to 2.0 (Greene et al., 1987a).

Anionic metal complexes such as SeO_4^{2-}, CrO_4^{2-}, and MoO_4^{2-}, bound most effectively at pH values near 2, were effectively stripped by adjusting the pH to a higher value.

Metal ions that showed little pH dependence in binding to algae were successfully stripped from the algal cells by the addition of specific ligands that form exceptionally stable complexes with these ions. Ag(I) binding to *Chlorella vulgaris*, *Chlorella pyrenoidosa*, and *Spirulina platensis* was effectively reversed by the addition of mercaptoethanol at pH 9 or above (Greene et al., 1987a). Under other conditions, however, Ag(I) bound to *Chlorella vulgaris* at pH 5 could be stripped at pH 2. This occurred when relatively high concentrations of Ag(I) were bound to the algal cells. Figure 12.5 demonstrates that as increasing amounts of Ag(I) were bound to the cells at pH 5, an in-

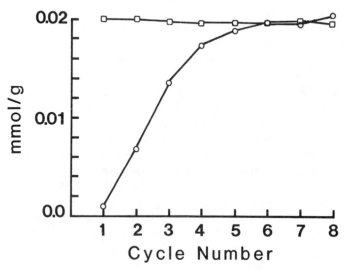

Figure 12.5 Binding and recovery of silver(I) from *Chlorella vulgaris*. Results are shown for a sequence of two reactions in which silver(I) was first bound to *Chlorella vulgaris* in a reaction at pH 5 and then stripped from the algae in a reaction at pH 2. This sequence, constituting a cycle, was repeated several times in succession. The binding reactions were performed by suspending washed algae at 5 mg/mL for 15 min (with shaking) in a solution which contained 0.1 mM silver(I) nitrate in 0.05 M sodium acetate at pH 5. The mixtures were then centrifuged, decanted, and the supernatants analyzed for silver. The amount of silver(I) bound to the algae during each binding reaction was calculated, and the results are shown in the upper curve (□). The stripping reactions were performed by suspending the algae pellets (following the above centrifugation) at 5 mg/mL in 0.05 M sodium acetate at pH 2 for 1 min with shaking, then centrifuging and decanting. This procedure was then repeated to ensure that stripping was complete. The lower curve (○) shows the total amount of silver(I) stripped from the algae by the two stripping solutions during each cycle.

creasing amount of Ag(I) was stripped at pH 2. These results suggest the presence of both primary (strong, pH-independent) binding sites, such as sulfhydryl groups, and secondary (weak, pH-dependent) binding sites, such as carboxylate groups for Ag(I) on the alga. Once the strong binding sites were saturated with Ag(I), additional binding of Ag(I) occurred at the weaker sites, and therefore stripping with dilute acid was effective. This example with Ag(I) points out that the classifications shown in Fig. 12.1 may be concentration-dependent.

Hg(II) binding to *Chlorella vulgaris* was effectively reversed by the addition of chloride ion at pH 2 or 5 or by the addition of mercaptoethanol at pH 2 or higher. Using a combination of pH gradient and mercaptoethanol, a scheme was used by Darnall et al. (1986a) to selectively and sequentially recover four different metal ions [Cu(II), Zn(II), Au(III), and Hg(II)] initially bound to polyacrylamide-immobilized *Chlorella vulgaris* (Fig. 12.6). In these experiments, a solution containing equimolar concentrations (0.1 mM) of $AuCl_4^-$, Cu(II), Zn(II), and Hg(II) was initially applied to a column of immobilized algae at pH 6.0. Binding of all four metal ions to the algal material was quantitative, since no detectable metal ions were present in the column effluent despite continued washing at pH 6.0. A pH gradient was then applied to the column, and as the column effluent gradually decreased from pH 6.0 to pH 2.0, first Zn(II) and then Cu(II) eluted from the column. However, gold and mercury remained bound to the column at pH 2.0. Subsequently, elution of the column with mercaptoethanol at pH 2.0 resulted in desorption of Hg(II), and then

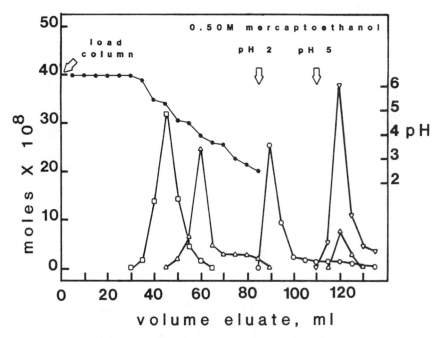

Figure 12.6 Binding and selective recovery of metal ions from *Chlorella vulgaris* immobilized in a polyacrylamide matrix. *Chlorella vulgaris* washed at pH 2.0 was immobilized in a polyacrylamide matrix by suspending the algae in a solution of acrylamide and N,N'-methylenebisacrylamide, followed by the addition of small amounts of ammonium persulfate and tetramethylenediamine to induce polymerization and crosslinking. The immobilized algae was passed through a 40-mesh sieve, washed over a 100-mesh sieve, and then packed into a glass column (5 cm by 0.7 cm). The column was washed extensively at pH 2 with 0.05 M acetic acid and then at pH 6.0 with 0.05 M sodium acetate. A solution (5 mL) containing 1.0×10^{-4} M each of mercury(II) acetate, zinc(II) acetate, copper(II) acetate, and hydrogen tetrachloroaurate(III) was loaded on the column with a flow rate of 0.1 mL/min. After metal-ion loading, the column was washed with 0.05 M sodium acetate at pH 6, and then pH was gradually decreased by washing the column with 0.05 M sodium acetate at a lower pH. Elution with mercaptoethanol in 0.05 M sodium acetate occurred at the indicated pH's (□) zinc, (△) copper, (○) mercury, (▽) gold, (●) pH. [*From Darnall et al., Environ. Sci. Technol. 20(2):206–208 (1986a).*]

elution with mercaptoethanol at pH 5.0 resulted in desorption of gold. A similar procedure was used by Greene et al. (1987b) to elute these metals from silica-immobilized *Chlorella vulgaris*. Acidic thiourea has been used to recover algal-bound gold (Greene et al., 1986b; Hosea et al., 1986; Darnall et al., 1988; Gee and Dudeney, 1988; Gardea-Torresdey, 1988; Kuyucak and Volesky, 1988). The algae can be re-used for Au binding many times thereafter.

Competition of ions for metal-binding sites

Chlorella vulgaris, Spirulina platensis, Chlorella pyrenoidosa, and *Cyanidium caldarium* bind Ca^{2+} and Mg^{2+} only very weakly (Greene et al., 1987b). This is one important advantage of these algal cells over ion-exchange resins in recovering heavy-metal ions from hard waters, since Ca^{2+} and Mg^{2+} may saturate cation-exchange resins and interfere with the binding of heavy-metal ions. Experiments with *Spirulina platensis,* which were performed at a constant ionic strength of 1.51 M (maintained with sodium nitrate) and at pH 5 to 6, indicated that 10,000 ppm Ca(II) produced a decrease of only 10 percent in Cu(II) binding at 6.4 ppm, 10 percent of Al(III) binding at 27 ppm, 21 percent of Cd(II) binding at 56 ppm, and 25 percent of Zn(II) binding at 6.5 ppm. Similarly, 10,000 ppm Mg(II) produced a decrease of 11 percent of Al(III) binding, 15 percent of Cd(II) binding, and 25 percent of Zn(II) binding.

Metal ions may compete with each other for binding sites on an algal surface (Crist et al., 1981; Nakajima et al., 1981). Another example is an experiment in which *Chlorella vulgaris* was sequentially exposed to a solution containing an equimolar mixture of nine metal ions at pH 5.0 (Greene et al., 1987a,b). The order of selectivity in binding of the nine different metal ions to the alga was

$$Al(III) \sim Ag(I) \gg Cu(II) > Cd(II) \geq Ni(II) \geq Pb(II) > Zn(II)$$
$$= Co(II) \geq Cr(III)$$

In the experiment described above, the binding of Ag(I) and Al(III) was essentially unaffected by the presence of all the other metal ions (Greene et al., 1987a,b). In contrast, binding of the other metal ions was decreased in multicomponent mixtures as compared to the binding of the same metal ion in the absence of other metal ions. One explanation for this behavior is that there are distinct classes of binding sites on the alga, which have preference for binding of either very hard or very soft metal ions. Al(III) ion was the hardest metal ion in the medium, and Ag(I) was the softest. Most of the other metal ions were borderline soft or borderline hard. Therefore, it would be expected that Ag(I) and Al(III) binding, which occurs to soft and hard sites, respectively, should not compete with each other, but also that binding of the borderline metal ions would compete substantially with each other because of similarities in their coordination chemistry.

Like Ag(I) ion, Hg(II) bound to *Chlorella vulgaris* rather independently of pH in the 2 to 5 range. Since Hg(II) is also a soft metal ion, it is predicted that competition for binding sites might occur between these ions when exposed to *Chlorella vulgaris*. Figure 12.7 shows that when algal cells are exposed repeatedly to equal volumes of fresh, 0.1-mM solutions of Ag(I) or Hg(II) at pH 5.0 or to a solution containing equimolar concentrations of the two metal ions, the algal cells have a higher binding capacity for Hg(II) than for Ag(I), and that the presence of Hg(II) drastically inhibits binding of Ag(I) to the alga. These results suggest that Hg(II) and Ag(I) compete for binding sites on *Chlorella vulgaris*, with Hg(II) being the more strongly bound. Other experiments have demonstrated that algal-bound Ag(I) is largely displaced from binding sites by the subsequent addition of Hg(II) (Greene et al., 1987a).

Specific examples of metal-ion binding to algae

Reports of gold binding to algal cells have recently generated much interest, and some researchers have investigated this phenomenon as

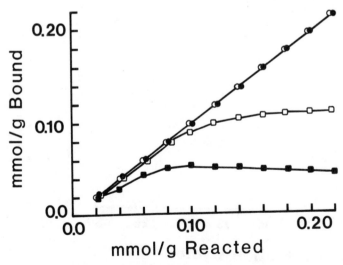

Figure 12.7 Sorption of Hg(II) and Ag(I) by lyophilized *Chlorella vulgaris*. Lyophilized *C. vulgaris* cells were washed and then suspended repeatedly for 15-min periods in solutions of 0.1 mM Ag(I) nitrate (□), Hg(II) acetate (○), or a solution containing 0.1 mM of each metal salt [Hg(II) (●), Ag(I) (■)].

it occurs in nature and under laboratory conditions (Dissanayake and Kritsotakis, 1984; Greene et al., 1986b; Hosea et al., 1986; Watkins et al., 1987; Gee and Dudeney, 1988; Kuyucak and Volesky, 1988; Darnall et al., 1988). Algal cells have a remarkable affinity for gold ions. The mechanism of gold binding, which has been studied with gold(I) and gold(III) complexes, is novel and complex (Greene et al., 1986b).

Figure 12.8 shows that gold binding to algae is reversible. A column containing silica-immobilized *Chlorella pyrenoidosa* was prepared, and a tetrachloroaurate(III) solution at pH 1.5 was passed through the column. Analysis of the column effluent indicated that gold binding was essentially quantitative despite continued washing with HCl. An acidic thiourea solution was then introduced into the column, and analysis of the column effluent indicated that there was nearly quantitative recovery of bound gold. Following a washing step, gold loading and stripping cycles were repeated more than 50 times with no loss of

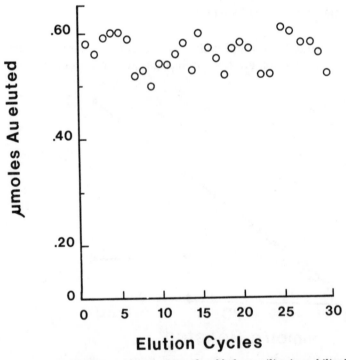

Figure 12.8 Binding and stripping of gold from silica-immobilized *Chlorella pyrenoidosa.* A small column containing silica-immobilized *Chlorella* was loaded with 10 mL of 0.100 m*M* gold(III) and was then stripped with 10 mL of 0.10 *M* thiourea. Values shown are micromoles of gold recovered on each cycle.

column performance. These data indicate that the immobilized alga have excellent recycling properties for $AuCl_4^-$ binding and recovery. Similar recycling experiments were performed in batch experiments using *Chlorella vulgaris* and a gold(I) complex, sodium gold(I) thiomalate.

Gold may be accumulated by algae from very dilute solutions, even when the waters contain a complex matrix such as seawater. High concentrations of gold were found in benthic algal mats offshore of Sri Lanka (Dissanayake and Kritsotakis, 1984). We have done laboratory experiments to determine gold binding at very low concentrations from different solutions. The experiments were performed by loading columns containing polyacrylamide immobilized *Chlorella* with low concentrations of gold spiked into an artificial seawater solution. Following the loading step, stripping with acidic thiourea was performed. Analysis of the stripping solution indicated that gold binding was nearly quantitative even at nanomolar concentrations from the artificial seawater solutions. It has not been determined, however, whether gold binding to immobilized algae from natural seawater will occur under these conditions.

An electrode which contains *Chlorella* in a carbon paste has recently been devised. The electrode can be used to quantitate gold in solution (Gardea-Torresdey et al., 1988b).

U(VI) binding to algae was investigated by Brierley and Brierley (1981), Nakajima et al. (1981, 1982), and by Greene et al. (1986a). The affinities of different algal species for U(VI) binding were found to depend on the species of uranyl ion complex in solution. For example, the presence of bicarbonate ion dramatically inhibited U(VI) binding to *Chlorella vulgaris* (Greene et al., 1986a). High concentrations of bicarbonate ion are normally found in waters derived from U ore mining and milling operations. Reduced U(VI) binding in the carbonate solution occurred as the pH was increased above pH 6. This was most likely the direct result of the increased stability of the U(VI) bicarbonate complex from pH 6 to pH 7. Other experiments indicated that U(VI) binding to *Chlorella vulgaris* from authentic mill waters was greatly improved by acidifying the waters from pH 8.0 to between pH 6.0 and pH 5.0 (Greene et al., 1986a). Nakajima et al. (1982) also used alkaline bicarbonate solutions to recover U(VI) bound to algae.

Copper binding to algae has been studied by different workers, and the recovery process has been successfully applied to the treatment of electroplating wastewaters and contaminated ground waters. Ferguson and Bubela (1974) found that Cu(II) binding to *Chlorella vulgaris, Ulothrix* sp., and *Chlamydomonas* was determined to be highly dependent on pH and that increased binding occurred as the pH was increased from 3 to 7. Darnall et al. (1986b) and Greene et al.

(1987b) have confirmed a similar pH dependence, using the algae *Chlorella vulgaris, Chlorella pyrenodosa, Spirulina platensis,* and others.

Cu(II) bound to free or immobilized algal cells was readily reversed by using an acidic solution (Darnall et al., 1986a). Figure 12.9 shows the cycling of an algal column in which loading of Cu(II) on the column was performed at pH 5.0, and stripping was performed with an acidic solution. The Cu(II) was quantitatively recovered from the algal column in just a few bed volumes of acid. Once regenerated, the column can be reused, and the acid regenerant solution can also be reused. It has been demonstrated that sulfuric acid solution containing 3800 ppm Cu(II) was effective at regenerating algal columns loaded with Cu(II), provided that its acidity was maintained near or below 1 (Greene et al., 1987a). Thus, large volumes of Cu(II) solution can be cleaned up, and reduced volumes of a highly concentrated Cu(II) solution can be obtained by reuse of the regenerating solution and recycling of the algal column.

The strong copper-binding characteristics of immobilized algae have made it possible to make an alga-modified carbon paste electrode

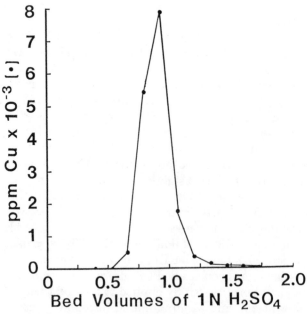

Figure 12.9 Copper stripping from silica-immobilized *Chlorella pyrenoidosa.* See text for details. (*From Greene et al., Algal ion exchangers for waste metal reclamation, American Chemical Society National Meeting Preprint Extended Abstract, New Orleans, Aug. 30–Sept. 4, 1987, 272(2):800–804.*)

which can be used to quantitate copper in aqueous solutions (Gardea-Torresdey et al., 1988a).

Mercury binding to algae has been well documented (Becker, 1983; Darnall et al., 1986a,b; Greene et al., 1987a). Although inorganic and organic mercury species may be biologically accumulated by living algae, killed algal cells are also effective at mercury binding. *Chlorella vulgaris* bound Hg(II) over a wide pH range, but binding was inhibited by the presence of chloride ion or by mercaptoethanol at pH 2 or above. The addition of these complexing agents was used to successfully recover algal-bound Hg(II).

Greene et al. (1987a,b) determined the available sulfhydryl content of *Chlorella vulgaris* by polargraphic titration of an algal cell suspension with sodium *p*-hydroxymercuriphenylsulfonate.

Al(III) binding to nonliving *Chlorella vulgaris, Chlorella pyrenoidosa* and *Spirulina* was studied by Darnall et al. (1986b) and Greene et al. (1987a). It was determined that there was a significant difference in the binding capacity of different algal organisms for Al(III) in acidic solutions. Based on the relative affinities of different algal organisms for Au(III) and Al(III) binding at pH 2, it was possible to separate and recover these ions by the sequential exposure of a solution to *Chlorella pyrenoidosa* and *Eisenia bicyclis*, respectively (Greene et al., 1987a; Gardea-Torresdey, 1988).

Mechanism of metal-ion binding to algae

The mechanism of metal-ion binding to nonliving algal cells may depend on the species of metal ion, the algal organism, and the solution conditions. Crist et al. (1981) presented evidence that metallic-cation binding to *Vaucheria* sp. occurred at least in part by an ion-exchange mechanism, and determined ratios of moles of hydrogen ions displaced per mole of metal cation adsorbed by the alga. The binding sites were thought to be amino and carboxyl groups as well as sulfates and imidazoles associated with polysaccharides and proteins. In experiments performed at initial pH values of 4.5, protons displaced by metal-ion adsorption gave the following ratios for H^+ displaced per divalent cation adsorbed: 1.2 (Cu^{2+}), 0.6 (Zn^{2+}), 0.59 (Mg^{2+}), and 0.30 (Sr^{2+}). The calculated ratios were found to vary with different metal-ion concentrations and initial pH values.

Greene et al. (1987b) demonstrated that the binding of Cu(II), Pb(II), Zn(II), Ni(II), Cd(II), and Cr(III) to *Spirulina platensis* was also accompanied by the liberation of protons. The experiments were performed by combining metal-ion solutions and algal suspensions that were initially at the same pH value, near pH 5.5, and subsequent determination of the quantity of metal ions adsorbed by the alga and the final pH value. In all cases the final pH values were substantially lower than the initial pH values, a result suggesting that metal-cation binding occurred at least in part by an ion-exchange mechanism in which some of the counter ions associated with the metal-binding sites were protons. The workers also determined ion-exchange ratios from the data. The importance of proton equilibrium in the mechanism of metal-cation binding was also shown by the pH dependence of binding, and by the fact that binding was reversible by lowering the pH.

Several metallic-anion complexes and oxoanions appear to bind electrostatically to algal cells, although other mechanisms are certainly possible. The occurrence of an electrostatic binding mechanism is consistent with the pH profiles of the metal ion binding to the alga, indicating that binding increases as the pH is lowered. Metal ions which follow this tend to include SeO_4^{2-}, MoO_4^{2-}, CrO_4^{2-}, $Au(CN)_2^-$, and $PtCl_4^{2-}$. Binding of anionic metal complexes probably occurs because protonated amines or imidazoles electrostatically attract the negatively charged metal ion.

Greene et al. (1986a,b), Hosea et al. (1986), and Watkins et al. (1987) have investigated the mechanism of gold complexes binding to *Chlorella vulgaris*. The gold complexes studied in the experiments were $AuCl_4^-$, $Au(CN)_2^-$, and sodium gold(I) thiomalate. In general, the results indicated that the mechanisms were different for each complex and also depended on the relative amount of gold complex initially present in the cell suspensions. Some of this evidence is described below.

Greene et al. (1986) determined that $AuCl_4^-$ (0.1 mM) binding to *Chlorella vulgaris* (5 mg/mL) was accompanied by the rapid liberation of 3 moles of chloride ion per mole of adsorbed gold. Algal-bound gold, initially applied at 0.067 mM/g as $AuCl_4^-$ and then extracted from the cells with 1.00 M sodium bromide showed no detectable $AuBr_4^-$ as determined by visible spectrophotometry. Since bromide forms strong complexes with gold(I) and gold(III) and will not oxidize or reduce gold in either oxidation state, these results suggest indirectly that the gold extracted by the sodium bromide was dibromoaurate(I), which does not absorb in the visible region. Therefore, it seems likely that algal-bound gold, initially applied in the +3 oxidation state, was converted to the +1 oxidation state. The x-ray absorption spectra of *Chlorella vulgaris* that contained ionic gold, initially applied as $AuCl_4^-$, provided direct spectral evidence that Au(III) was reduced to Au(I) under experimental conditions similar to those described above (Watkins et al., 1987). For *Chlorella vulgaris* samples that were prepared by reaction with low concentrations of $AuCl_4^-$ and contained 0.1% gold by dry weight of alga, the XANES (x-ray absorption near-edge structure) spectrum indicated that all algal-bound gold was in the $+1$ oxidation state. The XANES analysis on a sample of $AuCl_4^-$-treated *Chlorella vulgaris* that contained 7.5% gold by dry weight of alga indicated that about half of the gold remained in the $+3$ oxidation state and the other half in the $+1$ oxidation state. Thus, for the reaction of high concentrations of $AuCl_4^-$ with *Chlorella vulgaris*, the reduction of Au(III) to Au(I) was incomplete.

The EXAFS (extended x-ray absorption fine structure) region of the spectrum was used to determine the coordination environment of gold in samples prepared by the reaction of different gold complexes with *Chlorella vulgaris* (Watkins et al., 1987). The analyses for $AuCl_4^-$-treated alga indicated strong evidence for coordination of gold to nitrogen or oxygen. For algal samples that were prepared with $Au(CN)_2^-$ and *Chlorella vulgaris* and contained 0.08% gold by dry weight of alga, direct spectral evidence indicated that gold was coordinated to a sulfur atom. The mode of $Au(CN)_2^-$ binding was also determined to be concentration-dependent, since algae that contained 1.4% gold [applied as $Au(CN)_2^-$] contained a mixture of gold(I) species, with some of the gold(I) coordinated to sulfur. Other modes of gold binding to algae besides ligand substitution may include electrostatic interactions of the anionic gold complexes with positively charged functional groups such as protonated amines or imidazoles.

Greene et al. (1986b) and Hosea et al. (1986) found that algal-bound gold, initially applied as $AuCl_4^-$, was reduced to gold(0) slowly, over a period of days or weeks. The fraction of metallic gold formed on the alga was dependent on the amount of $AuCl_4^-$ initially applied to the alga, with increasing reduction to elemental gold occurring at high gold concentrations and over long periods of time. Gee and Dudeney (1988) reported that large crystals of elemental gold formed on *Chlorella vulgaris* cells when they were incubated in tetrachloroaurate(III) solutions. These results are consistent with theories that algal cells may have played a role in the transport and deposition of gold in the environment (Dissanayake and Kritsotakis, 1984).

Summary

If there is a single major conclusion to be drawn from the work on metal-ion binding to the cell walls of algae, it would be that the surface of algae is literally a mosaic of metal-ion binding sites. The lipid, carbohydrate, and protein components combine to create a spectrum of distinct binding sites—sites which differ in affinity and specificity. Both anions and cations can be bound. There are sites with high affinity for hard metal ions such as Al(III) and Fe(III), and there are sites with equally high affinities for such soft metal ions as Hg(II),

Ag(I), and Au(III). Selectivity in metal-ion binding can often be gained by judicious manipulation of solution parameters. Likewise, selectivity in stripping bound metal ions can be achieved with suitable parameter adjustment. The use of algal biomass is thus likely to enjoy increased popularity for the removal and recovery of heavy-metal ions from industrial waste or mining process streams.

References

Bailey, R. W., and Staehelin, L. A., The chemical composition of isolated cell walls of *Cyanidium caldarium, J. Gen. Microbiol.* **54**:269–276 (1968).

Becker, E. W., Limitations of heavy metal removal from wastewater by means of algae, *Water. Res.* **17**(4):459–466 (1983).

Bird, G. M., and Haas, P., On the nature of the cell wall constituents of *Laminaria sp*, mannuronic acid, *Biochem. J.*, **25**:403–411 (1931).

Bold, H. C., Alexopoulus, C. S., and Delevoryas, T., *Morphology of plants and fungi*, 4th ed., Harper and Row, New York, 1980.

Brierley, C. L., and Brierley, J. A., Biological processes for concentrating trace elements from uranium mine wastes, Technical Completion Report N. 140, New Mexico Water Resources Research Institute, N. Mex., 1981.

Ciferri, O., Spirulina, the edible microorganism, *Microbiol. Rev.* **47**:551–578 (1983).

Crist, R. H., Oberholser, K., Shank, N., and Nguyen, M., Nature of bonding between metallic ions and algal cell walls, *Environ. Sci. Technol.* **15**(10):1212–1217 (1981).

Darnall, D. W., Greene, B., and Gardea-Torresdey, J., Gold binding to algae, in P. R. Norris and D. P. Kelly (eds.), *Biohydrometallurgy*, Science and Technology Letters, Kew, Surrey, U.K., 1988, pp. 487–498.

Darnall, D. W., Greene, B., Henzl, M. T., Hosea, J. M., McPherson, R. A., Sneddon, J., and Alexander, M. D., Selective recovery of gold and other metal ions from an algal biomass, *Environ. Sci. Technol.* **20**(2):206–208 (1986a).

Darnall, D. W., Greene, B., Hosea, M., McPherson, R. A., Henzl, M., and Alexander, M. D., Recovery of heavy metals by immobilized algae, in R. Thompson (ed.), *Trace Metal Removal from Aqueous Solution*, The Royal Society of Chemistry, Burlington House, London, 1986b, pp. 1–24.

Dissanayake, C. B., and Kritsotakis, K., The geochemistry of Au and Pt in peat and algal mats offshore from Sri Lanka, *Chem. Geol.* **42**:61–76 (1984).

Ferguson, J., and Bubela, B., The concentration of Cu(II), Pb(II), and Zn(II) from aqueous solutions by particulate algal matter, *Chem. Geol.* **13**:163–186 (1974).

Filip, D. S., Peters, T., Adams, V. D., and Middlebrooks, E. J., Residual heavy metal removal by an algae-intermittant sand filtration system, *Water. Res.* **13**:305–313 (1979).

Gale, N. L., and Wixon, B. G., Control of heavy metals in lead industry effluents by algae and other aquatic vegetation, in *Proc. Intern. Conf. Management and Control of Heavy Metals in the Environment*, CEP Consultants Ltd., Edinburgh, U.K., 1979, pp. 580–583.

Gardea-Torresdey, J., Ph.D. dissertation, New Mexico State University, 1988.

Gardea-Torresdey, J., Darnall, D. W., and Wang, J., Bioaccumulation and measurement of copper at an alga-modified carbon paste electrode, *Anal. Chem.* **60**:72–76 (1988a).

Gardea-Torresdey, J., Darnall, D. W., and Wang, J., Bioaccumulation and voltammetric behavior of gold at alga-containing carbon paste electrodes, *J. Electroanal. Chem.* **252**:197–208 (1988b).

Gee, A. R., and Dudeney, A. W. L., Adsorption and crystallisation of gold at biological surfaces, in P. R. Norris and D. P. Kelly (eds.), *Biohydrometallurgy*, Science and Technology Letters, Kew, Surrey, U.K., 1988, pp. 437–452.

Gotelli, I. B., and Cleland, R., Differences in the occurrence and distribution of hydroxyproline-proteins among the algae, *Am. J. Bot.* **55**(8):907–914 (1968).

Greene, B., and Bedell, G., Algal cells or immobilized algae for metal recovery, in W. Junk (ed.), *An Introduction to Applied Phycology*, SBP Press (in press).

Greene, B., Gardea-Torresdey, J. L., Hosea, J. M., McPherson, R. A., and Darnall, D. W., Algal ion exchangers for waste metal reclamation, Am. Chem. Soc. Nat. Meeting Preprint Extended Abstract, **272**(2):800–804 (1987a).

Greene, B., Henzl, M. T., Hosea, J. M., and Darnall, D. W., Elimination of bicarbonate interference in the binding of U(VI) in mill-waters to freeze-dried *Chlorella vulgaris*, *Biotechnol. Bioeng.* **28**:764–767 (1986a).

Greene, B., Hosea, M., McPherson, R., Henzl, M., Alexander, M. D., and Darnall, D. W., Interaction of gold(I) and gold(III) complexes with algal biomass, *Environ. Sci. Technol.* **20**(6):627–632 (1986b).

Greene, B., McPherson, R., and Darnall, D. W., Algal sorbents for selective metal ion recovery, in J. Patterson and R. Pasino (eds.), *Metals Speciation, Separation, and Recovery*, Lewis, Mich., 1987b, pp. 315–332.

Hosea, M., Greene, B., McPherson, R., Henzl, M., Alexander, M. D., and Darnall, D. W., Accumulation of elemental gold on the alga *Chlorella vulgaris*, *Inorg. Chim. Acta* **123**:161–165 (1986).

Hunt, S., Diversity of biopolymer structure and its potential for ion-binding applications, in H. Eccles and S. Hunt (eds.), *Immobilization of Ions by Bio-Sorption*, Ellis Horwood, Chichester, U.K., 1986, pp. 15–46.

Jennett, J. C., Hassett, J. M., and Smith, J. E., Control of heavy metals in the environment using algae, in *Proc. Intern. Conf. Management and Control of Heavy Metals in the Environment*, CEP Consultants LTD, Edinburgh, U.K., 1979, pp. 210–217.

Karger, B. L., Snyder, L. R., and Horvath, C., *An Introduction to Separation Science*, Wiley, New York, 1973.

Kennedy, J. F., and Cabral, J. M. S., Immobilized living cells and their applications, *Appl. Biochem. Bioeng.* **4**:189–280 (1983).

Khummongkol, D., Canterford, G. S., and Fryer, C., Accumulation of heavy metals in unicellular algae, *Biotechnol. Bioeng.* **24**:2643–2660 (1982).

Krambeer, C., Bigger profits through improved wastewater treatment, *Finish. Manage.* **32**:36–37 (1987).

Kuyucak, N., and Volesky, B., New algal biosorbent for a gold recovery process, in P. R. Norris and D. P. Kelly (eds.), *Biohydrometallurgy*, Science and Technology Letters, Kew, Surrey, U.K., 1988, pp. 453–464.

Nakajima, A., Horikoshi, T., and Sakaguchi, T., Recovery of uranium by immobilized microorganisms, *Eur. J. Appl. Microbiol. Biotechnol.* **16**:88–91 (1982).

Nakajima, A., Horikoshi, T., and Sakaguchi, T., Studies on the accumulation of heavy metal elements in biological systems. XVII. Selective accumulation of heavy metal ions by *Chlorella regularis*, *Eur. J. Appl. Microbiol. Biotechnol.* **12**:76–83 (1981).

Northcote, D. H., Goulding, K. J., and Horne, R. W., The chemical composition and structure of the cell wall of *Chlorella pyrenoidosa*, *J. Biol. Chem.* **70**:391–397 (1958).

Olafson, R. W., Abel, K., and Sim, R. G., Prokaryotic metallothionein: Preliminary characterization of a blue-green alga (*Synechococcus* sp.) heavy metal-binding protein, *Biochem. Biophys. Res. Comm.* **89**(1):36–43 (1979).

Pearson, R., *Hard and soft acids and bases*, Dowden, Hutchinson, and Ross, Philadelphia, 1973.

Robinson, P. K., Mak, A. L., and Trevan, M. D., Immobilized algae: A review, *Process Biochem.* **21**(4):122–127 (1986).

Rochefort, W. E., Rehg, T., and Chau, P. C., Trivalent cation stabilization of alginate gel for cell immobilization, *Biotechnol. Lett.* **8**(2):115–120 (1986).

Segel, I. H., *Biochemical Calculations*, 2nd ed., Wiley, New York, 1976.

Siegel, B. Z., and Siegel, S. M., *The Chemical Composition of Algal Cell Walls*, CRC Critical Reviews in Microbiology, Chemical Rubber Company, Cleveland, 1973, pp. 1–26.

Tsezos, M., and Volesky, B., Biosorption of uranium and thorium, *Biotechnol. Bioeng.* **23**:583–604 (1981).

Volesky, B., Biosorbents for metal recovery, *Trends Biotechnol.* **5**:96–101 (1987).

Watkins, J. W. II, Elder, R. C., Greene, B., and Darnall, D. W., Determination of gold binding in an algal biomass using EXAFS and XANES spectroscopies, *Inorg. Chem.* **26**:1147–1151 (1987).

Whyte, J. N. C., Polysaccharides of the red seaweed *Rhodymenia pertusa*. 1. Water soluble glucan, *Carbohydr. Res.* **16**:220–224 (1971).

Wood, J. M., and Wang, H. K., Microbial resistance to heavy metals, *Environ. Sci. Technol.* **17**:582A–590A (1983).

Metal Immobilization Using Bacteria

Corale L. Brierley

Introduction

Years of research and development have yielded commercial processes using bacteria as biosorbents for accumulation of metals. Applications of these biosorbents include (1) removal of metals from aqueous, industrial effluents for pollution control, (2) remediation of contaminated surface waters, groundwaters, and lagoons, and (3) treatment of industrial process streams for resource recovery.

This chapter focuses on the commercial application of bacteria as biosorbents for metal's recovery and metal's reclamation. Topics discussed in the context of commercial use include

1. A brief summary of scientific principles for use of bacteria as biosorbents

2. Important characteristics of bacterial biosorbent materials

3. Applications' considerations

4. Process economics

Mechanisms of Metal Immobilization by Bacteria

The immobilization of metals by bacteria can occur actively or passively. In active immobilization, metal transformations or microbe-metal interactions are carried out by living, metabolically active microorganisms. In passive immobilization, metals are transformed by physical-chemical actions not necessarily requiring participation by living microorganisms. Mechanisms of active and passive metal immobilization have been extensively reviewed (Kelly et al., 1979; Brierley et al., 1985a; Hutchins et al., 1986; Lakshmanan et al., 1986). This section briefly defines important mechanisms of active and passive metal immobilization and provides examples of these mechanisms.

Active metal immobilization requires the metabolic functions of living microorganisms, because the immobilization often depends on expenditure of energy. The mechanisms by which bacteria actively immobilize metals are

1. Precipitation

2. Intracellular accumulation and complexation

3. Oxidation and reduction

4. Methylation and demethylation

Methylation and demethylation reactions are mechanisms that living bacteria use to transform metals in their environment (Thayer and Brinckman, 1982; Pettibone and Cooney, 1988). However, methylation is not considered at this time to be a commercially viable process of metal immobilization. Methylation can transfer some metals from aqueous and soil environments to the atmosphere, but the capture of these volatilized metals has not been seriously addressed from a commercial point of view. Demethylation reactions yield an inorganic metal; these metal ions can subsequently be involved in other microbe-mediated transformations.

Passive immobilization of metals occurs when

1. A solubilized metal is chelated by a substance produced and excreted by a microbial cell, or

2. A metal binds to a cell surface by physical-chemical reactions.

Precipitation

Precipitation of metals results when living bacteria produce and excrete a substance that chemically reacts with metals present in solution to produce an insoluble metal compound. The generally cited example of precipitation of metals is the production of hydrogen sulfide by sulfate-reducing bacteria and the subsequent formation of insoluble metal sulfides. Sulfate-reducing bacteria, of which the most prevalent genera are *Desulfovibrio* and *Desulfotomaculum,* live in anaerobic environments such as lake sediments, swamps, and anoxic soils. They oxidize organic matter and reduce sulfate to sulfide:

$$SO_4^{2-} + 10(H) \rightarrow H_2S + 4H_2O$$

The hydrogen sulfide (H_2S) reacts with metals to form water-insoluble, metal sulfide compounds. Precipitation of metals by the activities of sulfate-reducing bacteria has been documented to occur in the natural environment for removal of metals from streams and lakes (Jackson, 1978). This naturally occurring activity has been reproduced in purpose-built impoundments to remove soluble metals emanating from mining operations (Brierley and Brierley, 1980; Brierley et al., 1980). Researchers are now examining application of constructed wetlands to reduce both acid and metals that run off from abandoned mines and solid mine wastes associated with these abandoned mines. Wetlands, once constructed, require little or no maintenance and utilize the natural activities of plants and microorganisms to immobilize metals. An important mechanism for metal removal in wetland systems is the proliferation and activity of sulfate-reducing bacteria (Emerick and Cooper, 1987; Erickson et al., 1987).

Metals can also be precipitated from solution by HPO_4^{2-} derived from a phosphatase located on the bacterial cell surface. The phosphatase enzyme cleaves glycerol 2-phosphate liberating HPO_4^{2-}. Macaskie and Dean (1984, 1987a,b, 1988), working with *Citrobacter* sp., have demonstrated the cell-bound precipitation of cadmium, lead, and uranium by HPO_4^{2-}. *Citrobacter* sp., immobilized as a biofilm on reticulated foam, was found effective in removal of 90 percent of the uranium (1 mM uranyl nitrate, pH 6.9) provided at a flow rate of 1.25 mL/min. After passage of 40 L of uranyl nitrate, the amount of uranium precipitated by the phosphatase activity was 9 g uranium per gram bacterial dry weight (Macaskie and Dean, 1987b). Polyacrylamide gel-immobilized cells were comparably effective in uranium removal (Macaskie and Dean, 1988).

Intracellular and cell-surface accumulation

The active immobilization process of intracellular accumulation of metals by bacteria is a two-stage process. Metals are first bound passively to the cell surfaces by physical-chemical processes such as electrochemical attraction. Next the metals are accumulated inside of the cell usually by an energy-dependent transport system. The active transport system may be one used to convey metabolically essential elements such as potassium or magnesium across the cell membrane. The inability of the bacterial cell's transport system to differentiate between metabolically essential and nonmetabolic metals of the same charge can result in substantial intracellular accumulation of toxic, heavy metals. This phenomenon has been reviewed by Brierley et al. (1985a).

Significant intracellular accumulation of metals by bacteria was demonstrated by Charley and Bull (1979), using a mixed culture of bacteria consisting of *Pseudomonas maltophila, Staphylococcus aureus,* and a cornyeform organism, which accumulated over 300 mg Ag^+ per gram dry-weight biomass. Accumulation was observed at a rate of 21 mg $Ag^+/(h \cdot g$ biomass).

Uranium, radium, and cesium were observed to accumulate intracellularly in *Pseudomonas aeruginosa* (Strandberg et al., 1981). The researchers reported that in this case, however, metabolism was not required for uranium uptake. Upon accumulation these metals formed dense intracellular deposits that were observable with an electron microscope.

A chemoautotrophic bacterium of the genus *Thiobacillus* reportedly accumulated silver, forming precipitates of Ag_2S on its cell surface (Pooley, 1982). This metal sulfide compound represented up to 25 percent of the dry weight of the cell. Intracellular formation of a metal sulfide was also observed when growing cultures of *Klebsiella aerogenes* accumulated Cd ions and inorganic sulfide in a molar ratio close to unity. A subsequent intracellular formation of CdS was noted (Aiking et al., 1984).

Oxidation and reduction

Oxidation and reduction are the increase and decrease in valence state of a metal, respectively. Although oxidation and reduction generally require the functions of living microorganisms, some metal-ion reductions can occur passively by chemically reactive sites on the cell walls of microorganisms (Beveridge and Murray, 1976, 1980; Gee and Dudeney, 1988). Depending on the metal species, both oxidation and reduction result in metal immobilization. Several oxidation and reduction examples follow.

The oxidation of manganese is brought about by a limited number of bacterial species (Nealson and Ford, 1980; Adams and Ghiorse, 1987; Boogerd and de Vrind, 1987) and mature spores of marine *Bacillus* species (Mann et al., 1988). This oxidation by bacterial cells and *Bacillus* spores results in the formation of insoluble manganese oxide:

$$Mn^{2+} + \tfrac{1}{2}O_2 + H_2O \rightarrow MnO_2 + 2H^+$$

The mineral hausmannite (Mn_3O_4) is formed when manganese is oxidized by spore coats. However, formation of this oxidized species of manganese is not mediated by the bacterial spore, but rather it is chemical recrystallization of the precipitated amorphous manganese oxide (Mann et al., 1988). Microbially mediated oxidation of Mn^{2+} has been demonstrated to actively occur on a geological scale. Even in Arctic conditions rates were found to be sufficient to oxidize all of the Mn^{2+} in a lake under study (Johnston and Kipphut, 1988).

A partial reduction to a metal ion of a smaller net charge can occur by certain actively metabolizing bacteria. A partial metal reduction mediated by bacteria is the conversion of ferric iron to ferrous iron by species of the bacteria *Bacillus, Thiobacillus, Micrococcus, Rhodopseudomonas,* and *Pseudomonas* (Brock and Gustafson, 1976; Ehrlich, 1981; Lundgren et al., 1983). This reduction reaction is the result of electron transport reactions (Silverman and Ehrlich, 1964). *Alteronomas putrefaciens* has been demonstrated to couple growth with reduction of MnO_2 under anaerobic conditions. The investigators (Myers and Nealson, 1988) of this phenomenon suggested that Mn^{4+} served as the terminal electron acceptor.

Under anaerobic conditions, the acidothermophilic archaebacterium, *Sulfolobus* sp., has been demonstrated to reduce Mo^{6+} to Mo^{5+}. This metal-ion reduction requires actively metabolizing bacteria because the reduction appears coupled to the oxidation of S^0 to SO_4^{2-} (Brierley and Brierley, 1982).

The reduction of gold can occur passively. The reduction of gold from Au^{3+} to Au^0 has been observed in isolated cell walls of *Bacillus subtilis* (Beveridge and Murray, 1976) and nonmetabolizing whole cells of *Bacillus subtilis* (Gee and Dudeney, 1988). In both studies microscopic crystals of elemental gold were observed. It is probable that gold is bound to sites on and within the cell wall and these sites act as nucleation points for the reduction of the gold and growth of crystals.

Extracellular complexation

Extracellular complexing of metals occurs when bacteria produce metal-binding substances generated outside of the cell or excreted form the cell. These substances can be chelating agents, such as siderophores, or metal-binding extracellular polymers. The production of extracellular complexing agents requires an actively metabolizing cell or enzymatic production of the polymer outside of the cell. Once produced, however, the complexing agents immobilize metals passively. These complexing agents can in some cases be harvested and used repeatedly for metal removal.

A siderophore is a low-molecular-weight, iron-specific ligand. Siderophores are synthesized and excreted by bacteria for the purpose of capturing and supplying iron for metabolism (Lundgren and Dean, 1979).

Siderophores are generally derivatives of catechol or hydroxamate, possessing formation constants for ferric iron of 10^{30} to 10^{50}. Siderophores can be isolated from a number of bacteria, including species of *Pseudomonas, Actinomyces, Azotobacter,* and *Arthrobacter.*

DeVoe and Holbein (1985) patented the synthesis of a series of catechols and substituted catechols for the purpose of metal binding. These metal-binding "compositions" are covalently linked to solid supports such as porous glass beads. The catechols are stabilized against oxidation by substitution of electrophilic groups, such as NO_2^-, Cl^-, and Br^-. These substitutions also influence the metal-binding affinity among metals. Furthermore, they allow for the selective recovery of metals other than ferric iron (DeVoe and Holbein, 1985, 1986). The substituted and immobilized catechols have been commercially evaluated for removal of the actinides—plutonium, thorium, and uranium—from aqueous nuclear wastes and for treating industrial waste streams for removal of other toxic heavy metals. Removal efficiencies are reported to be greater than 99.9 percent with effluent concentrations less than 0.01 mg/L metal (Holbein et al., 1984; DeVoe and Holbein, 1985, 1986).

Extracellular polymers produced by bacteria consist of polysaccharides, proteins, and/or nucleic acids. Metals are immobilized by extracellular polymers by adsorption and flocculation. Extensively studied for their metal-binding action are the polymers produced by bacteria in sewage treatment facilities (Dugan, 1975; Brown and Lester, 1979, 1982a,b; Norberg, 1983; Norberg and Persson, 1984; Morper, 1986; Scott et al., 1986; Sterritt and Lester, 1986). These polymers, which bind soluble metal ions and flocculate insoluble metals by physical entrapment, are instrumental in removing toxic metals from industrial wastewaters discharged to sewage treatment facilities. Microbial production of polymers has been explored for treating industrial wastewaters. Bomstein (1972) patented the use of nucleoprotein derived from several bacteria, including the genera *Polyangium, Myxococcus, Bacillus, Leuconostoc, Flavobacterium, Micrococcus,* and *Alcaligenes,* as a flocculating agent for waste treatment.

Binding to the cell wall

There are at least three mechanisms for binding of metals on cell walls: (1) ion-exchange reactions, (2) precipitation, and (3) complexation. While most bacteria possess a cell wall structure to bind metals, the gram-positive bacteria, principally members of the genus *Bacillus,* have enhanced capacity for metal binding because of a thick network of peptidoglycan with attached macromolecules (Beveridge, 1984). In contrast gram-negative bacteria, which have more limited metal-binding capacity, have cell walls that are chemically and structurally much different than gram-positive cell walls. Gram-negative cell walls have a thin peptidoglycan layer between two membrane bilayers (Beveridge and Koval, 1981) Gram-positive cell walls and surfaces have a negative charge density owing to the presence of teichoic acid and teichuronic acid attached to the peptidoglycan network. The phosphodiesters of the teichoic acid and the carboxylate groups of the teichuronic acid contribute ion-exchange capacity to the cell wall (Hancock, 1986). This is the primary mechanism for metal binding to the cell wall. (See also Chap. 10.)

Bacterial metal-binding capacities are frequently higher than can be accounted for by ion-exchange processes alone. The increased binding capacity may come from a nucleation reaction. Nucleation is not well understood (Doyle et al., 1980; Hancock, 1986).

A third mechanism of metal binding that has been identified in bacterial cell walls is complexation. It is proposed that nitrogen and oxygen ligands in the wall contribute to complexation of transition-metal ions (Hancock, 1986).

The ability of bacteria to bind metals has been coupled with techniques for cell immobilization, technology utilized in mineral separations (e.g., high-gradient magnetic separation; Kolm et al., 1975), and other engineering designs to yield commercial biosorbent processes that are technically and economically competitive with existing processes for metals removal from aqueous solutions. The remainder of this chapter describes biosorbent applications and explores factors important to commercial viability of these applications.

Bacterial Metal Sorbents: Important Factors for Commercial Application

Commercial applications for biosorbents include metal removal from industrial waste effluents, industrial process waters, contaminated groundwaters and surface waters, landfill and soil leachates, nuclear waste streams, and process streams. The chemistry of waste and process streams from these sources is characteristically complicated. Often these aqueous streams have a mixed metal composition. Their concentration of metals varies, ranging from several mg/L to several thousand mg/L, and their pH values can be extremely acidic or basic and change over time. Organic and inorganic contaminants may be present and affect the chemistry and availability of the metals for accumulation and binding by biosorbents. The presence of corrosive and chemically aggressive substances may attack biosorbents, and accumulation of highly radioactive metals may affect the stability of biological materials. Oils and greases can foul adsorption materials, thus severely reducing effectiveness. Particulates, such as metallic metals, sands, clays, or sludge components from process or upstream treatments, reduce permeability in fixed-bed systems and themselves act as adsorbents for metals. Aqueous waste-flow volumes can have wide variability. For example, mining operations may have to treat millions of liters per day of effluent during spring runoff, but discharge volumes may diminish to almost nothing during dry periods. In some industrial applications, workers assigned to oversee and maintain waste treatment systems may not be completely knowledgeable about the nature of the wastes they are treating nor fully understand the operation of the in-place waste treatment system. Under these circumstances process upsets occur. A waste treatment system is expected to control effluents that differ substantially in pH, metal concentration, flow volume, and chemical composition from the normal waste stream that the waste treatment system was designed and installed to process. These characteristics of industrial waste and process streams mandate treatment technologies be compliant, versatile, and robust.

Technological capabilities for waste treatment processes must be further extended as a result of recent environmental regulations mandating lower metal concentrations in discharged effluents and prohibiting land disposal of hazardous wastes. These mandates dictate implementation of waste treatment technologies that can significantly lower metal content in aqueous streams and concentrate the metals in a form for subsequent reclamation. New waste treatment technologies must, therefore, be specifically developed to avoid solid hazardous waste production.

Coupled with technological demands are competitive pressures. Recent advances in alternative physical and chemical technologies place an additional burden on developers of biosorbent processes to create products that meet the chemical, environmental, technical, and economic challenges.

This section (1) identifies technical, environmental, and economic issues related to environmentally acceptable treatment of chemically complex waste and process streams with bacterial biosorbents, (2) defines why these issues are important, and (3) explores how these issues can be resolved. Discussion focuses on development of commercially applicable bacterial sorption materials that achieve environmental regulations, which impose rigorous effluent treatment standards, eliminate land disposal of toxic wastes, and promote waste reclamation and reuse or recycling.

Biomass: dead or alive?

Because living and nonliving, or nonmetabolizing, bacteria accumulate metals, either type of biological material can be employed for removal of metals from solution. Arguments for use of living systems are based on the consideration that (1) the biological adsorbent is now a renewable resource, consequently not requiring replacement when the material is metal loaded, and (2) products of metabolism, such as H_2S and HPO_4^{2-}, can be used in the metal immobilization process. Arguments for using nonliving, or nonmetabolizing, biological materials are supported by the fact that waste streams are often toxic to living systems, devoid of nutrients, and can possess extremely varying conditions over time—all situations that make the maintenance of living systems difficult, at best. Most biological metal-removal systems in, or approaching, commercialization employ nonliving or nonmetabolizing systems (Brierley et al., 1985; Darnall et al., 1986; Kuyucak and Volesky, 1988; Tsezos, 1988). The reasons for this are explored in more detail.

Industrial waste and process waters, contaminated groundwaters and surface waters, and leachates can be so chemically polluted as to be lethal to living microorganisms. The presence of toxic concentrations of metals and additives, such as surface-active agents, coupled with the variability of pH and salt concentrations in effluents and remedial sites preclude use of living systems in all but the most innocuous of commercial applications.

Also, to be considered in using living systems is the matter of supplying nutrients in a cost effective manner to maintain microbial viability in waste treatment plants. Unless inexpensive nutrients, such as partially digested sewage or decaying plant materials, are readily available, the maintenance of living microorganisms in waste treatment facilities is impractical. Moreover, nutrient-supplemented waste streams may support the growth and activity of microorganisms other than the bacteria which were originally intended. This contamination can result in the depletion of valuable nutrients and proliferation of undesirable species.

Although it is quite clear that nonliving, biological-based systems offer a competitive alternative to physical and chemical technologies in most applications, the use of such materials as metal sorbents has not been without contention because of the cost of obtaining suitable biomass material in needed quantities. Biomass can be produced specifically for biosorbent production through fermentation. However, the cost of such custom fermentation can be too high to produce an economically competitive waste treatment product. Most companies seeking to commercialize biosorbents for waste treatment have sought suitable biomass that is a by-product of pharmaceutical and enzyme manufacturing. These by-products, although readily available in large quantities at a reasonable price, may present unique problems in manufacturing biosorbents. The by-product biomass may contain extraneous recovery chemicals that can adversely affect metals sorption performance. The by-product microorganism is almost always a proprietary strain to the pharmaceutical or enzyme manufacturer. This factor can discourage the biotechnology firm from supplying the biomass for competitive reasons.

Exceptions to the employment of nonliving or nonmetabolizing bacterial systems for metals immobilization in aqueous systems are the use of naturally occurring microorganisms in purpose-built wetland systems for treating acid mine drainage (Emerick and Cooper, 1987; Erickson et al., 1987) and the use of the sulfate-reducing or phosphate-producing bacteria. The artificial wetlands approach to waste treatment depends on a complex ecological system composed of higher plants, mosses, and microorganisms to neutralize acid and remove metals from wastewaters. The wetland systems are being evaluated principally for the treatment of drainages from abandoned mining operations. The theory behind the wetland systems is to prepare an impoundment filled with organic mulch and to plant with emergent aquatic vegetation. Sulfate-reducing bacteria and microorganisms that actively accumulate and passively bind metal ions develop naturally in this ecological niche.

Immobilization of bacteria for product stability and engineering design

To employ bacteria or other microorganisms commercially in waste treatment, the biomass must be modified through immobilization to yield a product having increased mechanical strength, high particle porosity, hydrophilicity, increased resistance to chemicals, a particle size and density for proper engineering design (Tsezos, 1986), temperature stability, and regenerability (J. A. Brierley, unpublished data). To achieve immobilization, developers of biosorbent systems have attached bacteria to inert substrates such as reticulated foams, stainless steel wire, pan scourers (Macaskie and Dean, 1987a), cotton webbing, wooden supports, high-area alumina (Clyde, 1982), polyvinyl chloride (Tengerdy et al., 1981), and other inert supports (Jack and Zajic, 1977). To form biomass particles containing a minimum of inert materials and possessing considerable mechanical strength and stability against aggressive chemicals and stripping and regeneration agents, bacteria have been immobilized in macromolecules such as polyacrylamide gels, agar (Jack and Zajic, 1977; Nakajima et al., 1982), calcium alginate, cellulose acetate, glutaraldehyde, toluene diisocyanate (Nakajima et al., 1982), and other cross-linking agents (Serbus et al., 1973; Nemec et al., 1977; Votapek et al., 1978).

Unfortunately, high metal-loading and porosity of the bacterial biosorbent particle are nearly always sacrificed to achieve satisfactory particle strength and chemical resistance. Attachment of bacteria to an inert substance and cross-linking consume active sites where metal ions would normally bind. Addition of macromolecules and cross-linking chemicals can diminish the number and size of pores, thus preventing the contact of metal ions with active sites within the biosorbent particle.

Substantial market opportunities exist for use of biosorbents in the treatment of high-level radioactive wastes. Studies using bacteria for accumulation of radionuclides (Shumate and Strandberg, 1985; Strandberg and Arnold, 1988) demonstrate that microorganisms accumulate a variety of radionuclides. Little research, however, has been carried out on the effect of high-level radiation on the long-term stability of biosorbent products.

Almost certainly, continued research and development will yield biosorbent products that possess the full complement of desirable characteristics, including (a) high mechanical strength, (b) high porosity, (c) resistance to corrosive chemicals and radiation, (d) desirable particle size and density, (e) high metal-loading, and (f) ability to be repeatedly stripped of metals and regenerated for reuse. In the end any biosorbent product destined for commercial use must be evaluated in terms of performance and economics to ensure its competitiveness in the market.

Cationic and anionic metal removal

Environmentally regulated metals are present in industrial and nuclear waste and process waters and contaminated surface waters and groundwaters as both cations and anions. Bacterial sorption materials must effectively remove both types of metal species for these aqueous solutions to achieve regulatory standards. Because bacterial biosorbents are negatively charged, owing to the presence of carboxylate groups, cations are readily bound. However, removal of anions, such as CrO_4^{2-}, requires chemical pretreatment of the aqueous solution to convert the ions to a cationic species, chemical optimization of the solution to enhance anionic binding to bacterial biomass, or modification of the bacterial biosorbent to specifically bind anions. Chemical pretreatments add capital and operating costs to waste treatment, are often cumbersome, and are sometimes hazardous. Existing processes for removal of chromates from wastewaters employ a chemical pretreatment, involving reduction of Cr^{6+} to Cr^{3+} using Fe^{2+}, SO_2, or other reduced sulfur species. The resulting chemically reduced wastewaters are subsequently treated with caustic or lime to pH 7 to 9 to precipitate chromium as $Cr(OH)_3$. Iron and other metal cations, if present, are also precipitated. Resulting sludges settle poorly, dewater with considerable difficulty, and are usually of little economic value, necessitating their removal to a toxic waste landfill for final disposition. The sulfur-reduction processes are further characterized by the generation of noxious and toxic gases and require close control of operating parameters, especially pH, to operate efficiently.

Chromate removal from aqueous solution by binding of the anion to bacterial biomass can be enhanced by adjustment of the solution pH to 2.0 or less. Binding of CrO_4^{2-} at low pH occurs with bacteria (J. A. Brierley, unpublished results) and with other microbial sorbents (Darnall et al., 1986). The high concentration of H^+ present at low pH values imparts a positive charge to the microbial surface, resulting in electrostatic binding of certain anionic species.

Under some industrial conditions or when treating very large volumes of waste or process waters, it is impractical to do significant pH adjustment to allow removal of anions. Under these circumstances, biosorbents must be able to effectively accumulate oxyanions under prevailing conditions. This can best be accomplished by materials that have a cationic character. For bacterial sorbents to be commercially competitive they must be modified either genetically to increase the numbers and types of cationic groups normally present on the cell surface or chemically to achieve a cationic nature.

Efficiency of metal removal from solution

Removal efficiency is the ability of the bacterial biosorbent to remove metals from solution to achieve the desired regulatory discharge standard. Regulatory standards vary, depending on whether the water is discharged to a sewage treatment facility, to surface waters, or to groundwaters where it may be reused as drinking water. Also, discharge standards vary geographically and by industry. Given today's awareness of environmental contamination by past and present industrial activities, concern about potable water supplies and enhanced capability to measure and monitor increasingly small pollutant concentrations, it is prudent to state that metal-removal efficiencies are crucial to commercial viability of any new water treatment product. It is not sufficient to be able to report metal removals of 99.9 percent because this may not be adequate to achieve discharge standards. To meet today's rigorous performance standards, waste treatment processes must now achieve effluent discharge concentrations of 0.05 mg metal per liter or less. The problems with achieving high removal efficiencies are almost always related to removing metal ions when

1. The metal concentration in solution is initially very low.

2. There are other ions present that compete for binding sites on the biosorbent material.

3. Complexing or chelating agents are present that compete with the biosorbent.

4. Displacement of one metal with another metal occurs at the biosorbent binding site.

An example of treating aqueous solution for dilute metal concentration is the mining industry. The volumes of water produced may be so large (e.g., 8×10^6 L) as to preclude recycling of the water in mineral processing, thus necessitating discharge. In such cases the waters are often discharged to streams or flow to reservoirs that serve as drinking water supplies. The water discharged often has not been industrially used and represents groundwater or runoff waters that have been in contact with mineralized rock. These waters contain low metal concentrations, often less than 5 mg/L. However, discharge permits for such operations may specify limits of 0.05 mg/L for each metal present because of subsequent water usage. Treatment technologies to remove low metal concentrations are limited for technical and/or economical reasons. Adsorption technologies are usually considered for such applications; however, contact time between the aqueous solution and the adsorptive material must be increased to achieve the desired effluent standard.

Metal-removal efficiency by biosorbents is affected in the presence of high concentrations of alkali- or alkaline-earth metals, such as K^+, Na^+, Ca^{2+}, and Mg^{2+}. The problem is particularly acute when the target heavy-metal concentration is low. This combination of high earth-metal concentration and low target-metal concentration is observed in what is termed a *polishing* application. Industrial effluents may be treated with a combination of technologies to achieve the regulatory discharge limits. Most often used for "primary," or the first, treatment to remove large quantities of metals is addition of caustic, lime, or limestone that produces metal hydroxide precipitates. Lamella or filtration devices are used as solid-liquid separators. Although metal removal by this precipitation technique is significant, seldom does this precipitation technology achieve the desired effluent standard on a consistent basis. A polishing, or second treatment, step must be added. The polishing system must then treat an effluent not only containing a low metal concentration but a solution with a high concentration of Ca^{2+} or Na^+. To exacerbate treatment problems, the solution may have a high concentration of suspended solids due to incomplete liquid-solids separation. Although bacterial-based metal removal products are less prone than the nonselective ion-exchange resins to loading of alkali- or alkaline-earth metals, some alkali- or alkaline-earth-metal loading is observed.

Chelating agents, such as CN^-, EDTA, and NH_4^+, are often added to metal solutions for electroplating and other metal-finishing applications. These chelating agents then become part of the waste effluent to be treated. Chelated metals are a major waste treatment problem as few technologies can effectively remove chelated metals from solution. Biosorbents are no exception. Most often the metal solutions must be pretreated to destroy the chelate. Harsh pretreatments, such as strong oxidation, must be used to destroy some chelating agents.

In mixed-metal solutions nonselective biosorbents are desirable to achieve effluent standards. However, due to differing binding constants among metals, some metals will bind to the exchange sites more strongly than others. When the biosorbent nearly reaches its metal-loading capacity, some of the more weakly bound metals have a tendency to disassociate and be replaced by metals with higher binding constants. This phenomenon is termed *chromatographing* and results in reduced metal-removal efficiency. This problem, which also occurs with ion-exchange resins, can best be overcome by engineering an application's system that does not result in the biosorbent becoming fully metal-loaded.

Metal-loading capacity

Metal-loading capacity is defined as the amount of metal that is loaded onto a biosorbent on a dry-weight basis. Metal loading is important because it controls the size of the contactor unit, dictates the frequency of regeneration or replacement of the biosorbent, and influences the overall cost of waste treatment. The lower the metal loading, the larger and more capital intensive becomes the contactor unit. Replacement of the product or biosorbent regeneration becomes more frequent, increasing operating and, potentially, capital costs. The principal factors affecting metal loading are

1. Number and nature of the exchange, chelation, and nucleation sites on the bacterial material

2. Initial metal concentration in solution

3. Presence of chelating or complexing agents

Binding sites vary with the type of biomass used. As discussed earlier, the gram-positive bacteria have an abundance of carboxylate groups that result in high metal-loading capacity. This capacity, however, can be significantly reduced when the bacteria or their cell walls are immobilized either on an inert material such as glass beads or in a matrix such as polyacrylamide. If cross-linking agents are used to bind or immobilize microbial cells or cell fragments, excessive amounts of the agents may significantly reduce porosity of the resulting granule or bead, resulting in diminished metal loading.

Chelating agents adversely affect metal loading. The chelating agent effectively competes with the binding sites on the biosorbent, preventing binding of the metals with the exchange sites on the biomass. This problem is overcome by implementing pretreatment steps to destroy or destabilize the chelate.

The lower the initial metal concentration in solution, the lower will be the loading on the biosorbent. This is particularly noted when initial metal concentrations are less than 10 mg/L. Longer contact time between the biosorbent and the metal-containing solution diminishes this effect somewhat.

Biosorbent regeneration and reuse

Regeneration and reuse of the biosorbent are important economically and environmentally. From an economic standpoint, regeneration of the biosorbent lowers waste treatment costs for the user and can enhance profitability for the biosorbent producer. Regeneration of the biosorbent also affords an opportunity for reclaiming of the stripped metal into a form that is reusable by the waste generator, or conversion of the metal into a form that can be reintroduced to commerce through further processing and refining. Environmental regulations are now in force that severely limit disposal of solid wastes in hazardous waste landfills. The increased enforcement of these regulations will preclude landfill disposal of metal wastes and necessitate implementation of technologies that reclaim metals from wastes. To that end, biosorbent technologies must incorporate the capability for repeated regeneration and metal reclamation.

Engineering considerations

The engineering design of an application's equipment for biosorbents in metals treatment is important in customer use and economics. Customers seek waste treatment systems that solve their problems, are simple and easy to use, are compact, have a pleasing appearance, and are cost effective. Automation of the system can be a positive design parameter in selling the waste treatment system. Portable systems are important in certain commercial applications such as on-site remediation of contaminated groundwater or surface water.

Most biosorbent systems are designed similar to ion-exchange systems and utilize packed- or fixed-bed contactor systems. Aqueous streams can pass either upflow or downflow through the contactor. When the biosorbent in the contactor has reached its full loading capacity, another contactor containing fresh biosorbent product is brought on-line. The loaded biosorbent is either disposed of or regenerated for reuse. Regeneration can be carried out in the same contactor to avoid unnecessary handling of the biosorbent.

Another application's design that has been used for biosorbents (Brierley et al., 1985, 1988) is the fluid-bed or expanded-bed system. This system requires that the biosorbent have a significant density change between the metal-loaded and unloaded particle and a shrinkage factor upon metal loading. In this system the flow of the aqueous stream is directed upward through a column containing the biosorbent, expanding the biosorbent bed. The density of the biosorbent is balanced against the flow rate through the column so that it is contained within the column, and only clear, treated effluent flows out the top. As the biosorbent becomes fully loaded, it becomes heavier than the unloaded biosorbent, sinks to the bottom of the column, and is collected at the column base. It is then removed from the column and replaced with fresh biosorbent. The fully loaded biosorbent can be regenerated in a separate vessel.

The dispersed-bed system is utilized in situations with extremely large flow magnitudes. Again the biosorbent must demonstrate a density change and shrinkage factor upon metal loading to be employed in a dispersed-bed system. The primary advantage of the dispersed bed lies in its low biosorbent inventory per volume of wastewater treated. Dispersed-bed contactors can be utilized in series to handle large flow volumes and to ensure adequate contact time to achieve effluent standards. Wastewater flows directly from its source into the first contactor. The contactor is equipped with a centralized column through which air is forced. The rising air causes the water around it to be constantly vertically mixed. Biosorbent contained in the large contactor is therefore kept in constant motion. Through this method, the biosorbent particles are exposed to as much wastewater as possible. Residence time of the wastewater in each contactor is calculated for each specific industrial situation. Metals are removed from the waste stream as it flows through the first contactor and into the second. The number of contactors utilized in this system is determined for each individual situation. When the biosorbent becomes fully loaded in any given contactor, it is removed and replaced with fresh material.

Microorganisms, which have accumulated metals can be separated from solution using high-gradient magnetic separation (HGMS), a technique developed in the mining and mineral processing industries for the extraction of weakly magnetic colloids (Kelland, 1973; Kolm et al., 1975). For example, when bacteria accumulate uranium, they exhibit paramagnetism because of the slightly magnetic quality of the metal. The accumulation of uranium by bacteria is significant enough that HGMS can be used to separate the bacteria from solution. If the metal accumulated by the bacteria is only weakly magnetic, then a two-stage process can be implemented to make the microorganisms magnetic. This entails first conditioning the microorganisms by loading onto them a small amount of a magnetic ion. This is followed by placing the weakly magnetic organisms into the target metal solution and allowing further metal accumulation. Ideally the metal to be separated from the bacteria should be a sulfide metal. The sulfate-reducing bacteria, when cultured in metal-containing solutions, do yield precipitated metal sulfides, and these precipitates have been found to adhere to the cell walls of the bacteria. The work that has been carried out in this area has been directed toward using actively metabolizing sulfate-reducing bacteria coupled with HGMS for reclaiming metals from nuclear and other industrial effluents (Anonymous, 1988).

The application of biosorbents must be carefully examined, because most waste streams contain contaminants affecting the functioning of the biosorbent. Pretreatment systems may have to be installed to remove suspended matter that may plug fixed-bed contactors, destroy high cyanide concentrations that complex metals, adjust pH to optimize metal binding by the biosorbent, and remove high concentrations of organic contaminants that could adversely influence metals removal.

Economics of bacterial biosorbent products for metals treatment

The economics of biosorbents for waste treatment are complex and include, among others, these factors: cost of biomass, capital and operating costs of manufacturing, application's equipment costs, marketing and sales costs, pricing of competitive technologies, and market opportunities. Because of the performance characteristics of bacterial biosorbents, the most lucrative market opportunities for biosorbents appear to be in polishing, or secondary, treatment applications and metal removal from dilute concentration streams. This particular market niche makes biosorbents directly competitive with sorption technologies, such as ion exchange. Therefore, pricing strategies for biosorbents are most generally controlled by pricing of the nonselective ion-exchange resins. Of course, pricing of biosorbents can be substantially different, if special equipment or services are sold in conjunction with the biosorbent, from the pricing of ion-exchange resins, which are sold by their manufacturers as commodity chemicals.

Summary

Bacterial biosorbents have now entered the market for removing metals from aqueous industrial effluents, remediation of contaminated groundwaters and surface waters, and metal recovery from industrial process streams. The inherent ability of bacterial cells to accumulate, bind, precipitate, and transform metals has been coupled with cell immobilization and engineering systems to yield a robust and versatile technology for metal's treatment. First-generation bacterial biosorbent products are finding a receptive market for treating streams containing dilute concentrations of target metals. Increasing enforcement of strict effluent discharge standards and the mandating of regulations that prohibit land disposal of hazardous wastes are positive factors in utilization of biosorbents that are highly efficient in metal removal and allow for metal reclamation and biosorbent reuse. Economic evaluations of biosorbents demonstrate that this technology is competitive with existing technologies. Continued research and development is under way to produce next-generation biosorbents that incorporate features such as accumulation of anionic metal complexes, increased metal-loading capacity, greater particle stability, and heightened porosity, which will further enhance marketability.

References

Adams, L. F., and Ghiorse, W. C., Characterization of extracellular Mn^{2+}-oxidizing activity and isolation of a Mn^{2+}-oxidizing protein from *Leptothrix discophora* SS-1, *J. Bacteriol.* **169**:1279–1285 (1987).

Aiking, H., Stijnman, A., van Garderen, C., van Heerikhuizen, H., and van't Riet, J., Inorganic phosphate accumulation and cadmium detoxification in *Klebsiella aerogenes* NCTC 418 growing in continuous culture, *Appl. Environ. Microbiol.* **47**:374–377 (1984).

Anonymous, Biomagnetic separation and extraction process, *Min. J. London*, July (1988).

Beveridge, T. J., Bioconversion of inorganic materials: Mechanisms of the binding of metallic ions to bacterial walls and the possible impact on microbial ecology, in M. J. Klug and C. A. Reddy (eds.), *Current Perspectives in Microbial Ecology*, American Society for Microbiology, Washington, D.C., 1984, pp. 601–607.

Beveridge, T. J., and Koval, S. F., Binding of metals to cell envelopes of *Escherichia coli* K-12, *Appl. Environ. Microbiol.* **42**:325–335 (1981).

Beveridge, T. J., and Murray R. G. E., Uptake and retention of metals by cell walls of *Bacillus subtilis*, *J. Bacteriol.* **127**:1502–1518 (1976).

Beveridge, T. J., and Murray, R. G. E., Sites of metal deposition in the cell wall of *Bacillus subtilis*, *J. Bacteriol.* **141**:876–887 (1980).

Bomstein, R. A., Waste treatment with microbial nucleo-protein flocculating agent, U.S. Patent 3,684,706, 1972.

Boogerd, F. C., and de Vrind, J. P. M., Manganese oxidation by *Leptothrix discophora*, *J. Bacteriol.* **169**:489–494 (1987).

Brierley, C. L., and Brierley, J. A., Anaerobic reduction of molybdenum by *Sulfolobus* species, *Zbl. Bakt. Hyg., I. Abt. Orig.* **C3**:289–294 (1982).

Brierley, J. A., and Brierley, C. L., Biological methods to remove selected inorganic pollutants from uranium mine wastewater, in P. A. Trudinger, M. R. Walter, and B. J. Ralph (eds.), *Biogeochemistry of Ancient and Modern Environments*, Australian Academy of Science, Canberra, 1980, pp. 661–667.

Brierley, J. A., Brierley, C. L., Decker, R. F., and Goyak, G. M., Metal recovery, U.S. Patent 4,789,481, 1988.

Brierley, J. A., Brierley, C. L., and Dreher, K. T., Removal of selected inorganic pollutants from uranium mine waste water by biological methods, in C. O. Brawner (ed.), *First Intern. Conf. on Uranium Mine Waste Disposal*, Society of Mining Engineers of the American Institute of Mining, Metallurgical and Petroleum Engineers, New York, 1980, pp. 365–376.

Brierley, J. A., Brierley, C. L., and Goyak, G. M., AMT-BIOCLAIM®: A new wastewater treatment and metal recovery technology, in R. W. Lawrence, R. M. R. Branion, and H. G. Ebner (eds.), *Fundamental and Applied Biohydrometallurgy*, Elsevier, Amsterdam, 1985, pp. 291–303.

Brierley, C. L., Kelly, D. P., Seal, K. J., and Best, D. J., Materials and biotechnology, in I. J. Higgins, D. J. Best, and J. Jones (eds.), *Biotechnology Principles and Applications*, Blackwell Scientific, Oxford, 1985a, pp. 163–212.

Brock, T. D., and Gustafson, J., Ferric iron reduction by sulfur- and iron-oxidizing bacteria, *Appl. Environ. Microbiol.* **32**:567–571 (1976).

Brown, M. J., and Lester, J. N., Metal removal in activated sludge: The role of bacterial extracellular polymers, *Water Res.* **13**:817–837 (1979).

Brown, M. J., and Lester, J. N., Role of bacterial extracellular polymers in metal uptake in pure bacterial culture and activated sludge. I. Effects of metal concentration, *Water Res.* **16**:1539–1548 (1982a)

Brown, M. J., and Lester, J. N., Role of bacterial extracellular polymers in metal uptake in pure bacterial culture and activated sludge. II. Effects of mean cell retention time, *Water Res.* **16**:1549–1560 (1982b).

Charley, R. C., and Bull, A. T., Bioaccumulation of silver by a multispecies community of bacteria, *Arch. Microbiol.* **123**:239–244 (1979).

Clyde, R. A., Horizontal fermenter, U.S. Patent 4,351,905, 1982.

Darnall, D. W., Greene, B., Hosea, M., McPherson, R. A., Henzl, M., and Alexander, M. D., Recovery of heavy metals by immobilized algae, in R. Thompson (ed.), *Trace Metal Removal from Aqueous Solution*, The Royal Society of Chemistry, London, 1986, pp. 1–24.

DeVoe, I. W., and Holbein, B. E., Insoluble chelating compositions, U.S. Patent 4,530,963, 1985.

DeVoe, I. W., and Holbein, B. E., A new generation of solid-state metal complexing materials: Models and insights derived from biological systems, in R. Thompson (ed.), *Trace Metal Removal from Aqueous Solution*, The Royal Society of Chemistry, London, 1986, pp. 58–70.

Doyle, R. J., Matthews, T. H., and Streips, U. N., Chemical basis for selectivity of metal ions by the *Bacillus subtilis* cell wall, *J. Bacteriol.* **143**:471–480 (1980).

Dugan, P. R., Bioflocculation and the accumulation of chemicals by floc-forming organisms, EPA-600/2-75-032, September 1975, National Technical Information Service, Springfield, Va., 1975.

Ehrlich, H. L., The geomicrobiology of iron, in *Geomicrobiology*, Dekker, New York, 1981, pp. 165–201.

Emerick, J. C., and Cooper, D. J., Acid mine drainage in the west: The wetland approach, in *Proc. 90th National Western Mining Conf.*, Colorado Mining Association, Denver, 1987.

Erickson, P. M., Girts, M. A., and Kleinmann, R. L. P., Use of constructed wetlands to treat coal mine drainage, in *Proc. 90th National Western Mining Conf.*, Colorado Mining Association, Denver, 1987.

Gee, A. R., and Dudeney, A. W. L., Adsorption and crystallisation of gold at biological surfaces, in P. R. Norris and D. P. Kelly (eds.), *Biohydrometallurgy*, Science and Technology Letters, Kew Surrey, U.K., 1988, pp. 437–451.

Hancock, I. C., The use of gram-positive bacteria for the removal of metals from aqueous solution, in R. Thompson (ed.), *Trace Metal Removal from Aqueous Solution*, The Royal Society of Chemistry, London, 1986, pp. 25–43.

Holbein, B. E., DeVoe, I. W., Neirinck, L. G., Nathan, M. F., and Arzonetti, R. N., DeVoe-Holbein technology: New technology for closed-loop source reduction of toxic heavy metals wastes in the nuclear and metal finishing industries, in *Proc. Massachusetts Hazardous Waste Source Reduction Conf.*, 1984, p. 66.

Hutchins, S. R., Davidson, M. S., Brierley, J. A., and Brierley, C. L., Microorganisms in reclamation of metals, *Ann. Rev. Microbiol.* **40**:311–336 (1986).

Jack, T. R., and Zajic, J. E., The immobilization of whole cells, *Adv. Biochem. Eng.* **5**:125–145 (1977).

Jackson, T. A., The biogeochemistry of heavy metals in polluted lakes and streams at Flin Flon, Canada, and a proposed method for limiting heavy-metal pollution of natural rivers, *Environ. Geol.* **2**:173–189 (1978).

Johnston, C. G., and Kipphut, G. W., Microbially mediated Mn(II) oxidation in an ooligotrophic Arctic lake, *Appl. Environ. Microbiol.* **54**:1440–1445 (1988).

Kelland, D. R., High gradient magnetic separation applied to mineral beneficiation, *IEEE Trans. Magn. Mag.* **9**:307–316 (1973).

Kelly, D. P., Norris, P. R., and Brierley, C. L., Microbiological methods for the extraction and recovery of metals, in A. T. Bull, D. C. Ellwood, and C. Ratledge (eds.), *Microbial Technology: Current State, Future Prospects*, Cambridge University Press, Cambridge, 1979, pp. 263–308.

Kolm, H., Oberteuffer, J., and Kelland, D. R., High-gradient magnetic separation, *Sci. Am.* **233**:46–54 (1975).

Kuyucak, N., and Volesky, B., New algal biosorbent for a gold recovery process, in P. R. Norris and D. P. Kelly (eds.), *Biohydrometallurgy*, Science and Technology Letters, Kew Surrey, U.K., 1988, pp. 453–455.

Lakshmanan, V. I., Christison, J., Knapp, R. A., Scharer, J. M., and Sanmugasunderam, V., A review of bioadsorption techniques to recover heavy metals from mineral-processing streams, in *Proc. Second Ann. Gen. Meet. of Biominet*, CANMET Special Publication SP85-6, Canadian Government Publishing Centre, Ottawa, 1986, pp. 75–96.

Lundgren, D. G., Boucheron, J., and Mahoney, W., Geomicrobiology of iron: Mechanism of ferric iron reduction, in G. Rossi and A. E. Torma (eds.), *Recent Progress in Biohydrometallurgy*, Association Mineraria Sarda, Iglesias, Italy, 1983, pp. 55–70.

Lundgren, D. G., and Dean, W., Biogeochemistry of iron, in P. A. Trudinger and D. J. Swaine (eds.), *Biogeochemical Cycling of Mineral-Forming Elements*, Elsevier, Amsterdam, 1979, pp. 211–251.

Macaskie, L. E., and Dean, A. C. R., Cadmium accumulation by a *Citrobacter* sp., *J. Gen. Microbiol.* **130**:53–62 (1984).

Macaskie, L. E., and Dean, A. C. R., Use of immobilized biofilm of *Citrobacter* sp. for the removal of uranium and lead from aqueous flows, *Enzyme Microbiol. Technol.* **9**:2–4 (1987a).

Macaskie, L. E., and Dean, A. C. R., Uranium accumulation by a *Citrobacter* sp. immobilized as biofilm on various support materials, in O. M. Neijssel, R. R. van der Meer, and K. Ch. A. M. Luyben (eds.), *Proc. 4th Eur. Cong. on Biotechnology*, Elsevier, Amsterdam, 1987b, pp. 37–40.

Macaskie, L. E., and Dean, A. C. R., Uranium accumulation by immobilized biofilms of a *Citrobacter* sp., in P. R. Norris and D. P. Kelly (eds.), *Biohydrometallurgy*, Science and Technology Letters, Kew Surrey, U.K., 1988, pp. 556–557.

Mann, S., Sparks, N. H. C., Scott, G. H. E., and de Vrind-de Jong, E. W., Oxidation of manganese and formation of Mn_3O_4 (hausmannite) by spore coats of a marine *Bacillus* sp., *Appl. Environ. Microbiol.* **54:**2140–2143 (1988).

Morper, M., Anaerobic sludge—a powerful and low-cost sorbent for heavy metals, in H. Eccles and S. Hunt (eds.), *Immobilisation of Ions by Bio-Sorption,* Ellis Harwood, Chichester, U.K., 1986, pp. 91–117.

Myers, C. R., and Nealson, K. H., Bacterial manganese reduction and growth with manganese oxide as the sole electron acceptor, *Science* **240:**1319–1321 (1988).

Nakajima, A., Horikoshi, T., and Sakaguchi, T., Recovery of uranium by immobilized microorganisms, *Eur. J. Appl. Microbiol. Biotechnol.* **16:**88–91 (1982).

Nealson, K. A., and Ford, J., Surface enhancement of bacterial manganese oxidation: Implications for aquatic environments, *Geomicrobiology J.* **2:**21–37 (1980).

Nemec, P., Prochazka, H., Stamberg, K., Katzer, J., Stamberg, J., Jilek, R., and Hulak, P., Process of treating mycelia of fungi for retention of metals, U.S. Patent 4,021,368, 1977.

Norberg, A., Production of extracellular polysaccharide by *Zoogloea ramigera* and its use as an adsorbing agent for heavy metals, Ph.D. dissertation, University of Lund, Lund, Sweden, 1983.

Norberg, A. B., and Persson, H., Accumulation of heavy-metal ions by *Zoogloea ramigera, Biotech. Bioeng.* **26:**239–246 (1984).

Pettibone, G. W., and Cooney, J. J., Toxicity of methyltins to microbial populations in estuarine sediments, *J. Ind. Microbiol.* **2:**373–378 (1988).

Pooley, F. D., Bacteria accumulate silver during leaching of sulphide ore minerals, *Nature* **296:**642–643 (1982).

Scott, J. A., Palmer, S. J., and Ingham, J., Microbial metal adsorption enhancement by naturally excreted polysaccharide coatings, in H. Eccles and S. Hunt (eds.), *Immobilisation of Ions by Bio-Sorption,* Ellis Horwood, Chichester, U.K., 1986, pp. 81–88.

Serbus, C., Hora, K., Rezac, J., Pribil, S., Marvan, P., Krejdirik, L., and Stoy, A., Sorbent and method of manufacturing same. U.S. Patent 3,725,291, 1973.

Shumate, S. E. II, and Strandberg, G. W., Accumulation of metals by microbial cells, in M. Moo-Young (ed.), *Comprehensive Biotechnology,* Vol. 4, Pergamon Press, New York, 1985, pp. 235–247.

Silverman, M. P., and Ehrlich, H. L., Microbial formation and degradation of minerals, *Adv. Appl. Microbiol.* **6:**153–206 (1964).

Sterritt, R. M., and Lester, J. N., Heavy metal immobilisation by bacterial extracellular polymers, in H. Eccles and S. Hunt (eds.), *Immobilisation of Ions by Bio-Sorption,* Ellis Horwood, Chichester, U.K., 1986, pp. 121–134.

Strandberg, G. W., and Arnold, W. D., Jr., Microbial accumulation of neptunium, *J. Ind. Microbiol.* **3:**329–331 (1988).

Strandberg, G. W., Shumate II, S. E., Parrott, Jr., J. R., and North, S. E., Microbial accumulation of uranium, radium and cesium, in F. E. Brinckman and R. H. Fish (eds.), *Environmental Speciation and Monitoring Needs for Trace Metal-Containing Substances from Energy-Related Processes,* U.S. Department of Commerce, National Bureau of Standards, Washington, D.C., 1981, p. 27.

Tengerdy, R. P., Johnson, J. E., Hollow, J., and Toth, J., Denitrification and removal of heavy metals from waste water by immobilized microorganisms, *Appl. Biochem. Biotechnol.* **6:**3–13 (1981).

Thayer, J. S., and Brinckman, F. E., The biological methylation of metals and metalloids, *Adv. Organomet. Chem.* **20:**313–356 (1982).

Tsezos, M., Adsorption by microbial biomass as a process for removal of ions from process or waste solutions, in H. Eccles and S. Hunt (eds.), *Immobilisation of Ions by Bio-Sorption,* Ellis Horwood, Chichester, U.K., 1986, pp. 201–218.

Tsezos, M., The performance of a new biological adsorbent for metal recovery, Modeling and experimental results, in P. R. Norris and D. P. Kelly (eds.), *Biohydrometallurgy,* Science and Technology Letters, Kew Surrey, U.K., 1988, pp. 465–475.

Votapek, V., Marval, E., and Stamberg, K., Method of treating a biomass, U.S. Patent 4,067,821, 1978.

Engineering Aspects
of Metal Binding by Biomass

Marios Tsezos

Introduction

Several aspects of the microbiology of the interaction between metals and microorganisms have been presented in detail in previous chapters. The utilization of microbes in the leaching of metals has a long history and has already found several useful applications in industry (Brierley, 1982). The process of metal binding by biomass cannot claim such an extensive history.

The selective sequestering of metal ions and organic molecules from aquatic solutions by microbial biomass has been termed *biosorption* (Tsezos and Volesky, 1981; Friis and Myers-Keith, 1986; Kiff and Little, 1986; Tsezos and Seto, 1986; Yakubu and Dudeney, 1986; Bell and Tsezos, 1987; Kuyucak, 1987). Other terms, for example, *adsorption, bioadsorption,* or *bioaccumulation,* have also been used, but the term *biosorption* appears to have gained general acceptance, especially in view of the fact that the mechanism of biosorption has been shown to be different from that of simple adsorption or metabolically dependent accumulation (Beveridge, 1978; Tsezos and Volesky, 1982a,b; Tsezos, 1983; Tsezos and Keller, 1983; Friis and Myers-Keith, 1986; Kuyucak, 1987).

Over the last 15 years, biosorption of metals developed very rapidly from a laboratory curiosity to a full-scale industrial process (Brierley et al., 1986; McCready and Lakshmanan, 1986; Krambeer, 1987). As a result, in addition to the fundamental studies on the equilibrium, kinetics, selectivity, reversibility, and mechanism of biosorption, the study of engineering aspects of metal binding by microbial biomass became significant. This chapter summarizes the developments and progress made in this area.

Biomass Procurement

Propagation of biomass

The industrial application of biosorption requires that sufficient quantities of the appropriate microbial biomass become available for use. This biomass can be produced by fermentation for specific use as biosorbent, or it can be the by-product of other biochemical operations, such as brewing, pharmaceuticals production, or biological wastewater treatment (Nielsen and Hrudey, 1983; Brierley et al., 1986; Morper, 1986; Tsezos et al., 1987a). The possibility of harvesting this biomass from the ocean (marine algae) has also been proposed (Kuyucak, 1987).

Growth media effects

A comparison of the uranium biosorptive capacity of a *Penicillium* species grown on media based on a variety of carbon sources such as starch, lactose, maltodextrin, cheese whey, hexadecane, and so on, has suggested that the metal-binding properties of the microbial biomass are affected by medium composition (Tsezos et al., 1988a). Similar results have also been reported for uranium uptake capacity of 11 different strains of *Rhizopus* (Treen, 1981). As a result, when biomass is grown specifically for use as a biosorbent, the composition of each growth medium must be optimized with respect to the metal-binding capacity of the biomass and the cost of the media. The use of industrial by-products as raw material for fermentation media requires consideration of media cost. At the same time, the uniformity of medium composition and availability of ingredients must also be considered.

Considerably less control is available in regulating growth conditions for optimizing metal-binding capacity of the biomass whenever the biomass is obtained as an industrial by-product. It has been reported, however, that the radium-binding capacity of activated sludge remains quite constant, even among samples obtained from wastewater treatment plants in different cities (Tsezos and Keller, 1983).

Biomass characteristics

The age of the microorganisms may affect the metal uptake capacity of the microbial biomass. Recent work on this topic suggested that the optimum harvest time for *Penicillium* and *Rhizopus* is at the inflection point of the culture pH curve and that further aging of the biomass does not substantially improve the biosorptive uptake capacity of the biomass (Tsezos et al., 1988a).

Biomass is generally available in the form of individual cells or cell aggregates. The form depends on the characteristics of the biomass strain and the culture conditions. For example, *Rhizopus* and *Penicillium* species have been grown in fermentations either as a mycelial mat or as individual pellets of desirable particle size. The form in which biomass becomes available may facilitate the downstream processing of the biomass and its products (Treen, 1981).

Harvested biomass usually has the form of a dense slurry of living cells with considerable carryover of growth medium residuals. Washing the biomass reduces the medium residual that may interfere with subsequent processing and use of the biomass. Present industrial applications of biosorption make use of dead biomass. Dead biomass does not require nutrients and can be exposed to environments that cannot support life. Furthermore, dead microbial biomass has been shown to have, in general, as good, and usually better, biosorptive properties as the same biomass in a living state (Somers, 1963; Heldwein et al., 1977; Horikoshi et al., 1979, 1981; Nakajima et al., 1979, 1981; Sakaguchi et al., 1979). Several different methods are available for inactivating microbial biomass: for example, sterilization, heat treatment, and chemical treatment. Published information suggests that the way cells are inactivated affects their respective metal-binding capacity, and that heat treatment appears to be the best technique (Somers, 1963; Horikoshi et al., 1979, 1981; Nakajima, 1981; Tsezos et al., 1988a).

The preceding information suggests that standardization of the biomass procurement procedure and optimization of the procurement parameters which affect the metal-binding capacity of the biomass are the primary set of engineering parameters that need to be defined for a metal-recovery process based on biosorption.

Metal Uptake Properties of the Biomass

Equilibrium of biosorption

Extensive research on the equilibrium of biosorption of metals has shown that biosorption systems follow an adsorption-type isotherm (Tsezos and Volesky, 1982a,b; Tsezos and Keller, 1983; Yakubu and Dudeney, 1986; Kuyucak, 1987). The extent of biosorption of a metal ion by microbial biomass from a single-solute solution is a function of the equilibrium metal-ion concentration in solution, all other solution parameters such as pH and temperature remaining constant. The simple adsorption isotherm models of Langmuir and Freundlich have been shown to describe adequately the biosorption equilibrium over a concentration range, although breakpoints do appear on the isotherms (Tsezos and Volesky, 1981; Tsezos and Keller, 1983). The Freundlich model has the form

$$Q = KC^{1/n} \qquad (14.1)$$

where Q = metal uptake capacity of the biomass in units of weight of metal per unit weight of biomass
K = biosorption equilibrium constant indicative of the maximum biosorptive uptake
n = biosorption equilibrium constant indicative of the general shape (aggressiveness) of the isotherm
C = metal-ion solution equilibrium concentration

The Freundlich model has found wide use because of its simple form and because it can be linearized easily in the form

$$\text{Ln } Q = \text{Ln } K + \frac{1}{n} \text{Ln } C \qquad (14.2)$$

The model effectively describes the distribution of the metal ion between the solid phase (biomass) and the liquid phase (solution) at equilibrium.

A plot of Ln Q against Ln C usually yields a straight line with a slope of $1/n$ and intercept of Ln K. This is a quick but incorrect way of determining the value of the model parameters. A closer examination of Q reveals that Q is calculated from a mass balance involving the metal ion in batch contact with a known quantity of biomass (B) in a given volume (V) of metal-ion solution having an initial concentration C_0:

$$Q = \frac{(C_0 - C)V}{B} \qquad (14.3)$$

Therefore, C, the residual metal-ion solution concentration, is the true dependent parameter, not Q. A proper regression of the experimental data for determining the biosorption equilibrium parameter values should use the following expression, which is the combination of Eqs. (14.1) and (14.3) and necessitates a nonlinear regression (Tsezos and Seto, 1986):

$$\frac{(C_0 - C)V}{B} = KC^{1/n} \qquad (14.4)$$

Although the single-solute adsorption isotherm models describe adequately the biosorption equilibrium, the drawing of any mechanistic conclusions from the model fit alone should be avoided. In general, the fit should be interpreted as a simple mathematical representation of the biosorption equilibrium over a given concentration range. In multisolute solutions, which are encountered in most industrial applications, the ionic matrix of the solution is complex. The equilibrium of biosorption, as defined in single-solute solutions, may be affected significantly by the presence of other cations or anions in solution, depending on the chemical interaction of the other ionic species (co-ions) with the metal of interest and the biomass. Thus, whereas the biosorption of heavy metals is not affected significantly by the presence of alkali metals like sodium or potassium in solution (Polikarpov, 1966; Tsezos and Volesky, 1982a), the uptake of uranium and radium by microbial biomass is affected strongly by the presence of heavy-metal ions such as zinc, copper, or iron in solution; the effect is not, however, the same for all biomass types (Tsezos and Volesky, 1982a; Tsezos et al., 1986a). Radium uptake by *P. chrysogenum* is little affected by the presence of calcium but strongly affected by the presence of magnesium and barium. On the other hand, radium uptake by activated sludge biomass is affected strongly by the presence of calcium (Tsezos et al., 1986a). This is illustrated in Table 14.1, which summarizes the observed co-ion effect on radium uptake capacity of the two different biomass types at a common radium equilibrium solution concentration of 1000 pCi/L and pH 7.0 (Tsezos et al., 1986a).

The co-ion effect on the equilibrium of biosorption may have more than one cause. Zinc, copper, and iron(II) have been shown to reduce the uranium uptake capacity of *R. arrhizus* by competing with uranium for coordination at the cell wall chitin amino nitrogen (Tsezos, 1983). Iron as ferric iron, however, is most effective in reducing biomass metal-binding capacity by adsorbing on the biomass surface in the form of its hydrolysis products (Tsezos, 1983).

In addition to cations, the presence of anions can also affect the metal biosorptive uptake capacity of a biomass. The presence of carbonate ions in seawater suppresses the uranium biosorptive capacity of several microorganisms. Removal of the carbonate ions by pH manipulation has been shown to restore the biosorptive uptake (Sakaguchi et al., 1978; Horikoshi et al., 1979; Tsezos and Noh, 1984; Greene et al., 1986). This leads to the conclusion that the quantifica-

TABLE 14.1 Cumulative Co-Ion Effect on Radium Uptake by Biomass

Co-ions	Biomass radium uptake capacity (pCi/g)	
	P. chrysogenum	Activated sludge
None	40,000	37,000
$Ca^{2+} + Ba^{2+}$	36,000	16,000
$Mg^{2+} + Fe^{2+}$	1,900	7,200
Cu^{2+}	330	3,600

tion of the equilibrium of biosorption of a metal in industrial applications where solution ionic matrices are complex requires attention. The effect of co-ions depends on the biomass types used and the chemistry of the ions in solution.

A very important engineering parameter that influences the biosorption equilibrium is the solution pH, which affects the speciation of the ions in solution as well as the chemistry of the active sites on the biomass. The available literature suggests that, depending on the metal ion under consideration, an optimum pH range exists for its biosorptive sequestering. For example, in the case of uranium uptake by *R. arrhizus,* the optimum pH is 4, while for radium biosorption the optimum pH range is 7 and above (Tsezos and Volesky, 1981; Tsezos and Keller, 1983). The active sites of biomass biosorption, such as the cell wall chitin amino groups, cellular phosphate groups, and proteins, hydrolyze as solution pH changes, thus affecting the biomass metal-binding capacity (Beveridge, 1978; Tsezos, 1983; Friis and Myers-Keith, 1986; Gadd, 1986). The dependence of the metal-binding capacity of the microbial biomass on temperature has been shown to be small. In general, the limited existing information suggests an overall biosorption process that is slightly favored by increases in temperature (Horikoshi et al., 1981; Tsezos and Volesky, 1981; Nakajima et al., 1982).

Desorption equilibrium

In industrial applications of biosorption, the sequestered metal ions need to be recovered in a concentrated solution, simultaneously regenerating the biosorbent for additional use. This is the same operation as in the elution of conventional ion-exchange resins. The elution of the sequestered metal ions from the microbial biomass has not been studied as extensively as the respective uptake. Detailed studies on the elution of U from *R. arrhizus* have revealed that quantitative elution of the metal and the reuse of the microbial biomass are possible. They depend on the choice of the eluant and the elution conditions, as has been reported for the elution of other metals (Au, Ag, Al, etc.) from algal biomass (Tsezos, 1984; Krambeer, 1987; Kuyucak, 1987). Ammonium sulfate not only is a poor eluant for U, but the sulfate ions induce irreversible damage to the cell wall biosorption sites while reducing the potential for the reuse of the biomass in subsequent metal recovery cycles (Tsezos, 1984).

Another important elution parameter is the ratio of weight of metal-charged biomass to the volume of a given eluant that is used for the complete recovery of a bound metal. This is commonly referred to as the *solid-to-liquid (S/L) ratio.* The S/L ratio needs to be maximized so that complete metal recovery is achieved with a minimum eluant volume, yielding an eluate with the highest possible metal-ion concentration. The use of a proper eluant reverses the biosorption equilibrium because the biosorbed metal ion has a higher affinity for the eluant than for the biomass. This implies a stoichiometric relation between the metal ion of interest and the eluant and, consequently, a dependence of the optimum S/L ratio on the eluant composition. This dependence has been described for uranium and expressed mathematically by relating the maximum uranium concentration in the eluate to the S/L ratio and the composition of the eluant used (Tsezos, 1984).

Kinetics of metal biosorption

The proper engineering design of process equipment requires a good understanding of the process kinetics in addition to the equilibrium conditions. The kinetic information that has been reported in the literature on the rate of biosorption of metal ions is often difficult to interpret. This has primarily been done because the experimental procedures used to measure reported biosorption rates were not always well defined. Observed rates of biosorption depend strongly on the design of the experimental apparatus and on the experimental conditions used. In general, there is an observed overall rate of biosorption and an intrinsic rate of biosorption. Overall biosorption rates are the result of the intrinsic biosorption rate superimposed on the rate of transfer of the metal ion from the bulk solution to the actual biosorption sites (Tsezos et al., 1988b). The use of a well-mixed batch reactor to observe the rate of biosorption usually allows a direct observation of the intrinsic biosorption rate. A well-stirred reactor provides sufficient mixing so that the rate of transport of the metal ion in the bulk solution is not rate-limiting. Adequate mixing is commonly absent in simple shake-flask experiments, and the experimental kinetic results derived from such experiments commonly represent overall biosorption rates (Shumate et al., 1978; Nakajima et al., 1979; Norris and Kelly, 1979; Khummongkol et al., 1982; Tsezos and Volesky, 1982; Tsezos et al., 1986b, 1988b; Kuyucak, 1987).

Available information has shown that the intrinsic rate of biosorption is rapid and not likely to be rate-limiting in a biosorption-based process (Tsezos and Volesky, 1982a; Tsezos et al., 1986b, 1988b).

Process Design Considerations

Physical form of the biomass

All the existing or proposed industrial applications of biosorption use various forms of immobilized biomass (Brierley et al., 1986;

Krambeer, 1987). Microbial biomass in its natural form consists of small particles of low density, low mechanical strength, and low rigidity. The use of such particles in any conventional unit operation for contacting the biomass with a large volume of solution containing metal ions has been shown to not be practical. The main difficulty lies in the rapid and efficient separation of the biomass from the reaction mixture after contact (Treen, 1981; Brierley et al., 1986; Mellis, 1986; Krambeer, 1987; Tsezos et al., 1987a). Alternatively, immobilized microbial biomass could be produced in the form of particles of desirable size, mechanical strength, and rigidity while maintaining the native properties of the biomass. The simultaneous achievement of all the above objectives is not an easy task. The biosorptive equilibrium properties of the microbial biomass can be protected to a large extent during immobilization. The kinetic biosorptive characteristics of the native biomass are more difficult to protect. As will be discussed later, immobilized biomass usually has poorer kinetic characteristics as compared with the same biomass in a nonimmobilized form.

Satisfactory immobilized biomass types, also called *biosorbents,* resemble a conventional exchange resin in physical form. The principal applications of biosorbents are in the sequestering of metal values from complex, dilute process solutions and industrial waste solutions where the selectivity of biosorption provides a definite advantage over ion exchange. Typical examples are the treatment of wastewaters and the recovery of uranium from in situ biological leaching of ores (Brierley, 1986; McCready and Lakshmanan, 1986; Krambeer, 1987; Tsezos et al., 1987a). Most of the techniques for the production of biosorbents are proprietary, and discussion of the immobilization technologies is outside the scope of this chapter. Additional information on this topic is available in the references.

Kinetic characteristics of immobilized biomass

In order to analyze the kinetic characteristics of immobilized biomass, we first describe a generalized model of an immobilized biomass particle; we then describe a model for biosorptive uptake by the particle, using generalized batch-reactor mass transfer kinetics.

An immobilized biomass particle can be considered as basically composed of two parts, the biomaterial and the admixes (support material) used for immobilization. The mass transport properties of the two components of the biosorbent particle are not the same. Therefore, from a mass transfer viewpoint, the two components can be considered as two separate, sequential layers, each with its own mass transfer characteristics (Tsezos et al., 1988b). Figure 14.1 illustrates the con-

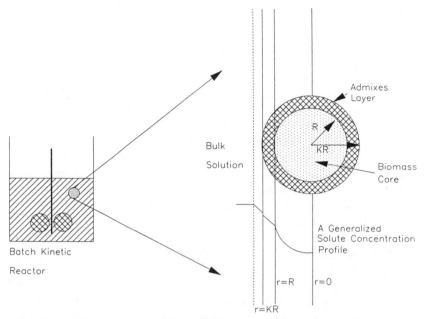

Figure 14.1 Conceptual representation of an immobilized biomass particle.

ceptual structure of such a spherical biosorbent particle.

When such an immobilized biomass particle is suspended in a solution containing a solute, essentially four consecutive mass transport steps are associated with the biosorption of the solute from the solution. The first step is the transport of the solute in the bulk solution. This step is usually very rapid because of mixing and convective flow. The second step involves the diffusion of the solute through the hydrodynamic boundary layer that surrounds the biosorbent particle. The third step is the diffusional transport of the solute across the layer of admixes. The fourth step is the diffusion of the solute in the biomaterial core. The actual biosorption of the metal ion on the active sites of the biomass is generally considered to be very rapid and thus insignificant in the context of the observed overall biosorption rate. Because the steps act in series, the slowest of the three will be the rate-limiting step. If the rates are comparable, the control of the overall rate may also be distributed between steps.

Important additional assumptions required for the development of the model are:

1. Uniform immobilized biomass particle size of spherical geometry.

2. Uniform layers of the admix components and the biomass.

3. No accumulation of solute by the admixes.

4. Local biosorption equilibrium exists in the pores of the biosorbent particle.
5. Accumulation of solute in the liquid inside the pores of the particle is negligible.

Using the Freundlich biosorption isotherm model to describe the biosorption equilibrium of the native biomass and working in a dimensionless domain, we can derive an analytical expression that describes the changing bulk solution concentration as a function of time for a well-mixed batch reactor. A detailed presentation of the model has been published (Tsezos et al., 1988b; Tsezos and Deutschman, 1989). The model allows an investigation of the effects of system parameters on the observed kinetics of biosorption by the immobilized biomass that are of engineering importance.

The particle size of the immobilized biomass particles has a significant effect on the observed overall biosorption kinetics. Figure 14.2 illustrates the simulated kinetic results for uranium uptake by immobilized microbial biomass of three different particle sizes (Tsezos and Deutschman, 1989). It is evident that small particle size improves the kinetic characteristics and, consequently, the performance of the immobilized biomass.

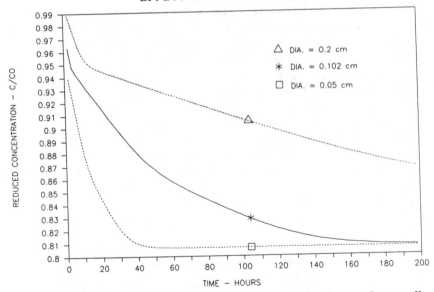

Figure 14.2 The effect of the immobilized biomass particle size on the overall biosorption rate.

The effect of initial metal-ion concentration on the observed kinetics is also significant. Higher initial concentrations establish stronger driving forces for mass transfer and result in faster kinetics than do lower metal-ion concentrations (Tsezos and Deutschman, 1989).

Figure 14.3 shows the effect of biosorptive metal-ion uptake capacity of native biomass on the observed overall kinetics (Tsezos and Deutschman, 1989). The higher metal-binding capacity of native biomass improves overall metal loading on the immobilized biomass and speeds up the kinetics. A second important conclusion that has been reached from modeling is that the metal-binding capacity of biosorbent which is actually available for engineering design is not necessarily the same as the intrinsic metal uptake capacity. This probably results from a reduction in the rate of metal uptake by the biosorbent as the immobilized biomass particle uptake capacity approaches saturation. A similar phenomenon has been observed with adsorption by activated carbon, and it is also possible for some other biosorbents to exhibit similar slow kinetics (Peel and Benedek, 1980).

Finally, the effective diffusivity of the metal ion in the pores of the immobilized biomass particles exerts an important influence on biosorption (Tsezos and Deutschman, 1989). A large value of the solute diffusivity in the biomass component of the particle has the effect of increasing the metal-ion uptake in a given period of time, compared

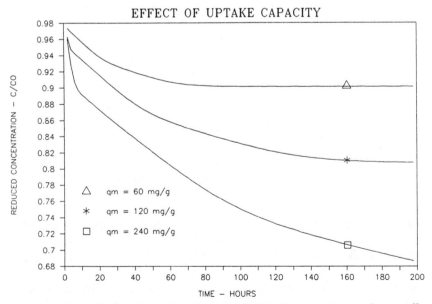

Figure 14.3 The effect of the native biomass metal-binding capacity on the overall biosorption kinetics.

with a low diffusivity. An interrelation between the overall mass transfer coefficient of the metal ion and the effective solute diffusivity in the biomass pores also exists (Tsezos and Deutschman, 1989), but a detailed discussion of this relation is beyond the scope of this chapter.

Conceptual flowsheets for biosorption applications

In the previous section reasons for the use of immobilized instead of native biomass were presented. In a separation process, such as biosorption, the best use of the biosorptive properties of the immobilized biomass particles is made by using a multistage, countercurrent scheme for contacting the biosorbent with the solution of interest. Such schemes have been extensively developed in the use of ion-exchange resins and activated carbon. Considerable experience has accumulated in their use by the metal- and pollution-control industries. This experience is an important advantage that these processes have, and it should be considered seriously in designing flowsheets for the application of biosorption. The most common of such schemes are the upflow or downflow packed-bed reactors and the continuous fluidized-bed reactors. Pilot scale results for both types of reactors applied to biosorption from industrial solutions have been reported (Brierley et al., 1986; Tsezos et al., 1989a,b). A schematic flowsheet for the fluidized-bed arrangement is shown in Fig. 14.4. The economics of each flowsheet need careful evaluation for each case. It is not possible at present to draw generalized conclusions.

The use of nonimmobilized biomass would necessitate an elaborate scheme of well-mixed reactors followed by separators, which is costly and less efficient than the schemes mentioned above (Mellis, 1986).

Conclusions

This chapter shows that the available understanding of the equilibrium, selectivity, reversibility, and kinetics of metal binding by microbial biomass has been developed to a considerable extent. Furthermore, it is evident that the above aspects of metal biosorption need to be defined separately for each metal ion and biomass pair, although some generalizations are possible. The process application of biosorption depends largely on the development of selective and efficient immobilized biomass biosorbents. Engineering considerations are therefore central in decisions concerning the commercial future of biosorption. Commercial biosorbents have recently appeared on the market, and future prospects are positive. The art of biosorption is, however, still young and faces a further trial period.

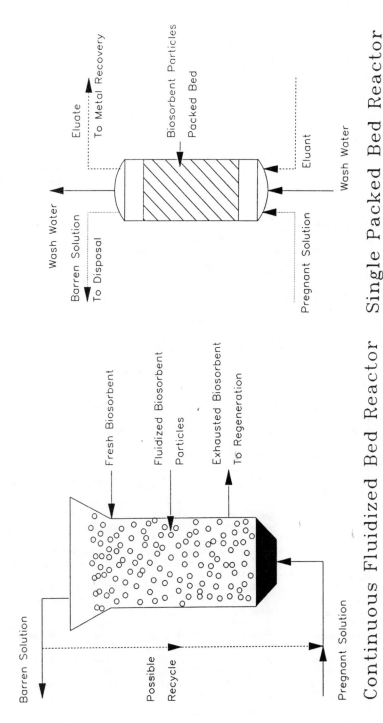

Continuous Fluidized Bed Reactor   Single Packed Bed Reactor

(a).   (b)

Figure 14.4 Examples of conceptual reactor arrangements for the application of biosorption.

337

References

Bell, J. P., and Tsezos, M., Removal of organic pollutants by biomass adsorption, *J. Water Pollut. Contr. Fed.* **59**:191–198 (1987).

Beveridge, T. J., The response of cell walls of *B. sultilis* to metals and microscopic stains, *Can. J. Microbiol.* **24**:89–104 (1978).

Brierley, C. L., Microbiological mining, *Sci. Am.* **247**:44–53 (1982).

Brierley, J. A., Goyak, G. M., and Brierley, C. L., Considerations for commercial use of natural products for metals recovery, in H. Eccles and F. Hunt (eds.), *Immobilization of Ions by Biosorption*, Ellis Horwood, Chichester, U.K., 1986, pp. 105–117.

Friis, N., and Myers-Keith, P., Biosorption of uranium and lead by *Streptomyces longwoodensis*, *Biotechnol. Bioeng.* **28**:21–28 (1986).

Gadd, G. M., The uptake of heavy metals by fungi and yeasts: The chemistry and physiology of the process and applications to biotechnology, in H. Eccles and F. Hunt (eds.), *Immobilization of Ions by Biosorption*, Ellis Horwood, Chichester, U.K., 1986, pp. 149–157.

Greene, B., Henzel, M. T., Hosea, J. M., and Darnall, D. W., Elimination of bicarbonate interference in the binding of U(VI) in mill-waters to freeze-dried *Chlorella vulgaris*, *Biotechnol. Bioeng.* **28**:764–767 (1986).

Heldwein, R., Tromballa, H. W., and Broda, E., Aufnahme von Cobalt, Blei und Cadmium durch Baeckerhefe, *Z. Allg. Mikrobiol.* **17**:299–308 (1977).

Horikoshi, T., Nakajima, A., and Sakaguchi, T., Studies on the accumulation of heavy metal elements in biological systems. IV. Uptake of uranium by *Chlorella regularis*, *Agric. Biol. Chem.* **43**:617–623 (1979a).

Horikoshi, T., Nakajima, A., and Sakaguchi, R., Studies on the accumulation of heavy metal elements in biological systems. IX. Uptake of uranium from sea water by *Synechococcus elongatus*, *J. Ferment. Technol.* **57**:191–194 (1979b).

Horikoshi, T., Nakajima, A., and Sakaguchi, T., Studies on the accumulation of heavy metal elements in biological systems. XIX. *Eur. J. Appl. Microbiol. Biotechnol.* **12**:90–96 (1981).

Khummongkol, D., Canterford, G. S., and Fryer, C., Accumulation of metals in unicellular algae, *Biotechnol. Bioeng.* **24**:2643–2660 (1982).

Kiff, R. J., and Little, D. R., Biosorption of heavy metals by immobilized fungal biomass, in H. Eccles and F. Hunt (eds.), *Immobilization of Ions by Biosorption*, Ellis Horwood, Chichester, U.K., 1986, pp. 71–80.

Krambeer, C., Bigger profits through improved waste water treatment, *Finish. Manage.* **32**:36–37 (1987).

Kuyucak, N., Algal biosorbents for gold and cobalt, Ph.D. thesis, McGill University, Montreal, 1987.

McCready, R. G. L., and Lakshmanan, V. I., Review of bioadsorption to recover uranium from leach solutions in Canada, in H. Eccles and F. Hunt (eds.), *Immobilization of Ions by Biosorption*, Ellis Horwood, Chichester, U.K., 1986, pp. 219–226.

Mellis Consultants, Technical and economical comparison of uranium recovery for bioleach solutions by the use of biomass, ion exchange and reverse osmosis, Report DSS-23440-5-9042, 1986.

Morper, M., Anaerobic sludge—a powerful and low cost sorbent for heavy metals, in H. Eccles and F. Hunt (eds.), *Immobilization of Ions by Biosorption*, Ellis Horwood, Chichester, U.K., 1986, pp. 91–104.

Nakajima, A., Horikoshi, T., and Sakaguchi, T., Studies on the accumulation of heavy metal elements in biological systems. XII. Uptake of manganese ion by *Chlorella regularis*, *Agric. Biol. Chem.* **43**:1461–1466 (1979).

Nakajima, A., Horikoshi, T., and Sakaguchi, T., Studies on the accumulation of heavy metal elements in biological systems. XVII. Selective accumulation of heavy metal ions by *Chlorella regularis*, *Eur. J. Appl. Microbiol. Biotechnol.* **12**:76–83 (1981).

Nakajima, A., Horikoshi, T., and Sakaguchi, T., Recovery of uranium by immobilized microorganisms, *Eur. J. Appl. Microbiol. Biotechnol.* **16**:88–91 (1982).

Nielsen, J. F., and Hrudey, S. E., Metal loadings and removal at a municipal activated sludge plant, *Water Res.* **17**:1041–1052 (1983).

Norris, P. R., and Kelly, D. P., Accumulation of metals by bacteria and yeasts, *Dev. Ind. Microbiol.* **20**:299–308 (1979).

Peel, R. G., and Benedek, A., Dual rate kinetic model of fixed bed adsorber, *J. Environ. Eng. Div., ASCE* **EE4**:797–813 (1980).

Polikarpov, G. C., *Radioecology of Aquatic Organisms,* North-Holland, New York, 1966.

Sakaguchi, T., Horikoshi, T., and Nakajima, A., Studies on the accumulation of heavy metal elements in biological systems. VI. Uptake of uranium from sea water by microalgae, *J. Ferment. Technol.* **56**:561–565 (1978).

Sakaguchi, T., Tsuji, T., Nakajima, A., and Horikoshi, T., Studies on the accumulation of heavy metal elements in biological systems. XIV. Accumulation of cadmium by green microalgae, *Eur. J. Appl. Microbiol. Biotechnol.* **8**:207–215 (1979).

Shumate, S. E. II, Strandberg, G. W., and Parrott, J. R., Jr., Biological removal of metal ions from aqueous process streams, *Biotechnol. Bioeng.,* Symp. No. 8, 1978, pp. 13–20.

Somers, E., The uptake of copper by fungal cells, *Ann. Appl. Biol.* **51**:425–437 (1963).

Treen, M., Biosorption of the uranyl ion by *Rhizopus*, M. Eng. thesis, McGill University, Montreal, Canada, 1981.

Tsezos, M., The role of chitin in uranium adsorption by *R. arrhizus, Biotechnol. Bioeng.* **25**:2025–2040 (1983).

Tsezos, M., Recovery of uranium from biological adsorbents—Desorption equilibrium, *Biotech. Bioeng.* **26**:973–981 (1984).

Tsezos, M., The performance of a new biological adsorbent for metal recovery: Modeling and experimental results, in P. R. Norris and D. P. Kelly (eds.), *Biohydrometallurgy,* Science and Technology Letters, Kew, Surrey, U.K., 1988, pp. 465–475.

Tsezos, M., ADM Inc., ORF, and Denison Mines Ltd., Optimization of fungal adsorption of uranium and cell immobilization and pelletization, *Final Report to CANMET,* contract DSS 23440-6-9052/01-SQ, 1988a.

Tsezos, M., Baird, M. I. H., and Shemlit, L. W., Adsorptive treatment with microbial biomass of Ra[226] containing waste waters, *Chem. Eng. J.* **32**:B29–B41 (1986a).

Tsezos, M., Baird, M. I. H., and Shemlit, L. W., The kinetics of radium biosorption, *Chem. Eng. J.* **33**:B35–B41 (1986b).

Tsezos, M., Baird, M. I. H., and Shemlit, L. W., The use of immobilized biomass to remove and recover radium from Elliott Lake uranium tailings streams, *Hydrometallurgy* **17**:357–368 (1987a).

Tsezos, M., Baird, M. I. H., and Shemlit, L. W., The elution of radium adsorbed by microbial biomass, *Chem. Eng. J.* **34**:B57–B64 (1987b).

Tsezos, M., and Deutschman, A., An investigation of engineering parameters for the use of immobilized biomass particles in biosorption, *J. Chem. Tech. Biotech.* (1989).

Tsezos, M., and Keller, D., Adsorption of radium 226 by biological origin adsorbents, *Biotech. Bioeng.* **25**:201–215 (1983).

Tsezos, M., McCready, R. G. L., and Bell, J. P., The continuous recovery of uranium from biologically leached solutions using immobilized biomass, *Biotech. Bioeng.* **34**:10–17 (1989a).

Tsezos, M., McCready, R. G. L., Salley, J., and Cuif, J. P., The use of immobilized biomass for the continuous recovery of uranium from biological leachate of Elliot Lake ore, *Int. Conf. Biohydrometallurgy,* Jackson Hole, Wyoming, Aug. 13–18, 1989b.

Tsezos, M., and Noh, S. H., Extraction of uranium from sea water using biological origin adsorbents, *Can. J. Chem. Eng.* **62**:559–561 (1984).

Tsezos, M., Noh S. H., and Baird, M. I. H., A batch reactor kinetic model for uranium biosorption using immobilized biomass, *Biotech. Bioeng.* (1988b).

Tsezos, M., and Seto, W., The adsorption of chloroethanes by microbial biomass, *Water Res.* **20**:7 (1986).

Tsezos, M., and Volesky, B., Biosorption of uranium and thorium, *Biotech. Bioeng.* **23**:583–604 (1981).

Tsezos, M., and Volesky, B., The mechanism of uranium biosorption, *Biotech. Bioeng.* **24**:385–401 (1982a).

Tsezos, M., and Volesky, B., The mechanism of thorium biosorption, *Biotech. Bioeng.* **24**:955–969 (1982b).

Yakubu, N. A., and Dudeney, A. W. L., Biosorption of uranium with *Aspergillus niger,* in H. Eccles and F. Hunt (eds.), *Immobilization of Ions by Biosorption,* Ellis Horwood, Chichester, U.K., 1986, pp. 183–200.

Fossil Fuel Processing

This section deals with another aspect of biobeneficiation, namely that applied to fossil fuels. The beneficiation process takes two forms in this instance. One is the microbial removal of inorganic and/or organic sulfur to lessen its contribution to air pollution during the combustion of these fuels. The other is in enhanced oil recovery (EOR) by microbial means. As the chapter on microbial coal desulfurization explains, a technology of microbial removal of pyritic sulfur has been developed, whereas a technology for removal of organic sulfur remains a challenge to be successfully met. An effective microbial technology for removal of organic sulfur from petroleum is a little closer to solution but still not ready for practical application. Microbially enhanced oil recovery (MEOR), while still in an experimental stage, offers promise as an effective technology.

Microbial Treatment
of Coal

Pieter Bos

J. Gijs Kuenen

Introduction

Background

Since the oil crises in 1973 and 1979 it has been realized that coal must play a more important role in the world's energy supply. Its reserves are far more important than those of any other fossil fuel (Schilling and Wiegand, 1987). In the immediate future, increase in the use of nuclear power as an alternative to fossil fuels is not very likely because of adverse public opinion. Although the environmental risks of coal use might initially have been underestimated in the discussions on alternative energy sources, awareness of the drawbacks of coal utilization, especially as related to its sulfur content, has increased coincidentally with the renewal of interest in its use. For this reason, coal research has moved especially in the direction of the development and improvement of clean-coal technologies. Most activities have been concentrated on the reduction of sulfur and nitrogen-oxide emissions, the main causes of acid rain.

There are different possibilities for reducing sulfur emissions. Sulfur compounds can be removed before combustion by using the traditional sink-float techniques if the inorganic sulfidic minerals are present as coarse particles (Palowitch and Deurbrouck, 1979). For the removal of finely distributed sulfidic minerals, physical separation techniques have been suggested, such as froth flotation (Zimmerman, 1979), high-gradient magnetic separation (Beddow, 1981), and electrostatic separation (Beddow, 1981). For the removal of both inorganic and organic sulfur, the possibilities of chemical methods have been studied. For an overview, see Morrison (1981). Morrison concluded that these chemical processes are able to remove inorganic sulfur from coal. He stated that the extent of organic sulfur removal is much less certain because of the inaccuracies involved in the determination of organic sulfur in coal. The removal of sulfur dioxide during combustion can be achieved by fluidized-bed combustion of coal-chalk mixtures (Podolski, 1984) or by simply injecting lime into the flame (Bortz and Flament, 1985). Flue-gas desulfurization (Prior, 1977) aims at the removal of sulfur dioxide after combustion. None of the techniques mentioned meets the problem of nitrogen-oxide emission, only a small part of which has its origin in the nitrogen content of coal. Nitrogen oxides are mainly produced in the flame from atmospheric nitrogen and oxygen. Reduction of nitrogen oxides can be achieved by lowering the flame temperature and by changes in the geometry of the burners.

Because coal is a solid material, its use presents other disadvantages, such as dust formation and transport and storage difficulties. Many studies have therefore been directed toward the possibilities of converting coal into gaseous or liquid fuels. With respect to gasification, even in situ processes are being considered (Jüntgen, 1987). Both coal gasification and coal liquefaction offer convenient possibilities for sulfur removal.

In parallel with the development of chemical and physical techniques for clean-coal technologies in the last decades, biotechnological alternatives such as microbial desulfurization, microbial coal liquefaction, and microbial gasification are also being explored. The biotechnological alternatives have been the subject of several selective or comprehensive reviews (Kargi, 1982b, 1984, 1986; Monticello and Finnerty, 1985; Olson and Brinckman, 1986; Finnerty and Robinson, 1986; Gouch, 1987; Kawatra et al., 1987; Klein et al., 1988). Most research has been in the field of microbial desulfurization of coal. The studies on processes aimed at the removal of inorganic sulfur compounds from coal by acidophilic sulfur and/or iron oxidizers are most advanced.

Coal as a substrate for microbial activity

A detailed treatise on coal structure and composition would not be appropriate within the framework of this chapter; however, a few aspects will be highlighted. First of all, it should be realized that "coal is merely a generic name and that the solids to which it is applied are more often more dissimilar than alike" (Berkowitz, 1979). The variation in coal types is related to the various biological materials which served as the starting material, the variations in the conditions during coal genesis, and the geological age. During the process of coal genesis the carbon content increases. Types of coal can be arranged according to their increasing carbon content, resulting in a series from peat, lignite, subbituminous coal, bituminous coal, to anthracite. Microscopy reveals a layered structure. Over the years a system has been developed to identify different types of layers (macerals) whose origin can be attributed to the various initial biological materials.

The chemical composition of coals can be characterized in several ways. The "proximate analysis" deals with the products formed by a destructive distillation (percent moisture, percent volatile matter, percent ash, and the remainder, which is designated as the fixed carbon content). The ash content reflects, more or less, the proportion of minerals in the coal. The "ultimate analysis" comprises the elemental analysis of coal (carbon, hydrogen, nitrogen, sulfur, and oxygen). These as well as other characteristics, such as the caloric value of a coal, are of practical importance. For details on coal analysis, see Karr (1978).

During coal genesis, which is initiated biologically and completed by complex chemical processes, the molecular structure of the starting material evolves more and more away from the original biological structures. Some model coal molecules have been depicted in the literature (Shinn, 1985). The gradual changes in the structure of the carbon skeleton during its geological aging process may be reflected in its increasing recalcitrance to aerobic microbial attack as the carbon content rises.

The organic carbon skeleton of coal contains other elements besides carbon, hydrogen, and oxygen, of which sulfur is often one of the most important. Until now there was no exact knowledge on the contribution of the different types of sulfur-containing organic structures. One might expect sulfur to be present in the form of organic sulfides (R_1—S—R_2), disulfides (R_1—S—S—R_2), thiol groups, or in thiophenic structures (Kargi, 1982b). Adequate and reliable analytical techniques are not available. Techniques used in this area, such as those described by Attar and Dupuis (1979) and Curie-point pyrolysis combined with mass spectroscopy (Boudou et al., 1987), are destructive in nature and may give rise to artifacts.

With respect to microbial activities, some minerals and, especially, metal sulfides, the most important being pyrite and marcasite (both with the molecular formula FeS_2 but with a different crystalline structure), are the most susceptible components in coal. This susceptibility is the origin of the environmental problem of acid mine drainage. If FeS_2-containing coals are exposed to moisture and oxygen, spontaneous FeS_2 oxidation starts, resulting in the production of ferric iron and sulfuric acid. Because the rates of spontaneous pyrite oxidation decrease as the pH falls, the process would stop at pH 4 if acidophilic pyrite oxidizing bacteria were absent. However, these bacteria are generally indigenous in coals and sulfidic mineral deposits, and they will take over from the spontaneous chemical oxidation process, reducing the pH to values even lower than 2. The ferric iron produced in these oxidation reactions will act as an oxidizing agent and will be able to solubilize other metal sulfides. The ferrous iron thus produced can be reconverted biologically to ferric iron. For this reason, the drainage waters of mines, coal and ore piles, and mine waste dumps often contain unacceptably high concentrations of heavy metals and, in combination with its low pH, are a menace to the environment (Lovell, 1983).

Although pyrite and, to a lesser extent, marcasite are the most important sulfur compounds in a large variety of coals, other inorganic sulfur compounds are also present. In weathered coals, low concentrations of inorganic sulfates (generally less than 0.1 percent) produced from mineral sulfides can be observed. Some coals have also been reported to contain elemental sulfur (less than 0.2 percent) (Greer, 1979).

In the ultimate sulfur analysis of coal according to ASTM standards (D3177-75 and D2492-68), the subdivision into pyritic, sulfate, and organic sulfur is more or less pragmatic. Organic sulfur is defined as the difference that can be found by subtracting the pyritic sulfur and the sulfur represented by inorganic sulfate from the total sulfur content. This means that the figures of organic sulfur analysis may also include inorganic sulfur species, such as elemental sulfur, polysulfides, polythionates, and the sulfur moiety in non-ferro metal sulfides.

The mineral composition of coals can be determined with techniques such as SEM-EDS by which one can determine not only the content of different minerals but also their particle size distribution and their location and association in the coal matrix (Vleeskens et al., 1985). For example, pyrite can be present as comparatively large crystals in fractures, finely distributed as single crystals (1 to 40 μm), or in framboid structures in the coal. It has been suggested that the larger pyrite particles were deposited in the coal long after coal genesis had started. Especially when coals were formed in a marine environment, the ferrous iron present in the seawater could have reacted with the huge amounts of hydrogen sulfide produced by sulfate-reducing bacteria (epigenetic pyrite). This would explain the extremely high sulfur content of some coals, which exceeds the normal sulfur content in biological materials. In contrast, the minute pyrite crystals (often with a diameter less than 1 μm) distributed in the coal matrix might have been formed mainly from the sulfur which was originally present in the starting material (syngenetic pyrite; Damberger et al., 1984). The syngenetic pyrite causes most of the problems in analysis. The ASTM pyrite analysis is based on a dissolution of pyrite in boiling concentrated nitric acid, followed by a determination of iron in the solution. For complete dissolution, the carbon matrix should be destroyed completely, but it is very likely that the ASTM procedure is not adequate when these finely disseminated pyrite crystals are present. Thus, part of the pyrite will be missed in the analysis of pyritic sulfur and will be included as part of the organic sulfur (Kos, 1982).

Microbial Desulfurization of Coal

This review will deal mainly with microbial coal desulfurization processes. Processes for the removal of inorganic sulfur compounds from coal will be discussed first. There are two different versions of this technique. In the first one, a complete oxidation of the inorganic sulfidic minerals is pursued (complete oxidation process). In the second, only an initial attack is used to change the surface characteristics of the sulfidic minerals. This change brings an effective application of physical separation techniques such as oil agglomeration and froth flotation within reach (microbial-assisted physical separation techniques). Finally, attention will be paid to the potential of microbiological processes aimed at the removal of organic sulfur.

Removal of inorganic sulfur compounds
from coal by complete oxidation processes

The principle behind the complete oxidation processes is based on the ability of microorganisms to oxidize reduced sulfur compounds present in the coal and convert them into water-soluble products which can be leached from the coal.

The organisms responsible for the oxidation of inorganic sulfur compounds such as pyrite belong to the group of acidophilic sulfur and/or iron-oxidizing bacteria. Of these the mesophilic, autotrophic organism *Thiobacillus ferrooxidans* is the best studied. They convert pyrite into ferric iron and sulfuric acid according to the reaction equation

$$4FeS_2 + 15O_2 + 2H_2O \rightarrow 2Fe_2(SO_4)_3 + 2H_2SO_4 \qquad (15.1)$$

The concept of removing inorganic sulfur from coal by its complete bacterial oxidation dates from about 30 years ago. Zarubina et al. (1959) and Silverman et al. (1963) were the first authors to suggest such a process. Being aware of the detrimental role of acidophilic sulfur and/or iron oxidizers in the drainage water of the mining industry, they suggested the exploitation of the pyrite-oxidizing capacity of organisms to leach out this compound from coal under controlled conditions. They realized that coals should be milled to such an extent that the bacteria can reach the pyrite surface, and that the pulverized coal should be mixed with water containing enough nutrients to allow biomass synthesis. Because most acidophilic pyrite oxidizers are obligately aerobic autotrophs, the presence of air is essential to supply them with oxygen as well as carbon dioxide, their carbon source. The original proposals on how the process should be run were almost identical with those for bacterial metal leaching processes.

It took more than a decade before serious studies on the feasibility of this concept appeared. The first and second oil crises have significantly stimulated research in this field. A survey by Bos et al. (1986) indicated that as of 1985 15 different countries were involved in research in this area. Most studies have been monodisciplinary, in which often either the technological or the microbiological aspects were insufficiently examined.

The compatibility of microbial coal desulfurization with other coal-using techniques. Although some studies suggest that microbial desulfurization of coal is a universal solution to the problem of acid rain, it is more realistic to consider microbial desulfurization of coal as one of the possibilities to reduce the emission of sulfur oxides. From the start it has been clear that apart from some apparent advantages, of which the selective removal of sulfidic minerals, heavy metals, and ash-forming components from coal without attacking the carbon matrix are the most important, the technique has a number of drawbacks: (1) the process is comparatively slow and will require huge reactor systems to treat a bulk product like coal, (2) it is unlikely that organic sulfur can be effectively removed. Although in recent years the possibility of removing organic sulfur has been regarded more optimistically, it is currently more realistic in evaluating the feasibility of microbial desulfurization processes to consider only inorganic sulfur removal, and (3) the microbial process produces a wastewater with a low pH and loaded with unacceptably high concentrations of iron and other metals. Neutralizing this waste by adding lime results in a product that cannot be easily discarded.

The economic feasibility of the process will also depend greatly on the site, the type of coal, governmental standards and regulations regarding permissible sulfur emissions, and on the possibility of integrating the process as a step in a sequence of other techniques. This means that, depending on the situation, other techniques such as fluidized-bed combustion or flue-gas desulfurization might be preferred over the microbial process. For example in the Netherlands, which is no longer a coal-producing nation, microbial desulfurization of coal has been directed toward pretreatment of imported coals which already have a low sulfur content (less than 1.0 percent total sulfur). This pretreatment step is aimed at making the coal more suitable for combustion in small-scale industrial power plants, thus avoiding the necessity of a flue-gas desulfurization installation. This option was realistic at the time when the Dutch government allowed coal to be used in that way if the sulfur contents did not exceed 0.8 percent. However, standards have changed since then. In 1990, only coals containing less than 0.35 percent total sulfur may be burned without the requirement for any sulfur dioxide emission reduction. Assuming that the microbial desulfurization process is only effective in the removal of inorganic sulfur species, application of this technology only makes sense if coals with an organic sulfur content of less than 0.35 percent are available. These types of coals are rare. The microbial process might be of interest if it can be combined with another sulfur-reducing technique such as the injection of lime into the flame (van Steenbergen, 1985). In the Dutch electric-power-producing industry, it was already decided at the beginning of the 1980s that for large coal-fired electricity plants, flue-gas desulfurization was obligatory.

Microbial desulfurization can also be considered as a potential treatment of high-sulfur coals at the mine site. Traditional coal preparation plants often produce quantities of low-quality coal. The sink-float techniques may not be effective if most of the pyrite crystals present in the coal are extremely small and finely distributed in the coal matrix. Microbial desulfurization of coal might be of interest for the upgrading of these coals. The value of coal depends on its sulfur content among other factors (Dugan, 1985).

Microbial desulfurization of coal is a wet process. If the coal is to be used in traditional installations, the product of the desulfurization process should be dried. In recent years, interest in coal-water mixtures has increased (Armson, 1986). These mixtures contain 70 percent pulverized coal and have almost the same appearance as heavy oils, which they can replace in oil-fed burners. A prerequisite for oil-water mixtures is that they should be clean, which means that they have a low sulfur and a low ash content. Microbial desulfurization could be a serious candidate for the cleaning step in the preparation of coal-water mixtures.

Pyrite-oxidizing bacteria used in desulfurization studies. Most authors using mesophilic bacteria (20 to 35°C) claimed to have used pure cultures of *T. ferrooxidans* or a mixed culture in which *T. ferrooxidans* dominated (e.g., Dugan and Apel, 1978; Detz and Barvinchak, 1979; Hoffmann et al., 1981; Jilek and Beranova, 1982; Myerson and Kline, 1983; Pocas, 1984). Others utilized an artificial mixture of *T. ferrooxidans* and *Thiobacillus thiooxidans,* based on the assumption that *T. ferrooxidans* oxidized pyrite indirectly by oxidizing the ferrous iron present in the water phase (Dugan and Apel, 1978; Rinder and Beier, 1983; Andrews and Maczuga, 1984). The ferric iron produced than acts as a chemical oxidizing agent for pyrite. Based on old data in the literature (Stokes, 1901; Garrels and Thompson, 1960), the formation of elemental sulfur from pyrite according to the equation

$$2Fe^{3+} + FeS_2 \rightarrow 3Fe^{2+} + 2S \tag{15.2}$$

has been taken for granted. Because *T. ferrooxidans* is considered to be a poor oxidizer of elemental sulfur, another organism such as *T. thiooxidans* should be present to remove the sulfur formed on the pyrite surface, which would otherwise stop the activity of *T. ferrooxidans*.

Although conclusive experimental evidence which can give a straightforward explanation is still not available, the present information suggests that mixed populations perform better than pure cultures in the process. In addition to the beneficial role of *T. thiooxidans* in the removal of elemental sulfur, a positive effect of acidophilic, oligotrophic heterotrophs, and facultative acidophilic autotrophs, such as *Acidiphilium cryptum* and *Thiobacillus acidophilus*, respectively, on the performance of *T. ferrooxidans* cultures has been described (Norris and Kelly, 1982; Harrison, 1984). Heterotrophic bacteria may be able to metabolize excretion products of *T. ferrooxidans* which could be inhibitory. Another explanation may be the production locally of additional carbon dioxide by heterotrophs growing on (trace) organic substrates. This would favor *T. ferrooxidans*, which probably has a low affinity for carbon dioxide (Torma et al., 1972). Kargi (1982a) found pyrite oxidation kinetics by an external carbon dioxide supply. These explanations are only tentative and need thorough checking for their validation. One deals here with complex ecosystems from which the real composition and the microbial interactions involved are not known in detail.

Because the coal samples in most cases are not sterilized, the contribution of the indigenous microflora present on the coal should not be neglected, but often is. Muyzer et al. (1987) recently used specific fluorescent antibodies raised against *T. ferrooxidans* to follow the growth of the organism on coal. They demonstrated that, although this organism can be isolated quite easily from coal samples during a microbial leaching process, its quantitative role in the leaching process is low. *T. ferrooxidans* is able to effectively colonize pyrite-bearing coal if it is inoculated as a pure culture into a sterile coal sample. However, if *T. ferrooxidans* is added to a coal sample containing a rich, indigenous population, it is not able to establish itself. Obviously, various other bacterial species are active in the pyrite solubilization. Given these results, the conclusion may be drawn that *T. ferrooxidans* is the organism that can be enriched and isolated from the samples most easily. Some authors mention *Leptospirillum ferrooxidans* (Helle and Onken, 1988; Norris, Barr, and Hinson, 1988) as an important member of the pyrite-oxidizing community. Further ecological studies are necessary to reveal the actual composition of the responsible microbial population.

In fact, little is known on the ecology of acidophilic, pyrite-oxidizing bacteria in coal. Harrison (1978) described a succession of different types of bacteria and fungi in an artificial coal heap. His studies gave a qualitative overview, but did not provide any quantitative data on the respective species. A similar study is that by Radway et al. (1987). It is expected that the application of techniques such as epifluorescence microscopy using acridine orange (Yeh et al., 1987) or combined immunofluorescence-DNA staining techniques (Muyzer et al., 1987) will be useful tools to unravel the composition of the complex microbial population.

Besides the mesophiles, also moderate thermophiles with an optimal temperature around 50°C (Murr and Mehta, 1982; Gökcay and Yurteri, 1983) and extreme thermophiles such as *Sulfolobus* species with an optimal temperature around 70°C have been studied or suggested for a microbial desulfurization process (Detz and Barvinchak, 1979; Kargi, 1982b, 1984, 1986; Murphy et al., 1985) The moderate thermophiles are similar to the representatives of the genus *Sulfobacillus* (Karavaiko et al., 1988) or the isolates described by Norris, Brierley, and Kelly (1980). In contrast to *T. ferrooxidans,* which is an obligate autotroph, the moderate and the extreme thermophiles are facultative autotrophs. Much less is known about the ecology of pyrite oxidation at elevated temperatures than at mesophilic temperatures. However, thermophiles have been isolated from coal-mining waste dumps in which spontaneous and biologically catalyzed pyrite oxidation gave rise to local hot spots (Karavaiko et al., 1988).

Suitability of coals. A survey of the literature shows that successful desulfurization has been achieved with a large variety of coal specimens. Most studies were performed with bituminous and subbituminous coals, although there are also reports on the desulfurization of lignites (Jilek and Beranova, 1982; Gökcay and Yurteri, 1983; Ruiz et al., 1984; Volsicky et al., 1985). Most reports deal only with one or a few different coal specimens. Detz and Barvinchak (1979), Hoffmann et al. (1981), and Bos et al. (1986) have reported on a larger variety of coal types.

Often details on the characteristics of the coal specimens involved are not provided. For this reason, it is in most cases impossible to indicate why some coal specimens give problems in the desulfurization tests. Some causes for poor leaching results mentioned in the literature are (1) the weathering of the coal prior to testing, (2) presence of toxic heavy metals, (3) a high pyrite concentration, (4) the presence in the coal of acid-neutralizing components such as carbonates, (5) inadequate leaching test procedures (suboptimal pH and a high ionic strength of the process water), and (6) presence of toxic organic carbon compounds.

1. *Weathering of the coal:*   Detz and Barvinchak (1979) reported on the suitability of 10 different coal types. The kinetics of the pyrite leaching varied by a factor of 20. From their report, it can be concluded that a correlation may exist between the degree of weathering and poor leaching results. Possibly the pyrite becomes inaccessible for microbial attack because the surface is covered with oxidation products. In the hands of the authors, the results obtained with weathered coal specimen also appeared to be less than those obtained with the original freshly milled specimens.

2. *Toxicity of heavy metals:*   Hoffmann et al. (1981) tested nine different coals. In only one specimen was the rate of pyrite removal less than expected. In the leachate from this sample they found higher concentrations of chromium and lead than in the leachates from other coals. However, this example is unique. Generally the pyrite-oxidizing microflora should be considered as tolerant of high heavy-metal concentrations.

3. *Pyrite concentration:*   With respect to the sulfur content of coal, high-pyritic and low-pyritic coals have been tested. Research in coal-producing countries has mostly been concentrated on the treatment of poor-quality coals with pyrite concentrations often as high as 3 to 6 percent (Detz and Barvinchak, 1979; Hoffmann et al., 1981). As mentioned before, in a coal-importing country such as the Netherlands most attention has been given to the possibilities of further reducing the pyritic content of low-sulfur coals (less than 2% pyrite; Kos et al., 1981, 1983; Bos et al., 1986, 1988). One should realize that the treatment of high-pyritic coals (including coal-mining wastes and coal fines) is accompanied by a number of problems related to extensive heat production in the process. Only a part of the pyrite oxidation energy will be fixed in the biomass. Most energy will be liberated as heat. The huge reactor systems necessary can be considered as more or less adiabatic. With high-sulfur coals, one can expect a rise in temperature to levels that exceed those which are optimal for mesophilic pyrite-oxidizing bacteria. This indicates that the use of moderate or extreme thermophiles would be profitable, because the economic feasibility of microbial coal desulfurization will be influenced in a nega-

tive way if cooling would be required. However, operation at elevated temperatures will also give rise to some technical difficulties. Higher temperatures will accelerate the kinetics of the precipitation of the undesirable jarosites, which are iron- and sulfate-containing compounds. The consequences of the treatment of high-sulfur coals have been discussed in more detail by Bos et al. (1988).

4. *Acid-neutralizing components in coal:*   Some types of coal contain high concentrations of carbonates. For effective pyrite removal by acidophiles, the pH should be controlled at below 2. In many shake-flasks experiments, the pH cannot be properly controlled. Although the pH can be adjusted from time to time, the rise in pH due to the dissolution of carbonates will result in undesirable precipitation of iron-containing compounds, such as jarosites, which will cover the pyrite surfaces (Kos et al., 1981). Moreover, in the period that the pH is too high, the kinetics of the bacterial pyrite oxidation will be slowed drastically. A remedy is to test coals in pH-controlled reactors.

5a. *Suboptimal pH:*   Although the optimal pH for ferrous iron oxidation by *T. ferrooxidans* is around 2.5, that for inorganic sulfur removal from coal is much lower. This is due to the increased risk of jarosite precipitation at higher pH values, which will diminish the inorganic sulfur reduction by the biological pyrite oxidation. Nevertheless, numerous reports claim an effective desulfurization at pH values above 2.5. For example, Rai (1985) demonstrated an effective pyrite removal (85 percent) in a 10 to 20 percent coal-water slurry within 7 to 14 days at a pH of 2.8. His results are misleading as far as inorganic sulfur removal is concerned because the coal was washed with 0.1 N hydrochloric acid to remove traces of adsorbed sulfate and iron before different sulfur compounds were analyzed. This procedure most likely succeeds in removing most of the jarosite precipitates from leached coal samples.

5b. *High ionic strength of the leachate:*   In leaching tests, the 9 K salts medium (Silverman and Lundgren, 1959) omitting the ferrous sulfate continues to be used (e.g., Kargi, 1984). This medium contains comparatively high concentrations of mineral salts. Because only the growth of autotrophs needs to be supported and low biomass concentrations are standard, the mineral concentrations far exceed the amount really essential. Dugan and Apel (1978) compared the 9 K salts medium with other mineral solutions. Among these, a solution containing 25 mM ammonium sulfate appeared to be the most effective. Hoffmann et al. (1981) tried to optimize the composition of the 9 K medium. They found that optimal desulfurization occurred at an N/P molar ratio of 90. This ratio varies greatly from the N/P molar ratio commonly found in microbial biomass. These data suggest that in the experiments of Hoffmann et al. an important part of the phosphate was supplied by the coal itself. Kos et al. (1983) demonstrated that for most coals the presence of minerals in the coal will meet the requirements of the organisms. Only the addition of a nitrogen source (0.5 mM), and sometimes a low concentration of phosphate (0.1 mM), is necessary if slurries with 20% coal (w/v) (containing 1% pyrite) in water are used. In the opinion of the authors, poor leaching results obtained in experiments using a water phase with a high ionic strength might be improved by reducing the concentration of mineral salts. During the last years, the interest in jarosites has increased. In the earlier reports on pyrite removal, the processes were followed by simple pyrite analysis before, during, and after the test. The results were sometimes controlled by following the concentration of ferric iron in the leachate. The possibility that other sulfur species could remain in a solid state in the coal were not considered seriously. To take account of this, it is essential to make complete iron and sulfur balances. Hoffmann et al. (1981) have tried to detect jarosites by using IR analysis. In their experiments they did not find these compounds. In contrast, Golomb and Beier (1977) report "Schlamm-bildung," which might refer to jarosites. These authors were using percolating systems with long residence times. Little is known about the kinetics of jarosite formation, but it can be extremely slow. Murr and Mehta (1982) also suggested the formation of jarosites, but they did not give any experimental data to support their statements. Although jarosite precipitation may not interfere with the microbial activity because it does not completely cover the mineral surfaces, its presence in coal after leaching is undesirable. At temperatures above 700°C, these compounds will decompose and contribute to the sulfur emission of the coal-fired installation.

6. *Toxic organic compounds:* Inhibition of the acidophiles by low concentrations of organic compounds, especially organic acids, has frequently been reported (Schnaitman and Lundgren, 1965; Tuttle and Dugan, 1976). However, suggestions that the presence of organic acids explains poor leaching results with coals are questionable. As was indicated earlier, the carbon skeleton of coal is almost insoluble, and weathering processes will not lead to a breakup of the macromolecules, but rather to an introduction of carboxylic groups.

Thus other explanations for the poor leaching results must be sought, in particular the accessibility of the pyrite. Most leaching tests are performed in shake flasks, with some authors using percolation columns (Rinder and Beier, 1983; Volsicky et al., 1985). Volsicky et al. discussed some characteristics which may influence the leaching results obtained with the percolating system. They stressed that the pyrite should be finely disseminated in the coal matrix. The dissolution of large pyrite crystals is too slow. They also noted that, for a good solubilization of the pyrite, the porosity of lignites must be high.

Characterization of the coals after leaching.   Most studies have focused on pyrite removal and provide data on the reduction of the different sulfur species present in the coal. Because the ferric iron produced from pyrite by microbial activity is a strong oxidizing agent, it will oxidize a variety of components present in the coal. Recently, Kawakami et al. (1988) indicated that not only pyrite and other sulfidic minerals are attacked by ferric iron, but also the carbon skeleton may be oxidized. A drastic change in surface characteristics may influence the suitability of the coal for different applications. The oxidation of the sulfidic minerals will lead to the dissolution of the metals bound to the sulfide. Kos et al. (1981), McCready and Zentilli (1985), and Bos et al. (1986) demonstrated that the microbial desulfurization process reduces the heavy-metal content in coal. This can be considered as an advantage because it prevents problems with heavy metals in the fly and bottom ash of coal-firing installations. On the other hand, one should realize that the presence of heavy metals in the wastewater of the process indicates that we are replacing one problem with another problem.

Because the process requires a water phase with a pH value less than 2, some of the mineral components of the coal will also go into solution. McCready and Zentilli (1985) were able to remove as much as 68 percent of the ash-forming components from their high-sulfur run-of-mine Brogan coal. Bos et al. (1987) reported that in some coals, the reduction may be of practical interest. As was discussed earlier, microbial desulfurization of coal may be used as a cleaning step in the preparation of coal-water mixtures. For the use of coal in the old-style oil burners, a low sulfur content, as well as a low ash content, is required.

For effective pyrite removal, fine milling of the coal is necessary. However, McCready and Zentilli (1985) stressed that the microbial desulfurization procedure leads to a further decrease of the particle size, probably smaller than required, due to the abrasive character of the solids suspension. This may be disadvantageous because the solids must eventually be dewatered to remove the pyrite oxidation products. Dewatering of solids becomes more problematic with decreasing particle size.

Reactor types for microbial desulfurization processes.   To judge the technical and economical feasibility of microbial coal desulfurization, data must not be limited simply to information on the suitability of a variety of coal types and the characteristics of the coals before and after treatment. Models must also be developed on how a microbial desulfurization process should be built. In principle, a choice can be made between (1) heap and percolation leaching or (2) leaching in a well-mixed suspension-tank reactor. Capital investments and energy requirements for heap and percolation leaching systems are low compared with those required for tank leaching. Although a critical comparison between heap and tank leaching in a given situation with a particular type of coal has not been published, it is most likely that circumstances will dictate which of the two will be most suitable. To evaluate the potentials of heap leaching and percolation systems, one must collect a set of data different from that for the evaluation of tank leaching.

Heap and percolation leaching.   For heap and percolation leaching, data must be available on coal and pyrite particle size distribution, porosity of the coal, the kinetics of jarosite precipitation, and the influence of these parameters on the overall performance of the leaching system. Because the use of a heap or a percolation system involves a variety of limitations (such as oxygen and carbon dioxide) and local differences in biomass concentrations, temperature, and pH, both the growth and pyrite oxidation kinetics of the microorganisms involved are not exploited fully. Therefore data on maximal growth and pyrite oxidation rates are less relevant. Apart from the slow diffusion of oxygen and carbon dioxide into these packed columns and heaps, the long residence time that will be necessary for decreasing pyrite concentration can also be caused by difficulties of the pyrite-oxidizing microflora in colonizing the coal. The microflora is almost completely immobilized by its strong tendency to adsorb to the solids. Because the percolation water contains low concentrations of bacteria, it is not an effective mediator to inoculate the coal particles buried deep in the heap or percolation column.

Beier (1987) presented a design for heap leaching of coal which, he claims, can achieve a pyrite removal of over 90 percent with an average cost of DM 54 per ton of coal. Others have also considered the use of heaps and percolating columns (Lazaroff et al., 1979; Pocas, 1984; McCready and Zentilli, 1985; Beyer, 1987; Tillet and Myerson, 1987). Their conclusions are less optimistic than those of Beier. McCready and Zentilli (1985) reported a 65% pyrite removal within 110 days. Beyer (1987) stated that in a percolating system, the pyrite removal that can be achieved is 10 to 15 percent less than that in suspension reactors, and pointed out that the residence time will be at least 70 days. Tillet and Myerson (1987) reported an expected residence time of years. Because of possible jarosite precipitation, long residence times would be unfavorable. Although the kinetics of jarosite formation at pH 2 are slow, one can expect that jarosite formation will interfere with leaching results in heaps and percolating columns. It should also be noted that environmental control of a coal heap leaching installation will be problematic.

Tank leaching.   The objective of tank leaching is to create process conditions which are optimal for microbial pyrite oxidation. For the design of a well-mixed suspension reactor, kinetic data on microbial growth and pyrite oxidation are necessary. Moreover, attention must be paid to mass transfer of gases and to mixing and sedimentation phenomena to create optimal conditions for microbial activity. For this reason we turn our attention to published data on the kinetics of microbial growth and pyrite oxidation in leaching systems. Following this the gas mass transfer and the mixing and sedimentation phenomena in suspension reactors will be discussed. Laboratory-scale, mechanically mixed, airlift and pachuca tank reactors have all been described in the literature. In this respect, the 2-inch pipeline loop of Rai (1985) should be mentioned. For a leaching process on an industrial scale, mixing with a mechanical device is not advisable because highly corrosive suspensions at a high pulp density and a low pH are involved. An air-stirred reactor offers the best prospects. For that reason, Detz and Barvinchak (1979) and Bos et al. (1986) suggested the use of mechanically mixed tank reactors in series and cascades of air-agitated pachuca tanks, respectively, for large-scale treatments. Others (Kargi and Cervoni, 1983; Beyer et al., 1986) have proposed the use of airlift (recycle) reactors.

Other reactor types.   Other reactor types which have been suggested are lagoons (Dina et al., 1982; Kargi, 1986; Sproull et al., 1986), or even the pipelines used for the transport of coal slurries (Rai, 1985). Vaseen (1985) suggested the use of a horizontal rotating-tube reactor. These options suffer more or less from the same disadvantages as heaps and percolation columns. The conditions offered to the active microbes are far from optimal, although the suggestion has been made that the slurry pipelines represent a plug-flow reactor type which is best for exploiting the first-order kinetics of pyrite oxidation (Andrews and Maczuga, 1984). However, Hoffmann et al. (1981) pointed out that a major drawback to a pressure conduit slurry pipeline reactor will be the need for numerous reaeration stations, which can be calculated from the overall stoichiometry of reaction equation (15.1).

Kinetics of bacterial growth.   The primary requirement for effective pyrite solubilization from coal is that sufficient biomass must be present. Knowledge of the kinetics of bacterial growth during the leaching of coal is thus important. To follow the growth of acidophilic, pyrite-oxidizing bacteria in suspensions of pulverized coal is problematic for several reasons. Because the pyrite-oxidizing microflora is autotrophic in nature, the cell yields on the available energy substrates are low, and one deals with minute bacterial biomass concentrations in the order of tenths of milligrams per liter of the coal suspension. Moreover, the cells of most pyrite-oxidizing bacteria show a strong tendency to attach to solid materials (Dispirito et al., 1983; Myerson and Kline, 1983; Wakao et al., 1984; Bagdigian and Myerson, 1986). This is true not only for mesophilic pyrite-oxidizing bacteria, but also for the extremely thermophilic *Sulfolobus* species (Murr and Berry, 1976; Chen and Skidmore, 1988). These authors demonstrated that pyrite oxidizing bacteria not only have a tendency to attach to solid particles, but that they have a preference for the surface of pyrite particles, and even for the dislocation sites of the pyrite crystals (Andrews, 1988).

Myerson and Kline (1983) tried to determine biomass concentrations in their leaching system by using a colorimetric protein determination. They found a cell yield of 0.4 g protein per gram of iron oxidized. Assuming a protein content of 70 percent in bacterial biomass, the yield expressed as grams of biomass per gram of iron will be almost 0.6 g biomass per gram of iron. This cell yield is highly unlikely. It is in the order of magnitude of cell yields of heterotrophic bacteria growing on organic substrates. *T. ferrooxidans* grown under optimal conditions in a chemostat shows a yield of 0.004 g of biomass per gram of ferrous iron (Hazeu et al., 1987). Huber et al. (1984) found that with their pyrite oxidizing, mixed population produced a yield between 0.044 and 0.066 g of biomass per gram of pyrite. On the other hand, yields which can be calculated from the data obtained by Yurteri and Gökcay (1987) by using a protein determination are probably far too low.

Other studies report on the use of most-probable-number (MPN) methods to enumerate the number of bacterial cells and to follow the growth kinetics (Beyer et al., 1986; Yurteri and Gökcay, 1987). The reliability of the method is doubtful because almost all of the cells are adsorbed to the surface of the particles. Assuming a particle size of 50 μm and a density of coal of 2.5 g/cm^3, 1 g of coal will contain 3000 particles. If a dilution series is started with a sample containing 1 g of coal, at a dilution of 10,000 times, the coal particles will be diluted out completely. Although numerous cells may be adsorbed to one coal particle, they will also be diluted out in one step. Thus, the results of the MPN test will reflect the particle size distribution of the coal sample more than it will give an indication of the number of cells. Another complication in the use of MPN methods might be the composition of the culture medium. If the 9 K medium of Silverman and Lundgren (1959) is used, one can expect that *T. ferrooxidans* will predominate because its viability in such a medium is comparatively high. Other microbes involved in pyrite oxidation might have growth problems with the unnecessarily high ferrous iron concentration present in the 9 K medium and they will be missed. Nevertheless, in spite of the obvious drawbacks of the MPN, the method sometimes yields useful information. Assuming a constant ratio between adsorbed and free cells in leaching systems, the application of MPN methods to the coal-particle–free-water phase might give information on the growth kinetics.

Another possible means of following the kinetics of bacteria growth during the leaching process is by measuring the carbon dioxide uptake rate. Because the bacteria are autotrophic, carbon dioxide uptake is directly related to the biomass production. Huber et al. (1983) were able to measure the growth rate of a mesophilic pyrite-oxidizing microflora on coal, which appeared to be around 0.05 per hour at pH 1.8. See Fig. 15.1 for their experimental setup and Fig. 15.2 for the results of a representative test. With this method Bos et al. (1986) were able to measure the growth rate of their pyrite-oxidizing mixed culture in a coal suspension as a function of the pH. Growth stops at pH 1.6.

Kinetics of pyrite oxidation.   The kinetics of pyrite oxidation can be followed in different ways. The most common is to measure the concen-

BATCH   REACTOR   SYSTEM

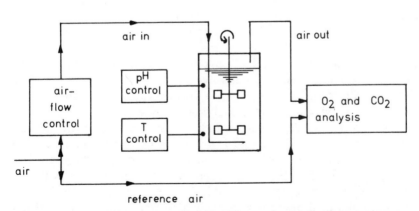

Figure 15.1   Schematic diagram of the 10-L batch reactor system used by Huber et al. (1983) to measure growth and pyrite oxidation rates during microbial desulfurization of coal. (*From Huber et al., Modelling design and scale up of a reactor for microbial desulphurization of coal, in G. Rossi and A. E. Torma (eds.), Progress in Biohydrometallurgy, Associazione Mineraria Sarda, Iglesias, 1983, pp. 279–289.*)

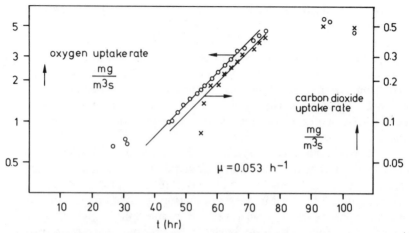

Figure 15.2 Logarithmic plot of oxygen and carbon dioxide uptake rates as measured in the 10-L batch reactor (see Fig. 15.1) during the microbial desulfurization of coal. (*From Huber et al., Modelling design and scale up of a reactor for microbial desulphurization of coal, in G. Rossi and A. E. Torma (eds.), Progress in Biohydrometallurgy, Associazione Mineraria Sarda, Iglesias, 1983, pp. 279–289.*)

tration of iron in the leachate as a function of time. The formation of sulfate can also be measured. Difficulties in representative sampling from coal suspensions during the leaching tests make pyrite analysis a less attractive tool. An elegant way to follow pyrite oxidation kinetics is to measure oxygen uptake rates (Huber et al., 1983) in the leaching system itself or by using a biological oxygen monitor.

It is generally assumed that in microbial leaching of pulverized coal by mesophilic bacteria, the kinetics of pyrite oxidation are first order with respect to the amount of pyrite present (Detz and Barvinchak, 1979). One would expect pyrite oxidation to follow a first-order reaction with respect to the available pyrite surface area because a heterogeneous oxidation reaction is involved which occurs at the interface of solid particles and the water phase. However, most experimental results obtained with pulverized coal do not allow a discrimination between first-order kinetics relative to pyrite surface and pyrite concentration. The most straightforward way to present the data on leaching rates is to list the first-order rate constants. However, these are frequently found in the literature as percentages of pyrite removal as a function of the residence time. Table 15.1 summarizes published kinetic data obtained under conditions which were found to be optimal for the system under study. These data are based on leaching experiments in which the coal was milled to such an extent that an almost complete removal of pyrite could be obtained (> 90 percent). In practice, most studies were done with pulverized coals to meet the require-

TABLE 15.1   Kinetic Data on the Oxidation of Pyrite in Coal

Authors	Published data on pyrite removal	Calculated k^* (1/h)
Chandra et al. (1979)	90% in 200 h	11.5×10^{-3}
Detz and Barvinchak (1979)	90% in 15 d	6.4×10^{-3}
Golomb and Beier (1977)	27.3% in 48 h	6.7×10^{-3}
Groudev and Genchev (1979)	70% in 28 d	1.8×10^{-3}
Hoffmann et al. (1981)	90% in 8 d	12.0×10^{-3}
Pocas (1984)	90% in 10 d	9.6×10^{-3}
Rossi and Salis (1977)	90% in 15 weeks	0.9×10^{-3}

$^*k$ = first-order rate constant.

ments for installations using powdered coal (< 100 μm). In those cases where the k value was not given, it was calculated from the data based on residence time and percentage pyrite removal. There is a significant variation between the different k values, which can be explained by the variations in the characteristics of the cultures used, the differences in the duration of the lag phase in the leaching test, by a limiting factor (biomass, oxygen, carbon dioxide), by inhibition of the oxidation (e.g., jarosite formation), or by the lack of a proper pH control during the leaching test. It can be concluded that for mesophilic microbial systems the maximal k is around 10^{-2} per hour, which means a pyrite removal of 90 percent within a residence time of 8 to 10 days.

The kinetics of pyrite removal from coal by mesophilic bacteria have also been followed as a function of different parameters, such as the pulp density of the suspension (e.g., Hoffmann et al., 1981; Rai, 1985; Beyer et al., 1986), the particle size distribution of the coal particles (e.g., Hoffmann et al., 1981; Vleeskens et al., 1985), and pH (e.g., Bos et al., 1986).

Most published data (e.g., Hoffmann et al., 1981; Beyer et al., 1986) indicate that the first-order kinetics remain constant at pulp densities up to about 20 to 25 percent pulverized coal in water (w/v). However, most experiments in this respect are performed in small-scale vessels, sometimes only in shake flasks. Inhomogeneous mixing of the coal particles in the test systems might have influenced the results. It is necessary to study the influence of pulp densities on the pyrite-removal kinetics under more rigidly controlled conditions. Because the investments for the reactor systems are the most important contributions to the overall costs of the process, the possibility of using higher pulp densities is most interesting. Andrews and Maczuga (1984) have indicated a pulp density as high as 50 percent in their projected desulfurization plant! In this respect, the expected difficulties to using extreme thermophiles at high pulp densities should be mentioned (Lawrence and Marchant, 1988). The extreme thermophiles are probably more sensitive to shear forces than are the mesophiles.

Studies by Bos et al. (1986) showed that their mesophilic pyrite-oxidizing bacteria stopped growing at pH 1.6. At lower pH values, pyrite oxidation continued at about the same rate to pH 1.4. The course of pH change in the different microbial cultures used in microbial desulfurization of coal depends on the cultures, samples, and conditions used. For example, moderate and extreme thermophiles which are active even at pH 1 have been described (Norris and Barr, 1988).

The data on moderate thermophiles suggest a pyrite-oxidizing rate comparable to that of mesophiles (Yurteri and Gökcay, 1988). Studies with extreme thermophiles suggest that their rates are considerably faster. Residence times of 5 days (Detz and Barvinchak, 1979; Kargi, 1982b, 1986) and even 3 days (Harsh et al., 1987) have been reported. However, at higher temperatures the kinetics of pyrite oxidation will deviate from simple first-order because of the increasing importance of nonbiologically catalyzed oxidation reactions which occur along with the biological reactions. As will be discussed below, this is particularly true for the spontaneous oxidation of sulfidic minerals by ferric ions. Although the sulfur-oxidizing capacity (Arkesteyn, 1980; Hazeu et al., 1987) of the acidophilic microflora must play an important role in the solubilization of pyrite, the ferric iron present in the leachate can also act as an oxidizer of the pyrite. In the literature, two possible reactions for this have been given (Stokes, 1901; Garrels and Thompson, 1960; Silverman, 1967):

$$FeS_2 + 2Fe^{3+} \rightarrow 3Fe^{2+} + S \qquad (15.3)$$

$$FeS_2 + 14Fe^{3+} + 8H_2O \rightarrow 15Fe^{2+} + 2SO_4^{2-} + 16H^+ \qquad (15.4)$$

In Eq. (15.3) the sulfur moiety of pyrite is oxidized to elemental sulfur, whereas in Eq. (15.4) it is oxidized completely to sulfate. Studies from Boogerd et al. (1988) have demonstrated that Eq. (15.3) is of no importance. Hardly any elemental sulfur can be detected as a result of the oxidation of pyrite by ferric iron. These findings seem to conflict with those of Beyer et al. (1987), who reported the formation of elemental sulfur during leaching tests with one of their four coal samples. They suggested that its formation might have been due to Eq. (15.3). Another possibility might be an incomplete oxidation catalyzed by the pyrite-oxidizing bacteria. It has been demonstrated that *T. ferrooxidans* can excrete elemental sulfur during the oxidation of reduced sulfur compounds under special conditions (Hazeu et al., 1988). The principle of incomplete bacterial oxidation of chalcopyrite in the presence of low concentrations of silver, as exploited in the process of Bruynesteyn et al. (1983), is probably related to this phenomenon.

Boogerd et al. (1988) showed that Eq. (15.4) increases in importance with increasing temperature. At temperatures of around 30°C the contribution to the overall kinetics is negligible. In accord with Arrhenius's law, the rates rise drastically with increasing temperature. At 45°C, pyrite oxidation by ferric iron reaches a level which has a significant influence on the overall kinetics (8 percent). At 70°C the influence is so important (43 percent) that simple first-order kinetics are no longer applicable. At these extreme temperatures, a significant concentration of ferrous iron is characteristically present in the leachate (Harsh et al., 1987), suggesting that the microbial ferrous iron oxidation is the bottleneck in the system.

Reactor design.   For optimal bacterial performance the following requirements should be considered in the reactor design: (1) biomass limitation must be prevented; (2) optimal conditions for microbial pyrite oxidation must exist everywhere in the reactor.

Although the first requirement is quite logical, it has been neglected in most designs, probably because sound data on growth kinetics in leaching processes are scarce. If one considers microbial desulfurization of coal in suspension reactors, continuous operation will be most suitable. Because of the strong adsorption of the pyrite-oxidizing bacteria to the solid particles in the suspension, inoculation of fresh incoming coal can only be achieved by extensive contact between the fresh coal particles and those overgrown with microbes. This means that a well-mixed tank reactor is needed.

The kinetics of pyrite oxidation indicate the suitability of another type of reactor. Andrews and Maczuga (1984), realizing that bacterial pyrite oxidation follows first-order kinetics, have advocated the use of a plug-flow reactor. One could think of a ditchlike reactor which is aerated and mixed by means of an air jet on the bottom. However, keeping the large scale in mind, one should also be aware of the risks of back-mixing that could disturb the plug-flow character of the reactor. An alternative might be a series of pachuca tank reactors which approximate a plug flow, as has been suggested by Detz and Barvinchak (1979), Bos et al. (1986), and Beyer et al. (1988).

One should note that in reactor design, two conflicting requirements have to be met. One is the prevention of bacterial biomass limitation, which requires a mixed-tank reactor, and the other is the kinetics of pyrite oxidation, which demands plug flow. This discrepancy can be solved by the suggestion of Huber et al. (1983), which involves a reactor configuration consisting of a mixed-tank reactor, which might again be a suspension reactor of the pachuca type, followed by a plug-flow reactor or a series of other pachuca tanks (see Fig. 15.3). The residence time in the first pachuca tank must be at least the reciprocal value of the maximal specific growth rate of the pyrite-oxidizing microflora (approximately 0.05 per hour). Thus a residence time of at least 20 h is required. Beyer et al. (1988) reported on experimental studies using cascades of pachuca tank reactors with a volume of 20 L each, and have demonstrated that this small-scale installation performs according to expectations. Although Pinches et al. (1988) have not studied pyrite removal from coal, their report on the performance of bacterial leaching reactors for the preoxidation of refractory gold-bearing sulfide concentrates demonstrates that the process works only if the residence time in the first part of their cascade of reactors surpasses 18 h. This is in agreement with the data of Huber et al. (1983).

To complete the reactor design, one needs, in addition to kinetic data on pyrite oxidation and bacterial growth, data on physical phenomena in suspension reactors. Bos et al. (1986, 1988) used the strategy which was outlined in the review by Sweere et al. (1987). First, a regime analysis based on characteristic time constants was used to identify the critical physical phenomena (gas mass transfer, mixing, and sedimentation in suspension reactors) which should be studied in detail. In the treatment of low-sulfur coals by mesophilic bacteria, it appeared that the mass transfer of oxygen and carbon dioxide from the gaseous phase into the liquid phase determined the performance of the reactor. In other words, the airstream required for supplying the

REACTOR CONFIGURATION

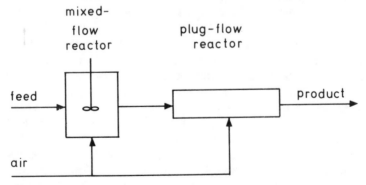

Figure 15.3 Reactor configuration for the desulfurization of coal proposed by Huber et al. (1983). (*From Huber et al., Modelling design and scale up of a reactor for microbial desulphurization of coal, in G. Rossi and A. E. Torma (eds.), Progress in Biohydrometallurgy, Associazione Mineraria Sarda, Iglesias, 1983, pp. 279–289.*)

organisms with enough oxygen and carbon dioxide is more than enough to keep the suspension well mixed and to prevent sedimentation of coal particles (Bos et al., 1986, 1988).

Microbially assisted physical desulfurization

Another approach for removing inorganic sulfur compounds from coal uses an initial attack of the pyrite by acidophilic bacteria to affect a change in the surface characteristics of the mineral. It is well known that the hydrophobicity of coal particles equals that of pyrite particles. This prevents the effective physical separation of pyrite from coal particles by separation techniques based on differences in hydrophobicity, such as oil agglomeration or (froth) flotation. Efforts to improve the efficiency of oil agglomeration and froth flotation by using additives to selectively change the surface characteristics of either the coal or the pyrite have not been very successful. The first suggestion that a change in the surface properties of pyrite by an initial attack by *T. ferrooxidans* could be combined with oil agglomeration came from Capes et al. (1973). A prerequisite for a high percentage of pyrite removal and low coal losses is extremely fine milling. In contrast to the complete oxidation process, the pyrite crystals must not only be accessible for microbial attack, they must be completely set free from the coal matrix. Butler et al. (1986) indicated that 90% pyrite removal can be obtained only if the particle size of the coal is less than 5 μm. Doddema (1983), studying the technical feasibility of microbially assisted oil agglomeration for low-sulfur coals, mentioned another important bottleneck. To achieve a fast initial attack, the coal slurry must be inoculated heavily with *T. ferrooxidans* cells. Because of the strong adsorption of the cells to the solid particles, it is not realistic to expect that this inoculation can be done by recirculating part of the process water. To keep the process running, *T. ferrooxidans* should be cultivated continuously on the pyrite concentrate produced in the agglomeration step. This cultivation will require additional reactor systems and will make the process economically less favorable.

Pooley and Atkins (1983) and Attia and Elzeky (1985) advocated the combination of an initial attack by acidophilic pyrite oxidizers and froth flotation. Dogan et al. (1986) compared the complete oxidation process with microbially assisted flotation using three different coals. Here, of course, the same set of disadvantages as mentioned above is also valid. In this case Townsley and Atkins (1986) were able to show that *T. ferrooxidans* could be replaced by heterotrophic organisms such as the yeasts *Saccharomyces cerevisiae* and *Candida utilis* and the bacteria *Escherichia coli* and *Pseudomonas maltophila* for pyrite removal. These microorganisms can be cultivated to higher densities and at a faster rate than *T. ferrooxidans*.

Based on these data, one can question whether the observed effects by acidophilic pyrite oxidizers can be attributed to the oxidative capacity of the cells or merely to the adsorption of biological material. The observation that bacterial conditioning can be achieved in 10 min (Elzeky and Attia, 1987) and even 2 min (Townsley et al., 1987) to obtain 90% pyrite removal is in agreement with the latter interpretation. Butler et al. (1986) drew similar conclusions from the observation that bacterial conditioning at pH 5 was optimal. At this pH value the oxidation rate of the pyrite oxidation by bacteria is extremely low.

Removal of organic sulfur from coal

The removal of inorganic sulfur from coal is considered to be only a partial solution to reduce the sulfur dioxide emission from coal-fired installations. This provides a strong incentive to study possibilities for also biologically removing organic sulfur compounds from coal. Although some authors have claimed to be able to remove organic sulfur from coal (Chandra et al., 1979; Hedrick, 1982; Isbister and Kobylinski, 1985; Kargi and Robinson, 1986; Kilbane, 1988), the authors of this review have not been convinced. By critically reading the relevant reports, one must conclude that a reliable analysis of the different sulfur compounds is missing. Not only are data on the organic sulfur analysis according to ASTM standards required, but direct analysis of elemental sulfur, inorganic sulfur other than pyrite, polysulfides, sulfate, and real organically bound sulfur are needed. As was stated previously, organic sulfur analysis by ASTM methods is error-prone. A calculation of the errors made in the routine analysis of pyrite, sulfate, and total sulfur will demonstrate that the reductions in organic sulfur reported by some authors are not significant (e.g., Chandra et al., 1979, 20 percent; Hedrick, 1982, 26 percent; Isbister and Kobylinski, 1985, depending on the type of coal, 19 to 57 percent). An example of such an error analysis is given in Table 15.2. Moreover, most reports on organic sulfur removal lack overall sulfur and carbon balances. Nevertheless, plant layouts and cost analyses for the com-

TABLE 15.2   Analysis of ROM Coal before and after Treatment with the DBT Degrading Organism CB1*

Sample	% Total S	% SO_4—S	% FeS_2—S	% Organic—S
ROM coal untreated	2.29 ± 0.05	0.2 ± 0.1	1.3 ± 0.05	0.79 ± 0.20
CB1 treated ROM coal	1.83 ± 0.05	0.2 ± 0.1	1.3 ± 0.05	0.53 ± 0.20

*Includes an error analysis by the authors of this chapter.
SOURCE: From Isbister and Kobylinski (1985).

bined removal of inorganic and organic sulfur from coal have been published (Kargi, 1986). Kargi suggested a two-step process in which pyrite is removed within 4 to 6 days by extreme thermophiles at 70°C and pH 2.5 in a first step. The second step is for organic sulfur removal and requires a residence time of 28 days. He assumed that only 40 percent of the organic sulfur can be eliminated. The operating costs were estimated to be around \$17 per ton of coal. The microbial inoculum used in the second step of the process should be pregrown on a dibenzothiophene-containing medium. Isbister and Kobylinski (1985) also proposed a two-step process scheme.

Some results have been published which support the idea that microorganisms might be able to oxidize organic sulfur compounds in coal. Radway et al. (1987), studying the microflora responsible for sulfur oxidation of a low-sulfur coal in a percolating system, reported the presence of heterotrophic, organic sulfur oxidizers in their leachates. They were using low-molecular-weight substrates to detect these organisms.

Numerous references deal with isolates which are able to attack organic sulfur-containing substrates such as dibenzothiophene (DBT; Kodama et al., 1973; Laborde and Gibson, 1977; Kargi and Robinson, 1984; Isbister and Kobylinski, 1985; Monticello et al., 1985; Stoner et al., 1988). Most of these studies demonstrate the formation of sulfate and breakdown products from the aromatic ring structures. Van Afferden et al. (1987) reported the existence of bacteria that might be able to use DBT as their sulfur source. The carbon skeleton of DBT remained untouched. Benzoate served as the carbon source. In addition to DBT, other model organic sulfur compounds have been used (Kargi, 1987).

Although organisms involved in DBT degradation are neutrophiles, Kargi and Robinson (1984) demonstrated the ability of the acidophilic thermophile *Sulfolobus acidocaldarius* to degrade DBT.

With our current knowledge of the breakdown of organic sulfur compounds, the use of recombinant DNA techniques to incorporate the genes for organic sulfur oxidation into acidophiles seems premature (Rhee and Wan, 1987). Although probably of scientific interest, these studies cannot yet be related to the development of a single-step process for the removal of both organic and inorganic sulfur compounds from coal, because it is not yet known which enzymatic systems should be selected.

Technical and economic feasibility of microbial desulfurization of coal

The extensive knowledge that has now accumulated leaves little doubt that a process for the removal of inorganic sulfur compounds of coal is technically feasible. However, in spite of optimistic reports on the microbial attack on organic sulfur in coal, the scientific basis is largely lacking, and for the time being its technical application is therefore doubtful. An urgent need exists for a critical evaluation of this research area. Special attention should be given to a reliable organic sulfur analysis in coal.

The question of whether microbial coal desulfurization is economically feasible has not yet been answered. However, complete oxidation process cost analyses have been published. A few examples will be given. Details of cost analyses on the microbially assisted processes are missing.

Detz and Barvinchak (1979) presented a process scheme for inorganic sulfur removal with a series of mechanically mixed reactors and estimated the costs of a 3,000,000-ton/year installation to be approximately $10 per ton. In their calculations they assumed a pyritic sulfur content in the coal of 2 percent.

Kargi (1982b, 1986) produced several schemes for a two-step process using *Sulfolobus acidocaldarius* with initial inorganic and then organic sulfur removal. He proposed shallow ponds with a depth of only 0.2 m as reactors. Preliminary economic analysis indicated costs of $17 per ton of coal in an installation fed with 1,000,000 ton/year. He claimed an almost complete removal of inorganic sulfur and a 40 percent reduction in organic sulfur.

Most cost estimates are based on data collected from simple leaching experiments in shake flasks that yielded kinetic data on pyrite oxidation as a function of various parameters, such as particle size, pH, temperature, composition of the process water, the inoculum size, and pulp density, as was described in the section on the kinetics of bacterial pyrite oxidation. Kinetics of bacterial growth was neglected in most feasibility studies. The same was true for the chemical engineering aspects of the process design. In the process scheme of Bos et al. (1986) the kinetics of bacterial growth and chemical engineering aspects were, however, taken into account. They proposed the use of a series of air-agitated suspension reactors (pachuca type). Their cost estimate was about $10 per ton of coal for a 1,000,000-ton/year installation.

Although these figures seem to be quite promising compared with, for example, the cost of flue-gas desulfurization ($20 to $40 per ton; Murphy et al., 1985), all of these estimates are based on small-scale experimentation. Studies on a larger scale are urgently needed to be able to evaluate the technical and economic feasibility of the complete oxidation process. As far as is known, a pilot-scale installation has not yet been built and tested.

Microbial Coal Liquefaction and Gasification

In recent years it was discovered that coals are not as inert microbiologically as was previously assumed. Fakoussa (1981) demonstrated that, although the microbial degradation rate is extremely slow, even bituminous coals can serve as a carbon substrate for some fungal and bacterial strains; they suggested that cell wall–bound, extracellular enzymes are responsible. The degradation mechanisms might be similar to that of wood- and cellulose-degrading microorganisms. Alternatively, coal degradation may be due to the release by microorganisms of extracellular enzymes or other (surface-active) cellular components into their environment. An extremely fine milling of coal is necessary for degradation.

Cohen and Gabriele (1982) demonstrated a much faster degradation of a heavily weathered lignite, leonardite, by some fungal species: *Polysporus versicolor* and *Portia monticola*. The faster rate may be explained by the greater resemblance of the structure of carbon compounds of lignites to the more geologically recent biological materials such as lignin. The macromolecules may be depolymerized by fungi possessing lignin-degrading enzymes, especially if the lignites such as leonardite are in a highly oxidized state. In contrast to nonweathered lignites which contain 20% oxygen, leonardite contains nearly 30% oxygen. Ward (1985) described some fungal isolates that degraded a weathered lignite. Scott et al. (1986) demonstrated the degradation of leonardite by allowing fungi to form a mat on a solid culture medium in a petri dish and then distributing pulverized lignite over the surface. The plates were incubated for several weeks. Some strains were able to produce liquid substances from the coal particles. Dahlberg (1986) reported similar results with leonardite.

To improve the susceptibility to attack of lignites which cannot be liquefied easily, pretreatment with oxidizing agents such as ozone, hydrogen peroxide, and nitric acid has been tested. A 48-h treatment with 8 N nitric acid of a nonsusceptible lignite transformed it into a material that could be liquefied by fungi almost completely. Scott et al. (1986) found that the product was water-soluble and that it had a relatively high molecular weight. The findings of Cohen and Gabriele (1982), Ward (1985), and Scott et al. (1986) have initiated other studies in the United States (Pyne et al., 1987; Quigley et al., 1987; Scott and Faison, 1987; Wyza et al., 1987). Wilson et al. (1986) and Pyne et al. (1987) were able to produce a cell-free extract from the fungus *Coriolus versicolor* which solubilized pretreated lignites. This activity was attributed by them to lignin-degrading enzymes (laccase). Efforts are being undertaken to characterize the products of the microbial liquefaction (Quigley et al., 1987).

Although we can speculate on the possible application and bioreactor configurations (Scott et al., 1986) of microbial coal liquefaction (bioreactor configurations have even been suggested by Scott et al., 1986), we cannot judge the technical and economic feasibility of the process at this moment. A constraint in the realization might be the expensive and unattractive pretreatment with nitric acid, which seems to be essential to making the lignites susceptible to microbial attack. In view of the low kinetic rates involved in the degradation of bituminous coal, it has been speculated that the liquefying organisms could be used underground in coal seams. The liquefied product could be recovered after proper incubation (Fakoussa, 1987).

The idea of microbial underground gasification is also highly speculative. The production of methane by anaerobic methanogens from peat was reported as early as in 1926 (Melin et al., 1926). It was suggested that the same mechanisms could produce methane from lignites, but sound experimental evidence to support this is not available. Nevertheless the possibility has already been integrated into futuristic plans (Leushner et al., 1985).

Concluding Remarks

A review of the recent literature on microbial treatment of coal easily gives the impression that biotechnology is substantially penetrating coal science and technology. Some authors suggest that microbial desulfurization of coal may be the final solution to the problem of acid rain and that microorganisms can readily convert coal into clean gaseous or liquid fuels. However, it is more realistic to consider microbial desulfurization as a possible means to reduce the sulfur content in coal, and this technique may be competitive in special situations with other techniques to reduce sulfur emissions. The possibilities of organic sulfur removal seem to be especially overestimated. It is too early to judge the technical and economic feasibility of coal liquefaction and coal gasification. One constraint in the realization of coal liquefaction might be the required pretreatment with oxidative agents such as concentrated nitric acid. Microbial coal gasification could be hampered by the extremely slow kinetics.

Acknowledgments

For critical reading and correction of the English text, the help of Dr. Lesley Robertson is gratefully acknowledged.

References

Andrews, G. F., The selective adsorption of Thiobacilli to dislocation sites on pyrite surfaces, *Biotechnol. Bioeng.* **31**:378–381 (1988).
Andrews, G. F., and Maczuga, J., Bacterial removal of pyrite from coal, *Fuel* **63**:297–302 (1984).

Arkesteyn, G. J. M. W., Contribution of microorganisms to the oxidation of pyrite, Thesis, Wageningen, 1980.

Armson, R., Water makes coal a brighter fuel, New Scientist, August 28, pp. 44–47 (1986).

Attar, A., and Dupuis, F., Data on the distribution of organic sulfur functional groups in coal, Am. Chem. Soc. Div. Fuel Chem. 24:166–177 (1979).

Attia, Y. A., and Elzeky, M. A., Biosurface modification in the separation of pyrite from coal by froth flotation, in Y. A. Attia (ed.), Coal Science and Technology 9. Processing and Utilization of High Sulphur Coals, Elsevier, Amsterdam, 1985, pp. 673–682.

Bagdigian, R. M., and Myerson, A. S., The adsorption of Thiobacillus ferrooxidans on coal surfaces, Biotechnol. Bioeng. 28:467–479 (1986).

Beddow, J. K., Dry separation techniques, Chem. Eng. 88:70–84 (1981).

Beier, E., Pyrite decomposition and structural alternations of hard coal due to microbe-assisted pyrite removal, in Proc. Biological Treatment of Coals Workshop, Vienna, Va., 1987, pp. 389–424.

Berkowitz, V., An Introduction to Coal Technology, Academic Press, New York, 1979.

Beyer, M., Microbial removal of pyrite from coal using a percolation bioreactor, Biotechnol. Lett. 9:19–24 (1987).

Beyer, M., Ebner, H. G., Assenmacher, H., and Frigge, J., Elemental sulfur in microbiologically desulfurized coals, Fuel 66:551–555 (1987).

Beyer, M., Ebner, H. G., and Klein, J., Influence of pulp density and bioreactor design on microbial desulphurization of coal, Appl. Microbiol. Biotechnol. 24:342–346 (1986).

Beyer, M., Höne, H.-J., and Klein, J., A multi-stage bioreactor for continuous desulphurization of coal by Thiobacillus ferrooxidans, in P. R. Norris and D. P. Kelly (eds.), Biohydrometallurgy, Proc. Intern. Symp. Warwick, Science and Technology Letters, Kew, Surrey, U.K., 1988, pp. 514–516.

Boogerd, F. C., Bos, P., and Huber, T. F., Microbial desulfurization of coal, Periodic report October 1, 1987–April 1, 1988, EEG contract EN3F-0036-NL, Delft University of Technology, 1988.

Bortz, S., and Flament, P., Recent IFRF fundamental and pilot scale studies on the direct sorbent injection process, Symp. Clean Combustion of Coal, Noordwijkerhout, The Netherlands, January 1985.

Bos, P., Huber, T. F., Kos, C. H., Ras, C., and Kuenen, J. G., A Dutch feasibility study on microbial coal desulfurization, in R. W. Lawrence, R. M. N. Brannion, and H. G. Ebner (eds.), Fundamental and Applied Biohydrometallurgy, Proc. 6th Intern. Symp. on Biohydrometallurgy, Elsevier, Amsterdam, 1986, pp. 129–150.

Bos, P., Huber, T. F., Luyben, K. Ch. A. M., and Kuenen, J. G., Feasibility of a Dutch process for microbial desulphurization of coal, Resour. Conserv. Recyc. 1:279–291 (1988).

Bos, P., Vleeskens, J. M., Kos, C. H., and Hamburg, G., Microbial desulphurization of coal: Characterization of the products, in J. A. Moulijn, K. A. Nater, and H. A. G. Chermin (eds.), Intern. Conf. on Coal Science, Elsevier, Amsterdam, 1987, pp. 431–434.

Boudou, J. P., Boulèque, J., Maléchaux, L., Nip, M., de Leeuw, J. W., and Boon, J. J., Identification of some sulphur species in a high organic sulphur coal, Fuel 66:1558–1569 (1987).

Bruynesteyn, A., Lawrence, R. W., Viszolyi, A., and Hackl, R., An elemental sulphur producing biohydrometallurgical process for treating sulphide concentrates, in G. Rossi and A. E. Torma (eds.), Progress in Biohydrometallurgy, Associazione Mineraria Sarda, Iglesias, Italy, 1983, pp. 151–168.

Butler, B. J., Kempton, A. G., Coleman, R. D., and Capes, C. E., The effect of particle size and pH on the removal of pyrite from coal by conditioning with bacteria followed by oil agglomeration, Hydrometallurgy, 15:325–336 (1986).

Capes, C. E., McIlhinney, A. E., Sirianni, A. F., and Puddington, I. E., Bacterial oxidation in upgrading pyritic coals, Can. Min. Metall. Bull. 66:88–91 (1973).

Chandra, D., Roy, P., Mishra, A. K., Chakrabarti, J. N., and Sengupta, B., Microbial removal of organic sulphur from coal, Fuel 58:549–550 (1979).

Chen, C. Y., and Skidmore, D. R., Langmuir adsorption isotherm for Sulfolobus acidocaldarius on coal particles, Biotechnol. Lett. 9:191–194 (1988).

Cohen, M. S., and Gabriele, P. D., Degradation of coal by the fungi *Polysporus versicolor* and *Poria monticola, Appl. Environ. Microbiol.* **44**:23–27 (1982).

Dahlberg, M., Some factors influencing the bioliquefaction of lignite, in *Proc. Biological Treatment of Coals Workshop*, Herndon, Va., 1986, pp. 172—193.

Damberger, H. H., Harvey, R. D., Ruch, R. R., and Thomas, J., Coal characterization, in B. R. Cooper and W. A. Ellingson (eds.), *The Science and Technology of Coal and Coal Utilization*, Plenum Press, New York, 1984, pp. 7–45.

Detz, C. M., and Barvinchak, G., Microbial desulfurization of coal, *Min. Congr. J.* **65**:75–86 (1979).

Dina, M., Shelly, J., and Fernandes, J. B., Demineralization and desulfurization in coal slurry pipeline, in H. G. Hedrick and J. Fernandes (eds.), *Proc. Symp. on Biological and Chemical Removal of Sulfur and Trace Elements in Coal and Lignite*, Louisiana Technical University, Ruston, 1982, pp. 119–120.

Dispirito, A. A., Dugan, P. R., and Tuovinen, O. H., Sorption of *Thiobacillus ferrooxidans* to particulate material, *Biotechnol. Bioeng.* **25**:1163–1168 (1983).

Doddema, H. J., Partial microbial oxidation of pyrite in coal followed by oil agglomeration, in G. Rossi and A. E. Torma (eds.), *Progress in Biohydrometallurgy*, Associazione Mineraria Sarda, Iglesias, 1983, pp. 467–478.

Dogan, Z. M., Ozbayoglu, G., Hicyilmaz, C., Sarikaya, M., and Ozcengiz, G., Bacterial leaching versus bacterial conditioning and flotation in desulfurization of three different coals, in R. W. Lawrence, R. M. N., Brannion, and H. G. Ebner (eds.), *Fundamental and Applied Biohydrometallurgy, Proc. 6th Intern. Symp. on Biohydrometallurgy*, Elsevier, Amsterdam, 1986, pp. 165–170.

Dugan, P. R., The value added to coal by microbial sulfur removal, in Y. A. Attia (ed.), *Coal Science and Technology. 9. Processing and Utilization of High Sulphur Coals*, Elsevier, Amsterdam, 1985, pp. 717–726.

Dugan, P. R., and Apel, W. A., Microbial desulfurization of coal, in L. E. Murr, A. E. Torma, and J. A. Brierley (eds.), *Metallurgical Applications of Bacterial Leaching and Related Microbiological Phenomena*, Academic Press, New York, 1978, pp. 223–250.

Elzeky, M., and Attia, Y. A., Coal slurries desulfurization by flotation using thiophilic bacteria for pyrite depression, *Coal Prep. J.* **5**:15–37 (1987).

Fakoussa, R. M., Kohle als Substrat fur Mikroorganismen: Untersuchungen zur Mikrobiellen Umsetzung nativer Steinkohle, Thesis, Bonn, 1981.

Fakoussa, R. M., Partial degradation of untreated, hard coal by microorganisms, in *Proc. Biological Treatment of Coals Workshop*, Vienna, Va., 1987, pp. 137–147.

Finnerty, W. R., and Robinson, M., Microbial desulfurization of fossil fuels: A review, *Biotechnol. Bioeng. Symp.* **16**:205–221 (1986).

Garrels, R. M., and Thompson, M. E., Oxidation of pyrite by iron sulfate solutions, *Am. J. Sci.* **258A**:57–67 (1960).

Gökcay, C. F., and Yurteri, R. N., Microbial desulphurization of lignites by a thermophilic bacterium, *Fuel* **62**:1123–1124 (1983).

Golomb, M., and Beier, E., Entpyritisierung von Steinkohle mit Hilfe von Bakterien, *Bergbau* **10**:419–422 (1977).

Gouch, G. R., *Biotechnology and Coal*, IEA Coal Research, London, 1987.

Greer, R. T., Organic and inorganic sulfur in coal, *Scanning Electron Microsc.* **1**:477–486 (1979).

Groudev, S. N., and Genchev, F. N., Microbial coal desulphurization: effect of the cell adaptation and mixed cultures, *Compt. Rend. Acad. Bulg. Sci.* **32**:353–355 (1979).

Harrison, A. P., Microbial succession and mineral leaching in a artificial coal spoil, *Appl. Environ. Microbiol.* **38**:861–869 (1978).

Harrison, A. P., The acidophilic thiobacilli and other acidophilic bacteria that share their habitat, *Ann. Rev. Microbiol.* **38**:265–292 (1984).

Harsh, D., Dybel, M., and Skidmore, D., Slurry concentration and mesh size in the development of microbial coal desulfurization, in *Proc. Biological Treatment of Coals Workshop*, Vienna, Va., 1987, pp. 231–241.

Hazeu, W., Batenburg-van der Vegte, W. H., Bos, P., van der Plas, R. K., and Kuenen, J. G., The production and utilization of intermediary elemental sulfur during the oxidation of reduced sulfur compounds by *Thiobacillus ferrooxidans, Arch. Microbiol.* **150**:574–579 (1988).

Hazeu, W., Schmedding, D. J., Goddijn, O., Bos, P., and Kuenen, J. G., The importance of the sulphur oxidizing capacity of *Thiobacillus ferrooxidans* during leaching of pyrite, in O. M. Neijssel, R. R. van der Meer, and K. Ch. A. M. Luyben (eds.), *Proc. 4th Eur. Congr. on Biotechnology*, vol. 3, Elsevier, Amsterdam, 1987, pp. 497–499.

Hedrick, H. G., Microbial methods for desulphurization, in H. G. Hedrick and J. Fernandes (eds.), *Proc. Symp. on Biological and Chemical Removal of Sulfur and Trace Elements in Coal and Lignite*, Louisiana Technical University, Ruston, 1982, pp. 73–104.

Helle, U., and Onken, U., Continuous bacterial leaching of a pyritic flotation concentrate by mixed cultures, in P. R. Norris and D. P. Kelly (eds.), *Biohydrometallurgy, Proc. Intern. Symp. Warwick*, Science and Technology Letters, Kew, Surrey, U.K., 1988, pp. 61–75.

Hoffmann, M. R., Faust, B. C., Panda, F. A., Koo, H. H., and Tsuchiya, H. M., Kinetics of the removal of iron pyrite from coal by microbial catalysis, *Appl. Environ. Microbiol.* **42**:259–271 (1981).

Huber, T. F., Kossen, N. W. F., Bos, P., and Kuenen, J. G., Modelling design and scale up of a reactor for microbial desulphurization of coal, in G. Rossi and A. E. Torma (eds.), *Progress in Biohydrometallurgy*, Associazione Mineraria Sarda, Iglesias, 1983, pp. 279–289.

Huber, T. F., Ras, C., and Kossen, N. W. F., Design and scale up of a reactor for the microbial desulphurization of coal: A kinetic model for bacterial growth and pyrite oxidation, in *Proc. Third Eur. Conf. on Biotechnology*, Vol. III, Verlag Chemie, Weinheim, 1984, pp. 151–159.

Isbister, J. D., and Kobylinski, E. A., Microbial desulfurization of coal, in Y. A. Attia (ed.), *Processing and Utilization of High Sulfur Coals. Coal Science and Technology 9*, Elsevier, Amsterdam, 1985, pp. 627–641.

Jilek, R.,and Beranova, E., Some experiments with bacterial leaching of brown coal, in *Proc. Intern. Conf. on Use of Microorganisms in Hydrometallurgy*, Hung, Acad. Sci. Local Comm., Pecs, 1982, pp. 167–174.

Jüntgen, H., Research for future in situ conversion of coal, *Fuel* **66**:443–453 (1987).

Karavaiko, G. I., Golovacheva, R. S., Pivovarova, T. A., Tzaplina, I. A., and Vartanjan, N. S., Thermophilic bacteria of the genus *Sulfobacillus*, in P. R. Norris and D. P. Kelly (eds.), *Biohydrometallurgy, Proc. Intern. Symp. Warwick*, Science and Technology Letters, Kew, Surrey, U.K., 1988, pp. 29–41.

Kargi, F., Enhancement of microbial removal of pyritic sulphur from coal using concentrated cell suspension of *Thiobacillus ferrooxidans* and an external carbon dioxide supply, *Biotechnol. Bioeng.* **24**:749–752 (1982a).

Kargi, F., Microbial coal desulfurization, *Enzyme Microb. Technol.* **4**:13–19 (1982b).

Kargi, F., Microbial desulfurization of coal, in A. Mizrahi and A. L. van Wezel (eds.), *Advances in Biotechnological Processes*, Vol. 3, Alan R. Liss, New York, 1984, pp. 241–272.

Kargi, F., Microbial methods for desulfurization of coal, *Trends Biotechnol.* **4**:293–297 (1986).

Kargi, F., Biological oxidation of thianthrene, thioxanthene and dibenzothiophene by the thermophilic organism *Sulfolobus acidocaldarius*, *Biotechnol. Lett.* **9**:478–482 (1987).

Kargi, F., and Cervoni, T. D., An airlift-recycle fermentor for microbial desulfurization of coal, *Biotechnol. Lett.* **5**:33–38 (1983).

Kargi, F., and Robinson, J. M., Microbiological oxidation of dibenzothiophene by the thermophilic organism *Sulfolobus acidocaldarius*, *Biotechnol. Bioeng.* **26**:687–690 (1984).

Kargi, F., and Robinson, J. M., Removal of organic sulfur from bituminous coal. Use of the thermophilic *Sulfolobus acidocaldarius*, *Fuel* **65**:397–399 (1986).

Karr, C. (ed.), *Analytical Methods for Coal and Coal Products*, Academic Press, New York, 1978.

Kawakami, K., Fujio, K., Kusunoki, K., Kusakabe, K., and Morooka, S., Kinetic study of coal slurry electrolysis. Oxidation and desulfurization of Illinois no. 6 coal by aqueous ferric chloride, *Fuel Process. Technol.* **19**:15–29 (1988).

Kawatra, S. K., Eisele, T. C., and Bagley, S., Coal desulfurization by bacteria, *Miner. Metall. Process.* **4**:189–192 (1987).

Kempton, A. G., Moneib, N., McGready, R. G. L., and Capes, C. E., Removal of pyrite from coal by conditioning with *Thiobacillus ferrooxidans* followed by oil agglomeration, *Hydrometallurgy*, **5**:117–125 (1980).

Kilbane, J. J., Sulfur-specific microbial metabolism of organic compounds, in *Proc. Bioprocessing of Coals Workshop*, Vienna, Va., 1988, pp. 156–166.

Klein, J., Beyer, M., van Afferden, M., Hodek, W., Pfeifer, F., Seewald, H., Wolff-Fischer, E., and Jüntgen, H., Coal in biotechnologie, in H.-J. Rehm and G. Reed (eds.), *Biotechnology*, vol. 6b, VCH Verlagsgesellschaft, Weinheim, 1988, pp. 497–567.

Kodama, K., Umehara, K., Shimizu, K., Nakatani, S., Minoda, Y., and Yamada, K., Identification of microbial products from dibenzothiophene and its proposed pathway, *Agric. Biol. Chem.* **37**:45–49 (1973).

Kos, C. H., A methods for the determination of pyrite in coal, *Recl. Trav. Chim. Pays-Bas* **101**:361–362 (1982).

Kos, C. H., Bijleveld, W., Grotenhuis, T., Bos, P., Poorter, R. P. E., and Kuenen, J. G., Composition of mineral salts medium for microbial desulphurization of coal, in G. Rossi and A. E. Torma (eds.), *Progress in Biohydrometallurgy*, Associazione Mineraria Sarda, Iglesias, 1983, pp. 479–490.

Kos, C. H., Poorter, R. P. E., Bos, P., and Kuenen, J. G., Geochemistry of sulfides in coal and microbial leaching experiments, in *Proc. Intern. Conf. on Coal Science, Düsseldorf*, 1981, pp. 842–847.

Laborde, A. L., and Gibson, D. T., Metabolism of dibenzothiophene by a *Beyerinckia* species, *Appl. Environ. Microbiol.* **49**:756–760 (1977).

Lawrence, R. W., and Marchant, P. B., Comparison of mesophilic and thermophilic oxidation systems for the treatment of refractory gold ores and concentrates, in P. R. Norris and D. P. Kelly (eds.), *Biohydrometallurgy, Proc. Intern. Symp. Warwick*, Science and Technology Letters, Kew, Surrey, U.K., 1988, pp. 359–374.

Lazaroff, N., Optimization of bacterial leaching of pyrite in coal, Final report 9506004, Department of Energy, Washington, D.C., 1979.

Leushner, A. P., Trantolo, D. J., Kern, E. E., and Wise, D. L., Biogasification of Texas lignite, in *Thirteenth Biennial Lignite Symp. on Technology and Use of Low-Rank Coals, Bismarck, N.D.*, Vol. 1, 1985, pp. 216–228.

Lovell, H. R., Coal mine drainage in the United States—an overview, *Water Sci. Technol.* **15**:1–25 (1983).

McCready, R. G. L., and Zentilli, M., Beneficiation of coal by bacterial leaching, *Can. Metall. Q.* **24**:135–139 (1985).

Melin, E., Norrbin, S., and Oden, S., Researches on the methane formation of peat, *Ingeniorsvetenskaps-akademiens, Stockholm, Handlingar* **53**:5–42 (1926).

Monticello, D. J., Bakker, D., and Finnerty, W. R., Plasmid-mediated degradation of dibenzothiophene by Pseudomonas species, *Appl. Environ. Microbiol.* **49**:756–760 (1985).

Monticello, D. J., and Finnerty, W. R., Microbial desulfurization of fossil fuels, *Ann. Rev. Microbiol.* **39**:371–389 (1985).

Morrison, G. F., Chemical desulphurization of coal, IEA Coal Research, London, 1981.

Murphy, J., Riestenberg, E., Mohler, R., Marek, D., Beck, B., and Skidmore, D., Coal desulfurization by microbial processing, in Y. A. Attia (ed.), *Coal Science and Technology 9. Processing and Utilization of High Sulphur Coals*, Elsevier, Amsterdam, 1985, pp. 643–652.

Murr, L. E., and Berry, V. K., Direct observations of selective attachment of bacteria on low-grade sulfide ores and other mineral surfaces, *Hydrometallurgy* **2**:11–24 (1976).

Murr, L. E., and Mehta, A. P., Coal desulfurization by leaching involving acidophilic and thermophilic microorganisms, *Biotechnol. Bioeng.* **24**:743–748 (1982).

Muyzer, G., de Bruyn, A. C., Schmedding, D. J. M., Bos, P., Westbroek, P., and Kuenen, J. G., A combined immunofluorescence-DNA-staining technique for enumeration of *Thiobacillus ferrooxidans* in a population of acidophilic bacteria, *Appl. Environ. Microbiol.* **53**:660–664 (1987).

Myerson, A. S., and Kline, P., The adsorption of *Thiobacillus ferrooxidans* on solid particles, *Biotechnol. Bioeng.* **25**:1669–1676 (1983).

Norris, P. R., and Barr, D. W., Bacterial oxidation of pyrite in high temperature reactors, in P. R. Norris and D. P. Kelly (eds.), *Biohydrometallurgy, Proc. Intern. Symp. Warwick*, Science and Technology Letters, Kew, Surrey, U.K., 1987, pp. 532–536.

Norris, P. R., Barr, D. W., and Hinson, D., Iron and mineral oxidation by acidophilic bacteria: Affinities for iron and attachment to pyrite, in P. R. Norris and D. P. Kelly (eds.), *Biohydrometallurgy, Proc. Intern. Symp. Warwick*, Science and Technology Letters, Kew, Surrey, U.K., 1987, pp. 43–59.

Norris, P. R., Brierley, J. A., and Kelly, D. P., Physiological characteristics of two facultatively thermophilic mineral-oxidizing bacteria, *FEMS Microbiol. Lett.* 7:119–122 (1980).

Norris, P. R., and Kelly, D. P., The use of mixed microbial cultures in metal recovery, in A. T. Bull and J. H. Slater (eds.), *Microbial Interactions and Communities*, Vol. 1, Academic Press, London, 1982, pp. 443–474.

Olson, G. J., and Brinckman, F. E., Bioprocessing of coal, *Fuel* 65:1638–1646 (1986).

Palowitch, E. R., and Deurbrouck, A. W., Wet concentration of coarse coal. 1. Dense medium separation, in J. W. Leonard (ed.), *Coal Preparation*, 4th ed., The American Institute of Mining, Metallurgical and Petroleum Engineers, New York, 1979, pp. 9.1–9.36.

Pinches, A., Chapman, J. T., te Riele, W. A. M., and van Staden, M., The performance of bacterial leach reactors for the pre-oxidation of refractory gold-bearing sulphide concentrates, in P. R. Norris and D. P. Kelly (eds.), *Biohydrometallurgy, Proc. Intern. Symp. Warwick*, Science and Technology Letters, Kew, Surrey, U.K., 1988, pp. 329–344.

Pocas, M. F., Kinetics of microbially catalyzed removal of sulphur from coals, *Actas Simp. Iberoam. Catal.* 9:1489–1490 (1984).

Podolski, W. F., Fluidized-bed combustion, in B. R. Cooper and W. A. Ellingson (eds.), *The Science and Technology of Coal and Coal Utilization*, Plenum Press, New York, 1984, pp. 263–305.

Pooley, F. D., and Atkins, A. S., Desulphurization of coal using bacteria by both dump and process plant techniques, in G. Rossi and A. E. Torma (eds.), *Progress in Biohydrometallurgy*, Associazione Mineraria Sarda, Iglesias, 1983, pp. 511–526.

Prior, M., *The Control of Sulphur Oxides Emitted in Coal Combustion*, IEA Coal Research, London, 1977.

Pyne, J. W., Stewart, D. L., Linehan, J. C., Bean, R. M., Powell, M. A., Lucke, R. B., Thomas, B. L., Campbell, J. A., and Wilson, B. W., Enzymatic degradation of low-rank coals by a cell-free enzymatic system from *Coriolus versicolor*, in *Proc. Biological Treatment of Coals Workshop*, Vienna, Va., 1987, pp. 174–188.

Quigley, D. R., Wey, J. E., Breckenridge, C. R., and Hatcher, H. J., Comparison of alkali and microbial solubilization of oxidized low-rank coal, in *Proc. Biological Treatment of Coals Workshop*, Vienna, Va., 1987, pp. 151–164.

Radway, J. C., Tuttle, J. H., Fendinger, N. J., and Means, J. C., Microbially mediated leaching of low-sulfur coal in experimental coal columns, *Appl. Environ. Microbiol.* 55:1056–1063 (1987).

Rai, C., Microbial desulfurization of coals in a slurry pipeline reactor using *Thiobacillus ferrooxidans*, *Biotechnol. Prog.* 4:200–204 (1985).

Rhee, K. H., and Wan, E. I., Overview of chemical and microbial coal-cleaning activities, in Y. P. Chugh and R. D. Caudle (eds.), *Processing and Utilization of High Sulfur Coals II*, Elsevier, New York, 1987, pp. 224–231.

Rinder, G., and Beier, E., Mikrobiologische Entpyritisierung von Kohlen in Suspension, *Erdöl Kohle Erdgas Petrochemie Brennst. Chem.* 36:170–174 (1983).

Rossi, G., and Salis, E., The microbial desulphurization of coal: Some results on the effect of the percentage of solids in the pulp, in *Proc. UMIST Conf. on Possibilities of Large Scale Microbial Leaching Processes*, 1977.

Ruiz, C., Gomez-Aranda, V., Inigo, B., and Gabalin, J. M., Microbiological removal of pyritic sulfur from Spanish lignites, *Afinidad*, 41:551–555 (1984).

Schilling, H.-D., and Wiegand, D., Coal resources, in D. J. McLaren and B. J. Skinner (eds.), *Resources and World Development*, Wiley, New York, 1987, pp. 129–156.

Schnaitman, C., and Lundgren, D. G., Organic compounds in the spent medium of *Ferrobacillus ferrooxidans*, *Can. J. Microbiol.* 11:23–27 (1965).

Scott, C. D., Microbial coal liquefaction, in *Proc. Biological Treatment of Coals Work-*

shop, Herndon, Va., 1986, pp. 128–140.

Scott, C. D., and Faison, B. D., Biological solubilization of coal in aqueous and nonaqueous media, in *Proc. Biological Treatment of Coals Workshop*, Vienna, Va., 1987, pp. 213–223.

Scott, C. D., Strandberg, G. W., and Lewis, S. N., Microbial solubilization of coal, *Biotechnol. Prog.* **2**:131–139 (1986).

Shinn, J. H., The structure of coal and its liquefaction products: A reactive model, in *Proc. Intern. Conf. on Coal Science*, Sydney, Australia, 1985, pp. 738–741.

Silverman, M. P., Mechanisms of bacterial pyrite oxidation, *J. Bacteriol.* **94**:1046–1051 (1967).

Silverman, M. P., and Lundgren, D. G., Studies on the chemoautotrophic iron bacterium *Ferrobacillus ferrooxidans*. I. An improved medium and a harvesting procedure for securing high cell yields, *J. Bacteriol.* **77**:642–647 (1959).

Silverman, M. P., Rogoff, M. H., and Wender, I., Removal of pyritic sulphur from coal by bacterial action, *Fuel* **42**:113–124 (1963).

Sproull, R. D., Francis, H. J., Krishna, C. R., and Dodge, D. J., Enhancement of coal quality by microbial demineralization and desulfurization, in *Proc. Biological Treatment of Coals Workshop*, Herndon, Va., 1986, pp. 83–94.

Stokes, H. N., Pyrite and marcasite, *Bull. U.S. Geol. Surv.* **182**:1–150 (1901).

Stoner, D. L., Microbial degradation of organosulfur compounds: implications for the microbial desulphurization of coal, in *Proc. Biological Treatment of Coals Workshop*, Vienna, Va., 1987, pp. 278–288.

Sweere, A. P. J., Luyben, K. Ch. A. M., and Kossen, N. W. F., Regime analysis and scale-down: Tools to investigate the performance of bioreactors, *Enzyme Microb. Technol.* **9**:386–398 (1987).

Tillet, D. M., and Myerson, A. S., The removal of pyritic sulfur from coal employing *Thiobacillus ferrooxidans* in a packed column reactor, *Biotechnol. Bioeng.* **29**:146–150 (1987).

Torma, A. E., Walden, C. C., Duncan, D. W., and Branion, R. M. R., The effect of carbon dioxide and particle surface area on the microbiological leaching of a zinc sulfide concentrate, *Biotechnol. Bioeng.* **14**:777–786 (1972).

Townsley, C. C., and Atkins, A. S., Comparative coal fines desulfurization using the iron-oxidizing bacterium *Thiobacillus ferrooxidans* and the yeast *Saccharomyces cerevisiae* during simulated froth flotation, *Process Biochem.* **21**:188–191 (1986).

Townsley, C. C., Atkins, A. S., and Davis, A. J., Suppression of pyritic sulfur during flotation tests using the bacterium *Thiobacillus ferrooxidans*, *Biotechnol. Bioeng.* **30**:1–8 (1987).

Tuttle, J. H., and Dugan, P. R., Inhibition of growth iron and sulfur oxidation in *Thiobacillus ferrooxidans* by simple organic compounds, *Can. J. Microbiol.* **22**:719–730 (1976).

van Afferden, M., Schacht, S., Beyer, M., and Klein, J., Microbial desulfurization of dibenzothiophene, *Am. Chem. Soc. Div. Fuel Chem.* **33**:561–572 (1988).

van Steenbergen, H. B., Eindrapport technisch-economische evaluatie project steenkoolzuivering. Inschatting van de vooruitzichten van voorgereinigde kool in de Nederlandse industrie, ESTS BV, IJmuiden, 1985.

Vaseen, V. A., Commercial microbial desulphurization of coal, in Y. A. Attia (ed.), *Processing and Utilization of High Sulfur Coals. Coal Science and Technology 9*, Elsevier, Amsterdam, 1985, pp. 699–715.

Vleeskens, J. M., Bos, P., Kos, C. H., and Roos, M., Pyrite association and coal cleaning: An optical image analysis study, *Fuel* **64**:342–347 (1985).

Volsicky, Z., Sebor, G., Dockal, M., and Puncmanova, J., Die Senkung des Schwefelgehaltes in nordböhmische Braunkohle durch Laugungsverfahren, *Aufbereit. Tech.* **26**:433–437 (1985).

Wakao, N., Mishina, M., Sakuray, Y., and Shiota, H., Bacterial pyrite oxidation. III. Adsorption of *Thiobacillus ferrooxidans* cells on solid surfaces and its effect on iron release from pyrite: the effect of organic additives. *J. Gen. Appl. Microbiol.* **30**:63–77 (1984).

Ward, B., Lignite-degrading fungi isolated from a weathered outcrop, *Syst. Appl. Microbiol.* **6**:236–238 (1985).

Wilson, B. W., Bean, R. M., Pyne, J., Stewart, D. L., and Fredrickson, J., Microbial beneficiation of low rank coals, in *Proc. Biological Treatment of Coals Workshop,* Herndon, Va., 1986, pp. 114–127.

Wyza, R. E., Desouza, A. E., and Isbister, J. D., Depolymerization of low-rank coals by a unique microbial consortium, in *Proc. Biological Treatment of Coals Workshop,* Vienna, Va., 1987, pp. 119–132.

Yeh, T. Y., Godshalk, J. R., Olson, G. J., and Kelly, R. M., Use of epifluorescence microscopy for characterizing the activity of *Thiobacillus ferrooxidans* on iron pyrite, *Biotechnol. Bioeng.* **30**:138–146 (1987).

Yurteri, R., and Gökcay, C. F., Bacterial desulphurization of Turkish lignites, in *Proc. Biological Treatment of Coals Workshop,* Vienna, Va., 1987, pp. 242–272.

Zarubina, N. N., Lyalikova, N. N., and Shmuk, E. I., Investigation of microbiological oxidation of coal pyrite, *Invest. Akad. Nauk. SSr. Otedl. Tekh. Mauk. Me. Toplivo.* **1**:117–119 (1959).

Zimmerman, R. E., Wet concentration of fine coal. Part 3: Froth flotation, in J. W. Leonard (ed.), *Coal Preparation,* 4th ed., The American Institute of Mining, Metallurgical and Petroleum Engineers, New York, 1979, pp. 10.75–10.104.

Microbial Desulfurization of Petroleum

Julia M. Foght
Phillip M. Fedorak
Murray R. Gray
Donald W. S. Westlake

Introduction

This chapter defines the sources and classes of organic sulfur found in petroleum and reviews reports in the literature pertaining to microbial degradation of model organic sulfur compounds. Evidence of in situ desulfurization of petroleum, based on biodegradation in reservoirs and oil-contaminated environments, and laboratory observations have stimulated studies on the bench-scale application of biodesulfurization technology. Proposed applications of strain selection and genetic engineering techniques to enhance biodesulfurization are examined. Finally, this chapter reviews process design and economic feasibility of the current petroleum biodesulfurization technology and suggests areas for future research and alternative process development.

Petroleum is generally considered to be a naturally occurring gaseous, liquid, or solid mixture predominantly composed of hydrocarbons. This definition includes conventional and heavy crude oils, mineral wax, asphalts (bitumens), and bituminous rock such as oil shales. Liquid synthetic crude oils resulting from upgrading of heavy oils, bitumens, and oil shales are also considered in this chapter.

Sulfur is the third most abundant element after carbon and hydrogen in petroleum, constituting approximately 0.04 to 5 weight percent in conventional crude oils (Speight, 1980). The sulfur content varies inversely with the API gravity of the petroleum so that light crude oils tend to have lower sulfur contents than heavy crudes and bitumens. Distillation of crude oils and analyses of the resulting fractions have shown that the amount of sulfur increases with the boiling range of the fraction and that the majority of the sulfur is found in the residues. For example, in an east Texas crude oil which contained 0.36 weight percent sulfur, 0.9 percent of the sulfur was in the gasoline and naphtha fraction, 1.3 percent was in the kerosene fraction, 15.4 percent was in the gas oil fraction, and 82.4 percent was in the residue (Speight, 1980).

As the world supply of low-sulfur petroleum decreases, there is increased interest in the use of heavy crudes, oils sands, and oil shales, which typically have higher sulfur contents than conventional crude oils. Rühl (1982) summarized the characteristics of several geographically diverse heavy oils and reported that the sulfur content ranged from 4.0 to 8.4 weight percent. Kerogen, the organic material in oil shales, generally has a lower sulfur content than heavy oils. For example, the sulfur content in 10 of 11 kerogens listed by Rühl (1982) ranged from 0.2 to 3.3 weight percent sulfur and the eleventh kerogen contained 8.5 weight percent sulfur. Oil shales have the unique problem of high nitrogen content, which may be several times higher than found in other petroleums (Harney, 1983).

The presence of sulfur in petroleums creates certain problems, which have been summarized by Rall et al. (1972) and Speight (1981). One of the most costly consequences of processing high-sulfur crude oils is corrosion of metals used in production and refining. Because sulfur is objectionable in refined petroleum products, it is normal refinery practice to remove essentially all sulfur from liquid fuels (gasoline and diesel). The only petroleum products which contain appreciable sulfur (above a fraction of a percent) are fuel oils for marine use and power generation, and asphalts for paving. Combustion of sulfur-containing fuels yields sulfur dioxide, which contributes to acid rain. Currently, the move to reduce sulfur dioxide emissions is the major impetus for desulfurization of fossil fuels.

To assess the feasibility of microbial desulfurization of petroleum, both the types of sulfur compounds present and the abilities of microorganisms to selectively metabolize these compounds must be considered. Although hydrogen sulfide and elemental sulfur are found in some crude oils (Rall et al., 1972), microbial removal of these inorganic species from petroleum will not be discussed because they are readily susceptible to chemical desulfurization technology. Low-molecular-weight aliphatic sulfur compounds such as methyl mercaptan and dimethyl sulfide also are readily removed chemically

(Speight, 1981) and are not suitable substrates for microbial treatment. Therefore, only the microbial removal of aromatic, alicyclic, and larger-molecular-weight aliphatic organic sulfur compounds will be considered.

The first major effort to identify sulfur compounds in crude oil was the 20-year API Research Project 48 (Rall et al., 1972) which focused on Wasson, Texas, crude oil (32.8° API, 1.85 weight percent sulfur). At the time Project 48 began, only 25 organic sulfur compounds had been identified in petroleum, including low-molecular-weight thiols, alkyl sulfides, cyclic sulfides, and thiophene. As a result of Project 48 investigations, an additional 176 organic sulfur compounds were isolated, identified, and grouped into 13 classes (Fig. 16.1). Although a few

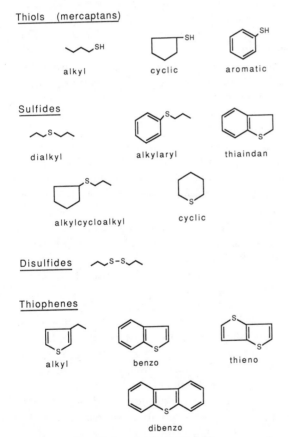

Figure 16.1   The 13 classes of organic sulfur compounds identified in the API Research Project 48. (Rall et al., 1972.)

disulfides were identified in Project 48, these are not regarded as being true constituents of crude oil because they are formed by the oxidation of thiols (Speight, 1981).

Thiophene derivatives are particularly abundant in crude oils with a high content of aromatics, resins, and asphaltenes (Tissot and Welte, 1984). Dibenzothiophenes (DBTs) are frequently the major thiophenic compounds in mature and altered high-sulfur crude oils. For example, DBT has been reported to constitute up to 70 percent of the organic sulfur in some Texas oils (Finnerty and Robinson, 1986). However, alkyl-substituted DBTs containing two or three additional carbon constituents are usually more abundant than unsubstituted DBT and can account for as much as 40 percent of organic sulfur in Middle East oils (Finnerty and Robinson, 1986). Based on sulfur composition, Ho et al. (1974) grouped oils into three categories. The first category was composed of immature oils which occur in young sediments of shallow depth and contain nonthiophenic sulfides and benzothiophenes. The second category included mature oils from deeper, older reservoirs, which contain more of the stable DBTs and fewer of the unstable benzothiophenes. The third category consisted of altered oils having a sulfur composition intermediate between the immature and mature oils, and which are found in shallow reservoirs subjected to water washing and microbial degradation.

Larger thiophene-containing ring systems are found in higher boiling fractions. For example, in a vacuum gas oil with boiling range from 350 to 500°C, Drushel and Sommers (1967) tentatively identified several classes of compounds containing four and five condensed rings (Fig. 16.2). Pioneering work by M. L. Lee and coworkers with synthetic fuels and shale oils (Later et al., 1981; Willey et al., 1981) has resulted in the separation, identification, and synthesis of S-heterocycles such as alkyldihydrobenzothiophenes, alkyldibenzothiophenes, benzonaphthothiophenes, phenanthrothiophenes, dinaphthothiophenes, and benzophenanthrothiophenes (Fig. 16.2).

Payzant et al. (1986) have isolated sulfide fractions from various petroleum samples. The sulfide content of the maltene fractions ranged from 0.3 to 16.1 weight percent. In addition, a number of novel cyclic sulfides including di-, tri-, and tetracyclic compounds were identified. Cyr et al. (1986) found a homologous series of hopane (hexacyclic) sulfides in petroleum. More recently, Payzant et al. (1989) identified a homologous series of monocyclic sulfides in nonbiodegraded petroleums. The structures of these cyclic sulfides are shown in Fig. 16.3.

Asphaltenes constitute a class of high-molecular-weight components of petroleum, based on their insolubility in aliphatic solvents such as n-pentane. They remain in distillation residues and contrib-

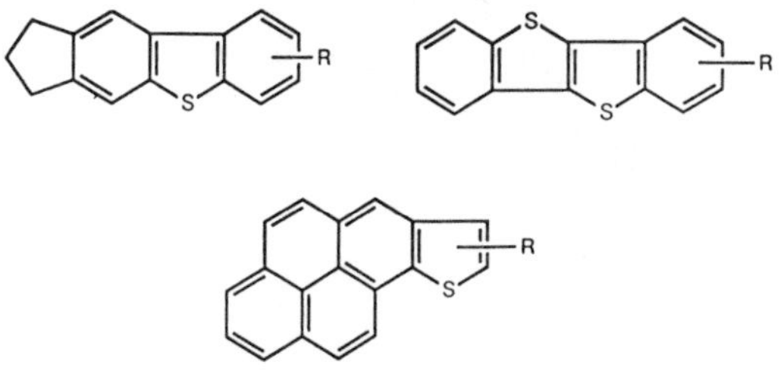

(a)

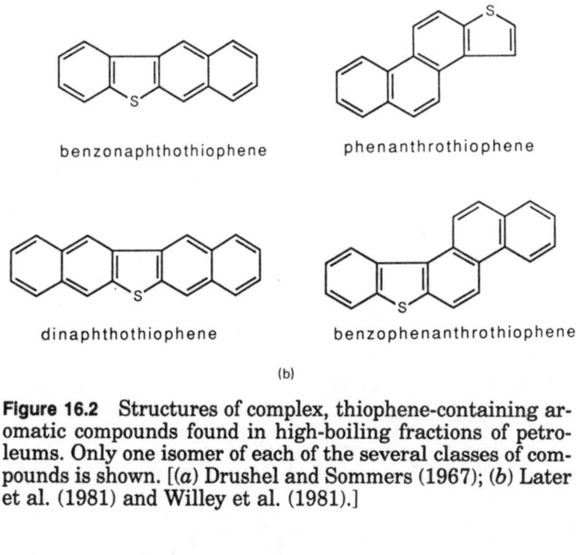

benzonaphthothiophene   phenanthrothiophene

dinaphthothiophene   benzophenanthrothiophene

(b)

Figure 16.2 Structures of complex, thiophene-containing aromatic compounds found in high-boiling fractions of petroleums. Only one isomer of each of the several classes of compounds is shown. [(a) Drushel and Sommers (1967); (b) Later et al. (1981) and Willey et al. (1981).]

ute to the sulfur content of the residues. The class comprises compounds with diverse chemical structures believed to resemble those shown in Fig. 16.2, but in complex, high-molecular-weight systems. In general, the aromaticity and the proportion of heteroatoms (sulfur, nitrogen, and oxygen) increase as the molecular weight of the asphaltenes increase (Speight, 1980; Cyr et al., 1987). Sulfur atoms in asphaltenes are present as sulfides and in thiophenic rings (Cyr et al., 1987) which are not readily accessible to microbial removal because of the large molecular size, complexity, and colloidal nature of the asphaltenes. However, although the sulfur content of the asphaltene fraction may be high, the maltene fraction may contain the majority of the sulfur in a low-asphaltene petroleum.

Chemical methods of petroleum desulfurization are currently practiced and have been reviewed by Speight (1981). Conventional

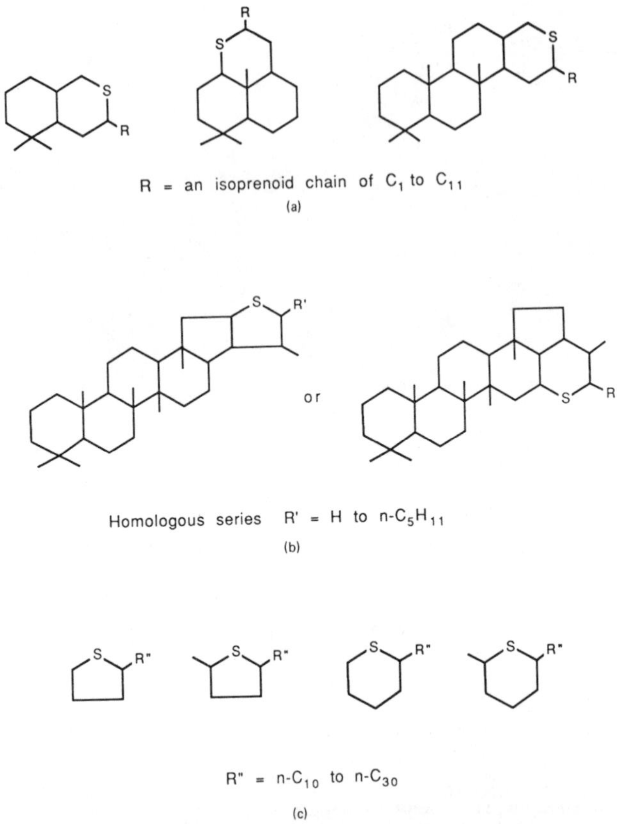

R = an isoprenoid chain of C_1 to C_{11}

(a)

or

Homologous series   R' = H to n-C_5H_{11}

(b)

R'' = n-C_{10} to n-C_{30}

(c)

Figure 16.3 Structures of a variety of cyclic sulfides identified in petroleums. [(a) Payzant et al. (1986), (b) Cyr et al. (1986), (c) Payzant et al. (1989).]

hydrodesulfurization employs catalytic reduction of organic sulfur to hydrogen sulfide. Process conditions require temperatures ranging from 270°C for naphthas to 430°C for residual fractions, and up to 14 MPa of hydrogen pressure (Speight, 1981). Sulfur removal of 50 to 100 percent (depending on feed and severity) is achieved with residence times on the order of 1 h. Carbon efficiency is 95 to 99 percent, where losses are due to the formation of light hydrocarbons.

Biodesulfurization, particularly of high-sulfur content petroleums, theoretically is an attractive alternative to hydrodesulfurization because of cost benefits from reduced temperature and pressure requirements and the potential for exquisite specificity of sulfur removal. In the past, various microbial systems were proposed for the desulfurization of oil, and some were patented (Strawinski, 1950, 1951; Zobell, 1953; Kirshenbaum, 1961). However, none of these processes has been used commercially (Malik, 1978). With the advent of new techniques in genetic engineering, and with an increased understanding of microbial metabolism of organic sulfur compounds in petroleum, biodesulfurization is being reevaluated as an alternative or addendum to conventional chemical desulfurization.

Recently, there have been two divergent experimental trends in aerobic biodesulfurization of fossil fuels: one has addressed biodesulfurization of petroleum, and the other has focused on removal of organic sulfur from coal. The former studies have emphasized oxidation of S-heterocycles to polar thiophene derivatives, which are subsequently removed from the oil by water washing. This approach has the advantage that many different microbes readily effect the oxidations, but the disadvantage of removing the valuable hydrocarbon moiety along with the heteroatom. The latter studies have stressed specific removal of the thiophenic heteroatom from DBT, yielding sulfate or sulfide and leaving the hydrocarbon skeleton intact. However, to date only a few organisms with this capability have been described, which limits the flexibility of study and process development.

This chapter emphasizes processes involving production of polar organic compounds from thiophenic sulfur, primarily because these phenomena have been reported most commonly in reference to petroleum biodesulfurization. Production of inorganic sulfur (sulfate, sulfides) from thiophenic sulfur is not discussed in the same detail, because these processes have been developed for use with coal (see Chap. 15).

Microbial Metabolism of Pure Organic Sulfur Compounds

An understanding of microbial metabolism of pure model organic sulfur compounds is necessary to develop a rational approach to microbial desulfurization. The term *metabolism* is used in this chapter to indicate alteration of an organic molecule, whether it is utilized as a carbon source or is metabolized fortuitously while the organism grows at the expense of another substrate (cometabolism). *Degradation* denotes the occurrence of extensive changes to the molecule, such as ring cleavage or loss of atoms. This section expands the review by Ensley (1984) of condensed thiophene metabolism, and concentrates on bacterial metabolism, because there is a lack of information about fungal metabolism of these compounds.

Thiophenes

Unsubstituted thiophene apparently is recalcitrant to microbial oxidation. There are no reports of aerobic thiophene utilization, and several reports of pure cultures failing to metabolize unsubstituted thiophene aerobically (Amphlett and Callely, 1969; Cripps, 1973; Kanagawa and Kelly, 1987). Although substituted thiophenes are not abundant in petroleum, some substituted thiophenes such as carboxythiophenes are susceptible to microbial metabolism and have been used as model compounds to study degradation of the thiophene nucleus. Amphlett and Callely (1969) described inducible metabolism of thiophene-2-carboxylate and thiophene-3-carboxylate to sulfide by a *Flavobacterium* sp. Cripps (1973) proposed a pathway for degradation of thiophene-2-carboxylate to carbon dioxide, sulfate, and cell mass by "Organism R1." Thiophene-2-carboxylate was utilized as sole carbon, sulfur, and energy source by Organism R1, whereas the isomer thiophene-3-carboxylate was a nonmetabolizable inducer of thiophene-2-carboxylate degradation. Sagardía et al. (1975) demonstrated that 2- and 3-methylthiophenes were metabolized by benzothiophene-adapted *Pseudomonas aeruginosa* "PRG-1." Recently, Kanagawa and Kelly (1987) showed that *Rhodococcus* spp., isolated from activated sludge, utilized thiophene-2-carboxylate, 5-methylthiophene-2-carboxylate, and thiophene-2-acetic acid as sole carbon and energy sources, but were incapable of growing on thiophene-3-carboxylate or methylthiophenes. Mutated *Escherichia coli* strains selected for growth on thiophene-2-carboxylate acquired the capability to metabolize thiophene-2-acetate, thiophene methanol, and thiophene methylamine (Abdulrashid and Clark, 1987). It would appear that, for some isolates, the position of thiophene substitution influences metabolism, while for others this is less important. Aerobic metabolism of substituted thiophenes is accomplished by cleavage of the thiophene nucleus with release of inorganic sulfur.

A single report on anaerobic metabolism of thiophene has appeared (Kurita et al., 1971). Bacterial cultures obtained from oil sludges produced hydrogen sulfide from thiophene when grown with polypeptone. Cell-free extracts also evolved hydrogen sulfide from thiophene anaerobically in the presence of methyl viologen, but the fate of the carbon moiety was not determined.

Benzothiophenes

The literature contains few reports of either aerobic or anaerobic benzothiophene metabolism. *P. aeruginosa* PRG-1 oxidized benzothiophene dissolved in light oil but could not use the compound as sole carbon source (Sagardía et al., 1975; Fuentes, 1984). The oxidized metabolites were not identified, although it was suggested that direct addition of oxygen to benzothiophene occurred. *Pseudomonas alcaligenes* DBT2 (Finnerty et al., 1983) and *Pseudomonas putida* CB1 (Isbister and Doyle, 1985) were reported to oxidize benzothiophene to uniden-tified products. Mixed culture enrichments from various water sam-ples were unable to metabolize benzothiophene unless supplied with naphthalene as an alternative carbon source (Bohonos et al., 1977). Analysis of degradation products from these cometabolizing cultures by gas chromatography–mass spectrometry showed oxidation of the thiophene ring at different positions, but no cleavage of either aro-matic ring.

Anaerobic degradation of benzothiophene by oil sludge bacteria yielded hydrogen sulfide and unknown carbon compounds (Kurita et al., 1971). Maka et al. (1987) reported that benzothiophene was de-graded anaerobically by mixed cultures from coal storage site soils, but the metabolites were not identified. Benzothiophene degradation by aquifer-derived methanogenic consortia has been observed (Godsy and Grbić-Galić, 1988), with thiophene-2-ol and 4-hydroxybenzene sulfonic acid identified as the ring cleavage products. In general, het-erocyclic compounds are more susceptible to anaerobic degradation than are their corresponding hydrocarbons (Grbić-Galić, personal communication).

Dibenzothiophene

DBT is the best-studied S-heterocycle and has been used as a model for studying petroleum biodesulfurization (Yamada et al., 1968; Hou and Laskin, 1976; Malik and Claus, 1976; Monticello et al., 1985). DBT usually is cometabolized by bacteria (Hou and Laskin, 1976; Kodama, 1977a; Monticello et al., 1985; Foght and Westlake, 1988), although some *Acinetobacter* and *Rhizobium* strains (Malik and Claus, 1976) and *Sulfolobus acidocaldarius* (Kargi and Robinson, 1984) have been reported to grow with DBT as sole carbon source.

Products of DBT cometabolism by *Pseudomonas jianii* were identi-fied, and a degradative pathway was proposed by Kodama et al. (1970, 1973). Subsequently, other genera were found to produce identical me-tabolites, presumably by similar pathways (Hou and Laskin, 1976; Malik and Claus, 1976; Laborde and Gibson, 1977; Monticello et al., 1985). A synthesis of proposed pathways is presented in Fig. 16.4 and Table 16.1, involving oxidation followed by cleavage of one benzene ring without release of the sulfur atom. The 3-carbon fragment liber-ated in the last enzymatic step has not been identified. Different pro-portions of metabolites accumulate in different species, but further pure culture aerobic degradation of the terminal metabolite, 3-hydroxy-2-formyl-benzothiophene (HFBT), has not been demonstrated.

The similarity of DBT cometabolism by bacterial genera isolated from geographically separated sources is notable: *Pseudomonas* spp.

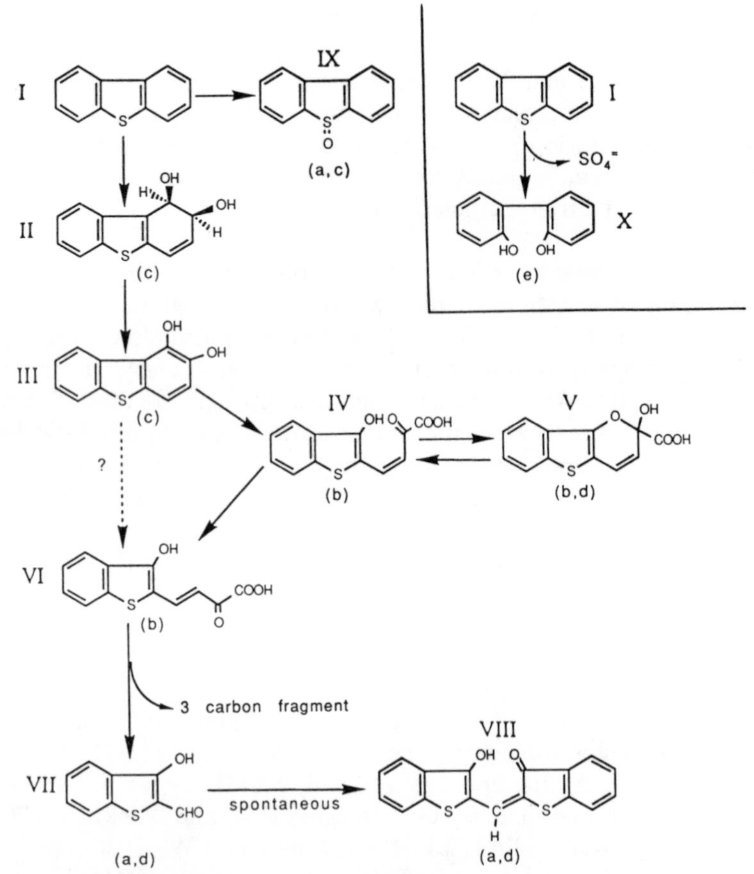

Figure 16.4 Synthesis of proposed pathways for aerobic degradation of dibenzothiophene by bacteria. For compound identification, see Table 16.1. Letters in parentheses refer to references: (a) Kodama et al. (1970); (b) Kodama et al. (1973); (c) Laborde and Gibson (1977); (d) Monticello et al. (1985); (e) Isbister et al. (1988).

TABLE 16.1   Identity of Dibenzothiophene (DBT) Metabolites Shown in Figure 16.4

Number	Compound name
I	Dibenzothiophene
II	(+)-cis-1,2-Dihydroxy-1,2-dihydrodibenzothiophene
III	1,2-Dihydroxydibenzothiophene
IV	cis-4-[2-(3-Hydroxy)-thianaphthenyl]-2-oxo-3-butenoic acid
V	Hemiacetal form of IV
VI	trans-4-[2-(3-Hydroxy)-thianaphthenyl]-2-oxo-3-butenoic acid
VII	3-Hydroxy-2-formylbenzothiophene (HFBT)
VIII	3-Oxo-[3'-hydroxy-thianaphthenyl-(2)-methylene]-dihydrothianaphthene
IX	Dibenzothiophene-5-oxide
X	2,2'-Dihydroxybiphenyl

isolated in Japan (Kodama et al., 1973), the United States (Monticello et al., 1985), and Canada (Foght and Westlake, 1988); a *Beijerinckia* sp. (Laborde and Gibson, 1977), a *Flavobacterium* sp. (Foght, 1985; Foght and Westlake, 1985), and a *Xanthomonas* sp. (Stoner et al., 1986) from the United States; and *Acinetobacter* and *Rhizobium* spp. from Germany (Malik and Claus, 1976) have been shown to produce similar metabolites of DBT.

A different mechanism of aerobic DBT metabolism, involving release of inorganic sulfur from the thiophene ring, was observed by researchers studying desulfurization of coal. Isbister et al. (1988) demonstrated with radiolabeled DBT that a genetically modified bacterium CB1 isolated from coal dump sites specifically liberated ^{35}S-sulfate from ^{35}S-DBT, producing 2,2′-dihydroxybiphenyl as an unassimilated organic residue (Fig. 16.4). The thermophilic, mixotrophic organism *S. acidocaldarius* released sulfate from DBT (Kargi and Robinson, 1986; Blount-Fronefield et al., 1986), although the organic residue was not determined.

Several *Pseudomonas* strains isolated from soil and strip-mine sites mineralized DBT-5,5-dioxide (DBT-sulfone), liberating sulfate (Ochman and Klubek, 1986). Foght and Westlake (1988) described *Pseudomonas* sp. HL7b, which metabolized DBT-5-oxide to unknown water-soluble products.

DBT also can be metabolized anaerobically. Köhler et al. (1984) incubated DBT dissolved in paraffin oil with *Desulfovibrio* spp. under sulfate-reducing conditions, and observed 15.9 percent desulfurization after 6 days' incubation. Maka et al. (1987) reported anaerobic degradation of DBT by mixed cultures from coal storage sites.

Dibenzylsulfide

Aerobic metabolism of dibenzylsulfide by a mixed bacterial culture generated several water-soluble products, one of which was identified as benzylmercaptoacetic acid (Babenzien et al., 1979). The pH of the medium dropped from 6.8 to 3.4 over 4 days' incubation, during which time 14.5 percent of the dibenzylsulfide was removed. Maintaining the medium at pH 6.8 increased the rate of dibenzylsulfide metabolism.

Köhler et al. (1984) observed anaerobic metabolism of dibenzylsulfide by *Desulfovibrio* sp. in a lactate medium. Cultures incubated with ^{35}S-dibenzylsulfide released ^{35}S-sulfide. Toluene and benzylmercaptan also were identified as metabolites of dibenzylsulfide.

Other organic sulfur compounds

Degradation of two alkylthiophenes (2-n-dodecyltetrahydrothiophene and 2-n-undecyltetrahydrothiophene) was demonstrated using bacterial and fungal isolates (Fedorak et al., 1988). This degradation involved oxidation and cleavage of alkyl side chains to produce, primarily, 2-tetrahydrothiophene carboxylic acid and 2-tetrahydrothiophene acetic acid. The fate of the thiophene ring was not ascertained, but it is believed that further biodegradation occurred.

Oxidation of phenyl sulfide and n-octyl sulfide by *P. putida* strains CB1 and CB2 has been reported (Isbister and Doyle, 1985; Isbister et al., 1988). Thioxanthene and thianthrene are metabolized to unidentified water-soluble products by *Pseudomonas* spp. (Finnerty et al., 1983; Foght and Westlake, 1988) and to sulfate plus unidentified organic residues by *S. acidocaldarius* (Kargi, 1987).

Thianaphthene, dibenzyldisulfide, butylsulfide, and octylsulfide dissolved individually in paraffin oil were incubated anaerobically with cultures of *Desulfovibrio* sp. (Köhler et al., 1984). Anaerobic metabolism led to small decreases (3 to 10 percent) in the amounts of the compounds recovered from the organic phase of the cultures.

Unfortunately, some organic sulfur compounds found abundantly in crude oils, such as alkyl-DBTs, are not commercially available as pure chemicals, and therefore their metabolism has not been investigated.

Removal of Organic Sulfur Compounds from Petroleum

Although metabolism of pure organic sulfur compounds has been demonstrated, degradation of these specific compounds in crude oil is influenced by biological factors, such as toxicity and enzyme induction, and chemical-physical properties of the oil, such as solubility and viscosity. Petroleum biodesulfurization has been assessed in comparative analytical studies using methods such as total sulfur analysis, gas chromatography, and mass spectrometry to observe disappearance of sulfur compounds rather than appearance of metabolites. These techniques have been applied to petroleum from biodegraded and nonbiodegraded reservoirs, petroleum spilled in the environment, and petroleum subjected to microbial attack under laboratory conditions. The results of these investigations are reviewed along with bench-scale studies on biodesulfurization of conventional and heavy crude oils.

In situ metabolism of organic sulfur compounds in petroleum

Evidence from petroleum reservoirs and crude oil spills suggests that the organic sulfur compounds vary in susceptibility to metabolism under environmental conditions. Metabolism is slow and selective, and the observed variability is due to the microbial population, ambient conditions, and physical-chemical properties of the petroleum. Deroo et al. (1974) observed that the decreasing thiophenic content (benzothiophenes, DBTs, and naphthobenzothiophenes) of western Canadian crude oils was paralleled by decreasing n-paraffinic and isoprenoid contents in the oils. They attributed these losses to in situ biodegradation and water washing of reservoir oils. In contrast, Burns et al. (1975) reported that in situ biodegradation of Beaufort Basin liquid hydrocarbons lowered the pour point of the oils but did not affect the sulfur content or viscosity.

In the late 1970s, two major oil spills occurred in the marine environment. The Amoco Cadiz spill impacted the coastline of France, and the Ixtoc blowout released crude oil into the Gulf of Mexico. A detailed chemical study of the fate of petroleum from the Amoco Cadiz spill revealed that the mixture of residual petroleum in the littoral zone was enriched in complex hydrocarbon components and aromatic organic sulfur compounds such as DBTs (Atlas et al., 1981). Gundlach et al. (1983) reported that the alkyl-DBTs persisted as major aromatic molecular markers 3 years after the Amoco Cadiz spill occurred. This enrichment appeared to result from selective metabolism of susceptible compounds, suggesting that DBTs are more recalcitrant than some other petroleum components. In contrast, enrichment of DBTs in the mousse generated from the Ixtoc spill was attributed to physical and photochemical processes (Patton et al., 1981), with no evidence for microbial degradation playing a role in the enrichment. Despite these observations of persistence, large-molecular-weight organic sulfur compounds cannot be totally recalcitrant to biodegradation because they have not accumulated in the environment. Their ultimate removal in situ probably depends on prolonged contact with microbial consortia.

In vitro metabolism of organic sulfur compounds in petroleum

Laboratory studies of petroleum degradation by pure and mixed cultures have demonstrated that petroleum biodesulfurization occurs in vitro. Walker et al. (1975, 1976) first reported the oxidation of S-heterocyclic compounds in conventional crude oils by microbes from an oil-polluted sediment sample. They concluded that compounds with a sulfur-containing aromatic nucleus were twice as resistant to biodegradation as their hydrocarbon analogues.

Fedorak and Westlake (1983) reported removal of alkylbenzo[b]thiophenes, DBT, and C_1- and C_2-DBTs from Prudhoe Bay crude oil by microbes in marine water samples. The order of susceptibility to metabolism was C_2-benzothiophenes > C_3-benzothiophenes and DBT > C_1-DBT > C_2-DBT. That is, susceptibility decreased with increasing molecular weight and substitution. Although many of these S-heterocyclic compounds were metabolized without nutrient supplementation of the medium, amendment with ammonium nitrate and phosphate increased the amount and number of compounds degraded. The pattern of S-heterocycle biodegradation in these experiments paralleled that of aromatic rather than n-alkane degradation. Similar results were reported for degradation of Prudhoe Bay crude oil by two greenhouse soil samples (Fedorak and Westlake, 1984). The presence of the oil dispersant Corexit 9527 was inhibitory to biodegradation of hydrocarbons and S-heterocyclic components of Prudhoe Bay crude oil (Foght et al., 1983). The low-molecular-weight S-heterocycles (C_2- and C_3-benzothiophenes, DBT, and C_1-DBT) in Rainbow and Redwater crude oils were metabolized within 14 days by mixed microbial populations (Westlake, 1983), but two larger compounds, one a benzonaphthothiophene isomer and the other unidentified, were resistant to microbial action.

A deep Kumak oil (2300 m depth), which had not undergone biodegradation in the reservoir, was subjected to microbial attack in vitro. The S-heterocycle content of residual oil from these cultures then resembled oil from the shallow Kumak pools (1350 m and 2150 m depth) which were considered to have undergone biodegradation in situ (Westlake, 1983). Similarly, Fedorak et al. (1988) demonstrated removal of n-alkyl monocyclic sulfides from Bellshill Lake oil by two bacterial cultures, and suggested that in situ biodegradation could be responsible for the absence of these compounds in biodegraded oils.

Bench-scale biodesulfurization studies

It is apparent from in situ and in vitro studies that S-heterocycles in oil are less susceptible to biodegradation than their hydrocarbon analogues. However, observation of sulfur depletion in biodegraded petroleum suggests that conditions could be optimized to develop a petroleum biodesulfurization process.

Eckart et al. (1980) assessed the ability of various aerobic mixed cultures to desulfurize Romashkino crude oil (1.69 weight percent sulfur). Two of the most active cultures were grown in sulfur-free mineral medium with oil as the sole source of carbon and sulfur. After 5 days of incubation at 30°C, petroleum losses were 25 percent and 22 percent by weight, and approximately 55 percent of the total sulfur was recovered in the aqueous phase.

Studies by Finnerty and coworkers have focused on the development of processes to oxidize the model compound DBT and other S-heterocycles in petroleum to polar products. In addition to oxidizing S-heterocycles, the organisms used in these studies oxidized aromatic hydrocarbons (benzene, naphthalene, anthracene, and phenanthrene) slowly, but were unable to grow at the expense of DBT or paraffinic components of oil. The cells were pregrown in an inducing medium, harvested, and suspended in a salt solution as a nongrowing biocatalyst. These cells effectively oxidized DBT dissolved in low-sulfur crude oil and removed indigenous DBT in a high-sulfur conventional crude oil, converting 100 percent of the DBT to polar products within 5 h at temperatures from ambient to 55°C (Finnerty et al., 1983; Finnerty and Hartdegen, 1984; Hartdegen et al., 1984). The biocatalyst appeared to be specific for S-heterocycles under the conditions tested, with no significant oxidation of hydrocarbons.

There are very few reports in the literature describing biodesulfurization of heavy crude oils (< 10° API). Finnerty and Hartdegen (1984) isolated an organism that reduced the viscosity of heavy oils by 95 to 98 percent. Subsequent treatment of the emulsion with a DBT-cometabolizing organism resulted in a significant increase in the release of water-soluble, organic sulfur products. Reduction of viscosity and increased surface area were believed to be responsible for the enhanced efficiency of desulfurization. Foght and Westlake (unpublished results) attempted to biodesulfurize heavy crude oil from Lloydminster (Saskatchewan) and Athabasca tar sands bitumen from Ft. McMurray (Alberta). Aerobic incubation with pure and mixed bacterial cultures of known oil-degrading capability yielded only minor changes in the gas chromatogram of sulfur compounds and small decreases in the total sulfur content, compared with conventional crude oils. This suggests that much of the organic sulfur in heavy crude petroleums is not readily available for microbial oxidation because of molecular complexity or reduced interfacial contact area due to viscosity.

Desulfurization of heavy oil distillate fractions using mixed, aerobic bacterial cultures was investigated by Eckart et al. (1981, 1982). Gas oil (1.2 to 2 weight percent sulfur), vacuum distillates (1.8 to 2 weight percent sulfur), and fuel oil (up to 4 weight percent sulfur) were used as the sole carbon and sulfur sources for the oil-degrading microorganisms (Eckart et al., 1981). The addition of an emulsifying agent was required to enhance desulfurization. After 5 to 7 days of incubation, sulfur removals were up to 20 percent from gas oil, up to 5 percent from vacuum distillates, and up to 25 percent from fuel oil. Supplementation of the mineral medium with yeast extract and glucose did not increase the extent of desulfurization. In another study (Eckart et al., 1982), approximately 30 percent of sulfur was removed from fuel-D-oil by a mixed population of bacteria. High-resolution mass spectrometry analysis showed removal of benzothiophene, DBT, and naphthobenzothiophene. However, this was accompanied by the removal of some other aromatic compounds and an approximate 60 percent decrease in the n-alkane content of the oil. NMR analysis showed that biodegradation caused a decrease in the degree of substitution and in the proportion of aromatic protons.

We have attempted to biodesulfurize Syncrude coker gas oil, the liquid product resulting from coking of Athabasca tar sands bitumen. Compared with the native bitumen, the coker gas oil has reduced viscosity, reduced average molecular weight, and equivalent total sulfur content (4.3 weight percent sulfur). Initial biodesulfurization attempts using pure and mixed population enrichment cultures were unsuccessful (unpublished results). The cultures metabolized the saturate fraction of the coker gas oil, but were unable to metabolize the aromatics and S-heterocycles in the coker gas oil (although they readily degraded these fractions in conventional crude oils). One pure culture, *Pseudomonas* sp. HL7b (Foght and Westlake, 1988) did metabolize the aromatics and some of the S-heterocycles in this oil. The gas chromatographic profile of S-heterocycles in the coker gas oil differs from that of several degradable conventional crude oils we have studied (unpublished observations). The apparent recalcitrance of this coker gas oil to biodesulfurization may result from the presence of inhibitory compounds or alkyl-substituted benzothiophenes and DBTs, which are prevalent in gas oils derived from heavy crude oils and bitumens (Clugston et al., 1976).

Although most studies on petroleum biodesulfurization have assessed aerobic bacterial processes, Eckart et al. (1986) have investigated the biodesulfurization of Romashkino crude oil under anaerobic conditions. Mixed cultures, in which *Desulfovibrio* spp. were the predominant organisms, were grown under sulfate-reducing conditions with lactate as a source of carbon and energy. Preliminary batch culture incubations for 7 days showed an increase in pH (to pH 8.5 to 9), a decrease in redox potential to approximately -280 mV, and a reduction in sulfur of approximately 8 percent. When pH was controlled at 7.2, desulfurization of 26 to 40 percent was observed after 48 h. Emulsifying agents were not required in these cultures. To optimize the process, the ratio of oil to nutrient solution was varied between 10 and 40 percent (v/v). After 48-h incubation, cultures with 40 percent oil showed less desulfurization than those containing 10, 20, or 30 percent oil. There was little difference in the amounts of desulfurization among the latter three cultures.

Degradation studies with model compounds and petroleum fractions indicate that microbial desulfurization can be selective for specific compounds such as DBT, and that removal of a variety of substituted sulfur compounds also can be achieved.

Enhancing Microbial Capabilities for Biodesulfurization

Selection of microbial strains for biodesulfurization

Materials from environments contaminated with petroleum or coal have been used as sources of desulfurizing bacteria, because prolonged contact with S-heterocycles may enrich for microorganisms possessing degradative activities. For example, oil-soaked soil (Sagardía et al., 1975; Monticello et al., 1985), oil sludges (Kurita et al., 1971), and coal dump sites (Isbister et al., 1988; Ochman and Klubek, 1986) have been used as sources of inocula for selection and isolation of desulfurizing bacteria. However, because some S-heterocycles are degraded cometabolically and do not sustain microbial growth (Kodama, 1977a; Monticello et al., 1985; Cripps, 1973), direct nutritional selection for S-heterocycle-oxidizing capability is not always possible. Instead, isolates have been selected for tolerance to toxic S-heterocycles such as benzothiophene (Sagardía et al., 1975) or DBT (Isbister et al., 1988) or by screening for production of colored metabolites from DBT (Finnerty et al., 1983; Foght and Westlake, 1988). In contrast, it may be possible to select microbes capable of releasing sulfate or sulfide from organic sulfur compounds by enrichment in a sulfur-limited medium with petroleum as the sole sulfur source. It is desirable to select isolates capable of metabolizing the S-heterocycles but not the economically valuable hydrocarbons in petroleum.

Screening for desulfurizing bacteria can be deliberately biased toward isolation of those with constitutive production of degradative enzymes versus those with inducible enzyme systems. For example, Foght and Westlake (1988) isolated *Pseudomonas* sp. HL7b, which constitutively produced the enzymes required for degradation of DBT by selecting colonies that immediately produced colored metabolites on DBT-sprayed agar plates. This characteristic has implications for economic desulfurization applications. For example, organisms requiring induction of S-heterocycle degradative enzymes with specific compounds such as naphthalene (Kodama, 1977b; Monticello et al., 1985) would be more expensive to produce than those synthesizing the enzymes during growth on an inexpensive carbon substrate.

Genetic engineering and mutagenesis of selected strains

Parallels have been drawn between the degradative pathways of DBT and aromatic hydrocarbons such as naphthalene (Monticello and Finnerty, 1985). Laborde and Gibson (1977) demonstrated that a

Beijerinckia sp. metabolized DBT in a manner analogous to naphthalene oxidation, via dihydrodiol and dihydroxy intermediates (Fig. 16.4). In fact, naphthalene dihydrodiol dehydrogenase purified from *P. putida* was capable of producing the same DBT intermediates as crude cell extracts of the *Beijerinckia* sp., suggesting that the degradative pathways and organisms share common oxidative steps and perhaps common broad-specificity enzymes. Despite the metabolic similarities, there have been numerous reports of plasmid-encoded aromatic hydrocarbon degradation but only one report of direct plasmid involvement in S-heterocycle degradation. Monticello et al. (1985) were the first to demonstrate plasmid-encoded DBT degradation in two *Pseudomonas* spp. The DBT-degradative genes from pDBT2 (the 55 Mdal plasmid of *P. alcaligenes* DBT2) were subcloned in a cosmid vector and transferred to an *E. coli* recipient (Finnerty and Robinson, 1986). The degradative genes were not expressed in the heterologous host, but DBT oxidation was restored upon mobilization of the construct into a cured, isogenic recipient (*P. alcaligenes* DM201). This manipulation apparently was facilitated by fortuitous clustering of the degradative genes in a 17-Mdal fragment of pDBT2. Perhaps surprisingly, aromatic hydrocarbon degradative plasmids demonstrated little homology with the DBT degradative genes (Finnerty and Hartdegen, 1984).

DBT metabolism by *Pseudomonas stutzeri* DBT3 was assumed to be encoded chromosomally (Monticello et al., 1985) because no plasmid was isolated. Location and cloning of chromosomal genes can be more difficult than with plasmid-encoded genes, but may result in enhanced degradative capability if the genes are cloned into high-copy-number vectors. It is difficult to speculate on the immediate potential for genetic engineering of S-heterocycle degradation by different genera without more information about the location and arrangement of degradative genes. Failure to conclusively demonstrate plasmid involvement in DBT metabolism (Kiyohara et al., 1983; Foght, 1985) could be due to chromosomal location or to a combination of chromosomal and plasmid loci for the degradative pathway genes.

Characteristics other than degradative capability can be introduced by genetic engineering. For example, heavy metals present in petroleum feedstocks contribute to chemical catalyst poisoning. Resistance to heavy metals is encoded on plasmids in many bacterial genera, and it might be possible to introduce heavy-metal resistance genes into biocatalyst cells or to select naturally resistant strains to reduce inhibition by heavy metals.

Chemical mutagenesis of isolates may enable selection of desired characteristics without resorting to intricate genetic engineering techniques. For example, Isbister and Doyle (1985) mutagenized a naturally occurring *P. putida* strain with diethyl sulfate, subsequently selecting for enhanced ability to oxidize DBT. Abdulrashid and Clark (1987) used successive rounds of ethyl methanesulfonate mutagenesis and selection to yield an *E. coli* strain capable of degrading thiophene-2-carboxylate and thiophene-2-methylamine. Thus, it is possible to mutagenize and select bacterial strains to enhance S-heterocycle degradative capabilities.

Transposon mutagenesis has been used both to "label" degradative genes with a selectable marker, such as antibiotic resistance, and to tailor oxidation products by truncating the degradative pathway. Finnerty and Robinson (1986) reported that Tn5 insertional mutagenesis of pDBT2 yielded a pool of mutants which were classified into groups according to the different DBT metabolites they accumulated. One group accumulated the first oxidation product, DBT-dihydrodiol, and the second group accumulated a ring cleavage product without formation of the terminal metabolite 3-hydroxy-2-formyl-benzothiophene (HFBT). This is significant, because it may be possible to minimize the oxidative steps required for efficient production of water-soluble compounds, rather than allowing the biocatalysts to waste reaction time by completing the pathway. In addition, accumulation of toxic polar metabolites (Monticello et al., 1985) may be prevented by engineering a truncated degradative pathway. It may also be possible to mutagenize genes so that nonspecific degradation of hydrocarbons is eliminated, precluding loss of economically valuable hydrocarbons from the oil. Site-specific mutagenesis is not currently feasible since sequences for S-heterocycle degradative genes have not been reported. However, this method could be useful in the future for altering pathway regulation and specificity of oxidation.

Horizontal and vertical expansion of S-heterocycle degradation pathways, analogous to manipulation of halogenated hydrocarbon degradation (Ramos and Timmis, 1987) is appealing but not yet practical for microbial desulfurization because information is limited regarding specificity and regulation of S-heterocycle metabolism. For example, when a DBT degradative plasmid (pDBT2) from *P. alcaligenes* DBT2 was introduced into *P. stutzeri* DBT3, the recipient accumulated the hemiacetal form of the DBT ring cleavage product rather than its usual metabolite HFBT (Fig. 16.4; Monticello et al., 1985). This suggests that pDBT2 encodes a regulatory function capable of influencing DBT degradation in trans in the recipient strain. However, there is no evidence yet that regulation of DBT degradation can be expanded to include degradation of other S-heterocycles in petroleum.

Consideration must be given to some of the problems inherent to genetic engineering of bacteria for enhanced desulfurization capability.

Stability of the engineered genes is crucial, and maintenance of the desired characteristics may require alternative selective pressure, such as antibiotic resistance to prevent loss of a cometabolic function. A high level of expression of the cloned genes must be achieved, which may be difficult in heterologous hosts. Because the S-heterocycles are virtually insoluble in water and their metabolites are removed by diffusion, the biocatalyst degradative enzymes should be located on the cell surface. This requires the host cell's membranes to be compatible with the oxidative enzymes, and the metabolites to be released into the aqueous phase for removal (unlike *P. aeruginosa* ERC-8 in which DBT metabolites were cell-associated; Hou and Laskin, 1976). These factors must be evaluated in addition to the physical consideration of biocatalyst preparation.

Reactor Design and Process Feasibility

Bioreactors: chemical considerations

Finnerty et al. (1983) described batch reactions in which DBT removal by a pre-grown biocatalyst was optimized with respect to cell concentration, pH, aeration, DBT concentration, agitation, and temperature. These optimum conditions were then applied to treatment of a low-sulfur crude oil supplemented with DBT and to a high-sulfur crude oil. Hartdegen et al. (1984) noted that the ratio of aqueous phase to petroleum phase in these batch reactions influenced the rate of desulfurization. When the amount of crude oil increased above 10 percent (v/v), the rate of desulfurization decreased dramatically. A ratio of 10 percent oil was considered too low for process application, but by increasing the oil volume to 50 percent, a compromise was reached with a desulfurization rate at 30 percent of optimum.

Hartdegen et al. (1984) suggested the use of flow-through bioreactors for desulfurization. The proposed bioreactor conformation consisted of packed-bed, immobilized cells sandwiched between the oil phase and the aqueous phase, separated by permeable membranes. This design would allow desulfurization to occur without mixing of oil and water phases, thereby avoiding potential bioemulsification problems. Removal of polar metabolites in the effluent stream might prevent inhibition of the biocatalyst. Hartdegen et al. (1984) recognized the need to optimize the reactive surface area of the bioreactor, since the oxidation process is most probably a cell surface phenomenon. This requires that substrate access to the cell surfaces be efficient and rapid.

Bioreactors: biological considerations

To produce a superior biocatalyst economically, several factors must be considered in addition to any pertinent genetic engineering constraints. First, pregrown cells should be used as nongrowing biocatalysts. This reduces the possibility that the cells will produce surface-active agents in response to growth on the oil, which could make separation of the product streams difficult, and reduces the possibility that the cells will grow at the expense of valuable hydrocarbons in the oil. Maximum activity of the biocatalyst cells should not require the presence of expensive inducers during growth (e.g., naphthalene, salicylate). To keep biocatalyst production costs low, the cells should grow on an inexpensive carbon source and have a high yield of active biomass per unit substrate.

The biocatalyst cells should not demonstrate substrate toxicity. Several investigators have shown that, fortunately, S-heterocycle toxicity may be reduced by dissolution in oil (Nakatani et al. 1968; Sagardía et al., 1975; Hou and Laskin, 1976). Therefore, compounds may be less toxic in an oil-desulfurization process than indicated by pure compound reactions. Furthermore, the use of selected hosts for cloned genes may allow production of substrate-tolerant biocatalysts.

Similarly, the biocatalyst cells should not be subject to product inhibition. Relief of inhibition by polar organic metabolites could be accomplished by using ion-exchange resins. Finnerty and Robinson (1986) reported that DBT oxidation increased 10-fold when ion-exchange resins were used to remove products from the reaction, and Laborde and Gibson (1977) found that resins were required for production of enzymatically active cell extracts. Alternatively, product inhibition could be prevented by adjusting the proportion of polar products formed, using strain selection, genetic engineering, or mutagenesis of the biocatalyst cells. Examples of such degradative pathway manipulation have been demonstrated, respectively, in different species of *Pseudomonas* (Monticello et al., 1985) by transferring a putative regulatory factor on a plasmid to a degradative strain (Monticello et al., 1985), and by transposon mutagenesis of degradative genes (Finnerty and Robinson, 1986).

The biocatalyst should have a high reaction rate as a function of gene copy number (e.g., by cloning degradative genes onto high-copy-number plasmids) or through cell surface characteristics (e.g., permeability, membrane protein conformations). This reaction rate should be measurable and predictable so that retention times can be calculated during the life of the biocatalyst. Also, the biocatalyst should specifically oxidize organic sulfur and S-heterocycles, without oxidizing economically valuable hydrocarbons at all, or at such a slow rate that the nonspecific reactions are insignificant with the operative retention time. Microbes with this specificity can be selected (Finnerty et al., 1983) or genetically engineered. At the same time, the biocatalyst should metabolize a broad range of S-heterocycles, ranging from thiophenes to alkyl-DBTs and larger molecules, so that maximum biodesulfurization is achieved.

Biocatalyst stability must be optimized with respect to storage and recycling. The cells should not lose activity when stored as wet cell mass (e.g., frozen at −20°C) or as lyophilized cells, and should be easily recyclable without loss of activity by autolysis or enzyme instability. The degradative genes should be stable within the cells so that the phenotype is not lost during pregrowth on a nonhydrocarbon medium. The cells and the degradative enzymes should be stable at the process temperatures (e.g., ambient to 55°C). Immobilization of the cells should be possible, achieving maximum available active surface area without disrupting enzymatic function.

Economic feasibility

Finnerty and Hartdegen (1984) presented a cost analysis for large-scale aerobic biodesulfurization of a 3.0 weight percent sulfur oil, using an immobilized whole-cell bioreactor producing polar organic metabolites. Low temperature and neutral pH conditions were assumed, with a water-to-oil ratio of 1:1, a 4- to 5-h residence time in the bioreactor, and a desired level of 0.7 weight percent sulfur in the final product. The calculations showed that biodesulfurization was not economically attractive compared with chemical desulfurization of high-sulfur crude oils or direct purchase of low-sulfur crude oils. The analysis assumed a substantial credit for conversion of waste-stream products to benzene and phenol, but a process to accomplish this step has not been demonstrated. Even if the sulfur and carboxylic acid groups were removed from the polar products, the remaining carbon structures would be alkylated mono- and dicyclic hydrocarbons which are worth less than benzene and phenol. The more likely scenario is that disposal of the polar sulfur species would add to the operating costs. Consequently, the report underestimates the net operating costs for biodesulfurization by at least 25 percent. Even with improvements in the process to achieve the stated objectives of a 1-h retention time and 12-month biocatalyst life, the microbial process would still be more expensive than the chemical process and more expensive yet than direct purchase of low-sulfur oil. Finnerty and Hartdegen's analysis assumed a price differential of $3.67 per barrel between high-sulfur residual oil and a low-sulfur product. At the time of this writing, the price differential fluctuates in the same range, even though the price per barrel is lower. In order to improve the economics of aerobic biodesulfurization as a competitive refinery process, the utilization of the polar organic sulfur compounds must be considered in conjunction with improvements to the microbial process.

Future Research and Alternative Process Development

Although biodesulfurization has been proposed in various forms for decades, much research is still required to permit economic development of the process. Several areas of potential interest and suggestions for new research directions are presented here.

To date, only a few bacterial species have been examined for biodesulfurization ability. Although the capabilities of fungal species and most archaebacteria have not been explored, they may possess the desired characteristics. Improvement to screening and analytical methods should enlarge the number of potential biodesulfurizing microbes available for evaluation.

An area of potential study is the use of mixed populations capable of degrading a broad range of S-heterocycles rather than a single population (genetically engineered or otherwise). A disadvantage is that population dynamics can be difficult to control. Another alternative is the use of immobilized enzymes rather than whole cells. Enzymes have been described which are capable of catalyzing reactions under nonpolar conditions (Deetz and Rozzell, 1988). However, this technology is not possible at the present time because of limited information about the S-heterocycle degrading enzymes, such as cofactor requirements.

This chapter has emphasized processes in which biodesulfurization is accomplished by producing polar organic metabolites because these have been the focus of recent reports. However, this is not the most attractive process because formation of water-soluble metabolites results in removal of both the sulfur and its associated structural carbon and hydrogen from the oil. The problems of recovery of these oxidized products in a useful form, or appropriate disposal of the wastes, are a significant drawback to the process. The higher the molecular weight of the oil, the greater is the loss of hydrocarbons due to selective desulfurization. In simple terms, the potential loss of hydrocarbons is given by

$$\% \text{ loss} = f_{C/S} \cdot \text{weight percent thiophenic sulfur in the oil}$$

where $f_{C/S}$ = ratio of hydrocarbon to sulfur in thiophenic species = $(AMW - 32)/32$
AMW = average molecular weight of the oil

For example, oxidation of DBT removes 4.75 g of hydrocarbon for every gram of sulfur removed. In a heavy oil with average molecular weight of about 400 and containing 5 percent thiophenic sulfur, complete conversion of sulfur would result in a loss of 57 percent of the oil into the aqueous phase! Clearly, aerobic biodesulfurization of high-sulfur oils may result in unacceptable conversion to aqueous species which have no demonstrated value. Even in lighter distillates, a loss of four to five parts of hydrocarbon for every part of sulfur removed is unattractive in comparison to chemical processes which give much higher carbon recovery for a given extent of sulfur removal. This observation does not invalidate previous research, but rather places the onus on researchers to isolate microbes (such as the coal-desulfurizing bacteria described by Isbister et al., 1988) with high-specificity enzymes for removing the heteroatom from thiophenic sulfur without oxidizing the hydrocarbon skeleton, or to develop techniques for economic conversion of the polar products to valuable feedstocks.

Theoretically, anaerobic biodesulfurization is an attractive alternative to aerobic processes for several reasons. First, the expense of providing aeration is eliminated in the anaerobic process. Second, because the rate of microbial transformation of valuable hydrocarbons under anaerobic conditions is much slower than under aerobic conditions, the quality of the desulfurized petroleum should not be reduced in the anaerobic process by nonspecific attack. Finally, although processing or disposal of polar products from aerobic biodesulfurization is an unresolved problem, hydrogen sulfide produced by anaerobic conversion of organic sulfur could be recovered readily by using existing chemical technologies. Very few studies of anaerobic biodesulfurization have been reported. In particular, the fate of the hydrocarbon backbone is unclear and, because this has implications for the economic feasibility of the process, requires study. However, if anaerobic conditions can effect conversion of S-heterocycles to hydrogen sulfide without oxidizing the hydrocarbon moiety, then the predicted lower rate of desulfurization would be balanced by more attractive recovery of hydrocarbons. The potential for anaerobic biodesulfurization requires further evaluation.

Another theoretical possibility is the introduction of microbes into suitable reservoirs to biodesulfurize petroleum in situ. Introduced microbes could be left in place over extended periods of time to slowly effect desulfurization under anaerobic conditions. Anecdotal reports of progressive souring of petroleum reservoirs suggest that such a process is possible. Ideal candidate reservoirs would have undergone primary production, and would be undergoing waterflooding to help flush additional oil out of the formation. Such a process would have significant advantages over biodesulfurization of produced petroleum. Under waterflood conditions, the interfacial contact area between the oil and the aqueous phase would be enormous, with high local ratios of water to oil. These conditions would enhance the transport of organic sulfur compounds to the active organisms in the aqueous phase. Because reservoirs have long production lives, the rate of desulfurization would be less important than in refinery biocatalysis. The low levels of sulfur removal demonstrated in the laboratory with actual oils (20 to 40 percent) would be acceptable in an in situ process, because even though surface processing would still be required the value of the oil would be enhanced.

Use of biodesulfurization in situ could also use the solubilization of thiophenic compounds to advantage. If carboxylic acids or sulfonates were produced, the sulfur compounds would be converted into surfactants. Such compounds are well known as mobilizing agents for oil recovery, where emulsification of a portion of the oil serves to enhance overall oil recovery. A combined biodesulfurization-biosurfactant formation process would balance the loss of hydrocarbon into the aqueous phase with improved recovery of oil from the reservoir. Such a process would, however, be subject to the same problems encountered with microbially enhanced oil-recovery processes (McInerney and Westlake, Chap. 17), such as slow dissemination of the introduced organisms, supplying an alternative carbon source for microbial utilization, and pore plugging by growing cells.

Conclusions

It is apparent from studies with bacterial cultures that a number of the organic sulfur compounds found in petroleum can be metabolized under aerobic or anaerobic conditions. A few compounds have been chosen, justifiably or not, as models for microbial degradation, but specific information is scarce regarding metabolism of most petroleum sulfur compounds. Reported bench-scale studies indicate that biodesulfurization of petroleum occurs. This is encouraging and is an incentive to address the biological uncertainties of the process and to optimize reaction conditions. Advances in genetic engineering techniques and applications may play a key role in making biodesulfurization of petroleum feasible. The potential for increasing the specificity of S-heterocycle oxidations and improving characteristics of biocatalysts by cloning and mutagenesis is exciting.

Current evaluations indicate that biodesulfurization of conventional petroleum feedstocks is not economical. Microbial treatment of high-sulfur petroleum such as heavy crude oils and bitumens may appear to be more feasible because of the disproportionately higher hydrodesulfurization costs associated with hydrogen demands and with chemical catalyst renewal due to coke deposition and heavy-metal poisoning. Paradoxically, the high-sulfur feedstocks are currently less amenable to biodesulfurization than conventional petroleums because low interfacial areas, high viscosity, and molecular complexity reduce accessibility and reactivity of the organic sulfur. Additionally, biodesulfurization of heavy crude oils through aerobic production of polar metabolites is not economically attractive because of the high carbon losses associated with high-molecular-weight sulfur. It is possible that the microbial process could be useful as an accessory treatment for coked bitumen and cracked or diluted heavy crudes in which the average molecular weight of the organic sulfur and viscosity are reduced. In this case, it is clear that biodesulfurization of asphaltene and resin fractions should be emphasized in research efforts and that processes be developed for petroleum in which the heteroatom is released from the organic skeleton to minimize hydrocarbon losses.

There are several unexplored areas of potential research, including anaerobic biodesulfurization and in situ processing. It is hoped that research efforts will be expended in the future towards commercial petroleum biodesulfurization process development.

Addendum

Since this chapter was written, additional technical reports have appeared supporting the alternative pathway of DBT degradation to biphenyl derivatives and sulfate. These have been reviewed by S. Krawiec (Dev. Ind. Microbiol., under review).

References

Abdulrashid, N., and Clark, D. P., Isolation and genetic analysis of mutations allowing the degradation of furans and thiophenes by *Escherichia coli*, *J. Bacteriol.* **169**:1267–1271 (1987).

Amphlett, M. J., and Callely, A. G., The degradation of 2-thiophenecarboxylic acid by a *Flavobacterium* species, *Biochem. J.* **112**:12p (1969).

Atlas, R. M., Boehm, P. D., and Calder, J. A., Chemical and biological weathering of oil, from the *Amoco Cadiz* spillage, within the littoral zone, *Estuarine Coastal Shelf Sci.* **12**:589–608 (1981).

Babenzien, H.-D., Genz, I., and Köhler, M., Oxydativer Abbau von Dibenzylsulfid. *Z. Allg. Mikrobiol.* **19**:527–533 (1979).

Bhadra, A., Scharer, J. M., and Moo-Young, M., Microbial desulphurization of heavy oils and bitumen, *Biotech. Adv.* **5**:1–27 (1987).

Blount-Fronefield, D. D., Gottlund, K., Krawiec, S., and Montenecourt, B. S., Correlation of dibenzothiophene disappearance and growth of *Sulfolobus acidocaldarius*, in *Proc. 86th Ann. Meeting American Society for Microbiology*, Abstract #Q-12, 1986.

Bohonos, N., Chou, T.-W., and Spanggord, R. J., Some observations on biodegradation of pollutants in aquatic systems, *Jap. J. Antibiot.* **30**(suppl):275–285 (1977).

Burns, B. J., Hogarth, J. T. C., and Milner, C. W. D., Properties of Beaufort basin liquid hydrocarbons, *Bull. Can. Pet. Geol.* **23**:295–303 (1975).

Clugston, D. M., George, A. E., Montgomery, D. S., Smiley, G. T., and Sawatzky, H., Sulfur compounds in oils from the western Canada tar belt, in T. F. Yen (ed.), *Shale Oil, Tar Sands and Related Fuel Sources*, Advances in Chemistry Series 151, American Chemical Society, Washington, D.C., 1976, pp. 11–27.

Cripps, R. E., The microbial metabolism of thiophene-2-carboxylate, *Biochem. J.* **134**:353–366 (1973).

Cyr, N., McIntyre, D. D., Toth, G., and Strausz, O. P., Hydrocarbon structural group analysis of Athabasca asphaltene and its g.p.c. fractions by [13]C n.m.r., *Fuel* **66**:1709–1714 (1987).

Cyr, T. D., Payzant, J. D., Montgomery, D. S., and Strausz, O. P., A homologous series of novel hopane sulfides in petroleum, *Org. Geochem.* **9**:139–143 (1986).

Deetz, J. S., and Rozzell, J. D., Enzyme-catalysed reactions in non-aqueous media, *Trends Biotechnol.* **6**:15–19 (1988).

Deroo, G., Tissot, B., McCrossan, R. G., and Der, F., Geochemistry of the heavy oils of Alberta, in L. V. Hills (ed.), *Oil Sands, Fuel of the Future, Can. Soc. Pet. Geol. Mem.* **3**:148–167 (1974).

Drushel, H. V., and Sommers, A. L., Isolation and characterization of sulfur compounds in high-boiling petroleum fractions, *Anal. Chem.* **39**:1819–1829 (1967).

Eckart, V., Hieke, W., Bauch, J., and Bohlmann, D., Microbial desulfurization of petroleum and heavy petroleum fractions. 2. Studies on microbial aerobic desulfurization of heavy petroleum fractions, *Zbl. Bakt. II. Abt.* **136**:152–160 (1981).

Eckart, V., Hieke, W., Bauch, J., and Gentzsch, H., Microbial desulfurization of petroleum and heavy petroleum fractions. 1. Studies on microbial aerobic desulfurization of Romashkino-crude oil, *Zbl. Bakt. II. Abt.* **135**:674–681 (1980).

Eckart, V., Hieke, W., Bauch, J., and Gentzsch, H., Microbial desulfurization of petroleum and heavy petroleum fractions. 3. The change of chemical composition of fuel-D-oil by microbial aerobic desulfurization, *Zbl. Mikrobiol.* **137**:270–279 (1982).

Eckart, V., Köhler, M., and Hieke, W., Microbial desulfurization of petroleum and heavy petroleum fractions. 5. Anaerobic desulfurization of Romashkino petroleum, *Zbl. Mikrobiol.* **141**:291–300 (1986).

Ensley, B. D., Jr., Microbial metabolism of condensed thiophenes, in D. T. Gibson (ed.), *Microbial Degradation of Organic Compounds*, Dekker, New York, 1984, pp. 309–317.

Fedorak, P. M., and Westlake, D. W. S., Microbial degradation of organic sulfur compounds in Prudhoe Bay crude oil, *Can. J. Microbiol.* **29**:291–296 (1983).

Fedorak, P. M., and Westlake, D. W. S., Degradation of sulfur heterocycles in Prudhoe Bay crude oil by soil enrichments, *Water Air Soil Pollut.* **21**:225–230 (1984).

Fedorak, P. M., Payzant, J. D., Montgomery, D. S., and Westlake, D. W. S., Microbial degradation of *n*-alkyltetrahydrothiophenes found in petroleum, *Appl. Environ. Microbiol.* **54**:1243–1248 (1988).

Finnerty, W. R., and Hartdegen, F. J., Microbial desulfurization of fossil fuels, in *The World Biotech Report 1984*, Vol. 1: *Europe*, Online Publications, U.K., 1984, pp. 611–622.

Finnerty, W. R., and Robinson, M., Microbial desulfurization of fossil fuels: A review, *Biotechnol. Bioeng. Symp.* **16**:205–221 (1986).

Finnerty, W. R., Shockley, K., and Attaway, H., Microbial desulfurization and denitrification of hydrocarbons, in J. E. Zajic, D. C. Cooper, T. R. Jack, and N. Kosaric (eds.), *Microbial Enhanced Oil Recovery*, PennWell, Tulsa, Okla., 1983, pp. 83–91.

Foght, J. M., Plasmids and the biodegradation of polycyclic aromatic hydrocarbons, Ph.D. Thesis, University of Alberta, 1985.

Foght, J. M., Fedorak, P. M., and Westlake, D. W. S., Effect of the dispersant Corexit 9527 on the microbial degradation of sulfur heterocycles in Prudhoe Bay oil, *Can. J. Microbiol.* **29**:623–627 (1983).

Foght, J. M., and Westlake, D. W. S., Degradation of some polycyclic aromatic hydrocarbons by a plasmid-containing *Flavobacterium* species, in *Proc. 85th Ann. Meeting American Society for Microbiology*, Abstract #Q-20, 1985.

Foght, J. M., and Westlake, D. W. S., Degradation of polycyclic aromatic hydrocarbons and aromatic heterocycles by a *Pseudomonas* species, *Can. J. Microbiol.* **34**:1135–1141 (1988).

Fuentes, F. A., Diauxic growth of *Pseudomonas aeruginosa* PRG-1 on glucose and benzothiophene, in *Proc. 84th Ann. Meeting American Society for Microbiology*, Abstract #N26, 1984.

Godsy, E. M., and Grbić-Galić, D., Anaerobic degradation pathways for benzothiophene in aquifer-derived methanogenic microcosms, in *Proc. 88th Ann. Meeting American Society for Microbiology*, Abstract #Q111, 1988.

Gundlach, E. R., Boehm, P. D., Marchand, M., Atlas, R. M., Ward, D. M., and Wolfe, D. A., The fate of *Amoco Cadiz* oil, *Science* **221**:122–129 (1983).

Harney, B. M., *Oil from Shale,* Dekker, New York, 1983.

Hartdegen, F. J., Coburn, J. M., and Roberts, R. L., Microbial desulfurization of petroleum, *Chem. Eng. Prog.* **80**:63–67 (1984).

Ho, T. Y., Rogers, M. A., Drushel, H. V., and Koons, C. B., Evolution of sulfur compounds in crude oils, *Am. Assoc. Pet. Geol. Bull.* **58**:2338–2348 (1974).

Hou, C. T., and Laskin, A. I., Microbial conversion of dibenzothiophene. *Dev. Ind. Microbiol.* **17**:351–362 (1976).

Isbister, J. D., and Doyle, R. C., A novel mutant microorganism and its use in removing organic sulfur compounds, Eur. Pat. Appl. EP 218,734, 1985.

Isbister, J. D., Wyza, R., Lippold, J., DeSouza, A., and Anspach, G., Bioprocessing of coal, in G. S. Omenn (ed.), *Environmental Biotechnology: Reducing Risks from Environmental Chemicals through Biotechnology,* Plenum Press, 1988, pp. 281–293.

Kanagawa, T., and Kelly, D. P., Degradation of substituted thiophenes by bacteria isolated from activated sludge, *Microb. Ecol.* **13**:47–57 (1987).

Kargi, F., Biological oxidation of thianthrene, thioxanthene and dibenzothiophene by the thermophilic organism *Sulfolobus acidocaldarius, Biotechnol. Lett.* **9**:478–482 (1987).

Kargi, F., and Robinson, J. M., Microbial oxidation of dibenzothiophene by the thermophilic organism *Sulfolobus acidocaldarius, Biotechnol. Bioeng.* **26**:687–690 (1984).

Kargi, F., and Robinson, J. M., Removal of organic sulfur from bituminous coal, *Fuel* **65**:397–399 (1986).

Kirshenbaum, I., Bacteriological desulfurization of petroleum, U.S. Patent 2,975,103, 1961.

Kiyohara, H., Sugiyama, M., Mondello, F. J., Gibson, D. T., and Yano, K., Plasmid involvement in the degradation of polycyclic aromatic hydrocarbons by a *Beijerinckia* species, *Biochem. Biophys. Res. Comm.* **111**:939–945 (1983).

Kodama, K., Co-metabolism of dibenzothiophene by *Pseudomonas jianii, Agric. Biol. Chem.* **41**:1305–1306 (1977a).

Kodama, K., Induction of dibenzothiophene oxidation by *Pseudomonas jianii, Agric. Biol. Chem.* **41**:1193–1196 (1977b).

Kodama, K., Nakatani, S., Umehara, K., Shimizu, K., Minoda, Y., and Yamada, K., Microbial conversion of petro-sulfur compounds. III. Isolation and identification of products from dibenzothiophene, *Agric. Biol. Chem.* **34**:1320–1324 (1970).

Kodama, K., Umehara, K., Shimizu, K., Nakatani, S., Minoda, Y., and Yamada, K., Identification of microbial products from dibenzothiophene and its proposed oxidation pathway, *Agric. Biol. Chem.* **37**:45–50 (1973).

Köhler, M., Genz, I.-L., Schicht, B., and Eckart, V., Microbial desulfurization of petroleum and heavy petroleum fractions. 4. Anaerobic degradation of organic sulfur compounds of petroleum, *Zbl. Mikrobiol.* **139**:239–247 (1984).

Kurita, S., Endo, T., Nakamura, H., Yagi, T., and Tamiya, N., Decomposition of some organic sulfur compounds in petroleum by anaerobic bacteria, *J. Gen. Appl. Microbiol.* **17**:185–198 (1971).

Laborde, A. L., and Gibson, D. T., Metabolism of dibenzothiophene by a *Beijerinckia* species, *Appl. Environ. Microbiol.* **34**:783–790 (1977).

Later, D. W., Lee, M. L., Bartle, K. D., Kong, R. C., and Vassilaros, D. L., Chemical class separation and characterization of organic compounds in synthetic fuels, *Anal. Chem.* **53**:1612–1620 (1981).

Maka, A., McKinley, V. L., Conrad, J. R., and Fannin, K. F., Degradation of benzothiophene and dibenzothiophene under anaerobic conditions by mixed cultures, in *Proc. 87th Ann. Meeting American Society for Microbiology,* Abstract #O-54, 1987.

Malik, K. A., Microbial removal of organic sulfur from crude oil and the environment: Some new perspectives, *Process Biochem.* **13**(9):10–12 (1978).

Malik, K. A., and Claus, D., Microbial degradation of dibenzothiophene, in *Fifth Intern. Fermentation Symp., Berlin,* Abstract #23.03, 1976.

Monticello, D. J., Bakker, D., and Finnerty, W. R., Plasmid-mediated degradation of dibenzothiophene by *Pseudomonas* species, *Appl. Environ. Microbiol.* **49**:756–760 (1985).

Monticello, D. J., and Finnerty, W. R., Microbial desulfurization of fossil fuels, *Ann. Rev. Microbiol.* **39**:371–389 (1985).

Nakatani, S., Akasaki, T., Kodama, K., Minoda, Y., and Yamada, K., Microbial conversion of petro-sulfur compounds. II. Culture conditions of dibenzothiophene-utilizing bacteria, *Agric. Biol. Chem.* **32**:1205–1211 (1968).

Ochman, M., and Klubek, B., Isolation of thiophene utilizing bacteria from soil and strip-mine spoil, in *Proc. 86th Ann. Meeting American Society for Microbiology*, Abstract #Q-11, 1986.

Patton, J. S., Rigler, M. W., Boehm, P. D., and Fiest, D. L., Ixtoc 1 oil spill: Flaking of the surface mousse in the Gulf of Mexico, *Nature* **290**:235–238 (1981).

Payzant, J. D., Montgomery, D. S., and Strausz, O. P., Sulfides in petroleum, *Org. Geochem.* **9**:357–369 (1986).

Payzant, J. D., McIntyre, D. D., Mojelsky, T. W., Torres, M., Montgomery, D. S., and Strausz, O. P., The identification of homologous series of thiolanes and thianes possessing a linear carbon framework from petroleums and their interconversion under simulated geological conditions, *Org. Geochem.* **14**:461–473 (1989).

Rall, H. T., Thompson, C. J., Coleman, H. J., and Hopkins, R. L., Sulfur compounds in crude oil, U.S. Department of the Interior, Bureau of Mines, Bulletin 659, 1972.

Ramos, J. L., and Timmis, K. N., Experimental evolution of catabolic pathways of bacteria, *Microbiol. Sci.* **4**:228–237 (1987).

Rühl, W., *Tar (Extra Heavy Oil) Sands and Oil Shales; Geology of Petroleum*, Vol. 6, Ferdinand Enke, Stuttgart, FRG, 1982.

Sagardía, F., Rigau, J. J., Martínez-Lahoz, A., Fuentes, F., López, C., and Flores, W., Degradation of benzothiophene and related compounds by a soil *Pseudomonas* in an oil-aqueous environment, *Appl. Microbiol.* **29**:722–725 (1975).

Speight, J. G., *The Chemistry and Technology of Petroleum*, Dekker, New York, 1980.

Speight, J. G., *The Desulfurization of Heavy Oils and Residua*, Dekker, New York, 1981.

Stoner, D. L., Means, J. C., and MacDonell, M. T., Degradation of dibenzothiophene by an estuarine Xanthomonad, in *Proc. 86th Ann. Meeting American Society for Microbiology*, Abstract #Q-10, 1986.

Strawinski, R. J., Method of desulfurizing crude oil, U.S. Patent 2,521,761, 1950.

Strawinski, R. J., Purification of substances by microbial action, U.S. Patent 2,574,070, 1951.

Tissot, B. P., and Welte, D. H., *Petroleum Formation and Occurrence*, 2nd ed., Springer-Verlag, New York, 1984.

Walker, J. D., Colwell, R. R., and Petrakis, L., Microbial petroleum degradation: Application of computerized mass spectrometry, *Can. J. Microbiol.* **21**:1760–1767 (1975).

Walker, J. D., Colwell, R. R., and Petrakis, L., Biodegradation of petroleum by Chesapeake Bay sediment bacteria, *Can. J. Microbiol.* **22**:423–428 (1976).

Westlake, D. W. S., Microbial activities and changes in the chemical and physical properties of oil, in E. C. Donaldson and J. B. Clark (eds.), *Proc. 1982 Intern. Conf. on Microbial Enhancement of Oil Recovery, Shangri-La, Afton, Oklahoma*, Bartlesville Energy Technology Center, Bartlesville, Oklahoma, 1983, pp. 102–111.

Willey, C. Iwao, M., Castle, R. N., and Lee, M. L., Determination of sulfur heterocycles in coal liquids and shale oils, *Anal. Chem.* **53**:400–407 (1981).

Yamada, K., Minoda, Y., Kodana, K., Nakatani, S., and Akasaki, T., Microbial conversion of petro-sulfur compounds. I. Isolation and identification of dibenzothiophene-utilizing bacteria, *Agric. Biol. Chem.* **32**:840–845 (1968).

ZoBell, C. E., Process of removing hydrocarbons from petroleum hydrocarbons and apparatus, U.S. Patent 2,641,564, 1953.

Microbially Enhanced Oil Recovery

Michael J. McInerney
Donald W. S. Westlake

Introduction

Oil, being an essential energy source, is both the lifeblood and a liability of many industrialized nations. The use of crude oil as an energy source has allowed many nations to develop a high standard of living. However, the increasing reliance upon foreign imports, especially in the United States, has decreased economic growth and employment and aggravated trade deficits. Continued economic growth will increase the demand for oil which cannot be met with existing production technologies or by new discoveries. Conventional production technologies are able only to recover about one-third of the oil originally in place in a reservoir. Thus, methods to recover the rest of the oil offer the most timely and cost-effective solution to reverse the decline in domestic production and increase the oil reserves of the United States (Department of Energy, 1985).

The long-term economic potential for enhanced oil recovery is large. In the United States, the total original oil in place in discovered reservoirs is estimated to be 488 billion barrels (Department of Energy, 1985). About 133 billion barrels have been produced through 1983 and 28 billion barrels are listed as reserves, that is, those which can be recovered by existing technologies. The difference, 327 billion barrels, cannot be recovered by conventional technology and is the target for enhanced oil-recovery (EOR) research and development. The actual EOR production in the United States is low and has never exceeded 5 percent of the total U.S. production, even though a variety of economic incentives has been provided to stimulate the development and application of EOR processes (Department of Energy, 1985). In contrast, 18 percent of the production of conventional oil in Alberta, Canada's main oil-producing province, involves EOR technology. It is projected that by the year 2010 one-third of the total oil production will be by EOR procedures (*Enhanced Oil Recovery Week,* April 25, 1988, Pasha Publications, Arlington, Va). The use of EOR technologies in these North American situations has been limited to a few very favorable reservoirs where reservoir conditions allow for accurate predictions of the amount of additional oil recovered. The major constraint against the general application of EOR is the unacceptable risk imposed by unpredictable technical performance. This lack of predictability stems from our inadequate knowledge of the reservoir, rock-fluid interactions, and the basic mechanisms by which EOR processes mobilize and displace oil.

Microbially enhanced oil recovery (MEOR) has some unique advantages which may offer an economically attractive collection of processes. However, the use of microorganisms may also be detrimental to oil recovery. Microbial corrosion processes, reservoir souring, loss of injectivity due to microbial plugging, and the biodegradation of crude oils are very serious problems which cause large economic losses to the petroleum industry. Although microbial activity is of economic concern to the petroleum industry, the microbiology of oil reservoirs remains for the most part an enigma, little studied due to the difficulty in obtaining uncontaminated samples. However, recent interest in the use of biodegradation for the cleanup of polluted groundwaters has increased our understanding of the actual and potential metabolic activities of subsurface microorganisms and the environmental factors which stimulate or limit metabolic processes in the subsurface (Suflita, 1987). Techniques to obtain uncontaminated subsurface material have been developed (Wilson et al., 1983) which have allowed us to greatly increase our understanding of shallow subsurface environments. Since diverse life-forms often exhibit similar metabolic pathways (Dagley, 1984), the metabolic principles gleaned from the study of surface and shallow subsurface habitats may be applied to deeper oil reservoirs. Knowledge about the distribution and transport of microorganisms and the predominant ecological processes which occur in shallow aquifers can be extrapolated to understand how microbial processes in deeper formations might be manipulated. An understanding of the microbial ecology of petroleum reservoirs is a prerequisite to the development of any EOR process whether it is a microbial process or not, since an inappropriate design may accelerate the detrimental activities of microorganisms.

Compared to other EOR processes, our understanding of MEOR is very limited. Thus, if MEOR is to become a viable technology, a greater knowledge of the properties and limitations of microorganisms in the deep subsurface is needed. One must first understand the major factor(s) which entrap oil and then develop a specific microbial process to recover that oil. There is no average oil reservoir, and the chemical and physical properties of oil reservoirs are quite variable, as are the factors which entrap oil. Thus, a generic microbial process will probably not be feasible when applied to a specific reservoir. Also, a microbial process must be addressed in terms which petroleum engineers, geologists, and executives can understand so that they can adequately assess the risks and potential benefits from the implementation of such a process. Herein we summarize our current understanding of the microbiology of oil reservoirs and analyze the potential for using microorganisms to enhance oil recovery.

Survey of the Microbiology of the Deep Subsurface

Microorganisms in their natural environment usually grow in consortia as a biomass consisting of various physiological and morphological types which are usually attached to surfaces. In evaluating the role that consortia play in an environment, scientists need to know the amount of biomass present, the nature of the organisms making up the consortia, and its biochemical capabilities and activities.

An assessment of some of the techniques available for enumerating and measuring the activity of aquatic bacteria is given in Costerton and Colwell (1979). The reported information supports the contention that standard microbiological methods are not satisfactory when studying in situ natural microbial populations and that new innovative methods are required. The numbers of organisms enumerated by viable plate count techniques for example are always lower than the numbers obtained by direct counting procedures, indicating the unsuitability of the standard plate count in environmental studies. Thus, the focus has shifted to the evaluation of the use of the measurement of unique biochemical parameters such as adenosine triphosphate (ATP), muramic acid, extractable lipids and the procaryote storage product poly-β-hydroxybutyrate in assessing population composition in natural environments (White et al., 1975).

The measurement of ATP concentration as an indicator of biomass revealed the presence of many more cells in seawater samples than were indicated by the plate count technique (Holm-Hansen and Booth, 1966). However, ATP is a universal indicator of life and thus does not differentiate between groups of organisms. Such a differentiation, however, is possible by use of *signature compounds* (White, 1983) such as muramic acid, whose content is an indicator of the amount of eubacterial organisms present in an environmental sample.

An assessment of the use of signature compounds (i.e., chemicals specific to a group of microorganisms) in the analysis of the composition of consortia biomass is presented by White (1983). The evaluation of metabolic activity and nutritional status of such consortia also can be investigated by measuring the incorporation of isotopically labeled substrates under carefully controlled conditions. White et al. (1983) used these techniques to compare the community structure of two aquifer sediments and an estuarine one. They concluded that sediment from the deeper aquifer (410 m) has a biomass one-half that of the shallow sediment sample and one-tenth that of the surface sediment. The community structure also was different because the deeper sediment was richer in anaerobes (i.e., there was a higher *cis*-vaccenic acid content), and gram-negative organisms formed a larger proportion of the biomass. A lack of long-chain polyenoic fatty acids was noted in aquifer sediments, indicating an absence of algae, protozoa, or fungi. The higher incidence of poly-β-hydroxybutyric acid content of cells from the deep aquifer suggests that cells in this environment were growing under conditions which led to unbalanced growth and an accumulation of this compound (White et al., 1983). The investigation of the community structure of soil profiles using this technique found that the vertical distribution of microorganisms differs greatly as a function of soil type (Federle et al., 1986). Improvements in the sensitivity of analytical methods and instrumentation together with an increasing knowledge of signature compounds will allow a biochemical approach to understand the composition and activities of microbial life at the consortium level.

The information obtained from many studies indicates that there are many different types of microbes present in deep subsurface populations including aerobes, facultative organisms, and anaerobes (Davis, 1967; McNabb and Dunlap, 1975; and Ghiorse and Wilson, 1988). This flora exhibits a broad range of physiological properties from heterotrophs (some with the capability of oxidizing some of the components of oil; others have been implicated in corrosion of metal) to methane-producing and sulfate-reducing bacteria. While it is important for our understanding of the role of microorganisms in the process of oil diagenesis to know the source of this flora, it is not necessary in order to exploit the physiological properties of bacteria in EOR procedures. Predominant in the literature on oil reservoir microbiology are the extensive reports of Russian scientists (e.g., Kuznetsova, 1960; Zinger, 1966) on the incidence of these organisms in subsurface waters associated with oil-bearing formations. The study of the interaction of aerobic hydrocarbon-degrading pseudomonads and sulfate-reducing bacteria (Kuznetsova and Gorlenko, 1965) was of particular interest because it established the importance of community association in microbial biochemical activities taking place in oil-bearing formations. This aerobe-anaerobe interaction was implicated by Ivanov and Belyaev (1983), using isotopic methods which showed that freshening stratal waters enhanced methane production by stimulating the aerobic degradation of oil.

Data have been obtained which indicate that the microbial populations of shallow aquifers can be activated or modified by inoculation and/or nutrient supplementation so as to degrade polluting chemicals. One of the earliest examples of this was the report of Raymond et al. (1976) on the stimulation of bacterial activity in groundwaters containing gasoline. Their method included supplementation of the groundwater with a source of nitrogen, phosphate, and oxygen to remove hydrocarbons from groundwater in a dolomite aquifer. A similar process was described by Werner (1985). The opportunities for the bioreclamation of petroleum-contaminated aquifers have been reviewed by Wilson and Ward (1987).

The data in the literature clearly support the idea that microbial activity can and does take place in relatively deep subsurface reservoirs. They also support the concept that growth can be stimulated or initiated through manipulation of the environment to make it more conducive to microbial growth. A comparison of rates of activity from literature data is impossible because of the different methods used for obtaining and reporting data. In many cases, the information made public is only a summary, the actual data being considered of a proprietary nature. One conclusion that can be drawn is that changes in reservoirs induced by microbial action take place at a slow rate even under nutrient-manipulated conditions requiring a period of months or years before an effect on a reservoir will be detected.

Factors Affecting Oil Recovery

An understanding of the multiphase flow properties of reservoir rock and the mechanisms which entrap oil is important to the successful development of any EOR process. Studies on the flow distribution of oil, gas, and water in the pores of the rock have been simulated by using a single layer of synthetic porous spheres sandwiched between two transparent plates (Craig, 1980). These studies showed that water and oil flow in what is termed *channel flow conditions,* where each fluid moves through its own tortuous network of interconnecting channels. In a preferentially water-wet formation, imbibition occurs where flow results from an increase in saturation of the wetting-phase water (Fig. 17.1; Craig, 1980). Initially, water saturation in the unaffected portions of the reservoir is low and exists as a thin film around the sand grains. In zones where oil and water are both flowing, part of

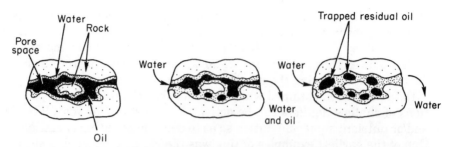

Figure 17.1 Visualization of oil production from a preferentially water-wet formation. Initially, oil is in a continuous phase in the pore space (left). After water enters the pore, through natural water drive or secondary waterflooding, oil is pushed out (center). After extensive waterflooding, oil is trapped in isolated droplets; only water will flow (right).

the oil exists in continuously interconnected channels, some of which have dead-end branches. After the reservoir has been extensively waterflooded, only isolated droplets of oil exist. In a preferentially oil-wet reservoir, flow is described as a drainage process where a decrease in the saturation of the wetting phase occurs. First, water forms continuous-flow channels throughout the largest pores. As the waterflood continues, water enters smaller and smaller pores until the flow of oil stops because of its decreased saturation in the flow channels. At residual oil saturation, oil exists in the smallest pores and as a thin film around the sand grains in the larger pores. This discussion illustrates the importance of wetability as a factor determining the nature of entrapped oil in the reservoir.

The rate of oil and water production from a well during primary and secondary recovery is governed by the capillary pressure in the vicinity of the wellbore (Donaldson, 1985; Craig, 1980). The relative fluid saturations of water and oil in this region are functions of the capillary pressure between these two fluids. An important factor governing the capillary pressure behavior is the pore entrance size distribution of the rock. If the pore entrance size distribution is decreased, there is a shift in the capillary pressure which causes oil production to stop at higher residual oil saturations (Fig. 17.2). Thus, a considerable amount of mobile oil which may be available a short distance from the well is not recovered because the shift in the pore entrance size distribution caused a higher water saturation in the vicinity of the wellbore. The migration of fine particles of clay and minerals, the precipitation of paraffins and asphaltenes, and compaction of the sand due to pressure drawdown can all decrease the pore entrance size distribution, thereby decreasing oil recovery. The stimulation of oil recovery observed after the growth of clostridia in wells (Hitzman, 1983) may be the result of a wellbore cleanout where precipitated material is removed by increased gas pressure, solvents and acids produced by

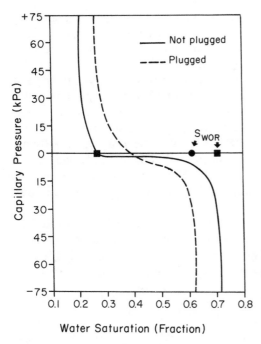

Figure 17.2 Idealized capillary pressure curves of Cleveland sandstone before and after plugging with particulates. S_{wor} is the water saturation at residual oil saturation (modified from Donaldson, 1985).

Water Saturation (Fraction)

the bacteria shifting the capillary pressure curve back to its originaldistribution (Donaldson, 1985).

The efficiency of oil recovery is defined by the equation (Craig, 1974)

$$E_r = E_d \times E_v \qquad (17.1)$$

where E_r = recovery efficiency expressed as fraction of original oil in place

E_d = microscopic oil displacement efficiency expressed as fraction of total volume of oil displaced in a unit segment of rock

E_v = volumetric sweep efficiency expressed as fraction of total reservoir volume contacted by recovery fluid

The microscopic displacement efficiency is a measure of how much oil is entrapped in small pores and dead-end pores (Fig. 17.1). Oil is entrapped in these narrow pores because of the capillary pressure which prevents the oil from flowing. The various forces which act to entrap oil are expressed in a dimensionless term called the *capillary number*, N_c (Taber, 1969):

$$N_c = \frac{\text{(Fluid velocity) (viscous forces)}}{\text{Interfacial tension}} \qquad (17.2)$$

An increase in the viscous forces or a decrease in the interfacial tension between the oil and brine will lead to further oil recovery. However, significant oil recovery occurs only when the capillary number is increased by 1000-fold or more (Reed and Healy, 1977). The only feasible way to achieve such a large increase in capillary number is to decrease the interfacial tension from about 10 to less than 0.01 mN/m. Micellar flooding techniques use surfactants to achieve low interfacial tensions and have given very high microscopic displacement efficiencies in laboratory studies (Van Poolen, 1980). Several microbially produced surfactants produce low interfacial tensions between oil and water (Finnerty and Singer, 1984a,b). However, the efficiencies of oil recovery by these biosurfactants in laboratory core experiments have not been well studied.

When EOR processes are implemented in the field, it is often the volumetric sweep efficiency which dominates the ultimate success of the process (Craig, 1974). The volumetric sweep efficiency is the product of the areal and vertical sweep efficiencies [Eq. (17.3); Craig, 1974]:

$$E_v = E_p \times E_i \qquad (17.3)$$

where E_p = areal sweep efficiency
E_i = vertical sweep efficiency

One factor resulting in poor volumetric sweep efficiency is the difference in mobilities of the oil and aqueous phases. Compared to oil, water moves rapidly through the reservoir. This results in an irregular waterfront with water pushing through the oil bank and reaching the production well before much of the oil is recovered. The relative mobilities of the aqueous and oleic phases are expressed in the mobility ratio [Eq. (17.4); Craig, 1974]

$$M = \frac{k_w \mu_o}{k_o \mu_w} \qquad (17.4)$$

where M = mobility ratio
k_w = relative permeability of water in waterflooded zone
k_o = relative permeability of oil in oil-saturated zone
μ_o = oil viscosity
μ_w = water viscosity

Mobility ratios less than 1.0 are favorable and result in a uniform displacement of the oil. Mobility ratios much greater than 1.0 are unfavorable, with water channeling through oil and leaving much of it behind.

Polymers such as xanthan gum are used to increase the viscosity of the aqueous phase, which decreases the mobility ratio and improves sweep efficiency.

A major factor resulting in poor sweep efficiency is permeability variation in the reservoir (Hutchinson, 1959; Craig, 1980). Variations in permeability are commonplace in petroleum reservoirs and dramatically affect the ultimate oil recovery. Water preferentially flows through the most permeable layers of rock, with little or no movement in the less permeable regions (Fig. 17.3). This results in early water breakthrough in

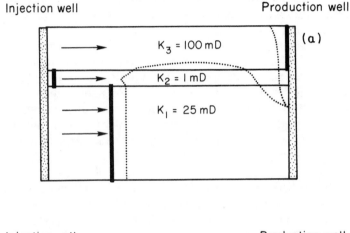

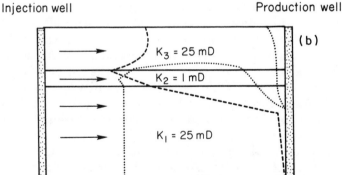

Figure 17.3 Waterflood patterns in an idealized reservoir composed of three different permeability layers in capillary contact with each other (Root and Skiba, 1965). K is the permeability of the sections in millidarcies (mD). (*a*) Dark lines are the simulated floodout pattern when the whole reservoir is open to injection and production. Dotted line is the floodout pattern when injection and production are allowed only in section 1. Note that cross-flow allows water to preferentially flow through section 3 even though this section is closed to injection and production. (*b*) Dashed line is the simulated floodout pattern when permeability of section 3 is reduced. Dotted line is as in (*a*).

the high-permeability regions. Further injection of water will follow this channel, bypassing the oil in the low-permeability regions. Because most oil reservoirs are composed of heterogeneous layers of rock, permeability is often the most important factor controlling sweep efficiency and the ultimate recovery of oil. A technology is needed to preferentially reduce permeability in the high-permeability zones without affecting the injectivity of fluids into the low-permeability zones. The reduction in permeability must occur deep within the reservoir in order to significantly improve the sweep efficiency (Fig. 17.3). Plugging the region near the wellbore will not dramatically increase sweep efficiency or oil recovery. Cross-flow between the layers will allow water to be redirected back into the high-permeability zone since this is the favored flow path (Root and Skiba, 1965; Silva and Farouq Ali, 1971).

Many plugging agents have been developed, including cement, colloidal clays, inert solids, paraffins, waxes, organic resins, and gels (Garland, 1966). The disadvantages of these agents are their inability of deeply penetrate the reservoir, migration of the agent when injection pressure is reduced, and the difficulty in determining the particle size needed. Crawford (1961, 1962) reported that injected bacteria tend to preferentially plug high-permeability zones, and that this plugging could correct permeability variation in oil reservoirs. Since bacteria multiply their mass at exponential rates, plugging could be enhanced under growth conditions. This increase in cell mass would divert flow to less permeable zones. Several studies (Jack and DiBlasio, 1985; Raiders et al., 1985, 1986a,b) have shown that the in situ growth of bacteria in sandstone preferentially plugs the high-permeability zone, which diverts fluid into the low-permeability zone, improving sweep efficiency and oil recovery.

The recovery of heavy oils is poor due to their high viscosity which prevents flow under normal reservoir conditions. The use of steam to decrease the viscosity of the oil and increase its mobility has been successful (Van Poolen, 1980). Several biosurfactants and emulsifiers have been shown to markedly reduce oil viscosity (Finnerty and Singer, 1984a,b) and improve the recovery of heavy oil from sand-packed columns. The emulsions which result after the use of steam or emulsifiers have large amounts of water which must be removed. Demulsifiers which can separate these two phases have been found in several different kinds of bacteria (Cairns et al., 1983).

Although physical and chemical factors do limit the recovery of oil from a reservoir, the diversity of the metabolic properties of microorganisms offers a wide range of possibilities for their use in EOR. The identification of the mechanism which has entrapped the oil is an essential first step in the development of a successful MEOR process.

Reservoir Conditions which Influence
Microbial Growth

A petroleum reservoir is composed of solid (rock), liquid (aqueous and oleic), and gaseous phases, all of which can influence the growth and metabolism of microorganisms. Data on the physical and chemical characteristics for 9 of the 10 top oil-producing states in the United States have been recently summarized (Clark et al., 1981; G. E. Jenneman, unpublished data). These sources provide the first comprehensive analysis of the limitations and potential applications of MEOR processes in existing reservoirs and have been extensively used in this section.

The solid phase of oil reservoirs is composed of rock, clays, and other insoluble minerals. Petroleum deposits are usually found in sedimentary rocks, but they can also occur in metamorphic and igneous rock formations. In the United States, sandstone is the most common oil-bearing rock, making up from 23 to 99 percent of reservoirs in all the states except Kansas and New Mexico, where limestone and dolomite predominate. The predominant minerals in sandstone and limestone are silicates and carbonates, respectively. In contrast the oil deposits in western Canada are primarily found in carbonate (limestone and dolomite) reef formations, with some production occurring in sandstone. In general, silicates and carbonates do not restrict microbial activity. However, other minerals and clays can dramatically influence microbial activity. The adsorption of bacteria to clay surfaces prevents the penetration and dispersion of bacteria in the reservoir. Clays swell when they absorb water. This restricts the movement of fluids in the pores. Consolidated rock has an enormous surface area which can act to concentrate nutrients and inhibitory compounds as well as provide a substratum for the attachment and growth of bacteria. The concentration of nutrients allows organisms to grow even when the concentration of nutrients in the liquid phases is very low.

Two important properties of consolidated rock are porosity and permeability. Oil, brine, and gas are trapped in the pore spaces inside of the rock. *Porosity* is a measure of this pore volume, which is expressed as a fraction of the total volume of the rock. *Permeability* is a measure of the ability of the porous material to transmit fluids. Porosities of reservoirs in oil-producing states in the United States range from 1 to 40 percent. Most reservoirs in Louisiana and Mississippi have porosities in the range of 26 to 40 percent, while 61 percent of the reservoirs in New Mexico have porosities less than 10 percent. Except for Oklahoma, New Mexico, and Wyoming, at least two-thirds of the average permeability values reported for reservoirs are above 100 millidarcies (mD). Most reservoirs in Oklahoma, New Mexico, and

Wyoming have average permeabilities of less than 100 mD; these states also have many reservoirs with low porosities.

The median pore entrance size distribution is an important parameter because of its relationship to the permeability of the rock (Craig, 1980). For cores of equal porosity, low-permeability cores will have median pore entrance size distributions skewed to the smaller pore entrance sizes while the reverse will be true for high-permeability cores. The relationship between permeability and pore entrance size is a complex one. Consolidated rock with a large proportion of small pores can have a high permeability if there are a few very large pores or fissures.

The pore entrance size is an important factor controlling the ability of bacteria to penetrate consolidated rock. Because bacteria have sizes in the range of pore entrance sizes found in sandstone, filtration tends to be the major factor restricting the transport of bacteria. Updegraff (1983) states that the pore entrance size should be about twice the diameter of the cell to allow passage of the bacteria through sandstone. Kalish et al. (1964) showed that larger cells or those which form aggregates or chains were filtered out more rapidly and caused greater permeability reductions than did smaller rods or cocci. We have made similar observations. The injection of over 100 pore volumes of a *Pseudomonas* strain into Berea sandstone cores caused little decrease in the permeability (Jenneman et al., 1984), but the injection of a larger *Bacillus* species resulted in almost 100 percent reduction in the permeability (Raiders et al., 1986a). The results of these and other studies (Raleigh and Flock, 1965; Jang et al., 1983a,b) all support the conclusion that filtration is the major mechanism which retards the penetration of bacteria through consolidated porous media and that little penetration occurs in less permeable rock where pore entrance sizes are small. Because of this, the injection of bacteria deep into a reservoir will be difficult since most of the bacteria will be filtered out within a short distance from the injection well. To avoid this problem, several workers have developed processes based on the use of spores or starved cells which have much smaller diameters. Many types of bacteria dramatically reduce the size of their cells to a diameter less than 0.3 μm in response to nutrient starvation (Torrella and Morita, 1981). The use of ultramicrobacterial forms for EOR has recently been studied (Lappin-Scott et al., 1988a,b; MacLeod et al., 1988). Suspensions of ultramicrobacteria were found to penetrate throughout the entire length of fused-glass-bead columns and sandstone cores without significantly reducing the injectivity or permeability. Thus, the use of ultramicrobacteria should prevent the accumulation of cells at the wellbore and allow for greater dispersal of bacteria in oil reservoirs.

Permeabilities less than 75 to 100 mD may represent the lower limit for effective microbial transport since pore entrance size distribution would contain pores which would restrict the passage of vegetative cells (Clark, 1981; Jenneman et al., 1985). However, there have been reports of bacterial transport in cores with permeabilities less than 75 mD (Hart et al., 1960; Kalish et al., 1964). In one case, transport occurred in a core with a permeability of 0.1 mD (Myers and McCready, 1966). Jenneman et al. (1985) found that rate of penetration of a *Bacillus* strain in nutrient-saturated cores was independent of permeability above 100 mD and rapidly declined for permeabilities below 100 mD.

The pore entrance size not only influences the microbial penetration, but it also influences microbial growth and metabolism. Zvyagintsev (1970) used rectangular capillary tubes to study the effect of pore size on growth rate and cell size. Cells of *Staphylococcus aureus* and *Saccharomyces vini* grew in large capillaries (400 μm thick by 150 μm wide), but no growth occurred and cells decreased in size in capillaries 5 μm thick and 3 μm wide. Zvyagintsev and Pitryuk (1973) found that decreasing capillary size from 500 by 250 μm to 8 by 4 μm decreased the growth rate of several bacterial species, and some species were unable to grow in these small capillaries. This was not due to a lack of nutrients, since similar results were obtained when the medium was continuously injected into the capillary.

Adsorption to surfaces of clays and other minerals is also important in limiting bacterial transport through porous medium. The relative surface properties of the solid matrix, the bacterial cell, and the liquid are important in determining the degree of adsorption (Absolom, 1983). Changing the surface properties of the solid surface does change the degree of adsorption of bacterial cells. Less retention and more effective transport of cells were observed in oil-saturated or sodium pyrophosphate-treated Berea sandstone cores compared to untreated cores (Jang et al., 1983a,b; Chang and Yen, 1985). Jenneman et al. (1986) found that steam sterilization increased the rate of penetration when compared to dry heat sterilization. Changes in permeability, porosity, and pore entrance size of the rock were not sufficient to explain the differences in the observed penetration times. However, electron dispersion spectroscopy and electron microscopy of autoclaved rock revealed changes in mineral composition of the surface and clay morphology. These data suggest that the increased rate of penetration in autoclaved cores was the result of a change in surface charge and a reduction in the surface area of clays.

Greater depths of penetration occur when bacteria are allowed to grow through the porous medium (Craw, 1908; Myers and McCready, 1966; Myers and Samiroden, 1967; Jenneman et al., 1984, 1985; Raiders, 1986a). The addition of a sucrose-based medium allowed bacteria to grow throughout the length of Berea sandstone cores. Semicontinuous injection of nutrients provided for better growth throughout the core and reduced the amount of growth which occurred at the face of the core. Similar results were obtained by Costerton and co-workers (Lappin-Scott et al., 1988a,b; MacLeod et al., 1988). Continuous injection of nutrient-containing brine results in preferential growth at the injection face (Shaw et al., 1985; Geesey et al., 1987).

Temperature and pressure both will increase with increasing depth of the formation. Thus, deeper formations will provide a less favorable environment for the growth of microorganisms. Temperature increases with depth at an average rate of 1 to 2°C per 100 ft, although major discontinuities in the temperature-depth profile have been noted. Temperatures of reservoirs vary greatly, but the vast majority of reservoirs (over 90 percent) have temperatures less than 85 to 99°C (Clark et al., 1981). Average temperatures range from 49 to 89°C. A variety of anaerobic bacteria grow at temperatures above 60°C, and some have been found which grow at temperatures of 95°C (Ljundahl, 1979). Thus, even deep formations may be targets for MEOR. The temperature-depth profile will restrict the use of microbes to reservoirs with depths less than 3500 m (temperature less than 100°C) and, in general, limit the use of most bacteria to reservoirs with depths less than 2500 m (temperature less than 85°C). This estimate is confirmed by the work of Philippi (1977), who concluded from a study of five reservoirs that the average cutoff temperature for bacterial degradation of crude oil was 66°C and ranged from 54°C to 77°C. Pressure gradients range from 0.43 to 1.0 psig per foot, depending on the geographical area. For most states, 90 percent of the reservoirs are estimated to have pressures less than 680 atm (Clark et al., 1981). Many states have reservoirs at depths less than 1000 m, which would be equivalent to about 200 atm. Pressures of 100 to 200 atm generally do not adversely affect bacterial growth, although decreased growth rates and yields may occur (Marquis and Matsumura, 1978). Pressure effects become critical when bacteria grow under suboptimal conditions of temperature, pH, and so on (Marqui and Matsumura, 1978; Marquis, 1982, 1983). This analysis suggests that temperature will be the more important parameter in restricting the application of MEOR processes. However, since pressure inhibits biochemical reactions involving volume changes, it will be an important consideration for processes based on in situ gas production.

The aqueous phase probably represents the most important part of the reservoir for MEOR, since it is the phase in which microbial growth and metabolism will occur. The water found in an oil reservoir is most often of marine origin and has been trapped in the rock from the geological past (connate water). During secondary and tertiary recovery operations, water from other formations is injected into the reservoir. These waters may provide a new source of nutrients and microbial species. Oil field brines are more highly mineralized than seawater or freshwater with regard to all minerals except magnesium, which is found at a concentration approximately equivalent to that in seawater (Clark et al., 1981). Oil field brines contain sufficient amounts of most elements to support microbial growth with the exception of phosphorous, nitrogen, and sometimes sulfur (Table 17.1). The pH of reservoirs varies from 3.0 to 9.9, with 66 percent of the reservoir having pH values between 6.0 and 8.0. This is a range which allows growth of most known microorganisms. Average NaCl concentrations in brines in Oklahoma and Mississippi are very high, usually in excess of 10 percent. In California, Colorado, and Wyoming, the average NaCl concentration is low. Sodium chloride comprises about 90 percent of the total dissolved solids of reservoir brines. Thus, tolerance to NaCl is one of the most important characteristics needed for organisms to be used for MEOR. Bacteria are known which grow at very high NaCl concentrations (approximately 32 percent) (Kushner, 1978). However, it would be preferable to have an organism which can grow over a wide range of salinities. Several halophilic and halotolerant anaerobic bacteria have recently been described (Rodriques-Valera, 1988).

McNabb and Dunlap (1975) suggest that sufficient carbon is available in the subsurface, although it may be in forms (e.g., soil humic acid components) which are not readily utilizable by bacteria. Significant levels of short-chain fatty-acid anions, especially acetate and propionate, have been detected in oil and gas wells in California and Texas (Carothers and Kharaka, 1978), which can support the growth

TABLE 17.1 Average Percentage Composition of Petroleum Reservoir Brine in Various States

Chemical	State								
	CA	CO	MI	NM	WY	OK	TX	LA	KS
Total solids	2.2	2.3	14.5	8.5	1.3	15.6	9.7	10.1	9.9
Chloride	1.2	1.2	8.8	5.0	0.6	9.4	5.2	5.8	6.0
Sodium	0.8	0.8	4.5	2.4	0.4	4.7	2.6	3.3	3.1
Potassium/sodium	0.5	0.4	4.0	2.7	0.3	4.1	3.7	3.7	2.5
Calcium	0.06	0.1	0.9	0.4	0.05	1.0	0.5	0.3	0.5
Sulfur	—*	—	—	—	—	0.04	—	—	—
Magnesium	0.02	0.02	0.11	0.15	0.01	0.18	0.12	0.06	0.16
Phosphate	$4.0^{\dagger}$	—	—	—	$2.0^{\dagger}$	$8.0^{\dagger}$	—	—	—
Nitrate	$3.0^{\dagger}$	—	—	$3.0^{\dagger}$	—	—	$2.0^{\dagger}$	—	—
Iron	0.001	0.006	—	0.005	0.006	0.02	0.007	0.008	0.006
Oxygen	$2.5^{\dagger}$	—	—	—	—	—	$0.2^{\dagger}$	—	—

*—; not available.
$^{\dagger} \times 10^{-4}$.
SOURCE: Adapted from Clark et al. (1981)

of a wide variety of organisms, including methanogenic and sulfate-reducing bacteria. Another carbon source in an oil reservoir is the crude oil itself. However, it will not support the growth of anaerobes because the initial microbial attack on hydrocarbons requires oxygen. Therefore, anaerobic growth in an oil reservoir depends on external carbon sources or the aerobic degradation of hydrocarbons under suboptimal conditions where oxidized carbon products will be formed. Subsurface microbial activity, as suggested by McNabb and Dunlap (1975), could be limited by the quality, rather than the quantity, of subsurface organic matter.

The concentration of oxygen is one factor which usually limits growth in the subsurface. This idea is supported by the observations that the efficient removal of hydrocarbons from shallow aquifers by microbial action requires aeration procedures. In addition, it has been observed that only sections of oil reservoirs in contact with oxygen-bearing aquifers (e.g., northwestern section of the Williston Basin in the southeastern corner of Saskatchewan, Canada) have chemical compositions indicative of having been subjected to bacterial degradation (Bailey et al., 1973a). Similarly degraded oils are also found in reservoirs undergoing waterflooding recovery procedures. Winograd and Robertson (1982) reported the presence of significant amounts of dissolved oxygen (2 to 8 mg/L) in a variety of aquifers ranging from 100 to 1000 m. Some of these aquifers were 10,000 years old and up to 80 km from recharge areas. These observations support the contention that subsurface environments, particularly those with low amounts of readily utilizable carbon sources, can contain significant levels of dissolved oxygen. In some locations (e.g., the Williston Basin) the level of oxygen is continually maintained by oxygenated percolation waters.

The oil phase is the third portion of the reservoir which must be considered for MEOR. The limitations imposed on MEOR by the oil phase are the toxicity of the light volatile fractions and the high densities of the heavy asphaltic crudes. Heavy oils are more difficult to recover because of the unfavorable mobility ratios between the brine and the oil. Crude oil with API densities less than 18° are considered heavy oils and are best recovered by steam injection. California and parts of Venezuela have many reservoirs with API gravities less than 18°; most reservoirs in the United States have crude oils with API gravities above 31°.

Based on the above considerations of the limitations imposed by a reservoir, it is possible to determine what percentage of the reservoirs could be used for MEOR processes in 9 of the top 10 oil-producing states in the United States (Clark et al., 1981; G. E. Jenneman, unpublished data). Criteria used for this analysis were NaCl less than 10 percent, pH between 4 to 9, permeability above 75 mD, API gravity of the oil above 18°, and temperature less than 75°C. At least 20 percent of the reservoirs in almost all of the states have favorable conditions which would make them potential targets for MEOR processes (Table 17.2). The parameters which seem to be the most limiting for MEOR are permeability and temperature.

Microbial Characteristics Which Make Them Suitable for MEOR

Bacteria constitute a very large, diverse group of unicellular life forms which encompass a wide range of nutritional, physiological, and biochemical characteristics that can be used in devising improved oil-recovery procedures. There are two ways in which microbes can be used in enhanced oil recovery (EOR) processes. Products (e.g., polysaccharides or surfactants) can be produced in fermentors and injected into the reservoir undergoing treatment. Alternatively, bacteria can be injected into the reservoir and a suitable nutrient mixture provided for growth in the reservoir. The ability to control the size, chemical composition, growth rate, and biochemical activities by altering the composition and concentration of the nutrients supporting growth will provide the basis for manipulating EOR procedures. Potentially, microorganisms could be used to improve both the microscopic oil displacement efficiency and the volumetric sweep efficiency. The production of solvents, surfactants, gases, and acids could change the wetability, decrease interfacial forces, and improve the mobility of oil, all of which would enhance the release of oil from rock surfaces and small pores. The formation of cells and extracellular polysaccharides in the larger pores and higher-permeability regions of the

TABLE 17.2   Percent of Reservoirs in Each State Potentially Treatable by MEOR

State	Salt 10%	pH 4–9	Permeability 75 mD	API gravity 17	Temp. 75°C	% Treatable*
Oklahoma	53[†]	97	31	99	96	15
Texas	53[†]	97	51	98	82	21
Louisiana	53[†]	97	93	99	75	36
Kansas	53[†]	97	39	99	98	19
California	100	98	82	84	80	54
Colorado	100	94	65	99	70	42
Mississippi	15	98	80	98	41	4
New Mexico	100	94	39	99	78	28
Wyoming	100	94	41	95	64	23

*Percentage calculated by considering each parameter as an independent variable.
[†]Average for four states.
SOURCE: From Clark et al. (1981).

reservoir will divert fluid flow to other unswept regions of the reservoir, improving the sweep efficiency.

Because most oil-bearing reservoirs are likely to be deficient in oxygen, one of the most important groupings of bacteria with regard to the choice of bacteria to use in reservoirs is based on their relationship to oxygen. Aerobic bacteria require oxygen for growth, whereas anaerobes do not and their growth is usually inhibited by its presence. A third group of bacteria, the facultative bacteria, can grow as aerobes if oxygen is present or as anaerobes if it is not. Anaerobically, they ferment, or, if suitable electron acceptors are present, they can respire. Under nutritional conditions which support balanced growth (i.e., all components of the cell are produced at a concentration which is required for cell duplication), aerobic bacteria will metabolize the carbon source primarily into new cell material and carbon dioxide. There is very little production of partially oxidized carbon compounds (e.g., organic acids) or extracellular polysaccharides. However, under unbalanced growth conditions, some aerobes will produce relatively large amounts of polysaccharides and surfactants which are useful in EOR processes. In contrast, anaerobic bacteria produce relatively large amounts of solvents such as acetone, butanol, and other alcohols, organic acids such as acetic, lactic, butyric, and large quantities of CO_2, CH_4, and N_2, depending on the availability of electron acceptors. The production of solvents during the growth of anaerobes can reduce the interfacial and surface tension forces, thus releasing oil from reservoir surfaces. Organic-acid production can result in the dissolution of carbonates in source rocks, enhancing the permeability of the reservoir. The solubilization of microbially produced gases in oil can result in the precipitation of the asphaltenes, leaving an oil with more desirable flow characteristics.

Microbial physiologists have long recognized that growth rate, size, and composition of bacterial cells are directly related to the richness of the medium in which the cells are grown (Mandelstam et al., 1982). For example, when bacteria are inoculated in a rich medium which supports fast growth rates, the cells are larger and contain more RNA per cell than when they are growing in minimal media which usually support only slow growth rates. The DNA content per cell also increases with the growth rate but to a lesser degree than RNA; hence, the DNA content decreases in relation to cell mass. Cell envelopes (i.e., the wall plus cytoplasmic membrane) are usually of constant thickness. Thus, the proportion of envelope to the total mass of the cell decreases with increasing growth rate and cell size, which results in a reduction of the ratio of surface area to volume of a cell as growth rate increases. This results in larger cells having a decreased ability to carry out exchange with their environment.

For EOR procedures involving "selective" plugging of the reservoir, the cell's response to the carbon-nitrogen ratio of the growth medium is important. A ratio of 10 parts carbon to 1 part nitrogen results primarily in cell production. If this ratio is increased, the cell will start to produce polysaccharides, which will enhance the cell's ability to reduce the permeability of the reservoir. Nutritional conditions which increase the growth rate of the cell are also important in this phase of oil recovery. A small increase in the radius of the cell, which results from nutrient enrichment, causes a relatively large increase in the volume occupied by that cell. For instance, doubling the radius of a sphere, a common bacterial shape, from 1 to 2 mm results in an eightfold increase in cell volume. In contrast, a similar increase in the radius of a rod, the shape of many of the bacteria with biochemical characteristics of possible use in EOR, results in a fourfold increase in cell volume. An increase in the length of a rod does not have as great an effect on cell volume as does an increase in its radius. However, increasing the length of a rod would allow the cells to span the width of a pore, which may accelerate the deposition of particulate and cellular material.

The relationship between cell size and growth rate can be exploited when bacteria are being prepared for inoculation into an oil reservoir (Lappin-Scott, 1988a,b; MacLeod et al., 1988). Growing the bacteria under conditions where their size (i.e., cell volume) is minimized increases the number of cells injected into the formation and their depth of penetration. Once the bacteria are in the formation, richer nutrient formulations can then be injected, which will result in faster growth rates and a corresponding increase in cell size and metabolic activity.

The low or undetectable oxygen content in most petroleum reservoirs indicates that anaerobic conditions primarily prevail in these environments. However, in cases where the introduction of oxygen into the reservoir occurs, hydrocarbon-degrading aerobes preferentially metabolize the lighter and more easily recoverable fractions of the oil (Bailey et al., 1973b). Also, anaerobes like the sulfate-reducing bacteria grow on by-products produced by the aerobic organisms and produce reduced sulfur compounds. Both processes detrimentally affect the economic value of the oil. Thus, aerobes are unsuitable for in situ EOR procedures.

The above discussion supports the prevailing idea that the selection of an organism for EOR has to be suited to the reservoir and possibly that section of the reservoir which is to be subjected to microbial enhancement procedures. McInerney (1983) suggests that such an organism would likely be a thermotolerant anaerobe with the ability to grow in the presence of high salt concentrations. The role of pressure cannot be ignored (Marquis, 1982, 1983). However, it is likely that the reservoir temperature would limit growth before pressure became the critical factor.

Neijssel and Tempest (1979) showed that the nature of the growth-limiting nutrient can have an effect on the metabolic products produced by *Klebsiella aerogenes*. While carbon-limiting conditions result in the most efficient conversion of carbon to cell material, no organic acids or extracellular polysaccharides are produced, both of which play an important role in enhanced oil recovery. Limiting amounts of either sulfate, ammonia, phosphate, or potassium result in the diversion of carbon from cell material to the production of organic acids and extracellular polysaccharides. Thus, the metabolic pathways expressed by this organism depend on a particular nutrient which limits growth. Under field conditions, the control of growth, cell size, and product formation by altering the concentration of the nitrogen source probably would be most economical and technically feasible.

To increase the microscopic displacement efficiency, the anaerobic microorganism should be capable of producing large amounts of gas, organic acids, and solvents from readily available, inexpensive, carbohydrate sources. Of the many bacterial types available, these requirements are readily met by the carbohydrate-fermenting members of the genus *Clostridium,* which produce CO_2, H_2, mixtures of acetic and butyric acids, and solvents like ethanol, *n*-butanol, acetone, and isopropanol from simple carbohydrate sources. A few species are capable of producing these acids and alcohols from cellulose. In addition, members of this genus produce endospores which penetrate consolidated rock more effectively than do the vegetative cells (Chang and Yen, 1985; D. O. Hitzman, U.S. Patent 3,032,472). The isolation of such organisms for EOR purposes was recently reviewed (Grula et al., 1983). Laboratory core experiments found that spore-forming bacteria physiologically similar to clostridia recovered about 32 percent of the original oil in place after the cores had been waterflooded to residual oil saturation (Bryant and Douglas, 1987). The production of gas, solvents, organic acids, and biosurfactants during the in situ growth of the bacteria in the cores were cited as the main mechanisms for enhanced oil recovery.

The successful utilization of clostridia or any other bacteria for EOR purposes will undoubtedly involve a redesign of the organism by modern genetic technologies. The ability to genetically manipulate clostridia has recently been discussed in reviews on the acetone-butanol fermentation (Jones and Woods, 1986a,b; Rogers, 1986). Genetic transfer systems, in particular DNA transformation techniques, have been developed, and genes from several different clostridial species have been expressed by means of their own promoters in *Escherichia coli*. These results indicate that our knowledge of *E. coli* cloning systems may be applied to clostridia to enhance reactions such as solvent production which could benefit EOR procedures.

Many organisms are known to produce biosurfactants (Cooper and Zajic, 1980; Zajic and Seffens, 1984; Singer, 1985). Several of these have been found to significantly reduce the interfacial tension between oil and brine to less than 0.01 mN/m (Finnerty and Singer, 1984a,b; Singer, 1985), making them potential candidates for EOR processes. However, these biosurfactants are produced by aerobic organisms which would make them unsuitable for in situ applications. Moreover, the effectiveness of these biosurfactants in high salt environments has only been studied in a few cases. Several anaerobic bacteria produce biosurfactants (LaRiviere, 1955; Cooper et al., 1980; Grula et al., 1983), although the degree of surface tension reduction is less than that observed for aerobically made biosurfactants, except for *Bacillus licheniformis* strain JF-2 (Javaheri et al., 1985). Strain JF-2 grows anaerobically under conditions that are found in many oil reservoirs, and the biosurfactant that it produces is effective under these conditions (Jenneman et al., 1983; Javaheri et al., 1985). However, it is not known whether the JF-2 biosurfactant generates low enough interfacial tensions to recover significant amounts of oil. Laboratory experiments show that oil recovery is increased by about 20 percent when strain JF-2 is grown in model sand-packed porous systems (Kianipey, 1986). Finnerty et al. (1985) found that the biosurfactant produced by an aerobic bacterium, strain H-13A, improved the recovery of heavy oil from sand-packed columns from 15 to 50 percent.

The production of polymers by microorganisms has been extensively studied (Kang et al., 1983; Kennedy and Bradshaw, 1984; Holzwarth, 1985). Xanthan gum is widely used to increase the viscosity of injection brine to provide better mobility control and sweep efficiency. The in situ application of microbial selective plugging processes requires that the organism grow and produce the polymer anaerobically at elevated temperatures and salinities. A variety of *Bacillus* species have been isolated which grow anaerobically and produce extracellular polymers at temperatures up to 50°C and salinities up to 10 percent NaCl (Zajic and Mesta-Howard, 1985; Pfiffner et al., 1986). These kinds of bacteria are effective in reducing permeability of Berea sandstone cores (Jenneman et al., 1984; Raiders et al., 1986a) and in improving sweep efficiency in laboratory model systems (Raiders et al., 1985, 1986a,b). Oil recoveries range from 8 to 35 percent of the oil originally in place after waterflooding. These bacteria are found in a variety of environments, particularly oil field brines, which suggests that the injection of these organisms into the reservoir may not be needed. All that may be required is the injection of nutrients to stimulate the growth of the indigenous populations. *Leuconostoc* species have been shown to effectively reduce permeability of fused-glass columns which have very high permeabilities ranging from 5 to 7 darcy (Jack et al., 1983; Jack and DiBlasio, 1985; Shaw et al., 1985). The transport and degree of permeability reduction can be controlled by altering the kinds of nutrients injected into the core. The use of a nutrient mixture which does not support extracellular polymer production allows the transport of the organisms throughout the core without significantly decreasing permeability. Injection of a nutrient mixture which supports extracellular polymer production leads to rapid reductions in permeability.

A variety of microorganisms have the ability to produce useful products and grow under the constraints imposed by the reservoir. Studies suggest that microbial growth and metabolism can be manipulated in the reservoir by altering the kinds and concentrations of nutrients used and the method of injection of the nutrient. If rapid plugging occurs at the injection face, permeability and injectivity can be restored by treating the wellbore with sodium hypochlorite followed by acid to degrade the cellular biomass and dissolve mineral precipitates (Clementz et al., 1982).

Evidence for Microbial Activity in Reservoirs

Recently, specialized drilling and sampling procedures for the collection of uncontaminated aquifer material from subsurface reservoirs down to a depth of about 300 m have been developed (Wilson et al., 1983). This allows the direct study of the microbial ecology of these habitats. Information on microbial activity in deep-subsurface, hydrocarbon-containing environments, however, is mainly based on a comparison of the chemical composition of hydrocarbons from reservoirs which have been subjected to waterflooding recovery procedures to the composition before treatment or by comparing the chemistry of parts of reservoirs which are in contact with groundwater aquifers with that of reservoirs which are free of water contact. Some data are also available from the analysis of core materials brought to the surface using drilling procedures.

The conclusive demonstration of an active and diverse microbial population in aseptically obtained aquifer material at depths to about 300 m (Ghiorse and Wilson, 1988) suggests that microorganisms may be found at even greater depths. Indeed, many data suggest that deep terrestrial subsurface reservoirs contain active and diverse populations of microorganisms. Olson et al. (1981) found sulfate-reducing and methanogenic bacteria in stratal waters from wells 1800 m deep. However, direct evidence for existence of microorganisms in consolidated porous media is lacking due to the difficulties in obtaining uncontaminated material. Despite these problems, many workers are convinced uncontaminated consolidated core samples have been obtained, and, thus, the microorganisms isolated from these samples are indeed indigenous to the reservoir (ZoBell, 1958; Kuznetsov et al., 1963; Davis, 1967). ZoBell (1958) found viable sulfate-reducing bacteria (SRB) in water and core samples taken from strata down to 3700 m near oil and sulfur deposits. Aseptic precautions were used to drill the well which included the use of alkaline (pH 11.5) drilling mud, biocides, and aseptic sampling procedures. No sulfate-reducing bacteria were found in the drilling mud and the absence of heterotrophic bacteria indicated that the samples were not contaminated by soils or other surface sources. Kuznetsov et al. (1963) cite the results of several Russian workers who detected the presence of viable bacteria in deep sedimentary rock. However, little or no information is given on the petrophysical properties and condition of the cores which is essential to the determination of the extent of contamination of these cores by drilling fluids. Microbial penetration can occur along fractures and high-permeability streaks (Kuznetsov et al., 1963; Jenneman et al., 1985). Weirich and Schweisfurth (1985) found viable bacteria in aseptically obtained samples of cores from a coal deposit 405 m deep, but not from samples taken from depths of 24 to 343 m. Their experiments showed that little penetration of the drilling fluids occurred.

An important question is whether the microorganisms found in subsurface reservoirs are indigenous or whether they were introduced into the reservoir during drilling and other operations. As discussed above, microorganisms can penetrate consolidated sandstones. ZoBell (1958) estimated a rate of movement of *Desulfovibrio* species of 0.5 to 13 m/year. This is fast enough to allow the dissemination of the organism throughout the reservoir during the normal operating life of the reservoir. Microorganisms could be introduced into oil reservoirs at the time of organic matter deposition; however, it is more likely that they would be introduced during maturation as spores and as viable cells from aquifers in porous strata that are in contact with the perimeters of reservoirs.

Evidence supporting the contention that microbial activity occurs in oil reservoirs is based on isotope fractionation data and on data obtained by comparing the chemical composition of oil pools which are derived from a common source, yet exist in different formations. Isotope enrichment studies (Thode et al., 1954; Feeley and Kulp, 1957) have implicated microbial activity in the formation of free sulfur and the carbon in the calcite cap rock in Gulf Coast salt domes. Recently, scanning electron microscopic studies by Sassen (1980) showed the intimate association of calcite, pyrite, elemental sulfur, gypsum, biode-graded oil, and microorganisms in samples of the cap rock from these salt domes.

Capillary gas chromatography (GC), with and without mass spectrometry, has been widely used in studying oil maturation processes. Oils which are thought to have been subjected to microbial action are deficient in low-molecular-weight compounds and in particular n-paraffins. The loss of these components has been frequently used as an indication that microbial action occurred. For example, Milner et al. (1977) presented GC profiles of oils from the Gulf Coast salt domes which showed a positive correlation between the presence of a degraded GC profile and the presence of a sulfur cap. These results support the isotopic data concerning the involvement of microbial activity in the formation of these oil-containing salt domes. Analyses by GC implicated microbial action in the alteration of the composition of oils in the Cretaceous Bell Creek field in Montana (Winters and Williams, 1969). Bailey et al. (1973a) presented isotopic fractionation and GC data on subsurface biodegradation of oils in a series of Mississippian oils in reservoirs of the northwest section of the Williston Basin formation in southeastern Saskatchewan. Using aerobic oil-degrading populations and laboratory conditions, Bailey et al. (1973b) converted unaltered oils from this basin into oils which had characteristics similar to in situ altered oils.

Losses also may be observed in the low-molecular-weight components of the aromatic and heterocyclic components of oil. Such changes are usually accompanied by changes in the physical properties (e.g., an increase in oil viscosity). Studies have implicated microbial activity in the conversion of conventional crude oils to the heavy oils and the bitumen beds of western Canada (Crawford, et al., 1977; Rubenstein et al., 1977). Changes in the thiophenic content (benzothiophenes, dibenzothiophenes, and naphthothiophenes) of western Canadian oils also has been attributed by Deroo et al. (1974) to microbial degradation and water washing. Burns et al. (1975) concluded that biodegradation occurred in a series of Beaufort Basin liquid hydrocarbons (Kumak field), which resulted in oils with lowered pour points but unchanged viscosities or sulfur content. Unaltered Kumak oils were biodegraded under laboratory conditions, yielding oils with properties similar to those degraded in situ oils (Westlake, 1983).

The potential of biodegradation in petroleum transformation in reservoirs has been reviewed by Milner et al. (1977). They conclude that convincing indirect evidence has been presented for this alteration process, but the problem of distinguishing between immature oils and biodegraded oils is not resolved. Tannenbaum and Aizenshtat (1984) applied material balance calculations in an assessment of the role of water washing and microbial degradation in the oil alteration process of an oil deposit in the Dead Sea. They attributed the faster removal of aromatics than of saturates at depths of 2000 m as being due to water washing by brines, while they attributed the loss of saturates from depths of 1500 m to the surface as being due to bacterial action initiated by the invasion of meteoric water.

Philippi (1977) reported that the average oil transformation cutoff temperature for oils from five different basins was 66°C. The abrupt change in temperature at which transformed oil is no longer found implicates microbial activity in oil maturation. Philippi (1977) also states that the increased optical activity of transformed oils is due to the fact that microbes utilize optical antipodes of racemic mixtures at different rates.

Sulfate-reducing bacteria have long been associated with corrosion and the souring of oil fields (ZoBell, 1958). However, there is no unequivocal evidence which supports a claim of involvement of SRB as initiators of oil degradation. Rather, the data suggest that the growth of SRB in a reservoir is the result of an interaction with aerobic, hydrocarbon-degrading bacteria. Kuznetsova et al. (1963) and Kuznetsova and Li (1964) reported that under field conditions sulfate reduction was stimulated in oil-bearing Devonian strata by the injection of freshwater containing sulfate and various microorganisms. Kuznetsova and Gorlenko (1965) concluded that sulfate reduction in reservoirs was a complicated process dependent in part on the presence and activity of aerobic bacteria of the genus *Pseudomonas*. Nazina et al. (1985) found sulfate-reducing bacteria to be widely spread in oil reservoirs. They used the alcohols and acids produced by aerobic hydrocarbon-oxidizing bacteria. Jobson et al. (1979) reported that a pure culture of SRB and mixed SRB cultures grew and produced sulfide from the residue of the degradation of oils by mixed populations of aerobic bacteria. Because these aerobic populations contained a mixture of gram-positive and gram-negative hydrocarbon-utilizing bacteria, a greater possibility existed for an association of aerobic petroleum-degrading bacteria and SRB than was suggested from the *Pseudomonas* association defined by Kuznetsova and Gorlenko (1965).

Reports from oil field engineers in northern Alberta (personal communication, Dr. M. Gray, Department of Chemical Engineering, University of Alberta) indicate that oils recovered from producing fields using waterflooding recovery procedures has turned sour (i.e., it has an increased H_2S content as compared to the original oil). Unfortunately, no data exist on the sulfur isotopic composition of the H_2S which should be enriched in the lighter isotope of sulfur if its production was due to microbial activity. However, fluids from such fields usually have a very diverse microbial population containing aerobes, facultative aerobes, and SRB (Westlake et al., 1986), and exhibit high corrosion rates unless biocides are used appropriately. Studies on the Southeast Vassar-Vertz reservoir in Paine County, Oklahoma, showed the souring of the reservoir during waterflooding (M. J. McInerney, W. Jenkins, R. Raiders, and R. K. Knapp, unpublished data). A brine high in sulfate from another aquifer is being used as the source for the water drive. The portions of the reservoir which have not been contacted by the waterflood do not contain sulfate and have less than detectable levels of sulfide and SRB. Portions of the reservoir which have been waterflooded contain sulfide and high numbers of SRB. Thus, the addition of the limiting nutrient—in this case, sulfate—stimulated the growth and metabolism of SRB.

Such studies provide only circumstantial evidence that the production of reduced sulfur compounds is of microbial origin. More definitive evidence, including isotopic studies with the above analyses, is needed to confirm such observations. Dockins et al. (1980) conducted such a study on Montana groundwater samples collected from depths of 10 to 260 m. They found the presence of SRB and the selection of the lighter isotope of sulfur during sulfate reduction, indicating a bacterial origin of the produced hydrogen sulfide. Salanitro (1985) found that the souring of an off-shore oil reservoir in the Gulf of Mexico was not the result of microbial activity based on isotopic studies and on the inability of isolated bacterial strains to grow at reservoir conditions. Abiotic reduction of sulfur to sulfide occurs at temperatures above 80°C, and the rate of this process increases with increasing temperature and pH (Belkin et al., 1985). Isotopic studies are essential to determine the origin of sulfide in thermophilic oil reservoirs.

Other potential sources of carbon for microbial growth in reservoirs undergoing waterflooding arise from the addition of biopolymers to improve the sweep efficiency of injected water in the reservoir (Holzwarth, 1985). Xanthan gums, the most commonly used biopolymer, are susceptible to hydrolysis by β-glucanases produced by a variety of microorganisms (Cadmus and Slodki, 1985; Sutherland, 1982). The products of such hydrolyses are partially oxidized carbon compounds which could serve as additional energy sources for indigenous bacterial flora. The in situ microbial degradation of xanthan gum was observed during a pilot test of a surfactant flood in the Loudon Field in Fayette County, Illinois. An increase in injectivity and a decrease in polymer concentration in the production wells coupled with large increases in bacterial numbers indicated that xanthan gum had been degraded after about 75 percent of the reservoir had been flooded (Bragg et al., 1983). Laboratory studies confirmed that the brine samples contained xanthan-degrading bacteria. Similarly, Grula and Sewell (1983) found that soil bacteria, reservoir isolates, and SRB released ammonia from polyacrylamide solutions in laboratory cultures. Thus, polyacrylamide might be used as a nitrogen source to support microbial growth in oil reservoirs. No organisms, however, were isolated which could cleave the carbon backbone of this polymer.

The results of published field trials also provide evidence for microbial activity in oil reservoirs (Hitzman, 1983). In these trials, either indigenous bacteria or injected bacteria were fed a carbohydrate-based nutrient, usually molasses, with the intention of stimulating in situ microbial growth and metabolism. In many of these trials, large increases in gas pressure, carbon dioxide content, bacterial numbers, short-chain monocarboxylic acids, alcohols, and sulfides as well as pH changes were observed at surrounding wells, located 100 to 400 m from the injection well which received the nutrients. In one test in Hungary it was estimated that an area of 60,000 m^2 was affected. These data show unequivocally that it is possible to stimulate the in situ growth of microorganisms in oil reservoirs.

The above information supports the contention that microbial activity can and does occur in oil reservoirs which are in hydrogeological communication with aquifers or are being subjected to waterflooding or an EOR process. Microbial growth in these reservoirs is often nutrient-limited, particularly for the availability of an electron acceptor in the case of aerobic metabolism. The injection of biopolymers and other organic compounds during waterflooding or an EOR process could support further microbial activity in reservoirs. The use of readily degradable compounds such as citrate, benzoate, and other organic-acid anions should be reconsidered in view of the above discussion and the fact that these compounds are excellent energy sources for sulfate-reducing bacteria (Pfennig and Widdel, 1981).

MEOR Field Trials

There has been an increasing number of field tests of MEOR processes in the last decade. The results of almost all microbial field trials have been summarized by Hitzman (1983). For the most part, a critical evaluation of these data is not possible because of inadequate reservoir characterization, sparse information on many of the trials, and the poor quality of the reservoirs in which the trials were conducted. Thus, it is difficult if not impossible to assess the efficiency of the process.

Since 1954, over 300 wells have been treated with microorganisms throughout the world; most of these trials have been in the United States (Hitzman, 1983). However, most of the wells were of the stripper well category, producing less than 10 barrels a day with minimal data available because of proprietary considerations. The trials conducted in the eastern European countries have been more comprehensive, but these trials are difficult to interpret because of translation problems with and omissions in the original manuscripts. A wide range of reservoir parameters were encountered in these trials with temperatures ranging from 22 to 97°C, depths between 50 and 1550 m, porosities of 11 to 36 percent, permeabilities from 10 to 8100 mD (highly fractured), oil types from heavy asphaltic to light paraffinic, and rock types of sandstone to limestone (Hitzman, 1983). Initially, the predominant species used were mixtures of aerobic and anaerobic bacteria with the exception of the Mobil test, which used *Clostridium acetobutylicum*. Later, the inocula consisted of stable mixtures of anaerobic bacteria. The size of the inoculum and nutrient addition increased as did the complexity of the treatment process. Most of the trials used molasses as the main nutrient with additional minerals added in certain trials.

Most of the microbial field trials can be classified as wellbore stimulations where the purpose of the treatment was to increase oil production from a particular well (Hitzman, 1983; Grula et al., 1985; Petzet and Williams, 1986). In this type of process, a well, close to its economic limit, is treated with a mixed anaerobic culture and a fermentable carbohydrate. The well is usually closed for 10 days to several months before fluid is allowed to flow. Hitzman (1983) states that these processes were most effective in carbonate wells with an API gravity of 15 to 30°, salinity less than 100,000 ppm, and a temperature around 35 to 40°C. Of the 24 wells treated by Petrogen Inc., 75 percent showed an increase in wellhead pressure and an increase in oil production for 3 to 6 months. The Hardin-Simmons University team has treated over 80 wells and obtained pressure increases in 64 of them (Petzet and Williams, 1986). More than 40 of these wells showed some increase in oil production. Complete oil production histories are not given, so it is difficult to assess the significance of these data or the cause of the increased oil production. Factors such as the interruption of oil production for 2 weeks to several months or the replacement of old equipment and piping may have contributed to the observed increases in oil production. Also, it is not possible to determine whether the injection of a specially designed mixture of bacteria is necessary. The addition of molasses to an anaerobic environment would enrich for organisms physiologically similar to the clostridial strains used in these inocula. It would be interesting to compare the results of a well-controlled field test where some wells receive both the inoculum of bacteria and molasses, others receive only molasses, and some are untreated.

The production of gas would pressurize the reservoir in the vicinity of the wellbore, and, when the well is opened, the depressurization would remove the entrapped particulates and bacterial cells. This might shift the capillary pressure distribution (Donaldson, 1985). The production of solvents and surfactants could alter the wetability of the rock, and, for carbonate reservoirs, acid production would dissolve some of the rock matrix. These factors would combine to change the relative permeability for oil and aid in its mobilization. Further studies are needed to confirm that these are the mechanisms which caused the increase in oil production. These studies suggest that using microorganisms for wellbore cleanout has the potential for developing into a widely used technology if the predictability of the process improves and more comprehensive tests are done using wells with complete production histories. This would provide the needed information to adequately assess the efficiency of the process and the economic risks and gains.

In MEOR processes, the purpose is to treat a significant portion of the reservoir in order to increase the ultimate recovery of oil from the reservoir—that is, to mobilize and recover residual oil. These processes usually involve the injection of cells and nutrients into the formation, the growth and proliferation of microorganisms throughout the reservoir, and the production of certain microbial product(s). In some cases, the injection of cells may not be needed since the addition of the appropriate nutrient mixture may select for the appropriate microbial activity. The analysis of data from several field trials shows that in situ microbial growth and metabolism does occur (see the section "Evidence for Microbial Activity in Reservoirs"). Bryant et al. (1988) reported that the surface tension in production wells decreased after the injection of a mixed bacterial culture containing *Bacillus licheniformis,* which is known to produce a biosurfactant anaerobically (Javaheri et al., 1985). Thus, it is possible to stimulate not only the in situ growth and metabolism of microorganisms but also the production of a specific metabolic product. However, it is not known whether the inoculated bacteria were needed or if a similar response would have occurred if only nutrients were added. Adequate controls have not been done to show that the inoculation of wells with bacteria is required or beneficial. The production of a specific metabolic product such as a biosurfactant would most likely require the injection of the appropriate strains because this property is not widely found among anaerobic bacteria.

By monitoring the concentrations of bacteria and nutrients in production wells, it is possible to estimate a mean proliferation rate of the bacteria and nutrients in the reservoir. In a highly permeable (3300 to 8100 mD) oil reservoir composed of medium to fine sand (Hodonin deposit, Czechoslovakia), the mean proliferation rates for nutrients and bacterial cells ranged from 1 to 8 m/day (calculated from data in Hitzman, 1983). In a less permeable reservoir (Petrova Ves deposit,

Czechoslovakia), the mean proliferation rates for nutrients and bacterial cells ranged from 1.7 to 3.6 and 1.4 to 3.6 m/day, respectively (calculated from data in Hitzman, 1983). These results indicate that permeability was an important factor affecting the movement of nutrients and cells. Nutrient movement followed preferential flow paths in the high-permeability zones. Wells that were separated from the injection wells by low-permeability zones were not affected, or the time required for breakthrough of the nutrients and cells was much longer. The bacterial populations in the reservoir followed the movement of the nutrients, which suggests that microbial penetration and proliferation throughout the reservoir can be achieved by growing the organisms throughout the formation (Jenneman et al., 1984).

In some reservoir-wide MEOR processes, additional oil production has occurred which has lasted for several months to several years (Hitzman, 1983; Lazar and Constantinescu, 1985). However, because of a lack of engineering and geological data on many of the treated reservoirs, it is not possible to determine whether other factors caused the increase in oil recovery. Such is the case for the 16-well microbial field trial in Last Chance, Colorado (Parkinson, 1985). In many of the field trials, the increase in oil production occurred after the increase in microbial activity, which suggests a causal relationship. One of the best-documented field tests was done by Mobil Oil Company in 1954 in the Upper Cretaceous Nacatoch Formation in Union County, Arkansas (Yarbrough and Coty, 1983). *Clostridium acetobutylicum* was injected and fed a 2 percent beet molasses solution. The rate of oil production increased 250 percent. Unfortunately, this section of the reservoir had been watered out and the residual oil saturation was very low (4.5 to 8.5 percent). Thus, the total amount of oil recovered was low. Currently, the U.S. Department of Energy is sponsoring a microbially enhanced waterflood field project in the Mink Unit of the Delaware-Childers field in Nowata County, Oklahoma (Bryant et al., 1988). To date, the oil recovery has increased 13 percent and the water-to-oil ratio has decreased. The total production from the field is low, and oil production from individual wells has not been measured. Thus, although these results are encouraging, it is difficult to assess the effectiveness of the process.

An encouraging note is that there has not been any serious loss of production or marked stimulation of H_2S production during any of the microbial field trials. Although it is not known whether residual oil was recovered in many cases, these trials do show that MEOR processes are controllable and do not result in many of the detrimental processes which were once anticipated. A major problem which has arisen is the loss of injectivity during the injection of certain grades of molasses. Care must be taken to use a high-quality feedstock or an on-line filtration system.

Conclusions

The potential for microbial processes to stimulate and enhance the recovery of oil is large. These processes may be the only economically viable EOR process in the near future. However, the lack of information on the effectiveness of MEOR has limited its application to marginal reservoirs where the financial loss which would be incurred should the project be unsuccessful is small. As is the case with other EOR processes, unpredictable technical performance has imposed an unacceptable risk which has restricted the general application of MEOR. Compared to other more studied EOR processes, it is not clear whether we really understand how and why microorganisms enhance oil recovery sufficiently to even begin to assess the technical performance of MEOR. There is clearly a need for more and better laboratory and field studies. Laboratory studies will allow us to test variables and understand mechanisms, while field work will tell us if we are proceeding along a realistic path and whether we have considered all the factors which may influence oil recovery and microbial growth in the subsurface. It is essential that these efforts combine the skills of all the pertinent disciplines: microbiology, engineering, geology, and chemistry. Only through a concerted effort of all of these disciplines will the needed information base be developed and the fears and disbelief of the industry (Shennan and Vance, 1987) be overcome. The utilization of genetic engineering technology to improve the performance of microbes in MEOR can only be intelligently applied when we understand how microbes effect the release of oil from reservoirs.

Many microorganisms have been isolated and characterized, which produce a variety of potentially useful products. It is possible to inject these organisms into the appropriate reservoir and stimulate their growth and metabolism or that of indigenous organisms by the addition of the appropriate nutrient source. Data from numerous field trials show that the appropriate products are produced and that additional oil is recovered. Well stimulation methods seem to be the most mature and economically viable processes. Reservoir-wide processes are just beginning to be tested in the field. Further sophistication of these trials involving reservoir computer simulations are needed to adequately test the performance of a given MEOR process and to assess its effectiveness.

Acknowledgments

We thank R. S. Bryant, G. E. Jenneman, and H. M. Lappin-Scott for allowing us to use unpublished information and materials submitted for publication.

References

Absolom, D. R., Lamberti, F. V., Policova, Z., Zingg, W., van Oss, C. L., and Neumann, A. W., Surface thermodynamics of bacterial adhesion, *Appl. Environ. Microbiol.* **46**:90–97 (1983).

Bailey, N. J. L., Jobson, A. M., and Rogers, M. A., Bacterial degradation of crude oil: Comparison of field and experimental data, *Chem. Geol.* **11**:203–221 (1973b).

Bailey, N. J. L., Krouse, H. R., Evans, C. R., and Rogers, M. A., Alteration of crude oils by waters and bacteria—evidence from geochemical and isotope studies, *Am. Assoc. Petrol. Geol. Bull.* **57**:1276–1290 (1973a).

Belkin, S., Wirsen, C. O., and Jannasch, H. W., Biological and abiological sulfur reduction at high temperature, *Appl. Environ. Microbiol.* **49**:1057–1061 (1985).

Bragg, J. R., Maruca, S. D., Gale, W. W., Gall, L. S., Wernau, W. C., Beck, D., Goldman, I. M., Laskin, A. I., and Naslund, L. A., Control of xanthan-degrading organisms in the Loudon pilot: Approach, methodology, and results, in *58th Annual Meeting of the Society of Petroleum Engineers, San Francisco*, Society of Petroleum Engineers, Dallas, 1983.

Bryant, R. S., Burchfield, T. E., Dennis, D. M., and Hitzman, D. O., Microbial enhanced waterflooding: Mink Unit project, in *Sixth Joint SPE/DOE Symp. on Enhanced Oil Recovery*, Tulsa, Okla., 1988.

Bryant, R. S., and Douglas, J., Evaluation of microbial systems in porous media for enhanced oil recovery, in *Symp. on Oilfield Chemistry*, Society of Petroleum Engineers, San Antonio, 1987.

Burns, B. J., Horgarth, J. T. C., and Milner, C. W. D., Properties of Beaufort basin liquid hydrocarbons, *Bull. Can. Petrol. Geol.* **23**:295–303 (1975).

Cadmus, M. C., and Slodki, M. E., Enzymic breakage of xanthan gum solution viscosity in the presence of salts, *Dev. Ind. Microbiol.* **26**:281–289 (1985).

Cairns, W. L., Cooper, D. G., and Kosaric, N., Bacterial-induced demulsification, in J. E. Zajic, D. C. Cooper, T. R. Jack, and N. Kosaric (eds.), *Microbial Enhanced Oil Recovery*, PennWell, Tulsa, Okla., 1983, pp. 106–113.

Carothers, W. W., and Kharaka, Y. K., Aliphatic acid anions in oilfield waters—implications for the origin of natural gas, *Am. Soc. Petrol. Geol. Bull.* **62**:2441–2453 (1978).

Chang, Y., and Yen, T. F., Effects of sodium pyrophosphate additive on the huff and puff nutrient flooding MEOR process, in J. E. Zajic and E. C. Donaldson (eds.), *Microbes and Oil Recovery*, Vol. 1, Bioresources Publ., El Paso, Tex., 1985, pp. 247–256.

Clark, J. B., Munnecke, D. M., and Jenneman, G. E., In situ microbial enhancement of oil recovery, *Dev. Ind. Microbiol.* **22**:695–701 (1981).

Clementz, D. M., Patterson, D. E., Aseltine, R. J., and Young, R. E., Stimulation of water injection wells in the Los Angeles basin by using sodium hypochlorite and mineral acids, *J. Petrol. Technol.* **34**:2087–2096 (1982).

Cooper, D. C., and Zajic, J. E., Surface active compounds from microorganisms, *Adv. Appl. Microbiol.* **26**:229–256 (1980).

Cooper, D. G., Zajic, J. E., Gerson, D. F., and Manninen, K. I., Isolation and identification of biosurfactants produced during the anaerobic growth of *Clostridium pasterianum*, *J. Ferment. Technol.* **58**:83–86 (1980).

Costerton, J. W., and Colwell, R. R., *Native Aquatic Bacteria: Enumeration, Activity and Ecology*, A.S.T.M. No. 695, American Society for Testing and Materials, Philadelphia, 1979.

Craig, F. F., Jr., Waterflooding, in *Secondary and Tertiary Oil Recovery Processes*, Interstate Oil Compact Commission, Oklahoma City, Okla., 1974, pp. 1–30.

Craig, F. F., Jr., *The Reservoir Engineering Aspects of Waterflooding*, Vol. 3, Society of Petroleum Engineers, Dallas, 1980.

Craw, J. A., On the grain of filters and the growth of bacteria through them, *J. Hygiene* **8**:70–73 (1908).

Crawford, P. B., Possible bacterial correction of stratification problems, *Prod. Monthly* **Dec.**:10–11 (1961).

Crawford, P. B., Water technology: Continual changes observed in bacterial stratification rectification, *Prod. Monthly* **Feb.**:12 (1962).

Crawford, R. J., Spyckerelle, C., and Westlake, D. W. S., Biodegradation of oil reservoirs, in O. P. Strausz and E. M. Lown (eds.), *Oil Sand and Oil Shale Chemistry*, Verlag Chemie, 1977, pp. 163–176.

Dagley, S., Introduction, in D. T. Gibson (ed.), *Microbial Degradation of Organic Compounds*, Dekker, New York, 1984, pp. 1–10.

Davis, J. B., *Petroleum Microbiology*, Elsevier, New York, 1967.

Department of Energy, *Federal Enhanced Oil Recovery Research: Increased Understanding of the 300-Billion-Barrel U.S. Residual Oil Resource and the Technologies to Produce It*, Office of Fossil Energy, Bartlesville Project Office, DOE/BC-85/6/SP, Bartlesville, Okla., 1985.

Deroo, G., Tissot, B., McGrossan, R. G., and Der, F., Geochemistry of the heavy oils of Alberta, in L. V. Hills (ed.), *Oil Sands, Fuel of the Future, Can. Soc. Petrol. Geol. Mem.* **3**:148–167 (1974).

Dockins, W. S., Olson, G. J., McFeters, G. A., and Turbak, S. C., Dissimilatory bacterial sulfate reduction in Montana groundwaters, *Geomicrobiol. J.* **2**:83–97 (1980).

Donaldson, E. C., Use of capillary pressure curves for analysis of production formation damage, in *SPE 1985 Production Operations Symp., Oklahoma City, Okla.*, 1985.

Federle, T. W., Dobbins, D. C., Thornton-Manning, J. R., and Jones, D. D., Microbial biomass, activity and community structure in subsurface soils, *Ground Water* **24**:365–374 (1986).

Feeley, H. W., and Kulp, J. L., Origin of Gulf coast salt-dome sulfur deposits, *Bull. Am. Soc. Petrol. Geol.* **5**:1802–1853 (1957).

Finnerty, W. R., and Singer, M. E., A microbial biosurfactant—physiology, biochemistry, and applications, *Dev. Ind. Microbiol.* **25**:31–40 (1984a).

Finnerty, W. R., and Singer, M. E., Microbiology and enhanced oil recovery, *Oil Gas Anal.* **4**:7–12 (1984b).

Finnerty, W. R., Singer, M. E., Ohene, F., and Attaway, H., The application of a biosurfactant to heavy oil displacement, in J. E. Zajic and E. C. Donaldson (eds.), *Microbes and Oil Recovery*, Vol. 1., Bioresources Pub., El Paso, Tex., 1985.

Garland, T. M., Selective plugging of water injection wells, *J. Petrol. Tech.* **18**:1550–1560 (1966).

Geesey, G. C., Mittelman, M. W., and Lieu, V. T., Evaluation of slime-producing bacteria in oil field core flood experiments, *Appl. Environ. Microbiol.* **53**:278–283 (1987).

Ghiorse, W. C., and Wilson, J. T., Microbial ecology of the terrestrial subsurface, *Adv. Appl. Microbiol.* **33**:107–173 (1988).

Grula, E. A., Russell, H. H., and Grula, M. M., Field trials in central Oklahoma using clostridial strains for microbial enhanced oil recovery, in J. E. Zajic and E. C. Donaldson (eds.), *Microbes and Oil Recovery*, Bioresources Pub., El Paso, Tex., 1985, pp. 144–150.

Grula, E. A., Russell, H. H., Bryant, D., Kenaga, M., and Hart, M., Isolation and screening of clostridia for possible use in microbially enhanced oil recovery, in E. C. Donaldson and J. B. Clark (eds.), *Proc. Intern. Conf. on Microbial Enhancement of Oil Recovery*, Bartlesville Energy Technology Center, U.S. Dept. of Energy, Bartlesville, Okla., 1983, pp. 43–47.

Grula, M. M., and Sewell, G. W., Microbial interactions with polyacrylamide polymers, in E. C. Donaldson and J. B. Clark (eds.), *Proc. Intern. Conf. on Microbial Enhancement of Oil Recovery*, Technology Transfer Branch, Bartlesville, Okla., 1983, pp. 129–134.

Hart, R. T., Fekete, T., and Flock, D. L., The plugging effect of bacteria in sandstone systems, *Can. Min. Metall. Bull.* **July**:495–501 (1960).

Hitzman, D. O., Petroleum microbiology and the history of its role in enhanced oil recovery, in E. C. Donaldson and J. B. Clark (eds.), *Proc. Intern. Conf. on Microbial Enhancement of Oil Recovery*, Technology Transfer Branch, Bartlesville, Okla., 1983, pp. 162–218.

Holm-Hansen, O., and Booth, C. R., The measurement of adenosine triphosphate in the ocean and its ecological significance, *Limnol. Oceanog.* **11**:510–519 (1966).

Holzwarth, G., Xanthan and scleroglucan: Structure and use in enhanced oil recovery, *Dev. Ind. Microbiol.* **26**:271–280 (1985).

Hutchinson, C. A., Jr., Reservoir inhomogeneity assessment and control. *Pet. Eng.* **Sept.**: B19–B26 (1959).

Ivanov, M. V., and Belyaev, S. S., Microbial activity in water flooded oilfields and its possible regulation, in E. C. Donaldson and J. B. Clark (eds.), *Proc. 1982 Intern. Conf. on the Microbial Enhancement of Oil Recovery*, Bartlesville Energy Technology Center, Bartlesville, Okla., 1983, pp. 48–57.

Jack, T. R., and DiBlasio, E., Selective plugging for heavy oil recovery, in J. E. Zajic and E. C. Donaldson (eds.), *Microbes and Oil Recovery*, Bioresources Pub., El Paso, Tex., 1985, pp. 205–212.

Jack, T. R., Thompson, B. G., and DiBlasio, E., The potential for use of microbes in the production of heavy oil, in E. C. Donaldson and J. B. Clark (eds.), *Proc. Intern. Conf. on the Microbial Enhancement of Oil Recovery*, Bartlesville Energy Technology Center, U.S. Dept. of Energy, Bartlesville, Okla., 1983, pp. 88–93.

Jang, L. K., Chang, P. W., Findley, J. E., and Yen, T. F., Selection of bacteria with favorable transport properties through porous rock for the application of microbial enhanced oil recovery, *Appl. Environ. Microbiol.* **46**:1066–1072 (1983a).

Jang, L. K., Sharma, M. M., Findley, J. E., Chang, P. W., and Yen, T. F., An investigation of transport of bacteria through porous media, in E. C. Donaldson and J. B. Clark (eds.), *Proc. 1982 Intern. Conf. on Microbial Enhancement of Oil Recovery*, Department of Energy, Bartlesville Technology Center, Bartlesville, Okla., 1983b, pp. 60–70.

Javaheri, M., Jenneman, G. E., McInerney, M. J., and Knapp, R. M., Anaerobic production of a biosurfactant by *Bacillus licheniformis* JF-2, *Appl. Environ. Microbiol.* **50**:698–700 (1985).

Jenneman, G. E., Knapp, R. M., McInerney, M. J., Menzie, D. E., and Revus, D. E., Experimental studies of in situ microbial enhanced oil recovery, *Soc. Petrol. Eng. J.* **24**:33–37 (1984).

Jenneman, G. E., McInerney, M. J., Crocker, M. E., and Knapp, R. M., Effect of sterilization by dry heat or autoclaving on bacterial penetration through Berea sandstone, *Appl. Environ. Microbiol.* **51**:39–43 (1986).

Jenneman, G. E., McInerney, M. J., and Knapp, R. M., Microbial penetration through nutrient saturated Berea sandstone, *Appl. Environ. Microbiol.* **50**:383–391 (1985).

Jenneman, G. E., McInerney, M. J., Knapp, R. M., Clark, J. B., Feero, J. M., Revus, D. E., and Menzie, D. E., A halotolerant, biosurfactant-producing *Bacillus* species potentially useful for enhanced oil recovery, *Dev. Ind. Microbiol.* **24**:485–492 (1983).

Jobson, A. L., Cook, F. D., and Westlake, D. W. S., Interaction of aerobic and anaerobic bacteria in petroleum biodegradation, *Chem. Geol.* **24**:355 (1979).

Jones, D. T., and Woods, D. R., Gene transfer, recombination and gene cloning in *Clostridium acetobutylicum*, *Microbiol. Sci.* **3**:19–22 (1986a).

Jones, D. T., and Woods, D. R., Acetone-butanol fermentation revisited, *Microbiol. Rev.* **50**:484–524 (1986b).

Kalish, O. J., Stewart, J. A., Rogers, W. F., and Bennett, E. O., The effect of bacteria on sandstone permeability, *J. Petrol Technol.* **16**:805–814 (1964).

Kang, K. S., Veeder, G. T., and Cottrell, I. W., Some novel bacterial polysaccharides of recent development, *Prog. Ind. Microbiol.* **18**:231–253 (1983).

Kennedy, J. F., and Bradshaw, I. J., Production, properties and applications of xanthan, *Prog. Ind. Microbiol.* **19**:319–371 (1984).

Kianipey, S., M.S. thesis, University of Oklahoma, Norman, 1986.

Kushner, D. J., Life in high salt and solute concentrations: Halophilic bacteria, in D. J. Kushner (ed.), *Microbial Life in Extreme Environments*, Academic Press, New York, 1978, pp. 317–368.

Kuznetsov, S. I., Ivanov, M. V., and Lyalikova, N. N., *Introduction to Geological Microbiology*, McGraw-Hill, New York, 1963.

Kuznetsova, V. A., Occurrence of sulfate-reducing organisms in oil-bearing formations of the Kuibyshev region with reference to salt composition of layer waters, *Mikrobiologiya* **29**:298–391 (1960).

Kuznetsova, V. A., and Gorlenko, V. M., Effects of temperature on the development of microorganisms from flooded strata of the Romashinko oil field, *Mikrobiologiya* **34**:274–278 (1965).

Kuznetsova, V. A., and Li, A. D., Developments of sulfate-reducing bacteria in the flooded oil bearing Devonian strata of Romashinko field, *Mikrobiologiya* **33**:276–280 (1964).

Kuznetsova, V. A., Li, A. D., and Tiforova, N. N., A determination of source contamination of oil-bearing Devonian strata of Romashinko oilfield by sulfate reducing bacteria, *Mikrobiologiya* **32**:581–585 (1963).

Lappin-Scott, H. M., Cusack, F., and Costerton, J. W., Nutrient resuscitation and growth of starved cells in sandstone cores: A novel approach to enhanced oil recovery, *Appl. Environ. Microbiol.* **54**:1373–1382 (1988a).

Lappin-Scott, H. M., Cusack, F., MacLeod, A., and Costerton, J. W., Starvation and nutrient resuscitation of *Klebsiella pneumonia* isolated from oil waters, *J. Appl. Bacteriol.* **64**:541–549 (1988b)

LaRiviere, J. W. M., The production of surface active compounds by microorganisms and its possible significance in oil recovery. I. Some general observations on the change of surface tension in microbial cultures, *Anton. van Leeuwenhoek J. Microbiol. Serol.* **21**:1–8 (1955).

Lazar, I., and Constantinescu, P., Field trial results of microbial enhanced oil recovery, in J. E. Zajic and E. C. Donaldson (eds.), *Microbes and Oil Recovery*, Bioresources Pub., El Paso, Tex., 1985, pp. 122–143.

Ljundahl, L. G., Physiology of thermophilic bacteria, *Adv. Microbiol. Physiol.* **19**:149–243 (1979).

MacLeod, F. A., Lappin-Scott, H. M., and Costerton, J. W., Plugging of a model rock system by using starved bacteria, *Appl. Environ. Microbiol.* **54**:1365–1372 (1988).

Mandelstam, J., McQuillen, J. K., and Dawes, I., *Biochemistry of Bacterial Growth*, Wiley, New York, 1982.

Marquis, R. E., Microbial barobiology, *Bioscience* **32**:267–271 (1982).

Marquis, R. E., Barobiology of deep oil formations, in E. C. Donaldson and J. B. Clark (eds.), *Proc. 1982 Intern. Conf. on the Microbial Enhancement of Oil Recovery*, Bartlesville Energy Technology Center, U.S. Department of Energy, Bartlesville, Okla. 1983, pp. 124–128.

Marquis, R. E., and Matsumura, P., Microbial life under pressure, in D. J. Kushner (ed.), *Microbial Life in Extreme Environments*, Academic Press, New York, 1978, pp. 105–159.

McInerney, M. J., Physiological types of microorganisms used for enhanced oil recovery, in E. C. Donaldson and J. B. Clark (eds.), *Proc. Intern. Conf. on Microbial Enhancement of Oil Recovery*, Bartlesville Energy Technology Center, U.S. Dept. of Energy, Bartlesville, Okla., 1983, pp. 38–42.

McNabb, J. F., and Dunlap, W. J., Subsurface biological activity in relation to ground water pollution, *Ground Water* **13**:32–44 (1975).

Milner, C. W. D., Rogers, M. A., and Evans, C. R., Petroleum transformation in reservoirs, *J. Geochem. Explor.* **7**:101–153 (1977).

Myers, G. E., and McCready, R. G. L., Bacteria can penetrate rock, *Can. J. Microbiol.* **12**:477–484 (1966).

Myers, G. E., and Samiroden, W. D., Bacterial penetration in petroliferous rock, *Producers Monthly* **31**:22–25 (1967).

Nazina, T. N., Rozanova, E. P., and Kuzentsov, S. I., Microbial oil transformation processes accompanied by methane and hydrogen-sulfide formation, *Geomicrobiol. J.* **4**:103–130 (1985).

Neijssel, O. M., and Tempest, D. W., The physiology of metabolite overproduction, *Sym. Soc. Gen. Microbiol.* **29**:53–82 (1979).

Olson, G. J., Dockins, W. S., McFeters, G. A., and Iverson, W. P., Sulfate-reducing and methanogenic bacteria from deep aquifers in Montana, *Geomicrobiol. J.* **2**:327–340 (1981).

Parkinson, G., Research brightens future of enhanced oil recovery, *Chem. Eng.* **92**(5):25–31 (1985).

Petzet, G. A., and Williams, R., Operators trim basic EOR research, *Oil Gas. J.* **84**(6):41–46 (1986).

Pfennig, N., and Widdel, F., Ecology and physiology of some anaerobes from the microbial sulfur cycle, in H. Bothe and A. Trebest (eds.), *Biology of Inorganic Nitrogen and Sulfur,* Springer-Verlag, Berlin, 1981, pp. 169–177.

Pfiffner, S. M., McInerney, M. J., Jenneman, G. E., and Knapp, R. M., Isolation of halotolerant, thermotolerant, facultative polymer-producing bacteria and characterization of the exopolymer, *Appl. Environ. Microbiol.* **51**:1224–1229 (1986).

Philippi, G. T., On the depth, time and mechanism of origin of the heavy to medium-gravity naphthenic crude oils, *Geochim. Cosmochim. Acta* **41**:33–52 (1977).

Raiders, R. A., Freeman, D. C., Jenneman, G. E., Knapp, R. M., McInerney, M. J., and Menzie, D. E., The use of microorganisms to increase the recovery of oil from cores, in *60th Annual Meeting of the Society of Petroleum Engineers*, Las Vegas, Nev., 1985.

Raiders, R. A., Maher, T. F., Knapp, R. M., and McInerney, M. J., Selective plugging and oil displacement in crossflow core systems by microorganisms, in *61st Annual Meeting of the Society of Petroleum Engineers*, New Orleans, La., 1986b.

Raiders, R. A., McInerney, M. J., Revus, D. E., Torbati, H. M., Knapp, R. M., and Jenneman, G. E., Selectivity and depth of microbial plugging in Berea sandstone cores, *J. Ind. Microbiol.* **1**:195–203 (1986a).

Raleigh, J. T., and Flock, D. L., A study of formation plugging with bacteria, *J. Petrol. Technol.* **17**:201–206 (1965).

Raymond, R. L., Jamison, V. W., and Hudson, J. O., Beneficial stimulation of bacterial activity in groundwaters containing petroleum products, *Am. Inst. Chem. Eng. Symp. Series* **73**:390–404 (1976).

Reed, R. L., and Healy, R. N., Some physical aspects of microemulsion flooding: A review, in D. O. Shah and R. S. Scheckter (eds.), *Improved Oil Recovery by Surfactant and Polymer Flooding*, Academic Press, New York, 1977, pp. 383–437.

Rodriques-Valera, F., *Halophilic Bacteria*, Vol. 1, CRC Press, Boca Raton, Fla., 1988.

Rogers, P., Genetics and biochemistry of clostridia relevant to development of fermentation processes, *Adv. App. Microbiol.* **31**:1–60 (1986).

Root, P. J., and Skiba, F. F., Crossflow effects during an idealized displacement process in a statified reservoir, *Soc. Petrol. Eng. J.* **5**:229–238 (1965).

Rubenstein, I., Strausz, O., Spyckerelle, C., Crawford, R. J., and Westlake, D. W. S., The origin of the oil sand bitumens of Alberta: A chemical and a microbiological simulation study, *Geochim. Cosmochim. Acta* **41**:1341–1353 (1977).

Salanitro, J. P., Are sulfate-reducing bacteria involved in sour gas production in oil field waterfloods? in *Annual Meeting Amer. Soc. Microbiol.*, Abstract no. 151, American Society for Microbiology, Washington, D.C., 1985, p. 155.

Sassen, R. F., Biodegradation of crude oil and mineral deposition in a shallow Gulf Coast salt dome, *Org. Geochem.* **2**:153–166 (1980).

Shaw, J. C., Bramhill, B., Wardlaw, N. C., and Costerton, J. W., Bacterial fouling in a model core system, *Appl. Environ. Microbiol.* **49**:693–701 (1985).

Shennan, J. L., and Vance, I., Microbial enhanced oil recovery techniques and offshore oil production, in E. C. Hill, J. L. Shennan, and R. J. Watkinson (eds.), *Microbial Problems in the Offshore Oil Industry*, Wiley, New York, 1987, pp. 73–91.

Silva, L. F., and Farouq Ali, S. W., Waterflood performance in the presence of stratification and formation plugging, in *46th Annual Meeting of the Society of Petroleum Engineers of AIME*, New Orleans, La., 1971.

Singer, M. E., Microbial biosurfactants, in J. E. Jazic, and E. C. Donaldson (eds.), *Microbes and Oil Recovery*, Bioresources Pub., El Paso, Tex., 1985, pp. 19–38.

Suflita, J. S., The biodegradation potential of ground water pollutants when oxygen is unavailable, in P. Churchill and R. Patrick (eds.), *Ground Water Contamination: Sources, Effects and Opinions to Deal with the Problem*, The Academy of Natural Sciences, Philadelphia, 1987, pp. 271–294.

Sutherland, I. W., An enzyme system hydrolyzing the polysaccharides of *Xanthomonas* spp. *J. Appl. Bacteriol.* **53**:385–393 (1982).

Taber, J. J., Dynamic and static forces required to remove a discontinuous oil phase from porous media containing both oil and water, *Soc. Petrol. Eng. J.* **9**:3–12 (1969).

Tannenbaum, E., and Aizenshtat, Z., Light oils transformation to heavy oils and asphalts—assessment of the amounts of hydrocarbons removed and the hydrological-geological control of the process, in *Exploration for Heavy Crude Oil and Bitumen*, AAPG Research Conference, Santa Maria, Calif., 1984.

Thode, H. G., Wanless, R. K., and Wallouch, R., The origin of native sulfur deposits from isotope fractionation studies, *Geochim. Cosmochim. Acta.* **5**:286–298 (1954).

Torrella, F., and Morita, R. Y., Microcultural study of bacterial size changes and microcolony and ultramicrocolony formation by heterotrophic bacteria in seawater, *Appl. Environ. Microbiol.* **41**:518–527 (1981).

Updegraff, D. M., Plugging and penetration of petroleum reservoir rock by microorganisms, in E. C. Donaldson and J. B. Clark (eds.), *Proc. 1982 Intern. Conf. on the Microbial Enhancement of Oil Recovery,* Department of Energy, Bartlesville Technology Center, Bartlesville, Okla., 1983, pp. 80–85.

Van Poolen, H. K., and Associates, *Fundamentals of Enhanced Oil Recovery,* PennWell, Tulsa, Okla., 1980.

Weirich, G., and Schweisfurth, R., Extraction and culture of microorganisms from rock, *Geomicrobiol. J.* **4:**1–20 (1985).

Werner, P., A new way for the decontamination of polluted aquifers by biodegradation, *Water Supply.* **3:**41–47 (1985).

Westlake, D. W. S., Microbial activities and changes in the chemical and physical properties of oil, in E. C. Donaldson and J. B. Clark (eds.), *Proc. Intern. Conf. on Microbial Enhancement of Oil Recovery,* Technology Transfer Branch, Bartlesville, Okla., 1983, pp. 102–111.

Westlake, D. W. S., Semple, K. M., and Obuekwe, C. O., Corrosion of ferric iron-reducing bacteria isolated from oil production systems, in S. C. Dexter (ed.), *Biologically Induced Corrosion,* National Association of Corrosion Engineers, Houston, 1986, pp. 195–200.

White, D. C., Analysis of microorganisms in terms of quantity and activity in natural environments, in J. H. Slater, R. Whittenbury, and J. W. T. Wimpenny (eds.), *Microbes in Their Natural Environments,* in *34th Symp. Soc. for General Microbiology,* Cambridge University Press, London, 1983, pp. 37–66.

White, D. C., Bobbie, R. J., Herron, J. S., King, J. D., and Morrison, S. J., Biochemical measurements of microbial mass and activity from environmental samples, in *Native Aquatic Bacteria: Enumeration, Activity and Ecology,* A.S.T.M. No. 695, American Society for Testing and Materials, Philadelphia, 1975, pp. 69–81.

White, D. C., Smith, G. A., Gehron, M. J., Parker, J. H., Findlay, R. H., Martz, R. F., and Frederickson, H. L., The groundwater aquifer microbiota: Biomass, community structure and nutritional status, *Dev. Ind. Microbiol.* **24:**201–211 (1983).

Wilson, J. T., McNabb, J. F., Wilson, B. S., and Noonan, M. S., Biotransformation of selected organic pollutants in ground water, *Dev. Ind. Microbiol.* **24:**225–233 (1983).

Wilson, J. T., and Ward, C. H., Opportunities for bioreclamation of aquifers contaminated with petroleum hydrocarbons, *Dev. Ind. Microbiol.* **27:**109–116 (1987).

Winograd, I. J., and Robertson, F. N., Deep oxygenated ground water: Anomaly or common occurrence? *Science* **216:**1227–1230 (1982).

Winters, J. C., and Williams, J. A., Microbial alteration of crude oil in the reservoir, in *Symp. Petrol. Transformations in Geologic Environments, Am. Chem. Soc. Meet., New York,* 1969, pp. E22–E31.

Yarbrough, H. F., and Coty, V. F., Microbially enhanced oil recovery from the Upper Cretaceous Nacatoch formation, Union County, Arkansas, in E. C. Donaldson and J. B. Clark (eds.), *Proc. 1982 Intern. Conf. on the Microbial Enhancement of Oil Recovery,* Department of Energy, Bartlesville Technology Center, Bartlesville, Okla., 1983, pp. 149–153.

Zajic, J. E., and Mesta-Howard, A. M., Properties of *Bacillus* from Conroe oil field (Texas) and other reservoir sources, in J. E. Zajic and E. C. Donaldson (eds.), *Microbes and Oil Recovery,* Bioresources Pub., El Paso, Tex., 1985, pp. 295–309.

Zajic, J. E., and Seffens, W., Biosurfactants, *Crit. Rev. Biotechnol.* **1:**87–107 (1984).

Zinger, A. S., Microflora of underground waters in the lower Volga region with reference to its use for oil prospecting, *Mikrobiologiya* **35:**305–311 (1966) (English translation).

ZoBell, C. E., Ecology of sulfate reducing bacteria, *Prod. Mon.* **22:**16–29 (1958).

Zvyagintsev, D. G., Growth of microorganisms in thin capillaries and films, *Mikrobiologiya* **39:**161–165 (1970).

Zvyagintsev, D. G., and Pitryuk, A. P., Growth of microorganisms in capillaries of various sizes under continuous flow and static conditions, *Mikrobiologiya* **42:**60–64 (1973).

Index